Monographs in Electrical and Electronic Engineering

Series editors: P. Hammond, T. J. E. Miller, and T. Kenjo

Monographs in Electrical and Electronic Engineering

21. *Brushless permanent-magnet and reluctance motor-drives* (1989) T. J. E. Miller
22. *Vector control of a.c. machines* (1990) Peter Vas
23. *Brushless servomotors: fundamentals and applications* (1990) Y. Dote and S. Kinoshita
24. *Semiconductor devices, circuits, and systems* (1991) Albrecht Möschwitzer
25. *Electrical machines and drives: a space-vector theory approach* (1992) Peter Vas
26. *Spiral vector theory of a.c. circuits and machines* (1992) Sakae Yamamura
27. *Parameter estimation, condition monitoring, and diagnosis of electrical machines* (1993) Peter Vas
28. *An introduction to ultrasonic motors* (1993) T. Sashida and T. Kenjo
29. *Ultrasonic motors: theory and applications* (1993) S. Ueha and Y. Tomikawa
30. *Linear induction drives* (1993) J. F. Gieras
31. *Switched reluctance motors and their control* (1993) T. J. E. Miller
32. *Numerical modelling of eddy currents* (1993) Andrzej Krawczyk and John A. Tegopoulos
33. *Rectifiers, cycloconverters, and a.c. controllers* (1994) Thomas M. Barton
34. *Stepping motors and their microprocessor controls. Second edition* (1994) T. Kenjo and A. Sugawara
35. *Inverse problems and optimal design in electricity and magnetism* (1995) P. Neittaanmäki, M. Rudnicki, and A. Savini
36. *Electric drives and their control* (1995) Richard M. Crowder
37. *Design of brushless permanent magnet motors* (1995) J. R. Hendershott and T. J. E. Miller
38. *Reluctance synchronous machines and drives* (1995) I. Boldea
39. *Geometry of electromagnetic systems* (1996) D. Baldomir and P. Hammond
40. *Permanent-magnet d.c. linear motors* (1996) Amitava Basak
41. *Vector control and dynamics of a.c. drives* (1996) D.W. Novotny and T. A. Lipo

Vector Control and Dynamics of AC Drives

D. W. Novotny and T. A. Lipo
Department of Electrical and Computer Engineering
University of Wisconsin-Madison
Wisconsin, USA

CLARENDON PRESS · OXFORD

Great Clarendon Street, Oxford OX2 6DP

Oxford University Press is a department of the University of Oxford.
It furthers the University's objective of excellence in research, scholarship,
and education by publishing worldwide in

Oxford New York

Auckland Cape Town Dar es Salaam Hong Kong Karachi
Kuala Lumpur Madrid Melbourne Mexico City Nairobi
New Delhi Taipei Toronto Shanghai

With offices in

Argentina Austria Brazil Chile Czech Republic France Greece
Guatemala Hungary Italy Japan South Korea Poland Portugal
Singapore Switzerland Thailand Turkey Ukraine Vietnam

Published in the United States
by Oxford University Press Inc., New York

First published 1996

A catalogue record for this book is available from the British Library

Library of Congress Cataloging in Publication Data
(Data available)

ISBN 978-0-19-856439-3

Printed and bound by CPI Group (UK) Ltd, Croydon, CR0 4YY

PREFACE

The material in this book represents the present state of evolution of the course notes for a graduate level course in the dynamics and control of AC drives. The period of evolution of this subject material reaches all the way back to the 1970's and is ongoing. Publication of the material at this time is not because the work is finished but rather a recognition that a formal, hard cover version of the material will be useful to our students and will contribute to the present state of the art.

While the primary audience for the text material has been electrical engineering graduate students, there were also many engineers in industrial positions and senior level students in elective courses who were exposed to various portions of the material. In particular, the concept of field orientation and the resulting ability to directly control torque was, and is, of great interest to a wide range of persons. This has influenced the means in which portions of the text have been organized with the goal being to produce a flexible body of tutorial material which can be used in a variety of ways.

Early on we adopted the concept of complex vector representation because of its great value as a visualization tool in clarifying the dynamic interactions between fluxes and currents in electric machines. This concept, despite having originated in the US, is widely used in Europe, Japan and elsewhere but is not commonly used here. Our use of the concept in the development of the material in this book grew until it is now central to almost the entire book. The power of the complex vector as a mathematical tool for simplifying polyphase electric machine modeling as well as a conceptual method for visualizing and interpreting results will become apparent to anyone reading Chapters 2, 3 and 4 who has tried to do some of the same things using real variable matrix methods.

Although the primary purpose of the text material is to treat vector control and field orientation, a number of other, more conventional topics are also treated as illustrations of the use of the basic machine model. These include the development of the standard steady state machine models, constant speed switching transients, trapped flux transient models, inverter–machine models and others. The material has been organized to allow a variety of ways it can be used by readers with differing backgrounds and needs. To illustrate some of these possibilities:

1) As a graduate level textbook, working through the material in the order presented.

2) For someone interested in the concept and steady state behavior of field oriented machines, skip directly to Chapter 5. The only prerequisite knowledge is steady state machine theory as presented in any introductory level book.

3) For someone interested in the dynamics of field orientation, skip directly to Chapter 6. The prerequisite knowledge is *d,q* machine modelling. This is available in Chapter 2 if needed.

4) For an introduction to complex variable analysis, Chapter 2 followed by Chapter 4 provide the basic material plus enough illustrations to clearly show the power of the method.

There are a number of example problems in the text and most chapters have a set of problems at the end to allow self study. These problems are taken from the course we have taught using the text material in note form. Answers are provided at the end of the book.

We dedicate this book to our many students who have helped shape the material through their participation in our courses over the past years.

Madison, WI, U.S.A. D.W.N. and T.A.L.
February, 1996

CONTENTS

1 Introduction to AC Drives *1*

1.1 Introduction 1

1.2 Review of DC Drives 2

1.3 Comparison of DC and AC Motor Drives 4

1.3.1 Commutator and Brushes 4
1.3.2 Power vs. Speed Range 5
1.3.3 Efficiency 5
1.3.4 Power Factor 5
1.3.5 Inertia 6
1.3.6 Protection 6
1.3.7 Motor Maintenance 6
1.3.8 Ruggedness 6
1.3.9 Standstill Performance 7
1.3.10 Size and Weight 7

1.4 Comparison of Synchronous and Induction Machines 7

1.5 Inverters for Adjustable Speed 9

1.6 The Six Step Voltage Stiff Inverter (VSI) 9

1.7 Six Step Inverter Speed and Slip Frequency Control Techniques . . 15

1.8 The Pulse Width Modulated Inverter (PWM) 18

1.8.1 Block Modulation 19
1.8.2 Sinusoidal PWM Modulation 20
1.8.3 Synchronous vs. Asynchronous Modulation 21
1.8.4 Pulse Dropping 23
1.8.5 Changeover to Six Step Mode 24
1.8.6 Losses Caused by Inverters 24
1.8.7 Torque Pulsations 25
1.8.8 Induction Motor Speed Control Using a PWM Inverter 25

1.9 The Current Stiff Inverter (CSI) 25

1.10 High Performance Drives vs. General Purpose Drives 31

References 33

2 d,q Modelling of Induction and Synchronous Machines *35*

2.1 Introduction 35

2.2 Winding Inductances 35

2.3 System Equations in the Stationary a,b,c Reference Frame 43
2.4 Determination of Induction Machine Inductances 45
2.5 Complex Vector Representation of Three Phase Variables. 48
2.6 Complex Variable Model of Three Phase Induction Machine 52
2.7 Turns Ratio Transformation 54
2.8 Transformation to a Rotating Reference Frame 56
2.9 Interpretation of Complex Vectors 61
2.10 Power Flow in the *d,q* Equivalent Circuit. 68
2.11 Example – Stator *a,b,c* and Rotor *d,q* Model (Stator Referred) . . . 70
2.12 The Electromagnetic Torque. 73
2.13 Analysis of Induction Motor Starting Performance Using *d,q*,0 Variables 78
2.14 Extension of *d,q*,0 Theory to Analysis of Salient Pole Synchronous Machines. 88
2.15 Extension of *d,q*,0 Theory to Analysis of Permanent Magnet Motors 102
References 103
Problems 105

3 d,q Models for Solid State Power Converters 109
3.1 Introduction. 109
3.2 *d,q* Model for Voltage Source Inverter 109
3.3 *d,q* Model for PWM Operation. 115
3.4 *d,q* Model for CSI System 116
3.5 Inverter *d,q* Models in a Synchronous Reference Frame 120
3.6 Examples of Inverter–Induction Motor Models 123
3.7 Fundamental Component Approximation for Steady State Operation 132
3.8 Duality of VSI and CSI Systems 144
References 145

Problems 146

4 Complex Vector Analysis of Induction Machines *151*

4.1 Introduction........ 151

4.2 Complex Vector Equivalent Circuit 152

4.3 Turns Ratio and Modified *d,q* Models 153

4.4 Steady State Equivalent Circuits........ 156

4.4.1 Sinusoidal Excitation in Complex Vector Form........ 156

4.4.2 Stator Referred Equivalent Circuits........ 158

4.5 Invariance of the Steady State Equivalent Circuit 164

4.6 Modified Equivalent Circuits 165

4.7 Electrical Transients at Constant Speed 170

4.7.1 Complex Vector Operational Impedances 171

4.7.2 Example – Sudden Application of Stator Current Source Excitation ... 172

4.7.3 General Solution for Switching Transients 177

4.8 The Constant Speed Eigenvalues 177

4.8.1 Example – Sudden Short Circuit of Unloaded Induction Machine 187

4.9 Transient Equivalent Circuits (Constant Rotor Flux Linkage)..... 190

4.9.1 Constant Rotor Flux Linkage Model 190

4.9.2 Transient Equivalent Circuit 191

4.9.3 Evaluation of Voltage Behind Transient Reactance........ 192

4.9.4 Example – Sudden Short Circuit of Unloaded Induction Machine 194

4.9.5 Comments on the Transient Equivalent Circuit........ 196

References 196

Problems 198

5 Principles of Vector Control and Field Orientation *203*

5.1 Introduction........ 203

5.2 DC Machine Torque Control........ 203

5.3 Requirements for Torque Control........ 206

5.4 Synchronous Machine Vector Control 208

5.4.1 CSI – Synchronous Machine (Commutatorless DC Machine)........ 208

5.4.2 Torque Control and Choice of γ 212

5.5 Synchronous Machine Steady State *d,q* Model 213
5.5.1 Steady State Conditions in *d,q* Variables 213
5.5.2 *d,q* Variable Vector Diagrams 214
5.6 Torque Control Implementations – Synchronous Machines. 217
5.6.1 Torque Control Using Field Orientation with a CSI. 217
5.6.2 Torque Control Using a CRPWM 218
5.6.3 Magnitude – Angle Resolver and Use in CSI Torque Control 220
5.6.4 Torque Control Requirements – Synchronous Machine. 221
5.7 Electrical Measurement of the Rotor Field Angle θ_r 222
5.8 Brushless DC Machines 222
5.8.1 Sine Wave Machines with Current Control 223
5.8.2 Trapezoidal Wave Machines with Current Control 224
5.8.3 Other Brushless Machines 225
5.9 Induction Machine Vector Control — Steady State 226
5.9.1 Conventional Equivalent Circuit Considerations. 226
5.9.2 Modified Equivalent Circuit. 228
5.9.3 Torque Control in Terms of $I_{s\phi}$ and I_{sT}. 230
5.9.4 Terminal Behavior in Terms of $I_{s\phi}$ and I_{sT} 233
5.10 Induction Machine Steady State *d,q* Model 234
5.10.1 Example – Induction Machine Field Orientation. 235
5.11 Implementation of Field Orientation in Induction Machines 240
5.11.1 Indirect Field Orientation 241
5.11.2 Direct Field Orientation 247
5.11.3 Torque Control Requirements 247
5.12 A Basic Difference Between Synchronous and Induction Machine Field Orientation. 249
References 250
Problems 252

6 Dynamics of Vector Control and Field Orientation 257
6.1 Introduction. 257
6.2 Dynamics of Induction Machine Field Orientation 257
6.2.1 Induction Machine *d,q* Model with Axis Rotation at an Angular Velocity ω_e 258
6.2.2 Rotor Flux Referred Equations. 258
6.2.3 Dynamic Response of Field Oriented Induction Machines 260

6.2.4 Block Diagram of Field Oriented Induction Machine 263

6.3 Indirect Controllers for Induction Machine Field Orientation 264

6.3.1 Indirect Controller with i_{ds}^{e*} and i_{qs}^{e*} as Inputs 265

6.3.2 Indirect Controller with λ_{dr}^{e*} and i_{qs}^{e*} as Inputs 265

6.3.3 Other Indirect Controllers 267

6.3.4 Indirect Control Using a CSI 268

6.3.5 Indirect Field Orientation Start Up Transient 269

6.4 Direct Controllers for Induction Machine Field Orientation 275

6.4.1 Direct Determination of Rotor Flux Angle 276

6.4.2 Measurement of Air Gap Flux 276

6.4.3 Voltage and Current Sensing 278

6.4.4 Implementation of Direct Field Orientation. 280

6.5 Induction Machine Field Orientation Using Air Gap Flux 282

6.5.1 Air Gap Flux Referred *d,q* Equations 283

6.5.2 Dynamic Response of Air Gap Flux Controlled Induction Machine. . . . 284

6.5.3 Steady State Interpretation of Air Gap Flux Control 286

6.5.4 Indirect Field Orientation Controller for Air Gap Flux 287

6.5.5 Direct Field Orientation Controller for Air Gap Flux 288

6.6 Dynamics of Synchronous Machine Vector Control and Field Orientation 290

6.6.1 *d,q* Model of Synchronous Machine 290

6.6.2 Vector Control and Angle Control 291

6.6.3 Dynamics of Synchronous Machine Field Orientation 293

6.6.3.1 Constant Field Current Operation (Constant Torque Region). . 293

6.6.3.2 Variable Field Excitation (Field Weakening Region) 295

6.6.4 Dynamic Response With $\gamma \neq 0$ (Angle Control) 296

6.6.5 Example – Synchronous Machine Field Orientation 298

6.7 Field Orientation Using Voltage as the Controlled Variable. 304

6.7.1 Stator Voltage Equations in Terms of Rotor Flux – Induction Machine . 304

6.7.2 Decoupling Equations for Field Orientation 305

6.7.3 Examples of Field Orientation Using Voltage Controlled Inverters 308

6.8 Stator Flux Based Field Orientation 309

6.8.1 Mathematical Model of a Stator Flux Oriented Induction Machine 310

6.8.2 Design of a Decoupler for Direct Stator Flux Oriented System 311

References 313

Problems 315

7 Current Regulation in Power Converters *317*

7.1 Introduction 317

7.2 Current Regulated Inverters 317

7.2.1 The Current Regulated CSI Inverter 318

7.2.2 The Current Regulated PWM Inverter 321

7.3 Hysteresis Regulators 322

7.4 Ramp–Comparison Controllers 331

7.5 Stationary Frame Regulators 335

7.6 Synchronous Frame Regulators 340

7.7 Feedforward Compensation 344

7.8 Augmented Feedforward Compensation 347

7.9 Augmented Feedforward Compensation with Decoupling 348

7.10 Predictive Control 349

7.11 Conclusion 351

References 351

Problems 352

8 Parameter Sensitivity and Saturation Effects in Indirect Field Orientation *353*

8.1 Introduction 353

8.2 Phasor Diagram for Detuned Operation 353

8.2.1 Field Orientation Torque Characteristics 355

8.2.2 Examples 361

8.2.3 Field Orientation Flux Characteristics 363

8.2.4 Influence of Saturation 364

8.2.5 Power Loss Considerations 365

8.2.6 Example — Detuned Induction Machine Field Orientation 369

8.3 Transient Response 377

8.3.1 Torque Response with Slip Gain Error 377

8.3.2 Block Diagram Representation of Slip Gain Error 383

8.3.3 Simulation of Transient Response with Slip Gain Error 384

8.4 Optimal Selection of Flux Level 388
8.4.1 Maximum Torque Per Stator Ampere 389
8.4.2 Maximum Drive Efficiency 390
8.4.3 Reduced Sensitivity to Controller Detuning 393
8.4.4 Optimal Utilization of the Converter 394
8.4.4.1 Converter Current Limits 394
8.4.4.2 Converter Current and Voltage Limits 395
References 397
Problems 398

9 Field Weakening Operation *401*
9.1 Introduction 401
9.2 Torque Demand and Availability of High Performance Drives 402
9.3 Induction Machine Control Strategy for Constant Horsepower Operation 405
9.4 Inverter Imposed Voltage and Current Limits 406
9.5 Maximum Torque Capability Curve Using the Conventional Method 409
9.6 Scheme for Achieving Maximum Torque Capability 410
9.6.1 Transition from Constant Torque to Constant Power Operation 411
9.6.2 Region #2 – Field Weakening with Constant Output Power 412
9.6.3 Region #3 – Field Weakening with Constant Speed*Power 413
9.7 Control System Implementation 414
References 416
Appendix - per Unit Representation 417
Answers to Problems *419*

Index 435

1 Introduction to AC Drives

1.1 Introduction

An important factor in worldwide industrial progress during the past several decades has been the increasing sophistication of factory automation. Manufacturing lines in an industrial plant typically involve one or more variable speed motor drives which serve to power conveyor belts, robot arms, overhead cranes, steel process lines, paper mills and plastic and fiber processing lines to name a few. Prior to the 1950s all such applications required the use of a dc motor drive since ac motors were not capable of true adjustable or smoothly varying speed since they inherently operated synchronously or nearly synchronously with the frequency of electrical input. The inherent disadvantages of dc drives, however, have prompted continual attempts to find better solutions to the problem. To a large extent, applications which require only a gradual change in speed are now being replaced by what can be called *general purpose* ac drives. In general, such ac drives often feature a cost advantage over their dc counterparts and, in addition, offer lower maintenance, smaller motor size, and improved reliability. However, the control flexibility available with these drives is very limited and their application is, in the main, restricted to fan, pump and compressor types of application where the speed need be regulated only roughly and where transient response and low speed performance are not critical.

Drives used in machine tools, spindles, high speed elevators, dynamometers, mine winders, rolling mills, glass float lines and the like have much more sophisticated requirements and must afford the flexibility to allow for regulation of a number of variables, such as speed, position, acceleration and torque. Such high performance applications typically require a high speed holding accuracy better than 0.5%, a wide range speed control of at least 20:1, and fast transient response, typically better than 50 rad/s for the speed loop. Until recently, such drives have almost exclusively been the domain of dc motors combined with various configurations of ac/dc converters depending upon the application. With suitable control, however, ac drives are more than a match for dc drives in a high performance application. It will be shown in this book that control of an ac machine is considerably more complicated than its dc motor counterpart. However, with continual advancement of microelectronics, these control complexities are rapidly being overcome. The gradual replacement of dc drives has already begun, for example in machine tool drives, and ac drives can be expected to continue to overtake them over the next decade. It is still too early to determine if dc drives will eventually be relegated to the history book as have nearly all other commutator machines such as rotary converters, amplidynes and the like. However, the next decade will surely witness a marked increase in the use of ac drives.

1.2 Review of DC Drives

Figure 1.1 shows three of the most common types of dc drives, consisting of a one, two and four quadrant drive respectively. In Figure 1.1(a) the one quadrant converter consists of a single bridge operating in a two pulse or three pulse mode depending upon the type of supply. The term "one quadrant' is used to denote that the bridge is capable of only one polarity of voltage and current. Hence, only motoring operation or acceleration in the forward direction can be obtained with a single quadrant drive. Adjustment of the dc armature voltage is obtained by phase control of the bridge which supplies a prescribed current to the machine and thus results in a prescribed torque. Braking torque is accomplished by the overhauling action of the load which drags the motor to a halt when the torque from the dc motor is removed.

In Figure 1.1(b) a more controlled form of braking is obtained by reversing the field current in the machine, which, in turn, reverses the counter emf. In this case the motor is supplied from a four or six pulse ac/dc converter. The polarity of the converter output voltage can then be reversed by means of phase control, thereby reversing the power flow so that operation is possible in two quadrants. Since the field winding has an inherently large inductance, the response of this system during the torque reversal is limited to about a few tenths of a second. Speed reversal can also be obtained by reversing the field as the motor reaches zero speed. However, since the torque becomes zero while the field current is reversed, speed performance near zero speed is adversely affected and a smooth reversal through zero speed is not possible with this system.

In the third type of motor drive, Figure 1.1(c), the motor is supplied by a double inverse parallel bridge type of arrangement. Both polarities of dc voltage and current are now possible if the field can again be reversed. In this case, torque can be reversed simply by first zeroing the current in the bridge providing motoring current and then increasing the current in the inverse parallel bridge. The response of the bridge is now limited only by the armature leakage inductance, the sampling theorem limit of the converter, and the commutation capability of the dc motor. Care must be taken upon reversing the dc current since a short circuit path exists if both the forward and reverse bridges are triggered simultaneously. Hence, the currents in both bridges must remain at zero for a few milliseconds to allow the forward conducting bridge to recover its

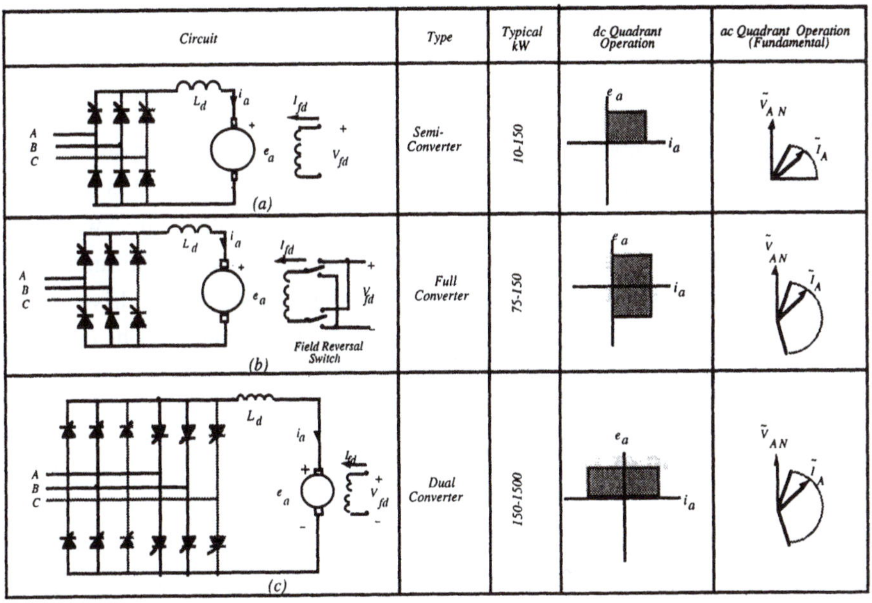

Circuit	Type	Typical kW	dc Quadrant Operation	ac Quadrant Operation (Fundamental)
(a)	Semi-Converter	10-150		
(b) Field Reversal Switch	Full Converter	75-150		
(c)	Dual Converter	150-1500		

Figure 1.1 Major dc motor drive types with both single phase and three phase supply. Additional components for three phase supply shown shaded

blocking ability before turning on the reverse conducting bridge. Speed reversal is easily accomplished by reversing the roles of the forward and reverse bridges. Smooth torque reversal is now possible since the current continues to flow in a single bridge as the motor reverses. Torque response of the system can now be improved to tens of milliseconds.

The basic layout of an ac drive employing a converter with a dc link is shown in Figure 1.2. In this case the line voltages and currents are first con-

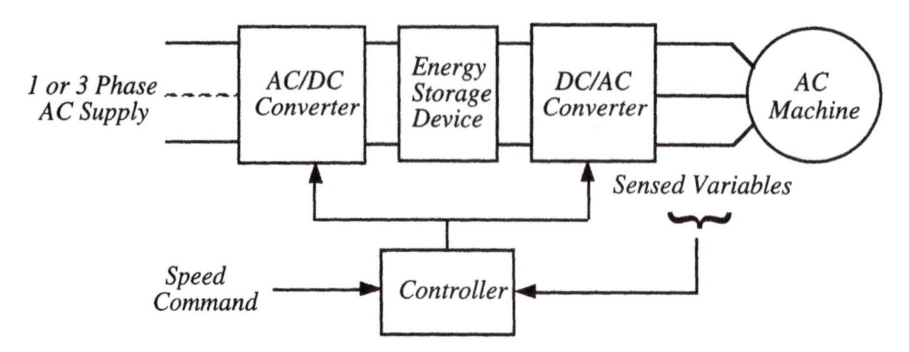

Figure 1.2 Basic ac motor drive employing a dc link

verted by a line side converter to dc quantities and subsequently by a machine side converter to the ac voltages and currents of variable frequency as required by the motor. Both converters consist of a bridge of solid state switching devices in the case of an ac drive while the machine side conversion function is accomplished by means of the commutator in the case of the dc drive. Note that the machine side converter together with the ac motor can be considered as the electromechanical equivalent of the dc motor.

1.3 Comparison of DC and AC Motor Drives

1.3.1 Commutator and Brushes [1]

DC motor armatures have two major weak points, the mechanical commutator and brush assembly. Neither of these components exist on squirrel cage induction motors. The absence of the mechanical commutator means that higher speeds are possible with the equivalent induction motor. Also, higher armature voltages can be used with an induction motor due to limitations in the voltage that can be supported between adjacent commutator segments in a dc machine armature. Transient response is also limited by the rate of rise of armature current which is set by the ability of the brush to complete the reversal of the current in the armature coil undergoing commutation. The maximum permissible rate of rise of current in modern solid frame dc motors is limited to 30 times rated current per second. For older motors this limit can be as low as 5 per unit per second. Although for modern laminated frame dc motors a rate of rise of 200 per unit per second is possible, even with this improved capability this feature can be the limiting factor affecting torque response in a dc motor. No such inherent limitation exists in an induction motor and the rates of current rise experienced in such a motor is limited only by the leakage inductance of the machine and the amount of voltage available to force the current from one value to another.

1.3.2 Power vs. Speed Range

Because of the presence of the commutator the speed of the dc machine is inherently limited. In general, for large motors it is not possible to obtain dc motors with a speed*power product greater than 2.6×10^6 (kW)(rpm). For example, it is difficult to locate a 1500 rpm, 1350 kW dc motor for purchase. Induction motors of this rating can be constructed for speeds reaching several times this rpm without great difficulty.

1.3.3 Efficiency

The efficiency of induction motors used for variable speed operation are generally comparable and frequently better than the equivalent dc motor efficiency, even though the presence of the rotor cage adds an additional loss component not encountered in a dc machine. Since the induction motor cage need not be designed to allow for a direct on line start when driven from a converter, the cage resistance can be selected solely to provide optimum running performance and minimum loss. Also, dc motor losses such as brush drop can be eliminated.

1.3.4 Power Factor

In general, the dc converter of a dc drive operates with an input fundamental component power factor which ranges from 0 to about 0.9 and increases approximately in proportion to the motor speed. Unity power factor cannot be achieved because the converter requires some voltage margin to allow for supply voltage dips, dynamic requirements and to prevent loss of commutation capability in the inversion mode. Although the induction motor always runs at lagging power factor, typically 0.95 at rated load for machines in the 300 kW range, the reactive power requirements of the motor are supplied by the dc link filter capacitor and inverter. With modern pulse width modulated inverters acting as the machine side converter, the input fundamental component power factor remains high, typically above 0.95, regardless of the motor speed.

1.3.5 Inertia

An ac machine of the same power rating and speed will normally have a lower inertia than its dc motor equivalent. In a dc machine, a parameter that affects commutation is the so called *reactance voltage* of the armature windings which relates to the commutating capability of the motor. The need to keep this parameter within certain bounds, constrains the length of core and hence results in an increased diameter for a given power and speed. The lower inertia of an induction motor translates to faster speed response for a machine with equal torque producing capability.

1.3.6 Protection

Monitoring and protection of a squirrel cage induction motor is simpler than for a dc motor. Without a direct on line starting capability and with the ability of the squirrel cage rotor to withstand much higher temperatures than the stator, simple stator temperature monitoring gives comprehensive thermal protection for the entire machine. Also, for high power applications it is usually necessary to use a dc circuit breaker to provide protection for large dc machines. These breakers are very expensive and require regular maintenance. However, DC circuit breakers need not be used for ac drive systems.

1.3.7 Motor Maintenance

An essential part of any economic comparison is the cost of keeping spares as well as the frequency of maintenance. A supply of brushes and brush holders are necessary stock items for dc motors. A dc motor must be regularly taken out of service to check or replace brushes and at less frequent intervals to resurface the commutator. Maintenance of these brushes is particularly burdensome in harsh environmental conditions. Except for the bearings, an induction motor is essentially maintenance free.

1.3.8 Ruggedness

Being on the rotating member, the rotor windings are the 'Achilles heal' of any motor. In this regard the squirrel cage rotor is clearly far more robust than the dc machine. Progressive degradation of insulation due to electrical stressing, thermal cycling and creep is eliminated in a squirrel cage winding. Over the years induction motors have been successfully used under conditions which could not be sustained by other electrical machine structures. If deterioration of the rotor does occur, such as a broken bar or end ring, the machine is often capable of continued operation, perhaps at reduced power, until maintenance can be scheduled. Modern computer data acquisition and signal processing techniques have made possible the early detection of broken bars and the like.

1.3.9 Standstill Performance

A normal dc machine cannot develop high torque for extended periods at standstill since the armature current then flows through a particular group of armature coils and commutator segments under this condition. Although special mill duty motors can be purchased to provide rated torque over several seconds they are essentially oversized relative to the induction motor which has no such limitation.

1.3.10 Size and Weight

Mainly because of the commutator assembly, a dc machine of the same torque capacity is significantly larger than the equivalent induction machine. As a result the 'footprint' of the motor is larger resulting in increased real estate to house the motor. Access doors must be larger and larger cranes need be used for maintenance. All of these factors lead to a greater overall cost for the installation.

Table 1.1 shows a comparison of the two motors for three widely different sizes corresponding to 6 MW, 75 kW and 1.5 kW [1]. The data for the two smaller machines are for standard squirrel cage machines while the larger machine is specially designed for a variable speed application. Machines specifically designed for variable speed would normally have a higher efficiency due to relaxation of the line starting requirement. Note the substantial improvement in efficiency and reduction in inertia available in the 6 MW cage induction machine

1.4 Comparison of Synchronous and Induction Machines

The synchronous motor is another machine which possesses many of the advantages of the induction motor. However, the synchronous machine needs slip rings, brushes and an insulated winding on the rotor except for small motors where the field winding can be replaced by permanent magnets.The synchronous machine can be operated at unity or even leading power factor resulting in a smaller kVA requirement for the machine side converter than for the induction motor. However, as we will see later in the book, operation at unity power factor reduces the torque transient response of the machine. Also, torque response is reduced due to the relatively high value of synchronous

Table 1.1 Comparison of Induction and DC Motors for Three Diverse Sizes

Machine Size	Large		Medium		Small	
Machine Type	Cage	DC	Cage	DC	Cage	DC
Rating(kW)	6000	6000	75	75	1.5	1.5
Speed (rpm)	60	60	1500	1500	1500	1500
Efficiency (%)	95.1	92	93.5	88	75	83
Inertia ($kg\ m^2$)	30000	60000	0.75	0.673	0.0039	0.0098
Length (cm)	1500	1500	86.3	102.2	32.25	41.6
Width (cm)	410	473	45	39	17	21
Weight (kg)	–	–	385	480	23	42.5

reactance which must be compared with the armature leakage inductance of the dc motor. Application of wound field synchronous motor drives are therefore limited to very high horsepower applications in which their overall cost favors the synchronous motor over an induction motor. An important factor in this cost trade off is the price advantage of a thyristor bridge which can be commutated simply by means of the synchronous motor emf versus a force commutated bridge. Since the switch itself must be relied upon to interrupt the current in an induction motor due to its inherent lagging power factor, expensive power switching devices such as bipolar transistors must be used which frequently gives a cost advantage in a high horsepower application to the synchronous motor drive. However, as better power switching devices such as gate turn off thyristors (GTOs) continue to decrease in cost, such situations are expected to diminish in the future.

The future for variable speed permanent magnet motor drives is, on the other hand, bright. However, issues concerning the cost of the magnets limit their application to relatively small values of horsepower.

1.5 Inverters for Adjustable Speed

The process of converting dc to ac power is called *inversion* and it is the *inverter* which creates the variable frequency from the dc source which is used to drive an induction motor at a variable speed. In general, two basic types of inverters exist which are totally different in their behavior. The so called Voltage Stiff (or Source) Inverter abbreviated as VSI is more common and this type of inverter creates a relatively well defined switched voltage waveform at the terminals of the motor. The dc bus is maintained as *voltage stiff* by the use of a large capacitor in the dc link. A *voltage source* inverter differs from the voltage stiff inverter in that the dc link is established with a dc source (battery for instance). The resulting motor current is then governed primarily by the motor load and the speed. The VSI is also termed a voltage source inverter. However, it is clear that the dc link itself is truly not a power source if the link voltage is supported only by an energy storage element such as a dc capacitor. The VSI is typically subdivided into two forms the so called *six step inverter* and the *pulse width modulated inverter.*

The second type of inverter, the Current Stiff (or Source) Inverter (CSI) provides a switched current waveform at the motor terminals. The dc bus is maintained as *current stiff* by use of a large inductor in the dc link. The voltage waveform is now governed primarily by the motor load and speed.

1.6 The Six Step Voltage Stiff Inverter (VSI) [2]

A simplified diagram of a basic three phase voltage stiff inverter bridge is shown in Figure 1.3. Note that the structure of this circuit follows the general depiction of an ac drive shown in Figure 1.2. The rectifier serves to establish a dc potential in much the same manner as in the dc motor drive. In addition, a relatively large electrolytic capacitor is inserted to 'stiffen' the link voltage and provide a path for the rapidly changing currents drawn by the inverter. It is the rms value of this ripple current which determines the value of the capacitor used. The capacitor is of appreciable size and cost, frequently 2,000 to 20,000

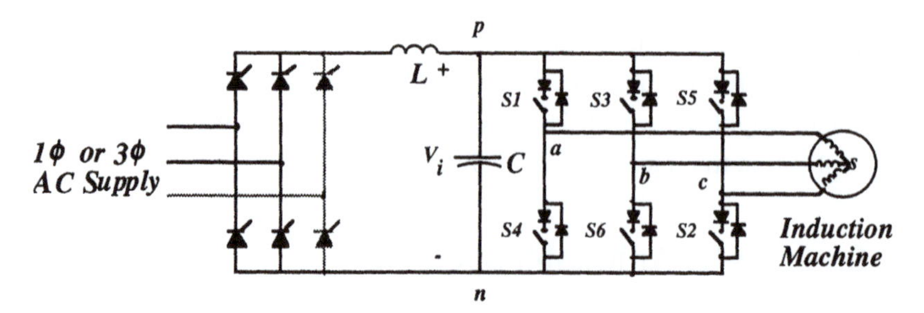

Figure 1.3 Basic three phase voltage stiff inverter (VSI)

microfarads and is a major cost item in the system. It is also usual to interpose some inductive reactance between the rectifier and the ac supply to limit fault current and to reduce the severity of the commutation dips produced by the rectifier. This series impedance is also helpful in attenuating the voltage spikes which may enter through the rectifier bridge due to switching or lightning strikes out in the utility system.

The inverter acts somewhat as the equivalent of the commutator assembly in a dc motor and converts the dc voltage to a variable frequency ac voltage. The inverter bridge is similar to the rectifier used in a dc motor drive except, because of the lagging power factor presented by the induction motor, the thyristors must be replaced with devices which are capable of being turned off as well as on. The choice at present is:

- Thyristors plus External Commutation Network
- Bipolar Junction Transistors (BJTs)
- MOS Field Effect Transistors (MOSFETs)
- Insulated Gate Bipolar Transistors (IGBTs)
- Gate Turn Off Thyristors (GTOs)
- MOS Controlled Thyristors (MCTs)

Since thyristors are not inherently capable of turn off, they require external commutation circuitry not shown in Figure 1.3. The types of commutation circuits are numerous but typically require one additional *commutation* thyristor and one capacitor (or pair of capacitors) per phase. Since the capacitor needs to have its polarity reversed in preparation for a commutation, a technique known as *resonant reversal* is used which requires at least one and often two resonant reversal inductors per phase. In addition, extra *snubbing* circuits are placed across and in series with each thyristor in order to keep them within their specified tolerances with respect to their turn on and turn off capabilities. The result is a complex circuit having numerous components and requiring a precisely timed firing sequence to prevent accidental short circuits. Thyristors have one important advantage over transistors which is their ability to withstand a substantial fault current for a brief time before protective devices (fuses or circuit breakers) operate. However, when used as an externally commutated switch they are presently of more interest for historical than for practical reasons.

Transistors have nearly completely replaced thyristors in inverter circuits below 500 kW. They are available in ratings to 1800 V and several hundred amperes and increases in the voltage rating have been occurring almost yearly. They have the great advantage of being able to be turned off as well as to turn on the current in the device. Hence, auxiliary components to accomplish turn off are not needed. Snubbing is still necessary if the transistors are to be used near their maximum ratings, which is generally an economic necessity. Of the transistor family, the bipolar junction transistor or BJT is the lowest cost and thus most widely used. The circuitry required to drive the transistor base is, however, relatively complex and is usually fabricated in a Darlington configuration to reduce the demand on the base current driver. A small reverse voltage must typically be applied to the base of a transistor in the off state and at turn off it is usually necessary to extract current from the base. IGBTs (insulated gate bipolar transistors) and MOSFETs (metal oxide semiconductor field effect transistors) are more recent additions to the transistor family and promise to replace BJTs in lower power applications. They do not suffer from second breakdown, a destructive loss mechanism in bipolar transistors that must be carefully controlled. Since both devices are turned off by field effect techniques, the current demand on the gate drive is minimal and they can even be turned off with an integrated circuit chip. Overall losses, parts count, and driver cost are markedly reduced with these devices resulting in an increasingly competitive product even though the devices remain more expensive than a BJT.

Gate turn off thyristors (GTOs) and MOS Controlled Thyristors (MCTs) are relative newcomers to the scene and promise to have the benefits of both thyristors and transistors. GTOs are both turned on and turned off by applying short gate current pulses. However, they require complex gate circuitry similar to BJTs. GTOs are presently available to 5000 volts and are becoming widely used in inverters operating at the utility medium voltage level (>1000 V). Such voltages are a necessary operating condition when the rating of the motor exceeds a few hundred horsepower.

In addition to the turn off element denoted as an ideal switch in Figure 1.3, each arm of the bridge normally contains an inverse parallel connected diode. These diodes which are called the *return current* or *feedback diodes*, must be provided to allow for an alternate path for the inductive motor current which continues to flow when the main power device is turned off. When regeneration occurs the roles of main power device and diode reverse. The diodes now return the regenerated power to the dc link while the thyristors carry the reactive current. The return of power to the dc link will raise the link voltage above its normal value and steps must be taken to absorb this regenerated power to prevent a dangerous link voltage buildup. Typically a resistor is switched in parallel with the dc link capacitor to absorb this energy or the input bridge is made bi–directional by adding a second inverse parallel bridge in much the same manner as Figure 1.1(c).

The basic operation of the six step voltage inverter can be understood by considering the inverter to effectively consist of the six ideal switches shown in Figure 1.3. While it is possible to energize the motor by having only two switches closed in sequence at one time, it is now accepted that it is preferable to have three switches closed at any instant since this strategy produces a higher output voltage under any operating condition [4]. This second pattern produces the voltage waveforms of Figure 1.4 at the terminals a, b and c referred to the negative dc potential n. The numbers written on the level parts of the waveforms indicate which switches are closed. This sequence of switching is in the order 561, 612, 123, 234, 345, 456 and back to 561. The line to line voltages V_{ac},V_{bc}, V_{ac} and line to neutral voltages V_{as},V_{bs}, V_{cs} then have the waveform shown in Figure 1.5. The line to line voltage contains an rms fundamental component of

$$V_{ll\,(rms\ fund)} = \frac{\sqrt{6}V_I}{\pi} \tag{1.6–1}$$

Thus a standard 460 V, 60 Hz induction motor would require 590 volts at the dc terminals of the motor to operate the motor at its rated voltage and speed. For this reason a 600 volt dc bus is quite standard in the US for inverter drives.

Although the induction motor functions as an active rather than a passive load, the effective impedances of each phase remain balanced. That is, so far as

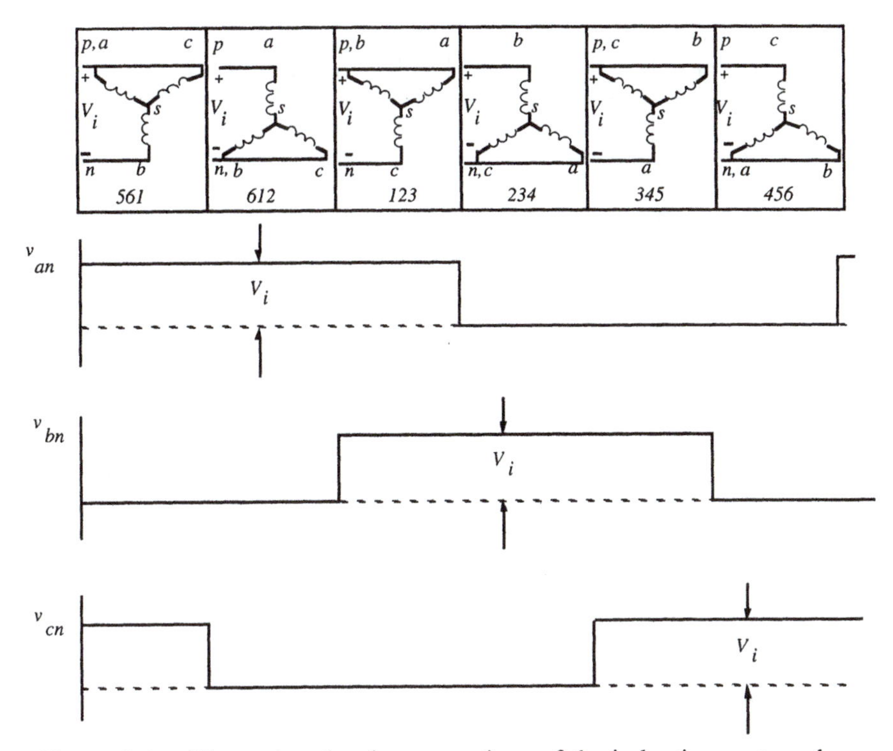

Figure 1.4 Illustrating the six connections of the induction motor when switched by a six step voltage stiff inverter

voltage drops are concerned, the machine may be represented by three equivalent impedances as shown in Figure 1.4 for the six possible connections. Note that a specific phase is alternately switched from positive pole to negative pole and that it is alternately in series with the remaining two phases connected in parallel or it is in parallel with one of the two phases and in series with the other. Hence the voltage drop across the phase is always 1/3 or 2/3 of the dc bus voltage with the polarity of the voltage drop across the phase being determined by whether it is connected to the positive or negative pole. A plot of one of the three motor line to line voltages is also given in Figure 1.5. Note that harmonics of order three and multiples of three are absent from both the line to line and the line to neutral voltages (and consequently absent from the currents). The presence of six *steps* in the line to neutral voltage waveform is the reason for this type of inverter being called a *six step* inverter. A Fourier analysis of these waveforms indicates a *square wave* type of geometric progression

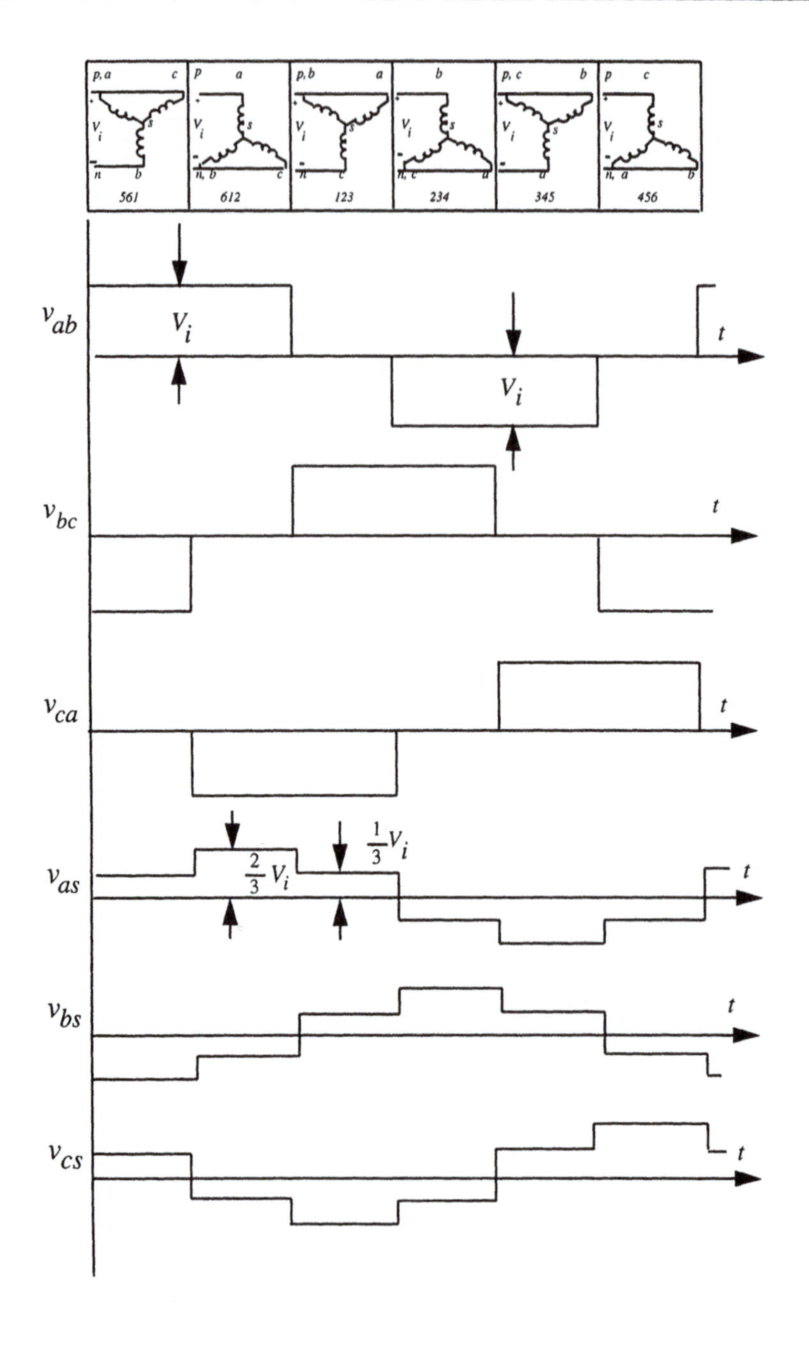

Figure 1.5 Line–to–line and line–to–neutral voltages across the induction motor phases when operating from a six step voltage inverter

of the harmonics. That is, the line to line and line to neutral waveforms contain 1/5th of the fifth harmonic, 1/7th of the seventh harmonic, 1/11th of the eleventh harmonic and so forth.

1.7 Six Step Inverter Speed and Slip Frequency Control Techniques

In general, the motor requires a nearly constant amplitude air gap flux for satisfactory working of the motor. Since the air gap flux is the integral of the voltage impressed across the magnetizing inductance and, assuming that the air gap voltage is sinusoidal,

$$\lambda_{ag} = \int v_{ag} dt = \int V_{ag} \sin\omega t dt$$

$$= -\frac{V_{ag}}{\omega} \cos\omega t$$

Hence, the motor requires a constant ratio of volts to hertz (plus a small additional boost to overcome stator resistance) and the dc voltage must be adjusted in almost direct proportion to frequency. A variable direct voltage can be readily obtained by using the controlled rectifier in Figure 1.3. In this case the inverter continues to operate in the so called six step mode giving the same motor voltage wave forms as shown in Figure 1.5 at all speeds. However, as the frequency is reduced the amplitude of the waveforms can be reduced in direct proportion (assuming ideal control) as shown in Figure 1.6. In general, the large dc link capacitor prevents the amplitude of the ac voltage from being increased as rapidly as the frequency, which is developed with practically no delay by simply feeding triggering pulses to the inverter switches. Hence, it is customary to slave the frequency to the dc link voltage as shown in Figure 1.7 to prevent the motor from ever receiving an inappropriate value of Volts/Hertz.

Of the two control 'handles', frequency control is by far the most critical as small changes in frequency produce large changes of slip frequency and, hence, large changes in current and torque. By slaving the frequency command to the dc bus voltage, the rate of change of frequency is generally limited to a value to which the motor can respond without drawing an excessive current or without regenerating. However this only solves half the problem since the slip frequency is also influenced by motor speed. A mechanical overload will quickly increase the slip frequency and hence motor current. This overload

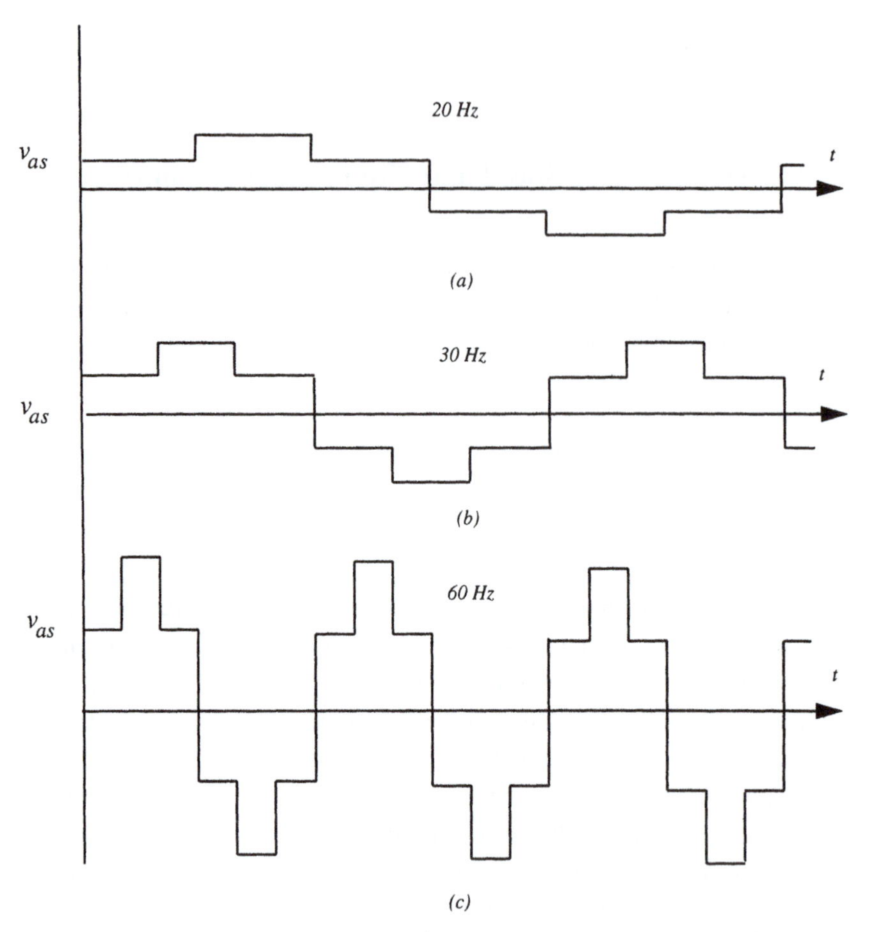

Figure 1.6 Six step line–to–neutral inverter output voltage waveforms at three different line frequencies

must be detected quickly and the inverter frequency reduced accordingly so that the slip frequency and current are kept low. An overload condition is most easily sensed by monitoring the dc link current. Similarly, an overhauling load will reverse the motor torque and cause regeneration and a rise of dc link voltage if the regenerated power exceeds the losses in the inverter. If there is no provision to accept this regenerated power, the motor frequency must be increased to prevent excessive regeneration. Thus, both link current and voltage contribute essential sensed variables in any control scheme.

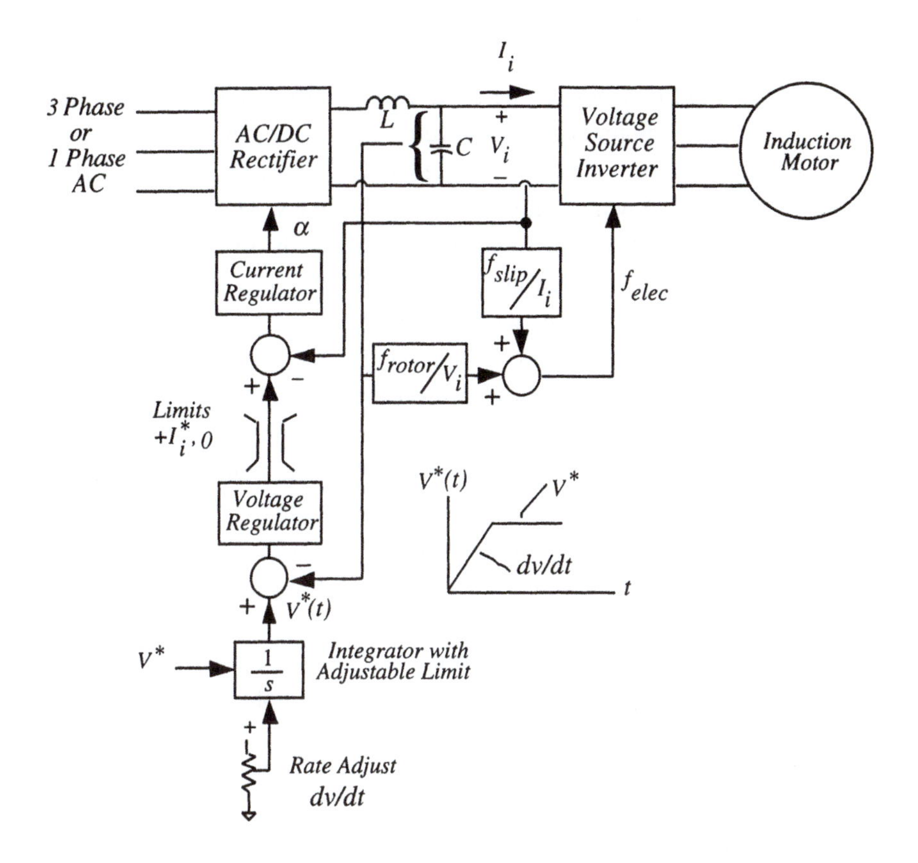

Figure 1.7 Open loop speed control system with limits on acceleration, frequency and current

The most common method for frequency control is to use a voltage controlled oscillator (VCO) to set the inverter frequency. This can be easily interfaced to the demanded speed signal via acceleration and deceleration limits which prevent excessive slip as a result of a sudden change of speed demand. In cases where speed regulation is of concern, improved regulation can be obtained by increasing the frequency as a function of dc link current (i.e. load) thereby adjusting the slip frequency while keeping the rotor speed roughly constant. Speed accuracy approaching 1% can be readily achieved in this manner. These controls are shown in Figure 1.7. Note that this type of motor drive is, in effect, an open loop speed controller since speed is not sensed explicitly. Such drives are well suited to applications in which speed regulation requirements are modest (>1–2%). Since a costly tachometer and associated hardware is not required, this type of controller is in widespread use in industry as a general purpose drive.

If desired, slip frequency can be controlled directly if a digital speed transducer is fitted to the motor shaft. The inverter frequency can then be generated by the digital addition of the shaft speed pulses and slip frequency pulses obtained from a VCO which determines slip frequency. This addition process must be carried out at a high frequency so that each inverter period of 60 degrees is comprised of a large number of shaft pulses to which are added a number of slip frequency pulses which is sufficiently large so that the inverter periods are of reasonable uniform duration. If regeneration is required the reversed sign of the speed error signal must cause the slip frequency VCO to subtract shaft pulses from the shaft transducer on a pulse for pulse basis to reduce the inverter frequency below synchronism and consequently produce regeneration. A typical number of pulses corresponding to full slip is 20–30. The addition/subtraction process must correctly take into account coincident or overlapping pulses. However, this is a relatively simple problem with modern digital electronics. Figure 1.8 shows such an arrangement.

It should be noted that precision speed control can be obtained without the need for a closed loop speed control if the induction motor is exchanged for a synchronous reluctance or permanent magnet motor. Speed holding precision is then totally dependent on the stability of the VCO. If the VCO is replaced by a crystal oscillator, divider and phase locked loop, then the precision can be as good as that of the crystal. This technique has found extensive application in the fiber spinning industry where a large number of motors must be speed–controlled with great precision and in exact synchronism with each other.

1.8 The Pulse Width Modulated Inverter (PWM) [2]

The PWM inverter again utilizes the voltage stiff dc link but combines both voltage control and frequency control within the inverter itself. It operates from a fixed voltage dc source, i.e. an uncontrolled diode rectifier or a battery (or perhaps both). In this case the power switches in the inverter are switched at a high frequency thus operating, in effect, as choppers. In general, modulation techniques fall into two classes: those which operate at a fixed switching ratio to the fundamental switching frequency (block or *picket fence* modulation), and those in which the switching ratio is continuously changing, usually

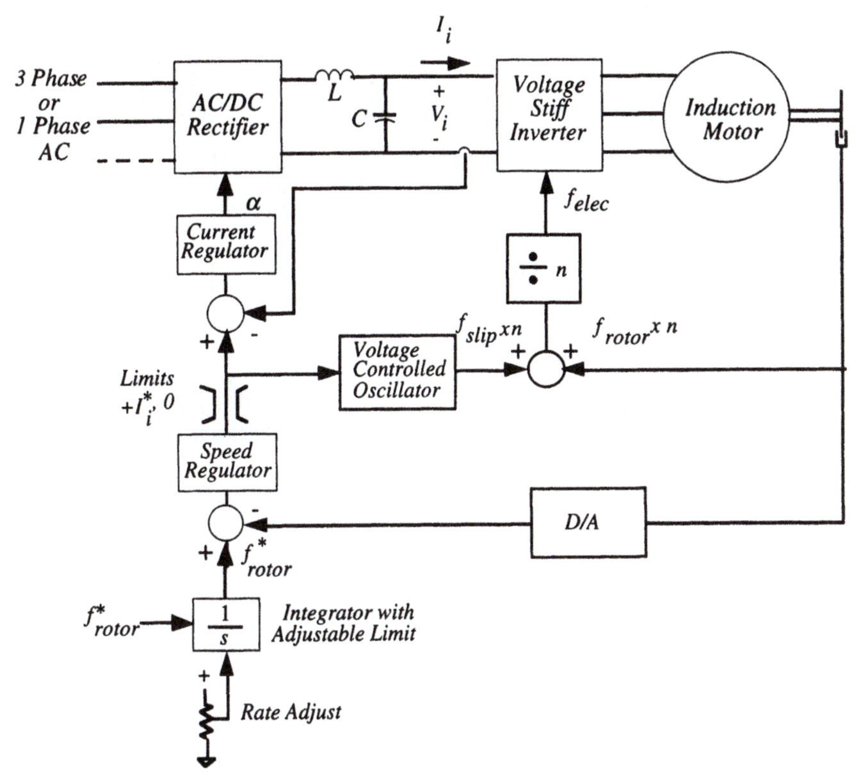

Figure 1.8 Closed loop speed control with digital addition of speed and slip frequency

sinusoidally, to synthesize a more nearly sinusoidal motor current (called *sinusoidal pulse width modulation* or sinusoidal PWM).

1.8.1 Block Modulation

Block modulation is the simplest type of modulation and is closest to simple six step operation. Instead of varying the amplitude of the motor voltage waveform by variation of the dc link voltage, it is varied by switching one or two of the inverter thyristors at a fixed switching ratio and adjusting the notches of the resulting pulses to control the motor voltage to suit the speed. The number of pulses is maintained constant over a predetermined speed range although the number of pulses may change discretely at several prescribed speeds.

A simple form of block modulation is shown in Figure 1.9 where the chopping is limited to the middle 60 degrees of each device conduction period resulting in minimum switching duty on the semiconductor switches. In spite

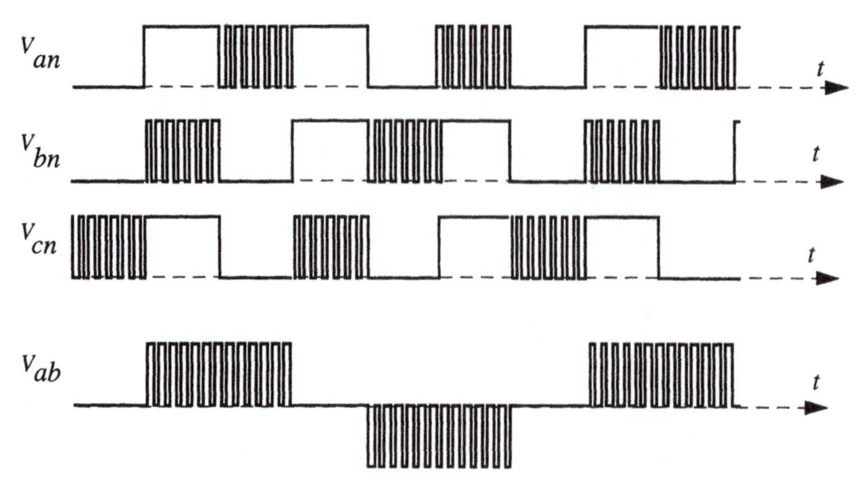

Figure 1.9 Typical block modulation waveforms showing voltages from line to negative dc bus and line to line voltage V_{ab}

of the similarities between block modulation and the six step mode, the torque pulsations at low speed are much less severe than for the six step inverter. The harmonics typical of a six step inverter are present also with block modulation, but there are also higher harmonics associated with the chopping frequency. The motor losses and motor noise, however, are significant compared to more elegant modulation algorithms and this method is no longer widely used.

1.8.2 Sinusoidal PWM Modulation

The objective of sinusoidal PWM is to synthesize motor voltages to produce currents as near to a sinusoid as economically possible. The lower order voltage harmonics can be greatly attenuated, leaving typically only 2 or 4 harmonics of substantial amplitude close to the chopping or carrier frequency. The motor now tends to rotate much more smoothly at low speed. Torque pulsations are virtually eliminated and the extra motor losses caused by the inverter are substantially reduced. To counterbalance these advantages, the inverter control is complex, the chopping frequency is high, typically 500–2500 Hz for GTOs and up to 5,000 or more for IGBT transistors, and inverter losses are higher than for the six step mode of operation. In one method for approximating a sine wave a high frequency triangular wave is compared with a fundamental frequency sine wave as shown in Figure 1.10. When the low frequency

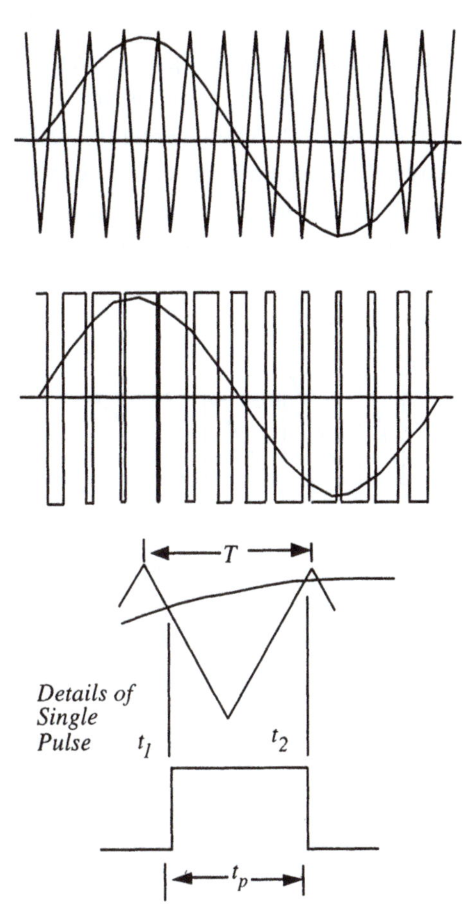

Figure 1.10 Illustration of sinusoidal pulse width modulation (PWM) technique

sine waves are used with 120 degrees phase displacement, the switching pattern for the six inverter devices ensues.

1.8.3 Synchronous vs. Asynchronous Modulation

If the carrier and modulation are to be synchronized, there must be a fixed number of carrier cycles in each modulation cycle. If this number is chosen to give a 'good' sinusoidal current waveform at low frequency, say at 1 Hz, then the ratio of carrier frequency to fundamental frequency will be in the neighborhood of 100–200. If the same ratio is then used at a motor frequency of 100 Hz, the chopping frequency becomes 10–20 kHz. This switching frequency is far too high for normal thyristor inverters and somewhat too high even for bipolar transistor inverters. Hence, a technique of 'gear changing' must be employed which reduces the switching ratio in steps as the fundamental frequency increases. As shown in Figure 1.11, a commercially available system employing a custom chip designed for such a purpose is now available [5]. Hysteresis must be included at each change in switching frequency ratio so that the system does not cycle continuously between two differing switching ratios at certain speeds.

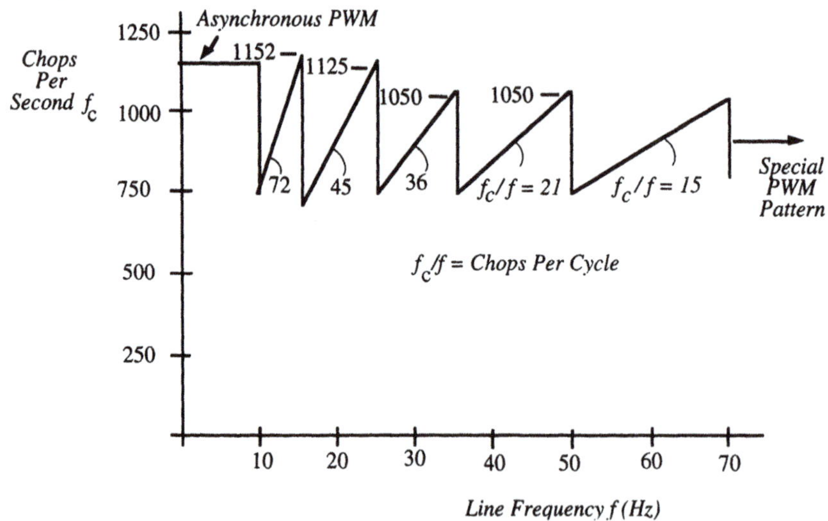

Figure 1.11 Illustration of 'gear changing' in a pulse width modulated inverter

Whereas the carrier frequency must be a fixed ratio of fundamental frequency at moderate and high speed to prevent undesirable *subharmonics* from appearing, asynchronous operation of the carrier frequency becomes acceptable at low speeds where the effect of the differing number of carrier cycles per modulation cycle is small. However, a change over to synchronous operation is required at some speed and some type of *phase locking* technique must be introduced to achieve the synchronization.

While Figure 1.10 provides a good picture of the modulation process, once the possibility of using a microprocessor is introduced, the 'analog' solution obtained by the intersection of a triangle wave and a sine wave becomes only one of many possibilities. This so called *natural modulation* can be replaced to advantage by *regular modulation* in which the modulating waveform is piece wise constant, i.e. sampled at the carrier frequency. An example of carrier and modulated waveforms for such a scheme is reproduced in Figure 1.12 for a

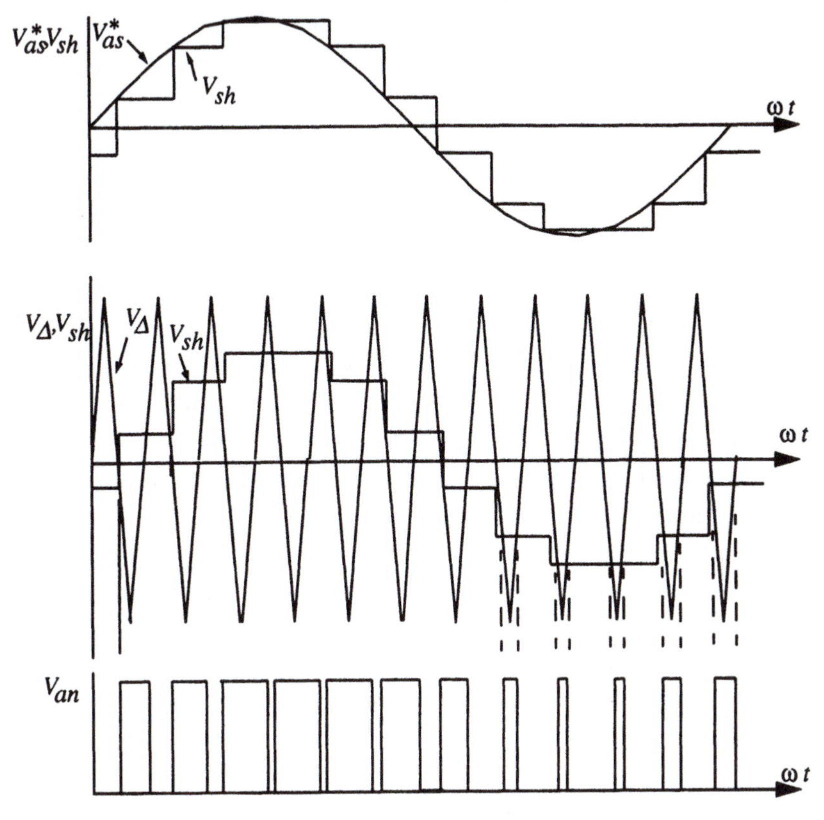

Figure 1.12 Switching pattern using regular sampling modulation

switching ratio of 9 [6].

1.8.4 Pulse Dropping

Referring to Figure 1.10 it can be noted that the center pulses of the PWM modulation are narrower than the pulses where the fundamental voltage passes through zero. These center pulses typically become progressively more and more narrow as the speed (and hence voltage) increases. When the conduction pulses become very narrow there is insufficient time to reverse bias the off going power device in force commutated systems. Hence, if the voltage across this device increases in a positive sense too soon, the device will again conduct and high losses will occur. Hence, if the modulation scheme calls for a pulse width below this minimum time, then this pulse must be omitted altogether and a cancelling adjustment made to a nearby pulse to correct for the change in the fundamental component. Once again, pulse dropping must be coupled with a small hysteresis to avoid cycling across the threshold condition.

1.8.5 Changeover to Six Step Mode

Without pulse dropping techniques it is apparent that sinusoidal modulation prevents the motor terminal phase voltage from reaching its maximum value (0.78 V_I) since 'notches' of minimum width imposed by the switching limitations of the solid state switching devices prevent full fundamental voltage from being reached. Operation at a reduced 'maximum' fundamental voltage is costly since the attendant motor must be designed for a lower voltage, implying in turn, higher currents for a given horsepower requirement. This increased current duty then impacts on the choice of switching device used in the inverter design and results in a more costly system. This disadvantage can be overcome if the modulation process is allowed to 'saturate' at high speeds, so that the waveform takes on first several forms of block modulation and, finally, full six step operation. During the intervening process when the transition is made from sinusoidal PWM through the various block modes to square wave operation, additional harmonics will exist which cause extra heating in the motor. However, since the motor is running near its rated speed, cooling is good and there is an overall benefit.

1.8.6 Losses Caused by Inverters [3]

The non–sinusoidal output of the inverter creates extra losses in the motor. The six step mode is worst with respect to voltage harmonic content with 20% of the 5th harmonic, 14% of the 7th and so forth. Sinusoidal PWM reduces the low order harmonics to very small proportions, but significant harmonics remain at or near at the carrier frequency. Frequency dependent losses such as skin effect and iron losses often result in motor losses being greater under PWM operation than with a six step inverter. Different manufacturers give varying degrees of derating, but typically the motor can give 80–90% of its nameplate output at its rated speed without an increase from its rated operating temperature. At lower speeds extra derating is required to counteract the poorer cooling. The inverter also has internal losses caused by device conduction voltage drop, losses in the auxiliary voltage supplies and gate/base drives, switching losses, losses in snubbers or other circuits to aid the switching capability of the power switching device and losses in the commutation circuits themselves (if any). The last three of these losses increase with inverter output frequency. An attempt to synthesize the best possible sine wave for the motor by selecting a high carrier frequency may well create more losses in the inverter than it saves in the motor. Overall efficiencies at full speed for an inverter fed induction motor drive lie between 70% for the smaller industrial drives to over 90% for large drives such as those used for traction.

1.8.7 Torque Pulsations

Torque pulsations, sometimes called *cogging torques*, are always present when the motor current is non–sinusoidal, but for all but the lowest speeds the rotor inertia effectively smooths them out to give a uniform speed. The six step voltage stiff inverter produces noticeable torque pulsations below about 3 Hz. However, with sinusoidal PWM these effects are virtually eliminated. In addition to torque pulsations which are an inherent result of the harmonics in the applied voltage waveform, a torque transient may occur whenever there is a sudden change of motor voltage or frequency. The latter can be avoided electronically, but considerable care must be taken in the design of the modulation scheme, particularly when changes in pulse number occur through pulse dropping, to ensure that the effective fundamental voltage before and after is unaltered.

1.8.8 Induction Motor Speed Control Using a PWM Inverter

.Figure 1.13 shows a speed controller based on use of a PWM inverter. Note that in this case the time delay introduced by the link capacitor can be avoided. While the voltage and frequency can now be changed almost instantly, acceleration and deceleration limits are again used to prevent changing the slip too rapidly, which as noted previously, could result in an over current condition

1.9 The Current Stiff Inverter (CSI)

Since induction motors have traditionally been designed to operate from a voltage source, the voltage stiff inverter was developed and used first since it is, in principle, an approximation of the waveform presented to the motor by the utility. The current stiff inverter on the other hand is very different in concept. Current fed inverter drives have been in use only for about twenty years. They have, however, several properties which make them attractive as well as an

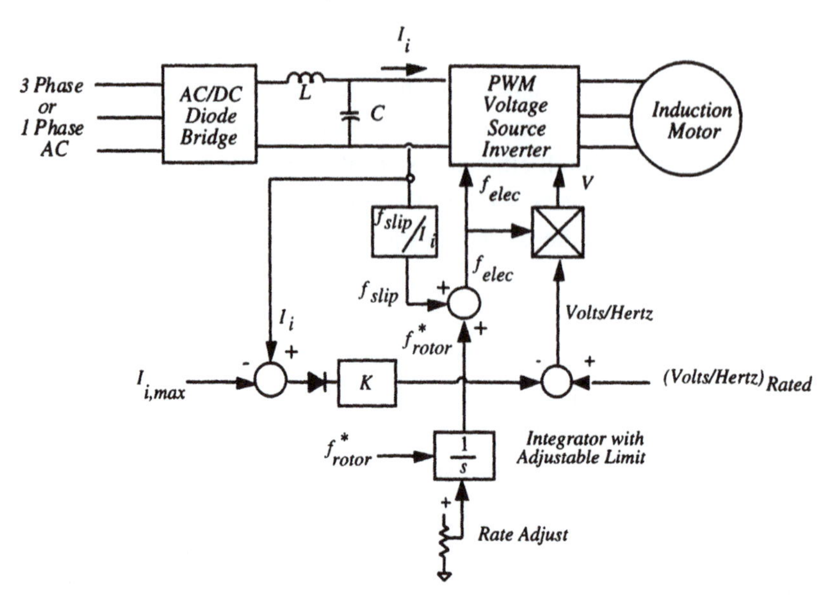

Figure 1.13 Basic induction motor speed control block diagram using a PWM inverter

inevitable number of undesirable effects. As the name implies the inverter switches of a CSI are fed from a constant current source. While a true constant current source can never be a reality it is reasonably approximated by a controlled rectifier or chopper with a current control loop as well as a large dc link inductor to smooth the current. The circuit is shown in Figure 1.14. In this case

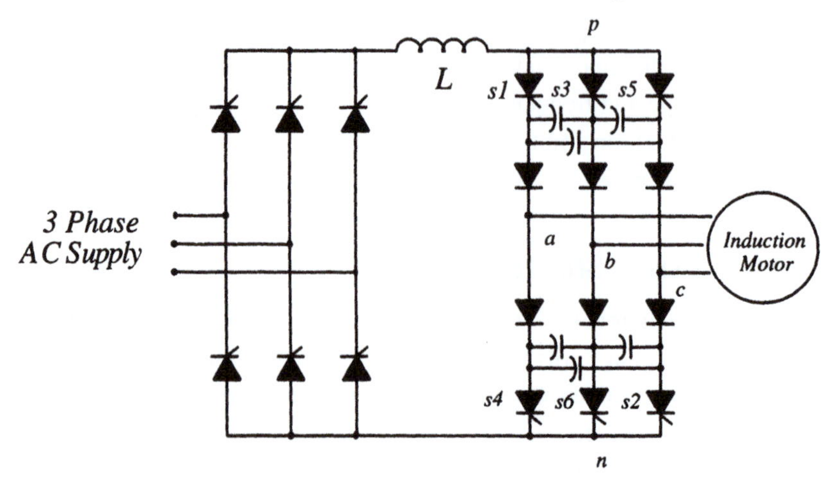

Figure 1.14 Schematic of current stiff inverter (CSI) drive

the current is switched sequentially into one of the three motor phases by the top half of the inverter and returns from another of the phases to the dc link by the bottom half of the inverter. Since the current is constant, there will be zero voltage drop across the stator winding self–inductance and a constant voltage drop across the winding resistances. Hence, the motor terminal voltage is not set by the inverter but by the motor. Since the motor is wound with sinusoidally distributed windings the consequent voltages which appear on the terminals of the motor are nearly sinusoidal. Ideally, the current waveform is an exact replica of the voltage waveform of a six step voltage stiff inverter. In practice, the motor currents cannot change instantaneously, and the transitions in the current waveform have a finite slope. During these transitions the current transfers from one inverter thyristor to the next with the aid of one of six commutating capacitors. A sketch of the motor voltage and currents is shown in Figure 1.15.

A very attractive feature of a CSI is its ability to ride through a commutation failure and to return naturally to normal operation. This feature is in marked contrast to most voltage fed inverters in which costly preventive measures are necessary. Another benefit of the CSI is its ability to regenerate into the utility supply by simply reversing the polarity of the rectifier dc output voltage. Again, this in contrast with the VSI where the current flow rather than dc voltage must be reversed, thereby requiring an additional inverse parallel connected six pulse bridge. This again is a desirable advantage over a VSI. Operation of the motor at negative slip automatically causes a reversal in the dc link voltage since the link current rather than voltage is the controlled variable. Hence, power is automatically regenerated into the ac supply.

An important limitation in the application of a CSI drive is the fact that open loop operation in the manner of a VSI is not possible. Figure 1.16 shows the torque/speed curve of a typical induction motor fed from both a voltage source and a current source. A marked 'peaking' exists in the case of the current source inverter. Inspection of this torque/speed curve suggests two possible operating points: one on the negatively sloped region which is usually regarded as stable and one on the positively sloped region which occurs at a speed below the point where the maximum torque is reached (breakdown torque) and is generally unstable (depending, however, upon the load torque vs. speed characteristics).

Careful examination of the point on the stable side of the torque speed curve reveals that continuous operation is not feasible here because the work-

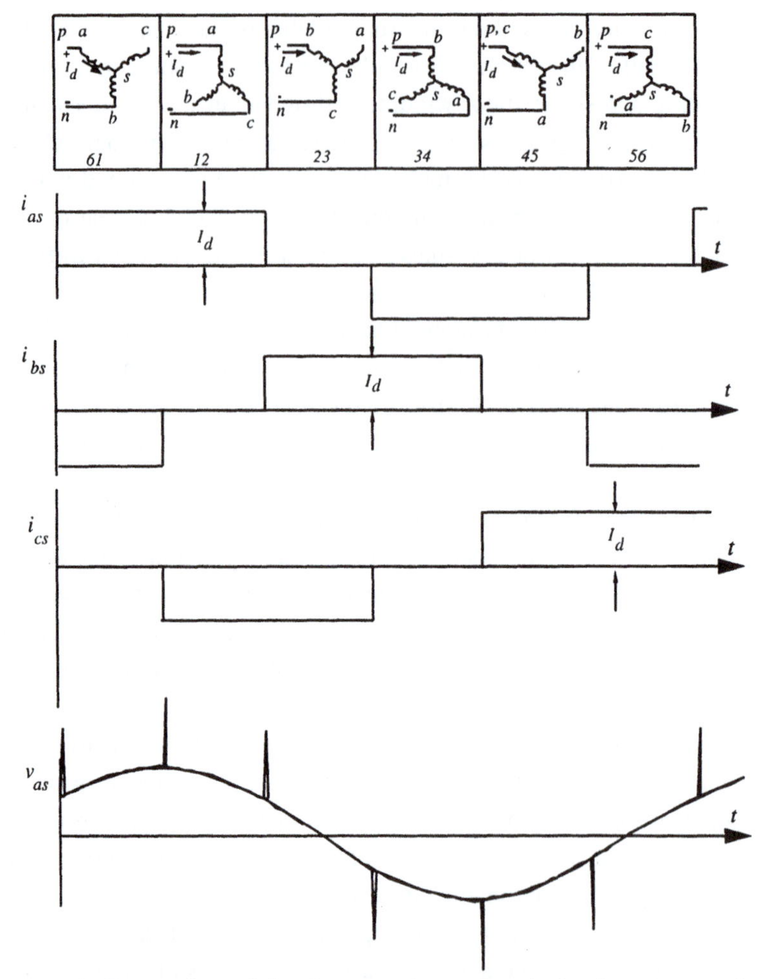

Figure 1.15 Illustration of the 6 connections of an induction motor when switched by a six step current stiff inverter showing motor line currents and one phase voltage

ing flux in the machine at this point is high, resulting in saturated operation and excessive magnetizing current and iron losses. On the other hand at the other point, the so called unstable point, the flux in the machine is near its rated value and losses are not excessive. (Note that this point corresponds to the intersection of the torque/speed curve for rated voltage with the torque/speed curve for rated current. At this point both voltage and current are thus at their rated values.). Unfortunately, being on the unstable side of the torque speed curve (for a

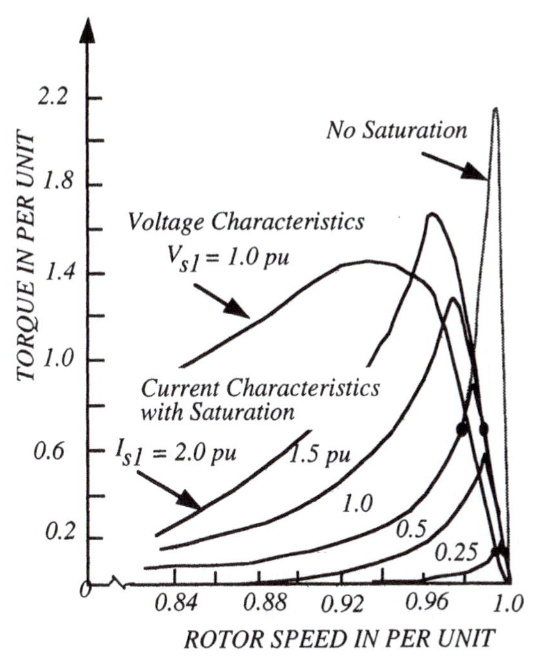

Figure 1.16 Induction motor torque–speed curves operating from a voltage source and a current source

current source) means that operation is impossible without some sort of feedback control which forces the machine to remain at this operating point. One system is a motor voltage control loop which regulates the motor voltage by controlling the input phase control rectifier (or dc link chopper). A representative block diagram of a system of this type is shown in Figure 1.17. Typically, an internal current loop is used as shown with the voltage error serving as a reference signal for the current regulator. Some IR drop compensation is often added and, since motor current and dc link current are proportional, this is easily accomplished. Additional compensating circuits are usually employed to improve system dynamics.

The size and cost of the ac commutating capacitors and the dc link inductor are the major disadvantage of this inverter. The capacitors are large because they must absorb the total energy stored in the leakage field of the winding when the current is commutated. In order to keep this energy interchange at a minimum, motors of special design are advantageous. In particular, these motors are designed to have as small a leakage reactance as possible. Unfortunately, this requirement is at odds with conventional motor design since small leakage reactances imply an over designed motor from the point of view of

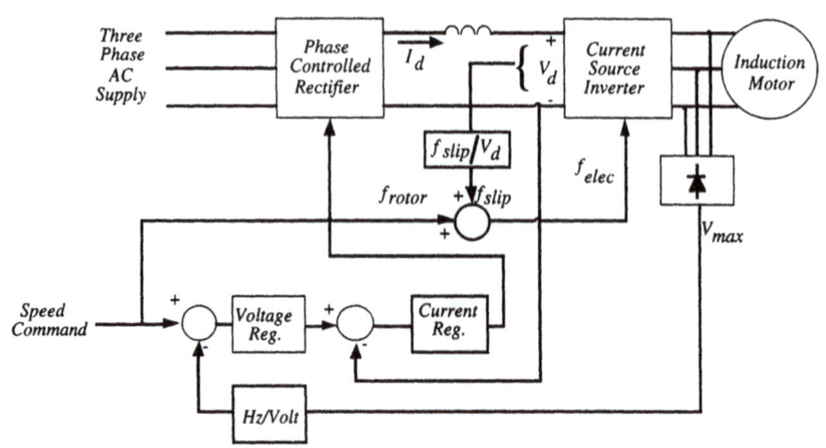

Figure 1.17 Block diagram of CSI–induction motor drive with motor voltage feedback

continuous rating. Finally, the use of thyristors rather than transistors or turn off thyristors in this circuit can be noted. While the CSI could potentially employ switches with turn off capability, they would not enhance the performance of the converter significantly since the rate of change of motor current and, thus, the voltage rise across the motor would remain fixed by the capacitors.

Turn off devices can be used in a current stiff link configuration if the capacitors are freed from their commutating requirements and simply placed across the terminals of the machine. Figure 1.18 shows a motor drive configu-

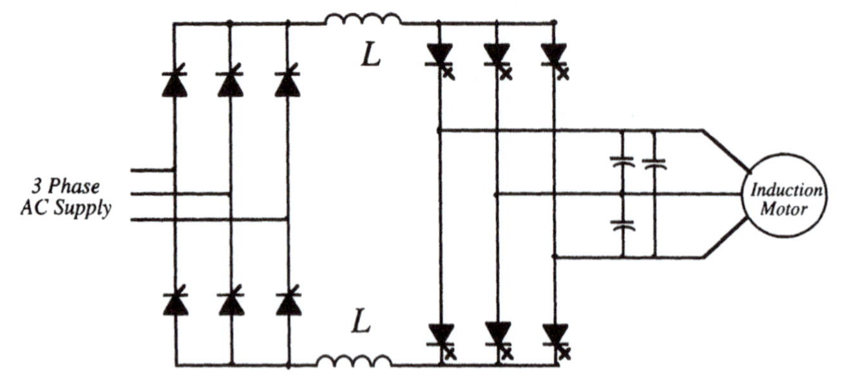

Figure 1.18 Current stiff inverter drive employing turn off devices (GTO is shown as an illustration)

ration which is becoming increasingly popular in high horsepower applications. In this case, the three top switches are sequentially switched, one at a time, to feed the link current into the motor/capacitor network and one of the bottom three switches returns the current to the link. The capacitors are now selected to absorb only the harmonic currents and need not be sized to absorb the total energy in the commutated phase during switching. The motor current is much more sinusoidal than the square wave current impressed on the motor in the conventional scheme. Pulse width modulation is again used to suppress the low frequency 5th and 7th harmonic torque pulsations which are inherent in six step operation. A major disadvantage of this scheme is the potential for resonance between the capacitors and the motor inductances. Care must be taken to avoid impressing current harmonics into the motor/capacitor network which will excite one of the system resonant frequencies. This possibility can be avoided by careful use of pulse width modulation. However, since the motor parameters must be known to implement such an approach, the drive is presently not popular for general purpose applications.

1.10 High Performance Drives vs. General Purpose Drives

At present, the vast majority of ac drives manufactured in this country and abroad concern applications which do not require fast transient response. Accordingly, the design of control circuits to service such ac drives are based on the so called 'per phase' T equivalent circuit of the induction motor [7],[8]. Such an equivalent circuit, however, is derived assuming steady state conditions. In the case of an induction machine, the resulting equivalent circuit assumes that the output power can be represented by an equivalent rotor resistor of $(1-S)R_2/S$ where S is the per unit slip. While this resistor is present in the equivalent circuit during transients, it is a fictitious resistor which does not truly represent the electrical behavior of the motor. Thus, the electrical time constants as calculated from this equivalent circuit are incorrect. Design of speed controllers such as Figure 1.7 can still proceed since speed is set in open loop fashion and the motor voltages and currents are not used except to perform a limiting function. Hence, the electrical time constants do not markedly influence the control gains.

When a closed loop speed control, such as Figure 1.18, was implemented on this basis, however, it became quite apparent that the motor time constants now began to influence the control gains [9]. When speed regulation requirements are not tight this control still yields adequate speed regulation of a few per cent. However, continued increase in gain to improve speed regulation and improve transient response leads to an instability and performance degradation at low speed which can not be predicted from the per phase equivalent circuit Clearly, additional factors are involved in control of an induction machine. These questions were largely answered by K. Hasse in his classic Ph.D thesis in 1969 which also pointed the way to improved controllers [10]. In this work, the essential need for the control of the spatial position of rotor flux and stator current was identified. It was emphasized that the electromagnetic torque is fundamentally the cross product of flux and current in much the same manner as other ac and dc machines. This is in contrast to the conventional per phase equivalent circuit approach which, in effect, views torque production as only frequency dependant and ignores spatial considerations. That is, torque production becomes equivalent to the power dissipated in a resistance inversely proportional to per unit rotor slip [i.e. $(I_2)^2R_2/S$].

The requirement for accurate spatial positioning of the stator ampere turns with respect to the rotor flux has led to an emphasis on precise regulation of the stator current based on information as to the spatial position of the rotor flux density. This form of modulation, termed current regulated pulse width modulation or CRPWM will be discussed in Chapters 4 and 6. This type of approach to ac motor control is variously called *field oriented control* or *vector control* depending upon whether the motor variables are viewed as distributed field quantities (flux densities and MMFs) or circuit variables, (flux linkages and currents). A simplified block diagram of a high performance drive is shown in Fig. 1.19. Note the presence of the inner current loop which is notably absent

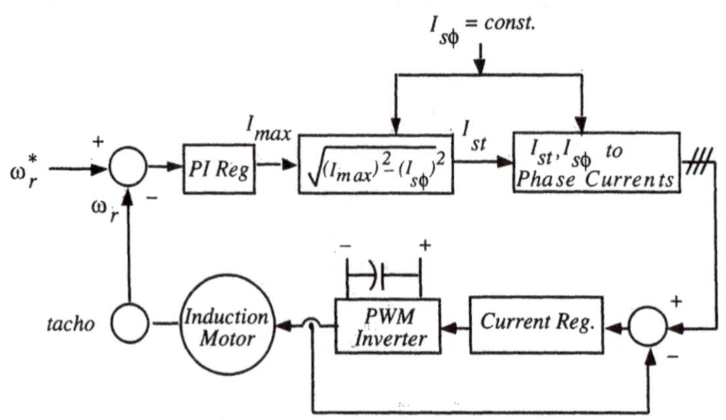

Figure 1.19 Simplified block diagram of high performance induction motor drive incorporating field oriented control

from the general purpose drive controller. The essence of field oriented control is that both the flux producing component and torque producing component of the stator current are identified and then expressed as equivalent three phase current commands. Regulation is then used to ensure that the stator currents are faithful reproductions of the commanded values.

While the principle of field oriented control is relatively straightforward, a familiarity with motor behavior beyond that afforded by the simpler per phase equivalent circuit is clearly necessary. In the next chapter we return to the basics to derive a true transient model of the induction machine. In subsequent chapters we use this model to develop a keen understanding of the principles involved in the design of a field oriented controllers for both induction and synchronous machines.

References

[1] N.D. Matheson, "Evaluation of Improved Induction Motor Control Schemes", Master of Eng. Science Thesis, University of Sydney, Sydney Australia, 1988.

[2] R.M. Davis, "Inverter–Fed Induction Machines", Proc. of Drives/Motors/Controls'82, 29 June–1 July 1982, Leeds U.K., pp. 66–75.

[3] D.W. Novotny, "A Comparative Study of Variable Frequency Drives for Energy Conservation Applications", University of Wisconsin Report #ECE–81–4, 1981.

[4] T.A. Lipo and F.G. Turnbull, "Analysis and Comparison of Two Types of Square Wave Inverter Drives", IEEE Trans. on Industry Application, Vol. IA–11, No. 2, March/April 1975, pp. 137–147.

[5] J.A. Houldsworth and W.B. Rosink, "Introduction to PWM Speed Control System for 3–Phase AC Motors", Electronics Components and Applications, vol. 2., No. 2, February 1980.

[6] S.R. Bowes, "New Sinusoidal Pulsewidth–Modulated Inverter", Proc. IEE, vol. 122, November 1975, pp. 1279–1285.

[7] C.J. Amato, "Variable Speed with Controlled Slip Induction Motor", in Conf. Record of IEEE Industrial Static Power Conversion Conf., Nov. 1–3, 1965, pp. 181–189.

[8] J.T. Salihi, P.D. Agarwal and G.J. Spix, "Induction Motor Control Scheme for a Battery Powered Electric Car (GM – Electrovair I)", in Conf. Rec. IEEE Industry and General Applications Society Annual Meeting, October 1967, pp. 107–114.

[9] A.B. Plunkett, J.D. D'Atre and T.A. Lipo, "Synchronous Control of a Static AC Induction Motor Drive", IEEE Trans. on Industry Applications, Vol. IA–15, No. 4, July/August 1979, pp. 430–437.

[10] K. Hasse, "Zur Dynamik Drehzahlgeregelter Antriebe Mit Stromrichtergespeisten Asynchron–Kurzschlublaufermaschinen", Ph.D. Dissertation, Tech. Hochschule Darmstadt, July 17, 1969.

Problems

Problem 1-1 Effect of Gating Delay on Inverter Waveform

Modern IGBT inverters which operate in the six step mode of Fig. 1.5, utilize what can be termed 180 degree switching. That is, gate pulses are applied to each of the two transistors making up one inverter leg for 180 electrical degrees. In effect, therefore, either one or the other of the two transistors is always on. Other types of switching have been used in the past, for example 150, 120 and 90 degree switching in order to prevent “shoot throughs”, i.e. both top and bottom transistors in conduction at the same time. The sketch below illustrates, for example, 120 degree switching for a three phase inverter. If the inverter load is assumed to be a simple balanced set of wye connected resistors, plot one line to line voltage and one phase voltage for 150, 120 and 90 degree switching.

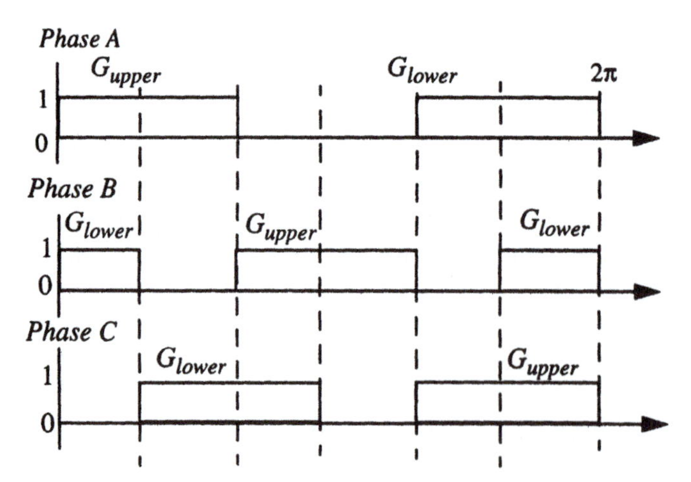

Gate pules to upper and lower transistor switches of each leg of a three phase inverter.

2 *d,q* Modelling of Induction and Synchronous Machines

2.1 Introduction

The induction motor is truly the workhorse of the industrial world. Until recently, however, they were limited only to driving loads which required nearly constant speed (or at best several discrete speeds) since the efficiency of the motor suffered drastically as the speed was lowered below the rated speed. With the advent of variable frequency power supplies this problem has been largely overcome. However, costs associated with the inverter have thus far prevented across the board replacement of dc motors by ac induction motor drives. Analysis of induction machines is typically presented by means of the so called per phase equivalent circuit. However, this model proves inadequate when applied to dynamic conditions as encountered in a variable frequency drive.

We will begin our study by first determining the coupled circuit equations of the induction machine. While the result will be a formidable set of equations we will learn that by using a complex vector representation these equations reduce quite readily to a manageable form. The theory is then extended to the synchronous machine including the effects of non–uniform air gap and non–symmetric rotor windings typical of such a machine.

2.2 Winding Inductances

Let us begin by considering a simplified representation of the winding distribution of one phase of a three phase induction machine as shown in Figure 2.1. For purposes of analysis we will assume that a uniform air gap induction machine can be modelled initially as two concentric cylinders separated by a thin air gap of constant radial length g. It will be assumed that the surface of the stator is equipped with sinusoidally wound three phase windings. We will neglect the effects of slotting so that the windings can be visualized as being of negligible depth and existing in the air gap. Also, harmonic components of flux

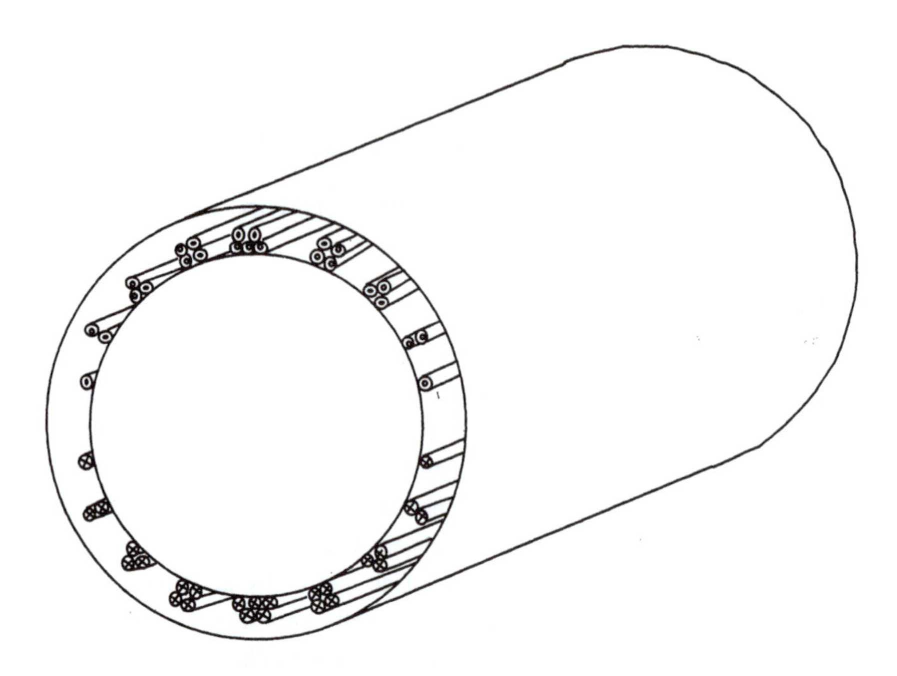

Figure 2.1 Placement of conductors of one stator or rotor phase around the air gap of a two pole machine. The details of the slot shape are not shown for clarity.

arising from the placement of the actual conductors in the discrete slots (harmonic leakage flux) will be neglected.

In general, the rotor of the machine may be either a winding placed in slots (wound rotor machine) or consist of a simple set of shorted bars placed in the rotor slots (squirrel cage winding). In the case of the wound rotor machine the rotor windings can be idealized in the same manner as the stator windings. When the machine is equipped with a squirrel cage rotor the corresponding idealizing assumptions are less evident. However, it is important to note that the flux density in the air gap, being *spatial* in nature, is sinusoidally distributed spatially even for arbitrary *temporal* (time dependant) values of stator winding current. Hence, rotor currents are always induced which are also sinusoidally distributed *spatially* even if the rotor consists of a shorted cage. As a result, we can replace the actual winding distribution (actually an n phase concentrated winding) with an equivalent three phase sinusoidally distributed winding. For further discussion of the procedure for replacing the squirrel cage by an equivalent sinusoidally distributed winding see Langsdorf [1].

In the following discussion the permeability of the stator and rotor iron is assumed to be infinite and saturation and iron losses, as well as end winding and slotting effects are neglected. For convenience, we will assume a two pole construction. However, the theory can be readily adapted to multi–pole machines at a later stage. A visual representation of the winding distribution for a typical phase of either the stator or the rotor is shown in Figure 2.2. We

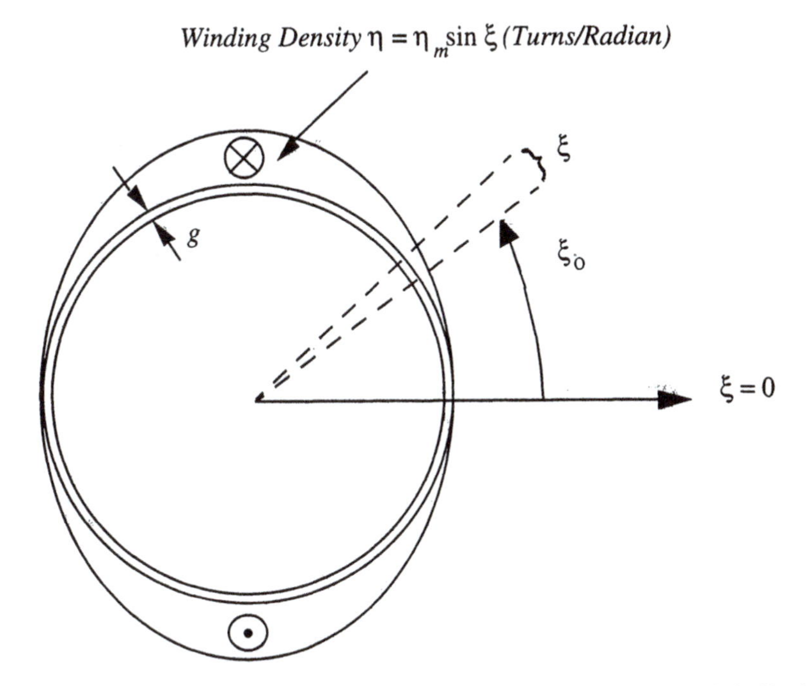

Figure 2.2 Idealized induction machine illustrating sinusoidal distribution of one phase winding

will designate the winding as phase '*a*'. If the winding is sinusoidally distributed, we can represent this fact by a winding density distribution of the form:

$$\eta(\xi) = \eta_m \sin\xi \tag{2.2–1}$$

where we have located the reference point for the angle ξ at the magnetic axis of the winding, i.e. the maximum value of the fundamental component of air gap flux density produced by the winding. Since η_m is not normally specified explicitly, it is first necessary to relate the peak value of this winding density distribution η_m to the number of turns per pole pair. If the number of turns is N_x then, clearly

$$\int_0^{\pi} \eta_m \sin\xi d\xi = N_x \tag{2.2–2}$$

which results in

$$\eta_m = \frac{N_x}{2} \tag{2.2–3}$$

The number of turns N_x is called the *effective number of turns* and, in general, is related to the actual total number of turns N_t by

$$N_x = k_p k_d k_s N_t \tag{2.2–4}$$

where k_p, k_d, and k_s are the winding pitch, distribution and skew factors respectively.

Let us now locate a small portion of the winding defined by the differential dN. If this differential portion of the winding is located at ξ_o, we can write,

$$dN(\xi_o) = \frac{N_x}{2}\sin\xi_o d\xi \tag{2.2–5}$$

We evaluate first the effects of such a small incremental winding acting by itself. For this purpose consider a line integral around the path 12341 shown in Figure 2.3 By Ampere's Law we can write,

$$\oint \bar{H}\cdot d\bar{l} = \int_1^2 \bar{H}_{12}\cdot d\bar{l}_{12} + \int_2^3 \bar{H}_{23}\cdot d\bar{l}_{23} + \int_3^4 \bar{H}_{34}\cdot d\bar{l}_{34} + \int_4^1 \bar{H}_{41}\cdot d\bar{l}_{41} = I_x dN(\xi_o) \tag{2.2–6}$$

Since B must be continuous across the air–iron boundary and since $\mu_i >> \mu_o$,

$$H_{iron} = \frac{B_{iron}}{\mu_{iron}} << H_{air} = \frac{B_{air}}{\mu_{air}} \tag{2.2–7}$$

Hence, H_{iron} can be neglected or, alternatively, a ‘saturation factor’ can be introduced to later account for H_{iron}. If the air gap is small and uniform, the field intensity is constant over the paths 12 and 34, so that eqn (2.2–6) becomes

$$H_{12}(\xi_o + \zeta)g + H_{34}(\xi_o - \zeta)g = dN(\xi_o)I_x \tag{2.2–8}$$

where the functional dependance of H_{12} and H_{34} on ξ_o and ζ are explicitly shown. By symmetry $H_{12} = H_{34}$. Thus, in general, for an arbitrary value of ζ,

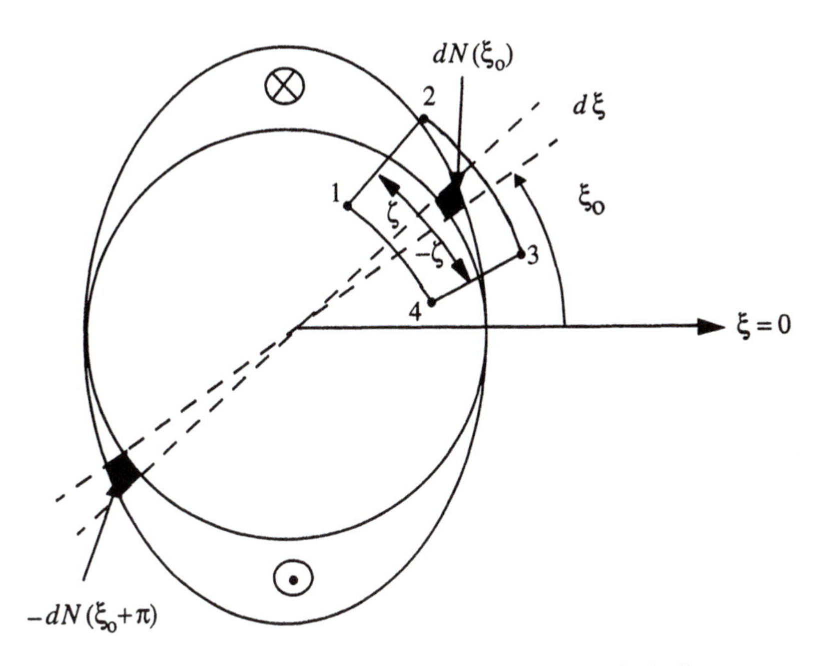

Figure 2.3 Differential number of winding turns of winding x

$$H(\zeta + \xi_o) = \frac{dN(\xi_o)I_x}{2g} \quad \text{for} \quad 0 < \zeta < \pi \tag{2.2–9}$$

and

$$H(\zeta + \xi_o) = -\frac{dN(\xi_o)I_x}{2g} \quad \text{for} \quad \pi < \zeta < 2\pi \tag{2.2–10}$$

We have dropped the subscript '12' for convenience. From symmetry it is clear that the field intensity is uniform (i.e. constant) around the gap and, assuming that positive polarity corresponds to the field intensity H directed from rotor to stator, it is negative during the interval $\pi < \zeta < 2\pi$. This solution is, in fact, the field produced by what is termed a *concentrated full pitch coil* and corresponds to a what is essentially a square wave of flux density in the air gap.

Since all field lines are perpendicular to the surface of interest the differential flux crossing the gap at any point along the air gap is

$$\begin{aligned} d\phi &= \bar{B} \cdot d\bar{A} \\ &= \mu_o \bar{H} \cdot d\bar{A} \\ &= \mu_o H dl\,(r d\zeta) \end{aligned} \tag{2.2–11}$$

which becomes, with the aid of eqn (2.2–5) and (2.2–9),

$$d\phi = \mu_o N_x I_x \left(\frac{r}{4g}\right) \sin\xi_o d\xi\,(d\zeta dl) \quad \text{for} \qquad 0 < \zeta < \pi \tag{2.2–12}$$

and is the negative of this value for ζ elsewhere.

Referring to Figure 2.4 it is important to observe that each packet of flux $d\phi$ links a different number of turns. The number of turns linked by a differential packet of flux located at a position ζ relative to ξ_o is

$$N = \int_{\xi_o - \zeta}^{\xi_o + \zeta} \frac{N_x}{2} \sin u\,du \tag{2.2–13}$$

where u is a dummy variable of integration. Equation (2.2–13) can be evaluated as

$$N = N_x \sin\xi_o \sin\zeta \tag{2.2–14}$$

Assuming no fringing, the flux is evenly distributed throughout the length l of the machine. The total number of flux linkages linked by all such packets of flux $d\phi$ is

$$d\lambda_{xm} = \mu_o \frac{N_x I_x}{4g} r \sin^2\xi_o d\xi \int_0^l dl \int_0^\pi N_x \sin\zeta d\zeta \tag{2.2–15}$$

$$= \mu_o \frac{N_x^2 I_x}{2g} rl \sin^2\xi_o d\xi \tag{2.2–16}$$

where we have made use of the symbol 'λ' to denote flux linkage (weber turns) rather than flux ϕ (webers). Note that the differential is still used since this expression accounts only for the flux linkages produced by the differential current sheet element located at ξ_o.The subscript '*xm*' for the flux linkage variable is used to remind us that this flux corresponds to the *magnetizing* component rather than the total (i.e. *self*) component of flux linkages.

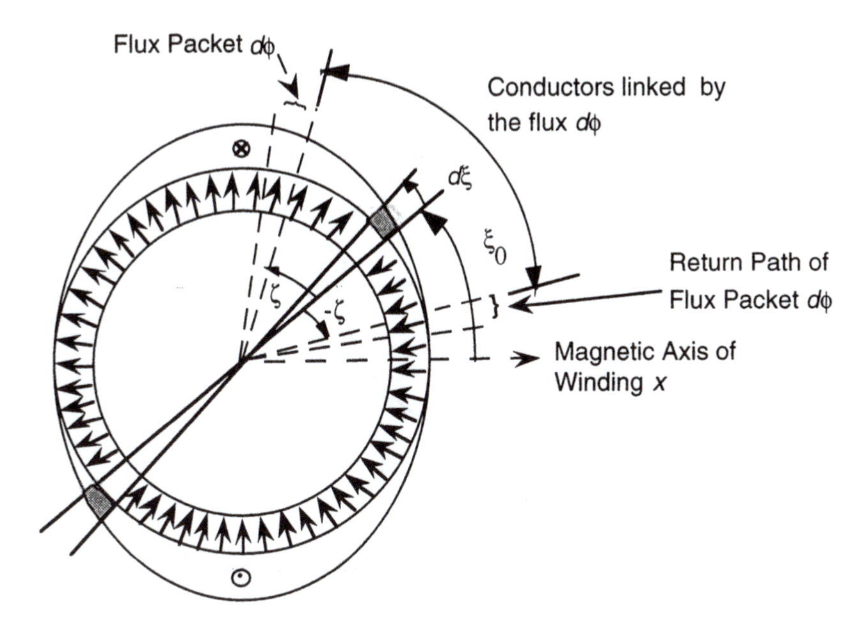

Figure 2.4 Showing winding portion linked by an arbitrary flux packet located at $\zeta + \xi_0$

The total flux linkage is the sum of the linkages from all such differential windings. The subscript on the variable ξ_o can now be dropped. The result is

$$\lambda_{xm} = \int_0^{\pi} \mu_o \frac{N_x^2 I_x}{2g} (rl) \sin^2\xi d\xi \tag{2.2–17}$$

$$= \mu_o N_x^2 I_x \left(\frac{rl}{g}\right)\left(\frac{\pi}{4}\right) \tag{2.2–18}$$

Thus the magnetizing inductance associated with this winding is

$$L_m = \frac{\lambda_{xm}}{I_x} = \mu_o N_x^2 \left(\frac{rl}{g}\right)\left(\frac{\pi}{4}\right) \tag{2.2–19}$$

In general, current in phase a also links the other windings located on either the stator or the rotor. Let us now calculate the flux linking one of the other stator windings whose magnetic axis is displaced counterclockwise from the first by an angle α as shown in Figure 2.5. In this case the number of turns N linked by the flux packet $d\phi$ is given by a modification of eqn (2.2–14) which becomes

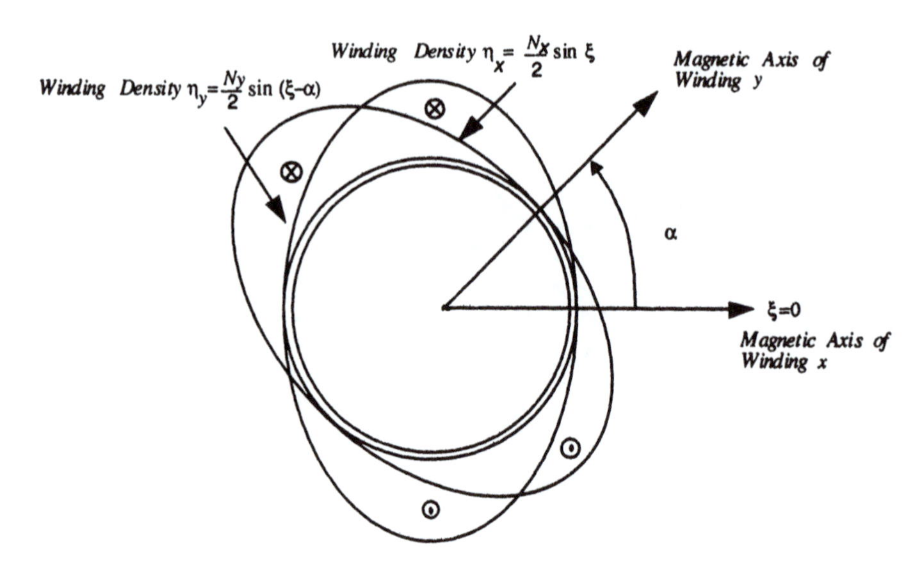

Figure 2.5 Sinusoidal current sheets for two windings displaced by an angle α

$$N = N_y \sin(\xi_o - \alpha)\sin\zeta \tag{2.2–20}$$

Hence, the total number of flux linkages of winding 'y' linked by the flux packet $d\phi$ located at $\xi_o + \zeta$ is

$$d\lambda_{xy} = \mu_o \frac{N_x I_x}{4g} rl \sin\xi_o d\xi \int_0^{\pi} N_y \sin(\xi_o - \alpha)\sin\zeta d\zeta \tag{2.2–21}$$

$$= \mu_o \frac{N_x N_y I_x}{2g} rl \sin\xi_o \sin(\xi_o - \alpha)\, d\xi \tag{2.2–22}$$

The total flux linkages of winding 'y' due to a current in winding 'x' is therefore,

$$\lambda_{xy} = \int_0^{\pi} \mu_o \frac{N_x N_y I_x}{2g} rl \sin\xi \sin(\xi - \alpha) d\xi$$

$$= \mu_o N_x N_y I_x \left(\frac{rl}{g}\right)\left(\frac{\pi}{4}\right)\cos\alpha \tag{2.2–23}$$

The mutual inductance between winding x and winding y is therefore

$$L_{xy} = \frac{\lambda_{xy}}{I_x} = \mu_o N_x N_y \left(\frac{rl}{g}\right)\left(\frac{\pi}{4}\right)\cos\alpha \tag{2.2–24}$$

From an analogous derivation it is not difficult to show that reciprocity is maintained. That is

$$L_{yx} = \frac{\lambda_{yx}}{I_y} = L_{xy} \tag{2.2–25}$$

2.3 System Equations in the Stationary *a,b,c* Reference Frame

The general result which we have obtained can now be used to calculate the mutual inductance between any pair of stator or rotor windings. Let us now proceed to calculate the state equations of our idealized three phase induction machine. In particular, we will assume the winding configuration shown in Figure 2.6. In this case the winding placement is only conceptually shown with the center line of the equivalent inductors directed along the magnetic axes of the windings. Again an elementary two pole machine is considered. 'Balanced' three phase windings are assumed for both the stator and the rotor. That is, all three stator windings designated as the *as*, *bs* and *cs* windings are assumed to have the same number of effective turns, N_s, and the *bs* and *cs* windings are symmetrically displaced from the *as* winding by ±120°. The subscript 's' is used to denote that these windings are *stator* or *stationary* windings. The rotor windings are similarly arranged but have N_r turns. These windings are designated by *ar, br, cr* in which the second subscript reminds us that these three windings are *rotor* or *rotating* windings.

The voltage equations describing the stator and rotor circuits can be written conveniently in matrix form as

$$\boldsymbol{v}_{abcs} = r_s \boldsymbol{i}_{abcs} + p\boldsymbol{\lambda}_{abcs} \tag{2.3–1}$$

$$\boldsymbol{v}_{abcr} = r_r \boldsymbol{i}_{abcr} + p\boldsymbol{\lambda}_{abcr} \tag{2.3–2}$$

where p represents the operator d/dt and $\boldsymbol{v}_{abcs}$, $\boldsymbol{i}_{abcs}$ and $\boldsymbol{\lambda}_{abcs}$ are $3x1$ vectors defined by

$$\boldsymbol{v}_{abcs} = \begin{bmatrix} v_{as} \\ v_{bs} \\ v_{cs} \end{bmatrix}; \qquad \boldsymbol{i}_{abcs} = \begin{bmatrix} i_{as} \\ i_{bs} \\ i_{cs} \end{bmatrix}; \qquad \boldsymbol{\lambda}_{abcs} = \begin{bmatrix} \lambda_{as} \\ \lambda_{bs} \\ \lambda_{cs} \end{bmatrix} \tag{2.3–3}$$

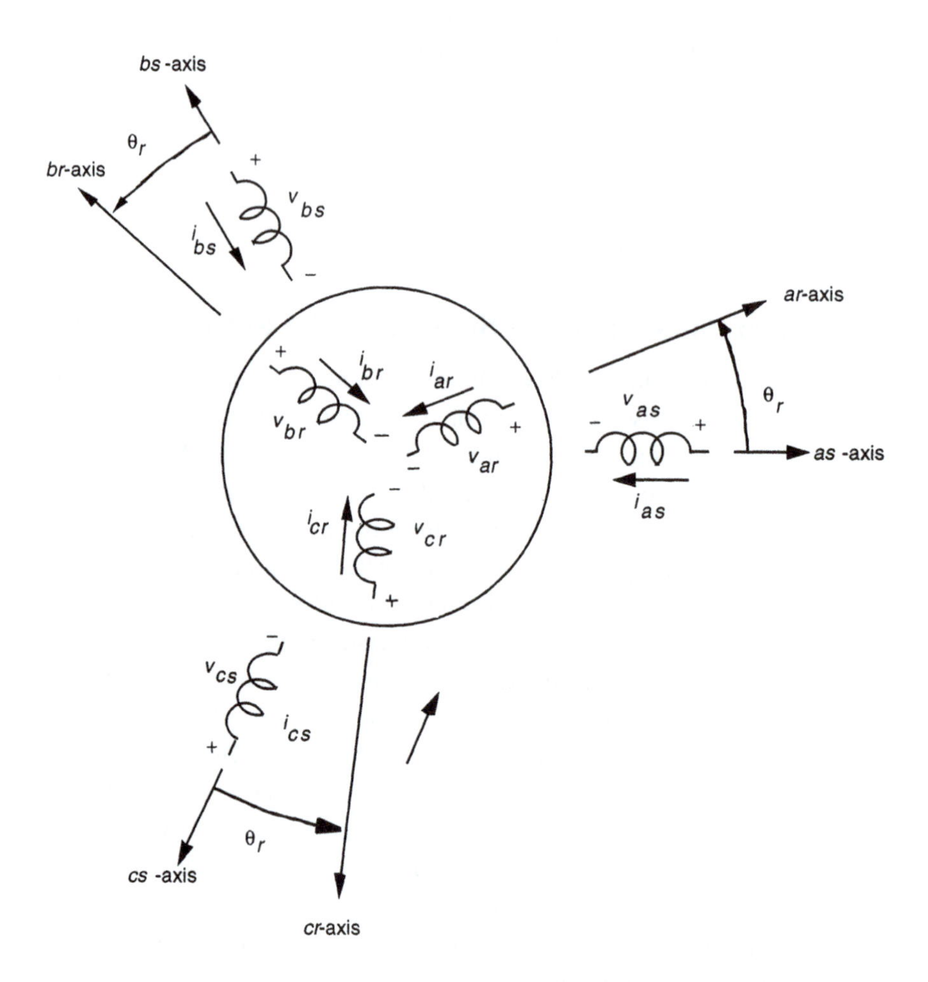

Figure 2.6 Magnetic axes of a three phase induction machine

Similar definitions apply for the rotor variables $\boldsymbol{v}_{abcr}$, $\boldsymbol{i}_{abcr}$ and $\boldsymbol{\lambda}_{abcr}$.

In general, coupling clearly exists between all of the stator and rotor phases. The flux linkages are therefore related to the machine currents by the following matrix equations.

$$\boldsymbol{\lambda}_{abcs} = \boldsymbol{\lambda}_{abcs(s)} + \boldsymbol{\lambda}_{abcs(r)} \tag{2.3–4}$$

$$\boldsymbol{\lambda}_{abcr} = \boldsymbol{\lambda}_{abcr(s)} + \boldsymbol{\lambda}_{abcr(r)} \tag{2.3–5}$$

where

$$\boldsymbol{\lambda}_{abcs\,(s)} = \begin{bmatrix} L_{as} & L_{abs} & L_{acs} \\ L_{abs} & L_{bs} & L_{bcs} \\ L_{acs} & L_{bcs} & L_{cs} \end{bmatrix} \boldsymbol{i}_{abcs} \tag{2.3–6}$$

$$\boldsymbol{\lambda}_{abcs\,(r)} = \begin{bmatrix} L_{as,\,ar} & L_{as,\,br} & L_{as,\,cr} \\ L_{bs,\,ar} & L_{bs,\,br} & L_{bs,\,cr} \\ L_{cs,\,ar} & L_{cs,\,br} & L_{cs,\,cr} \end{bmatrix} \boldsymbol{i}_{abcr} \tag{2.3–7}$$

$$\boldsymbol{\lambda}_{abcr\,(s)} = \begin{bmatrix} L_{ar,\,as} & L_{ar,\,bs} & L_{ar,\,cs} \\ L_{br,\,as} & L_{br,\,bs} & L_{br,\,cs} \\ L_{cr,\,as} & L_{cr,\,bs} & L_{cr,\,cs} \end{bmatrix} \boldsymbol{i}_{abcs} \tag{2.3–8}$$

$$\boldsymbol{\lambda}_{abcr\,(r)} = \begin{bmatrix} L_{ar} & L_{abr} & L_{acr} \\ L_{abr} & L_{br} & L_{bcr} \\ L_{acr} & L_{bcr} & L_{cr} \end{bmatrix} \boldsymbol{i}_{abcr} \tag{2.3–9}$$

Note that as a result of reciprocity, the inductance matrix in the third flux linkage equation, eqn (2.3–8), above is simply the transpose of the inductance matrix in the second equation, eqn (2.3–7).

2.4 Determination of Induction Machine Inductances

While the number of inductances defined above is large, the task of solving for all of these inductances is straightforward. The self inductance of stator phase *as* is obtained by simply setting $\alpha = 0$, and by setting N_x and N_y in eqn (2.2–24) to N_s or, alternatively, by setting $N_x = N_s$ in eqn (2.2–19). Whereby,

$$L_{am} = \mu_o N_s^2 \left(\frac{rl}{g}\right)\left(\frac{\pi}{4}\right) \tag{2.4–1}$$

The subscript m is again used to denoted the fact that this inductance is a *magnetizing* inductance. That is, it is associated with flux lines which cross the air gap and link rotor as well as stator windings. In general, it is necessary to add a relatively small, but important, leakage term to eqn (2.4–1) to account for leakage flux. This term accounts for flux lines which do not cross the gap but instead close with the stator slot itself (slot leakage), in the air gap (belt and harmonic leakage) and at the ends of the machine (end winding leakage). Hence, the total self inductance of phase *as* can be expressed.

$$L_{as} = L_{ls} + L_{am} \tag{2.4–2}$$

Since the windings of the *bs* and *cs* phases are identical to phase *as*, it is clear that the magnetizing inductances of these windings are the same as phase *as* so that, also

$$L_{bs} = L_{ls} + L_{bm} \tag{2.4–3}$$

$$L_{cs} = L_{ls} + L_{cm} \tag{2.4–4}$$

It is apparent that L_{am}, L_{bm} and L_{cm} are equal making the self inductances also equal. It is therefore useful to define the stator magnetizing inductance

$$L_{ms} = \mu_o N_s^2 \frac{rl}{g}\left(\frac{\pi}{4}\right) \tag{2.4–5}$$

so that

$$L_{as} = L_{bs} = L_{cs} = L_{ls} + L_{ms} \tag{2.4–6}$$

The mutual inductance between phases *as* and *bs*, *bs* and *cs*, and *cs* and *as* are derived by simply setting $\alpha = 2\pi/3$ and $N_x = N_y = N_s$ in eqn (2.2–24). The result is

$$L_{abs} = L_{bcs} = L_{cas} = -\mu_o N_s^2 \frac{rl}{g}\left(\frac{\pi}{8}\right) \tag{2.4–7}$$

or, in terms of eqn (2.4–5),

$$L_{abs} = L_{bcs} = L_{cas} = -\frac{L_{ms}}{2} \tag{2.4–8}$$

The flux linkages of phases *as*, *bs*, and *cs* resulting from currents flowing in the stator windings can now be expressed in matrix form as

$$\boldsymbol{\lambda}_{abcs(s)} = \begin{bmatrix} L_{ls} + L_{ms} & -\frac{L_{ms}}{2} & -\frac{L_{ms}}{2} \\ -\frac{L_{ms}}{2} & L_{ls} + L_{ms} & -\frac{L_{ms}}{2} \\ -\frac{L_{ms}}{2} & -\frac{L_{ms}}{2} & L_{ls} + L_{ms} \end{bmatrix} \cdot \boldsymbol{i}_{abcs} \tag{2.4–9}$$

Let us now turn our attention to the mutual coupling between the stator and the rotor windings. Referring to Figure 2.6 we can see that the rotor phase *ar* is displaced from the stator phase *as* by the electrical angle θ_r where θ_r in this case is a variable. Similarly, the rotor phases *br* and *cr* are displaced from *bs* and *cs* respectively by θ_r. Hence, the corresponding mutual inductances can be obtained by setting $N_x = N_s$, $N_y = N_r$, and $\alpha = \theta_r$ in eqn (2.2–24).

$$\begin{aligned} L_{as,\,ar} = L_{bs,\,br} = L_{cs,\,cr} &= \mu_o N_s N_r \frac{rl}{g}\left(\frac{\pi}{4}\right)\cos\theta_r \\ &= \frac{N_r}{N_s} L_{ms} \cos\theta_r \end{aligned} \tag{2.4–10}$$

The angle between the *as* and *br* phases is $\theta_r + 2\pi/3$, so that

$$L_{as,\,br} = L_{bs,\,cr} = L_{cs,\,ar} = \frac{N_r}{N_s} L_{ms} \cos\left(\theta_r + 2\pi/3\right) \tag{2.4–11}$$

Finally, the stator phase *as* is displaced from the rotor *cr* phase by the angle $\theta_r - 2\pi/3$. Therefore,

$$L_{as,\,cr} = L_{bs,\,ar} = L_{cs,\,br} = \frac{N_r}{N_s} L_{ms} \cos\left(\theta_r - 2\pi/3\right) \tag{2.4–12}$$

The above inductances can now be used to establish the flux linking the stator phases due to currents in the rotor circuits. In matrix form,

$$\boldsymbol{\lambda}_{abcs\,(r)} = \frac{N_r}{N_s} L_{ms} \begin{bmatrix} \cos\theta_r & \cos\left(\theta_r + 2\pi/3\right) & \cos\left(\theta_r - 2\pi/3\right) \\ \cos\left(\theta_r - 2\pi/3\right) & \cos\theta_r & \cos\left(\theta_r + 2\pi/3\right) \\ \cos\left(\theta_r + 2\pi/3\right) & \cos\left(\theta_r - 2\pi/3\right) & \cos\theta_r \end{bmatrix} \begin{bmatrix} i_{ar} \\ i_{br} \\ i_{cr} \end{bmatrix} \tag{2.4–13}$$

The total flux linking the stator windings is clearly the sum of the contributions from the stator and the rotor circuits, eqns (2.4–9) and (2.4–13),

$$\boldsymbol{\lambda}_{abcs} = \boldsymbol{\lambda}_{abcs\,(s)} + \boldsymbol{\lambda}_{abcs\,(r)} \tag{2.4–14}$$

It is not difficult to continue the process to determine the rotor flux linkages. In terms of previously defined quantities, the flux linking the rotor circuits due to rotor currents is

$$\boldsymbol{\lambda}_{abcr(r)} = \begin{bmatrix} L_{lr} + \left(\frac{N_r}{N_s}\right)^2 L_{ms} & -\frac{1}{2}\left(\frac{N_r}{N_s}\right)^2 L_{ms} & -\frac{1}{2}\left(\frac{N_r}{N_s}\right)^2 L_{ms} \\ -\frac{1}{2}\left(\frac{N_r}{N_s}\right)^2 L_{ms} & L_{lr} + \left(\frac{N_r}{N_s}\right)^2 L_{ms} & -\frac{1}{2}\left(\frac{N_r}{N_s}\right)^2 L_{ms} \\ -\frac{1}{2}\left(\frac{N_r}{N_s}\right)^2 L_{ms} & -\frac{1}{2}\left(\frac{N_r}{N_s}\right)^2 L_{ms} & L_{lr} + \left(\frac{N_r}{N_s}\right)^2 L_{ms} \end{bmatrix} \begin{bmatrix} i_{ar} \\ i_{br} \\ i_{cr} \end{bmatrix}$$

(2.4–15)

where L_{lr} is the rotor leakage inductance. The flux linking the rotor windings due to currents in the stator circuits is

$$\boldsymbol{\lambda}_{abcr(s)} = \frac{N_r}{N_s} L_{ms} \begin{bmatrix} \cos\theta_r & \cos(\theta_r - 2\pi/3) & \cos(\theta_r + 2\pi/3) \\ \cos(\theta_r + 2\pi/3) & \cos\theta_r & \cos(\theta_r - 2\pi/3) \\ \cos(\theta_r - 2\pi/3) & \cos(\theta_r + 2\pi/3) & \cos\theta_r \end{bmatrix} \begin{bmatrix} i_{as} \\ i_{bs} \\ i_{cs} \end{bmatrix}$$

(2.4–16)

Note that the matrix of eqn (2.4–16) is the transpose of eqn (2.4–13).

The total flux linkages of the rotor windings are again the sum of the two components defined by eqns (2.4–15) and (2.4–16), that is,

$$\boldsymbol{\lambda}_{abcr} = \boldsymbol{\lambda}_{abcr(r)} + \boldsymbol{\lambda}_{abcr(s)} \qquad (2.4\text{–}17)$$

2.5 Complex Vector Representation of Three Phase Variables

It is apparent that the extensive amount of coupling between the six circuits makes the analysis of this machine a rather formidable task. However, we are now in a position to determine if there is any simplification that can be expected in these coupled equations. The approach to simplification involves the topic of *axes transformations* or *reference frame theory* and constitutes an essential aspect of machine analysis. A very useful and compact form of the machine equations for uniform air gap machines is obtained by using complex variable notation. This concept was introduced by Lyon [2] and was initially treated as a formal transformation in the same sense as the *d,q* transformation typically used for synchronous machine analysis. In this section the concept is introduced simply as a different type of notation rather than a totally new transformation of the basic phase variable machine equations.

For those readers with long experience using real variable transformations, the additional abstraction associated with complex notation may seem unnecessary and perhaps counterproductive. However, in cases where the symmetry of the *d,q* model is maintained, complex notation is ideally suited to exploit this symmetry. It provides a compact notation, easy algebraic manipulation, very simple graphical interpretations and often a very logical and direct development of concepts which are difficult to develop using real variable analysis. While it is certainly true that the use of complex variables can be avoided thereby gaining the extra degree of generality of *d,q* theory, there is, in the case of induction machines, a great deal of manipulative and conceptual value in their use.

Consider, first of all, the equation describing the instantaneous position of the stator air gap MMF. If phase *as* is sinusoidally distributed with a maximum value located at $\xi = \alpha$, then the MMF in the gap resulting from current flowing in phase *as* is

$$F_{as} = \frac{N_s}{2} i_{as} \cos(\beta) \tag{2.5–1}$$

where $\beta = \xi - \alpha$. Similarly, if currents flow in phases *bs* or *cs* which are spatially displaced from phase as by 120 electrical degrees, then the respective air gap MMFs are:

$$F_{bs} = \frac{N_s}{2} i_{bs} \cos(\beta - 2\pi/3) \tag{2.5–2}$$

$$F_{cs} = \frac{N_s}{2} i_{cs} \cos(\beta + 2\pi/3) \tag{2.5–3}$$

The total air gap MMF resulting from all three currents flowing simultaneously is

$$F_{abcs} = \frac{N_s}{2} [i_{as} \cos\beta + i_{bs} \cos(\beta - 2\pi/3) + i_{cs} \cos(\beta + 2\pi/3)] \tag{2.5–4}$$

It is not difficult to show that if the currents, i_{as}, i_{bs}, and i_{cs} form a balanced sinusoidal three phase set of currents, then eqn (2.5–4) describes an MMF wave having a constant amplitude and rotating at a constant angular velocity.

It is useful to introduce the Euler relation wherein

$$e^{j\theta} = \cos\theta + j\sin\theta \tag{2.5–5}$$

Solving for cos θ and sin θ, we have,

$$\cos\theta = \frac{e^{j\theta} + e^{-j\theta}}{2} \tag{2.5–6}$$

$$\sin\theta = \frac{e^{j\theta} - e^{-j\theta}}{2j} \tag{2.5–7}$$

Using eqn (2.5–6), the MMF for three phase excitation becomes

$$F_{abcs} = \frac{N_s}{2}\left[i_{as}\frac{(e^{j\beta} + e^{-j\beta})}{2} + i_{bs}\frac{(e^{j(\beta-2\pi/3)} + e^{-j(\beta-2\pi/3)})}{2} + i_{cs}\frac{(e^{j(\beta+2\pi/3)} + e^{-j(\beta+2\pi/3)})}{2}\right] \tag{2.5–8}$$

Collecting terms,

$$F_{abcs} = \frac{N_s}{4}\{ [i_{as} + i_{bs}e^{-j2\pi/3} + i_{cs}e^{j2\pi/3}]\, e^{j\beta} + [i_{as} + i_{bs}e^{j2\pi/3} + i_{cs}e^{-j2\pi/3}]\, e^{-j\beta}\} \tag{2.5–9}$$

Defining

$$\underline{a} = e^{j2\pi/3} \tag{2.5–10}$$

and noting that

$$\underline{a}^2 = \underline{a}^{-1} = e^{j4\pi/3} = e^{-j2\pi/3} = \underline{a}^{\dagger} \tag{2.5–11}$$

where † denotes the complex conjugate, eqn (2.5–9) becomes

$$F_{abcs} = \frac{N_s}{4}\{ [i_{as} + \underline{a}^2 i_{bs} + \underline{a} i_{cs}]\, e^{j\beta} + [i_{as} + \underline{a} i_{bs} + \underline{a}^2 i_{cs}]\, e^{-j\beta}\} \tag{2.5–12}$$

If we define

$$\underline{i}_{abcs} = \frac{2}{3}(i_{as} + \underline{a}i_{bs} + \underline{a}^2 i_{cs}) \quad (2.5\text{–}13)$$

then

$$\underline{i}^{\dagger}_{abcs} = \frac{2}{3}(i_{as} + \underline{a}^2 i_{bs} + \underline{a}i_{cs}) \quad (2.5\text{–}14)$$

Equation (2.5–12) can now be written in the compact form

$$F_{abcs} = \frac{3}{2}\frac{N_s}{4}(\underline{i}_{abcs}e^{-j\beta} + \underline{i}^{\dagger}_{abcs}e^{j\beta}) \quad (2.5\text{–}15)$$

Note that whereas the MMF F_{abcs} remains a real variable, it is represented as the effects of two complex variables $\underline{i}_{abcs}$ and $\underline{i}^{\dagger}_{abcs}$. Quantities defined in the manner of eqn (2.5-13) are called *complex space vectors.* They should not be confused with *complex phasors* which are normally used to represent sinusoidally alternating quantities in the steady state.

The inverse relations for finding the physical currents from the complex vector is obtained by expressing eqn (2.5-13) in complex form

$$\begin{aligned}\underline{i}_{abcs} &= \frac{2}{3}\left[i_{as} - \frac{1}{2}(i_{bs} + i_{cs}) + j\frac{\sqrt{3}}{2}(i_{bs} - i_{cs})\right] \\ &= i_{as} + j\frac{1}{\sqrt{3}}(i_{bs} - i_{cs})\end{aligned} \quad (2.5\text{–}16)$$

if the sum of the three stator currents equals zero (no zero sequence component). Hence,

$$i_{as} = Re\,[\underline{i}_{abcs}] \quad (2.5\text{–}17)$$

Similarly,

$$\begin{aligned}\underline{a}^2\underline{i}_{abcs} &= \frac{2}{3}(\underline{a}^2 i_{as} + \underline{a}^3 \underline{i}_{bs} + \underline{a}^4 i_{cs}) \\ &= \frac{2}{3}(i_{bs} + \underline{a}i_{cs} + \underline{a}^2 i_{as}) \\ &= i_{bs} + j\frac{1}{\sqrt{3}}(i_{cs} - i_{as})\end{aligned} \quad (2.5\text{–}18)$$

So that,

$$i_{bs} = Re\,[\underline{a}^2\,\underline{i}_{abcs}] \quad (2.5\text{–}19)$$

Finally,

$$i_{cs} = Re\,[\underline{a}\underline{i}_{abcs}] \tag{2.5–20}$$

Note that an additional term must be added to these expressions if the three phase currents contain a zero sequence component.

2.6 Complex Variable Model of Three Phase Induction Machine

Consider now the representation of the machine equations in complex variable form. Transformation of the voltage equations is, indeed, almost trivial. If we multiply the second row of the stator voltage matrix equation, eqn (2.3–1), by '$\underline{a}$' and the third row by $\underline{a}^2$, add the three rows and multiply the entire result by 2/3 we immediately obtain,

$$\underline{v}_{abcs} = r_s\underline{i}_{abcs} + p\underline{\lambda}_{abcs} \tag{2.6–1}$$

where

$$\underline{v}_{abcs} = \frac{2}{3}(v_{as} + \underline{a}v_{bs} + \underline{a}^2 v_{cs}) \tag{2.6–2}$$

and so forth for $\underline{i}_{abcs}$ and $\underline{\lambda}_{abcs}$. The complex vector expression for the voltage equation of the rotor circuits is developed in a similar manner and written in complex form as

$$\underline{v}_{abcr} = r_r\underline{i}_{abcr} + p\underline{\lambda}_{abcr} \tag{2.6–3}$$

The stator flux linkages were identified in Section 2.4 as having two components, $\underline{\lambda}_{abcs(s)}$ resulting from stator currents (eqn(2.4–9)) and $\underline{\lambda}_{abcs(r)}$ arising from rotor currents (eqn (2.4–13)). Upon multiplying the second row of eqn (2.4–9) by $\underline{a}$ and the third row by $\underline{a}^2$ and adding the three rows, the stator flux linkages resulting from stator currents can be written as

$$\lambda_{as(s)} + \underline{a}\lambda_{bs(s)} + \underline{a}^2\lambda_{cs(s)} = \left[L_{ls} + L_{ms}\left(1 - \frac{\underline{a}}{2} - \frac{\underline{a}^2}{2}\right)\right](i_{as} + \underline{a}i_{bs} + \underline{a}^2 i_{cs}) \tag{2.6–4}$$

It is not difficult to demonstrate that

$$1 - \frac{\underline{a}}{2} - \frac{\underline{a}^2}{2} = \frac{3}{2} \tag{2.6–5}$$

so that eqn (2.6–4) reduces to

$$\underline{\lambda}_{abcs\,(s)} = \left(L_{ls} + \frac{3}{2}L_{ms}\right)\underline{i}_{abcs} \tag{2.6–6}$$

The second component of stator flux linkages arises from the rotor current. By repeated use of eqn (2.5–6), eqn (2.4–13) can be written as

$$\begin{bmatrix} \lambda_{as\,(r)} \\ \lambda_{bs\,(r)} \\ \lambda_{cs\,(r)} \end{bmatrix} = \frac{N_r}{N_s}\frac{L_{ms}}{2}\left\{\begin{bmatrix} e^{j\theta_r} & \underline{a}e^{j\theta_r} & \underline{a}^2e^{j\theta_r} \\ \underline{a}^2e^{j\theta_r} & e^{j\theta_r} & \underline{a}e^{j\theta_r} \\ \underline{a}e^{j\theta_r} & \underline{a}^2e^{j\theta_r} & e^{j\theta_r} \end{bmatrix} + \begin{bmatrix} e^{-j\theta_r} & \underline{a}^2e^{-j\theta_r} & \underline{a}e^{-j\theta_r} \\ \underline{a}e^{-j\theta_r} & e^{-j\theta_r} & \underline{a}^2e^{-j\theta_r} \\ \underline{a}^2e^{-j\theta_r} & \underline{a}e^{-j\theta_r} & e^{-j\theta_r} \end{bmatrix}\right\} \cdot \begin{bmatrix} i_{ar} \\ i_{br} \\ i_{cr} \end{bmatrix} \tag{2.6–7}$$

Again, multiplying the second row of this matrix equation by $\underline{a}$ and the third row by $\underline{a}^2$ and adding the three rows results in

$$\lambda_{as\,(r)} + \underline{a}\lambda_{bs\,(r)} + \underline{a}^2\lambda_{cs\,(r)} = \frac{N_r}{N_s}\frac{L_{ms}}{2}\,[3\,(i_{ar} + \underline{a}i_{br} + \underline{a}^2 i_{cr})e^{j\theta_r} \tag{2.6–8}$$

$$+ \quad (1 + \underline{a} + \underline{a}^2)\,(i_{ar} + i_{br} + i_{cr})e^{-j\theta_r} \quad]$$

However, it is readily shown that

$$1 + \underline{a} + \underline{a}^2 = 0 \tag{2.6–9}$$

so that in terms of complex space vectors, eqn (2.4–13) can be written in the equivalent compact form

$$\underline{\lambda}_{abcs\,(r)} = \frac{N_r}{N_s}\frac{3L_{ms}}{2}\underline{i}_{abcr}e^{j\theta_r} \tag{2.6–10}$$

where

$$\underline{\lambda}_{abcs\,(r)} = \frac{2}{3}(\lambda_{as\,(r)} + \underline{a}\lambda_{bs\,(r)} + \underline{a}^2\lambda_{cs\,(r)}) \tag{2.6–11}$$

and

$$\underline{i}_{abcr} = \frac{2}{3}(i_{ar} + \underline{a}i_{br} + \underline{a}^2 i_{cr}) \tag{2.6–12}$$

The total stator flux linkages are therefore

$$\underline{\lambda}_{abcs} = \left(L_{ls} + \frac{3}{2}L_{ms}\right)\underline{i}_{abcs} + \frac{N_r}{N_s}\frac{3L_{ms}}{2}\underline{i}_{abcr}e^{j\theta_r} \tag{2.6–13}$$

In a like manner it can be easily shown that the rotor flux linkages can be expressed in the form

$$\underline{\lambda}_{abcr} = \left[L_{lr} + \frac{3}{2}\left(\frac{N_r}{N_s}\right)^2 L_{ms}\right]\underline{i}_{abcr} + \frac{N_r}{N_s}\frac{3L_{ms}}{2}\underline{i}_{abcs}e^{-j\theta_r} \quad (2.6\text{–}14)$$

2.7 Turns Ratio Transformation

It is rather apparent that the turns ratio N_s/N_r causes these equations to be needlessly cumbersome. The appearance of the turns ratio is, of course, not unexpected since the induction motor is simply another type of coupled magnetic circuit and is related to the transformer. In fact we can view the induction motor as equivalent to a transformer with a short circuited and rotating secondary. We can get rid of the explicit dependance of these expressions on the turns ratio by referring the rotor circuits to the stator in much the same manner as utilized for a transformer. Equations (2.6–3), (2.6–13) and (2.6–14) can be manipulated to form

$$\left(\frac{N_s}{N_r}\right)\underline{v}_{abcr} = \left(\frac{N_s}{N_r}\right)^2 r_r\left[\left(\frac{N_r}{N_s}\right)\underline{i}_{abcr}\right] + p\left[\left(\frac{N_s}{N_r}\right)\underline{\lambda}_{abcr}\right] \quad (2.7\text{–}1)$$

$$\underline{\lambda}_{abcs} = \left(L_{ls} + \frac{3}{2}L_{ms}\right)\underline{i}_{abcs} + \frac{3L_{ms}}{2}\left[\frac{N_r}{N_s}\underline{i}_{abcr}\right]e^{j\theta_r} \quad (2.7\text{–}2)$$

$$\left(\frac{N_s}{N_r}\right)\underline{\lambda}_{abcr} = \left[\left(\frac{N_s}{N_r}\right)^2 L_{lr} + \frac{3}{2}L_{ms}\right]\left[\left(\frac{N_r}{N_s}\right)\underline{i}_{abcr}\right] + \frac{3L_{ms}}{2}\underline{i}_{abcs}e^{-j\theta_r}$$

(2.7–3)

We can now define the primed variables

$$\left(\frac{N_s}{N_r}\right)\underline{v}_{abcr} = \underline{v}'_{abcr} \quad (2.7\text{–}4)$$

$$\left(\frac{N_s}{N_r}\right)\underline{\lambda}_{abcr} = \underline{\lambda}'_{abcr} \quad (2.7\text{–}5)$$

$$\left(\frac{N_r}{N_s}\right)\underline{i}_{abcr} = \underline{i}'_{abcr} \quad (2.7\text{–}6)$$

$$\left(\frac{N_s}{N_r}\right)^2 r_r = r'_r \tag{2.7–7}$$

$$\left(\frac{N_s}{N_r}\right)^2 L_{lr} = L'_{lr} \tag{2.7–8}$$

Also, since the 3/2 term appears in all of the flux linkage expressions it is useful to define the *magnetizing inductance*

$$L_m = \frac{3}{2}L_{ms} = \frac{3}{2}N_s^2\left(\mu_o\frac{rl}{g}\right)\frac{\pi}{4} \tag{2.7–9}$$

Combining the above results the equations for stator and rotor circuits become, finally

$$\underline{v}_{abcs} = r_s\underline{i}_{abcs} + (L_{ls} + L_m)\,p\underline{i}_{abcs} + L_m\,p\,(\underline{i}'_{abcr}e^{j\theta_r}) \tag{2.7–10}$$

$$\underline{v}'_{abcr} = r'_r\,\underline{i}'_{abcr} + (L'_{lr} + L_m)\,p\underline{i}'_{abcr} + L_m p\,(\underline{i}_{abcs}e^{-j\theta_r}) \tag{2.7–11}$$

The rotor voltage $\underline{v}'_{abcr}$ is, of course, identically zero for the case of a squirrel cage machine, but has been carried through this analysis for generality.

Employing the normal rules for differentiation, these two equations can be written in alternative form as

$$\underline{v}_{abcs} = r_s\underline{i}_{abcs} + (L_{ls} + L_m)\,p\underline{i}_{abcs} + L_m\,(p\underline{i}'_{abcr})\,e^{j\theta_r} + j\omega_r L_m\,\underline{i}'_{abcr}\,e^{j\theta_r} \tag{2.7–12}$$

$$\underline{v}'_{abcr} = r'_r\,\underline{i}'_{abcr} + (L'_{lr} + L_m)\,p\underline{i}'_{abcr} + L_m\,(p\underline{i}_{abcs})\,e^{-j\theta_r} - j\omega_r L_m\,\underline{i}_{abcs}e^{-j\theta_r} \tag{2.7–13}$$

where $\omega_r = p\theta_r = \dfrac{d\theta_r}{dt}$.

It is important to note the enormous simplification afforded by the complex space vector representation. It is a tedious but educational exercise to confirm that eqns (2.7–12) and (2.7–13) generate the six scalar equations denoted by eqns (2.3–1) and (2.3–2).

2.8 Transformation to a Rotating Reference Frame

While the complex vector approach to writing the machine equations results in a compact form, the essential sinusoidal coupling between the stator and rotor circuits with rotor position remains. This coupling can essentially be eliminated, however, if the stator and rotor equations are referred to a common frame of reference. This common frame of reference can be non–rotating in which case it is associated with the stator and is called the *stator* or *stationary* reference. Alternatively the d,q axes can be made to rotate with the same angular velocity as the rotor circuits, and is termed the *rotor* reference frame. On some occasions it may even be useful to rotate these axes synchronously with one of the complex vectors denoting stator or rotor voltage, current or even flux. In this case the axes are said to be a*ttached* to the vector. In the most general case the axes need not be specified resulting in an *arbitrary* or *freely rotating* reference frame [3].

Consider, for example, a rotating set of d,q axes as defined by Figure 2.7. Figure 2.7 also shows axes of reference corresponding to the magnetic axes of the three phase stator and three phase rotor circuits. Variables along the a,b and c stator axes can be referred to the q– and d–axes by the expressions

$$f_{qs} = \frac{2}{3}\left[f_{as}\cos\theta + f_{bs}\cos\left(\theta - \frac{2\pi}{3}\right) + f_{cs}\cos\left(\theta + \frac{2\pi}{3}\right)\right] \tag{2.8–1}$$

$$f_{ds} = \frac{2}{3}\left[f_{as}\sin\theta + f_{bs}\sin\left(\theta - \frac{2\pi}{3}\right) + f_{cs}\sin\left(\theta + \frac{2\pi}{3}\right)\right] \tag{2.8–2}$$

where the symbol f is used to represent any of the three phase stator circuit variables such as voltage, current or flux linkage. It should be noted here that the coefficient 2/3 in these two equations is somewhat arbitrary. The choice of 2/3 is usually selected so as to maintain the same length of the voltage and current vectors for sinusoidal steady state as we shall see in Chapter 4. Another choice is to use the coefficient $\sqrt{2/3}$ in which case the power as calculated in the d–q coordinate system is the same as in the a,b,c system. In this case the transformation is said to be *power invariant.* Since there are three phases, then in general, it is necessary to define a third new variable to obtain a unique transformation. This third new variable is typically defined as the *zero sequence component*,

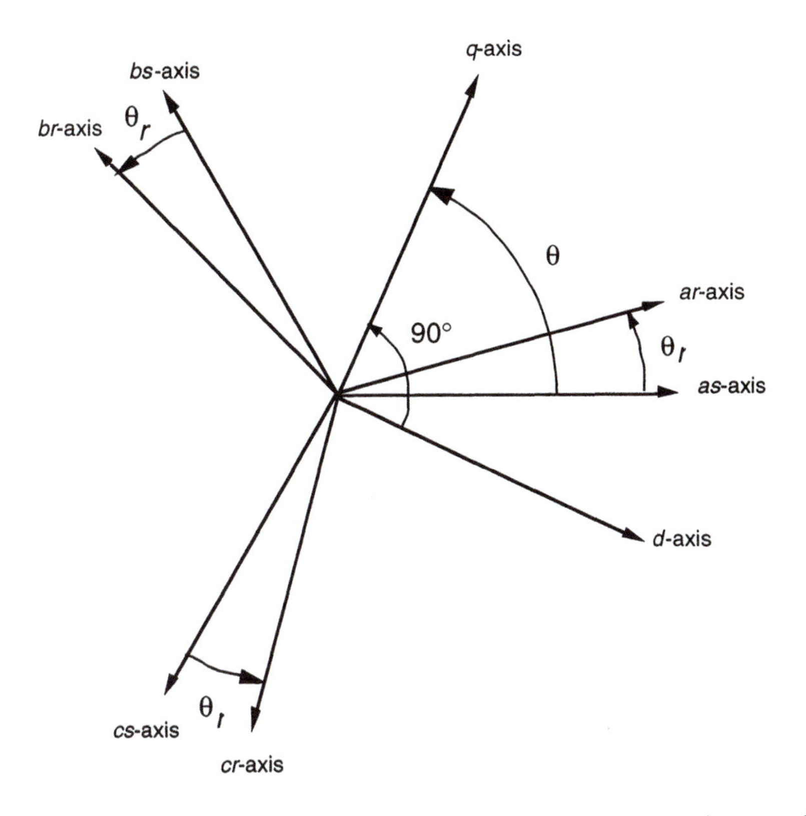

Figure 2.7 Location of rotating d,q axes relative to the magnetic axes of the stator and rotor phases

$$f_{0s} = \frac{1}{3}[f_{as} + f_{bs} + f_{cs}] \tag{2.8–3}$$

In this context the use of the symbol *0* in the subscript of the new variable is used to designate a zero component *normal* to the d,q plane

In practical applications such as those considered in this book, the machine is connected either in delta or in wye without a neutral return path. Thus, the three stator currents sum to zero. As a result, the other significant three phase variables such as the phase voltages, the stator flux linkages and rotor currents also sum to zero. Hence, it is necessary to concern oneself only with projections on the d– and q–axes. If eqn (2.8–2) is multiplied by j and the result subtracted from eqn (2.8–1) we have immediately, from Euler's Equation,

$$\underline{f}_{qds} = f_{qs} - jf_{ds} = \frac{2}{3}[f_{as}e^{-j\theta} + f_{bs}e^{-j(\theta - 2\pi/3)} + f_{cs}e^{-j(\theta + 2\pi/3)}] \tag{2.8–4}$$

which can be written as,

$$\begin{aligned}\underline{f}_{qds} &= \frac{2}{3}e^{-j\theta}[f_{as} + \underline{a}f_{bs} + \underline{a}^2 f_{cs}] \\ &= e^{-j\theta}\underline{f}_{abcs}\end{aligned} \tag{2.8–5}$$

This expression, combined with the zero sequence definition in eqn(2.8–3), defines a general rotating axis complex vector transformation. When the zero sequence quantity is zero, which is the usual case in machine analysis, the single expression in eqn (2.8–5) completely describes the transformation. Note that the rotation associated with the general *d,q* transformation is separately described by the exponential function $e^{-j\theta}$. The $\underline{a}$ operator, in effect, handles the transformation from three phase to two phase *d,q* variables.

In a similar manner, rotor variables can also be transformed to the rotating axes by a complex exponential function. In this case, however, since the rotor is itself already rotating, a relative rotational angle must be used rather than the angle θ. It is common to choose the angle between the *q*–axis of the rotating *d–q* frame and the *a*–phase magnetic axis of the rotor circuits. The equation of transformation relating rotor variables to equivalent variables rotating at an arbitrary rotational speed $d\theta/dt$ becomes

$$\begin{aligned}\underline{f}_{qdr} &= \frac{2}{3}e^{-j(\theta - \theta_r)}[f_{ar} + \underline{a}f_{br} + \underline{a}^2 f_{cr}] \\ &= e^{-j(\theta - \theta_r)}\underline{f}_{abcr}\end{aligned} \tag{2.8–6}$$

Equations (2.8–5) and (2.8–6) can now be used to transform the machine complex vector equations to a rotating reference frame. Multiplying eqn (2.7–10) by $e^{-j\theta}$ yields

$$e^{-j\theta}\underline{v}_{abcs} = r_s e^{-j\theta}\underline{i}_{abcs} + (L_{ls} + L_m)e^{-j\theta}p\underline{i}_{abcs} + L_m e^{-j\theta}p(\underline{i}'_{abcr}e^{j\theta_r}) \tag{2.8–7}$$

However, by the chain rule of differentiation,

$$x\frac{dy}{dt} = \frac{d}{dt}(xy) - \frac{dx}{dt}y$$

Thus, eqn (2.7–10) can also be written in the form

$$\begin{aligned} e^{-j\theta}\underline{v}_{abcs} &= r_s e^{-j\theta}\underline{i}_{abcs} + (L_{ls} + L_m)\,p\,(e^{-j\theta}\underline{i}_{abcs}) + L_m p\,[\underline{i}'_{abcr}e^{-j(\theta-\theta_r)}] \\ &\quad + j\omega\,[\,(L_{ls} + L_m)\,e^{-j\theta}\underline{i}_{abcs} + L_m\underline{i}'_{abcr}\,e^{-j(\theta-\theta_r)}] \end{aligned} \tag{2.8–8}$$

where

$$\omega = p\theta = \frac{d\theta}{dt} \tag{2.8–9}$$

Upon multiplying through by 2/3 and making use of eqns (2.8–5) and (2.8–6) we quickly arrive at

$$\underline{v}_{qds} = r_s\underline{i}_{qds} + (L_{ls} + L_m)\,p\underline{i}_{qds} + L_m p\underline{i}'_{qdr} + j\omega\,[\,(L_{ls} + L_m)\,\underline{i}_{qds} + L_m\underline{i}'_{qdr}\,] \tag{2.8–10}$$

In a similar manner it is not difficult to show that in a rotating d,q coordinate system, the rotor circuit equations can be represented by

$$\begin{aligned} \underline{v}'_{qdr} &= r'_r\,\underline{i}'_{qdr} + (L'_{lr} + L_m)\,p\underline{i}'_{qdr} + L_m p\underline{i}_{qds} \\ &\quad + j(\omega - \omega_r)\,[\,(L'_{lr} + L_m)\,\underline{i}'_{qdr} + L_m\underline{i}_{qds}] \end{aligned} \tag{2.8–11}$$

As a final step, the zero sequence component must be determined. If the three rows of eqn (2.3–1) are summed and the resulting equation simplified by means of eqns (2.4–9) and (2.4–13), it is not difficult to show that the result is

$$v_{as} + v_{bs} + v_{cs} = (r_s + pL_{ls})\,(i_{as} + i_{bs} + i_{cs})$$

Multiplying this equation by $1/3$ and making use of the definition of the zero sequence quantity, eqn (2.8–3), we have finally,

$$v_{0s} = (r_s + pL_{ls})\,i_{0s} \tag{2.8–12}$$

In a similar manner it is easily shown that if a zero sequence current flows in the rotor, the equation describing its behavior is given by

$$v'_{0r} = (r'_r + pL'_{lr})\,i'_{0r} \tag{2.8–13}$$

It is important to note that while the form of these zero sequence equations is correct, the parameters of the particular result we have obtained is, in fact, due to our assumptions. In particular, the inductance of the zero sequence circuits is not generally equal to the stator and rotor per phase leakage inductances. In practice, the stator and rotor leakage flux components consist of both components which either cross or enter the air gap (zig zag and tooth top leakage flux) and components which do not involve the air gap (slot, harmonic, skew and end winding leakage flux). When the stator (or rotor) is excited with a zero sequence current, their gap dependant flux components change (i.e. reduce) drastically to continue to satisfy Gauss' Law while the non–air gap dependant fluxes are hardly affected. Hence, the zero sequence inductance is typically less than the per phase leakage inductance being roughly 0.8 to 0.95 of this value.

Replacing p by d/dt and collecting our results to this point, eqns (2.8–10) to (2.8–13) can be written in scalar form, for the stator:

$$v_{ds} = r_s i_{ds} + \frac{d\lambda_{ds}}{dt} - \omega\lambda_{qs} \tag{2.8–14}$$

$$v_{qs} = r_s i_{qs} + \frac{d\lambda_{qs}}{dt} + \omega\lambda_{ds} \tag{2.8–15}$$

$$v_{0s} = r_s i_{0s} + \frac{d\lambda_{0s}}{dt} \tag{2.8–16}$$

where

$$\lambda_{ds} = L_{ls} i_{ds} + L_m (i_{ds} + i'_{dr}) \tag{2.8–17}$$

$$\lambda_{qs} = L_{ls} i_{qs} + L_m (i_{qs} + i'_{qr}) \tag{2.8–18}$$

$$\lambda_{0s} = L_{ls} i_{0s} \tag{2.8–19}$$

For the rotor:

$$v'_{dr} = r'_r i'_{dr} + \frac{d\lambda'_{dr}}{dt} - (\omega - \omega_r)\lambda'_{qr} \tag{2.8–20}$$

$$v'_{qr} = r'_r i'_{qr} + \frac{d\lambda'_{qr}}{dt} + (\omega - \omega_r)\lambda'_{dr} \tag{2.8–21}$$

$$v'_{0r} = r'_r i'_{0r} + \frac{d\lambda'_{0r}}{dt} \tag{2.8–22}$$

where

$$\lambda'_{dr} = L'_{lr} i'_{dr} + L_m (i_{ds} + i'_{dr}) \tag{2.8–23}$$

$$\lambda'_{qr} = L'_{lr}\, i'_{qr} + L_m (i_{qs} + i'_{qr}) \tag{2.8–24}$$

$$\lambda'_{0r} = L'_{lr}\, i'_{0r} \tag{2.8–25}$$

The equivalent circuit of an induction machine in a freely rotating reference frame is shown in Figure 2.8. Note that we have proceeded in a general way so that voltages may also be applied to the rotor circuits. In such machines the rotor is wound with coils in the same manner as the stator. Such machines are normally termed *wound rotor induction machines*. If the machine has a wound rotor with slip rings the rotor voltages are typically defined by external circuit conditions. For example, consider the case of a three slip ring machine connected to external resistors $R_x v'_{dr} = -(N_s/N_r)^2 R_x i'_{dr}$, and $R_x v_{qr}' = -(N_s/N_r)^2 R_x i'_{qr}$ with $v'_{0r} = 0$. If the machine is a squirrel cage construction the rotor voltages v'_{dr} , v'_{qr} and v'_{0r} are identically zero.

In the vast majority of cases the machine is connected in delta or wye such that the neutral current does not flow. In this case the neutral axis voltages v_{0s} and v'_{0r} remain identically zero. Hence, from eqns (2.8–19) and (2.8–25), the currents i_{0s} and i'_{0r} are identically zero and solution of these equations are always trivial. The four remaining equations are often assembled in a single matrix equation expressed as

$$\begin{bmatrix} v_{ds} \\ v_{qs} \\ v'_{dr} \\ v'_{qr} \end{bmatrix} = \begin{bmatrix} r_s + L_s p & -\omega L_s & L_m p & -\omega L_m \\ \omega L_s & r_s + L_s p & \omega L_m & L_m p \\ L_m p & -(\omega - \omega_r) L_m & r'_r + L'_r p & -(\omega - \omega_r) L'_r \\ (\omega - \omega_r) L_m & L_m p & (\omega - \omega_r) L'_r & r'_r + L'_r p \end{bmatrix} \cdot \begin{bmatrix} i_{ds} \\ i_{qs} \\ i'_{dr} \\ i'_{qr} \end{bmatrix} \tag{2.8–26}$$

where $L'_r = L'_{lr} + L_m$ and $L_s = L_{ls} + L_m$

2.9 Interpretation of Complex Vectors

Since the use of complex space vectors is not a familiar concept it is useful to pause in this analysis to consider the complex vector concept more fully. From Section 2.8 the equation relating stator *a,b,c* variables to stator *d,q* variables can be expressed as

$$\underline{f}_{qds} = \frac{2}{3} e^{-j\theta} [f_{as} + \underline{a} f_{bs} + \underline{a}^2 f_{cs}] \tag{2.9–1}$$

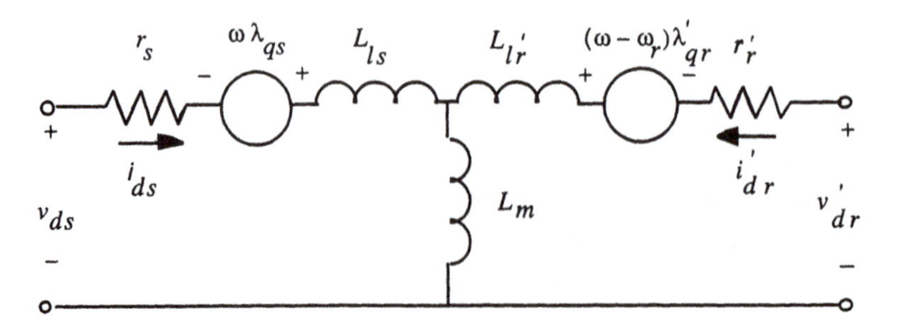

d– axis equivalent circuit

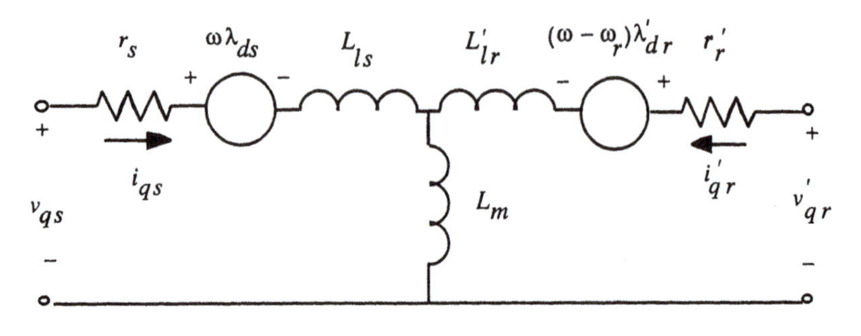

q– axis equivalent circuit

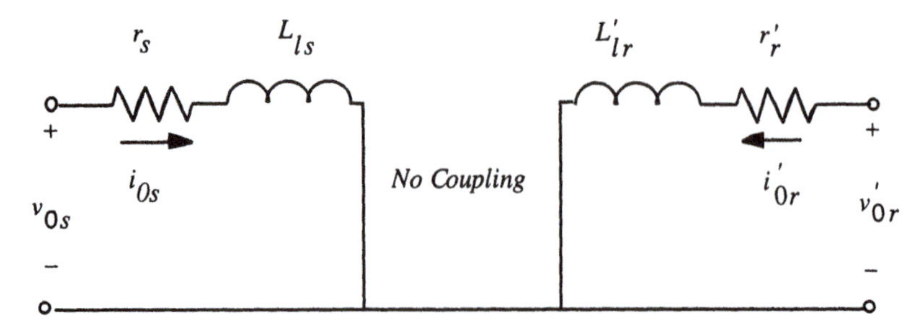

0– axis equivalent circuit

Figure 2.8 The $d,q,0$ equivalent circuits of three phase induction machine

When the *d,q reference frame* does not rotate we have $\theta = 0$. In this case the reference frame is said to be *fixed in the stator.* Using a superscript *s* to denote this fact, the equation relating *a,b,c* variables to *d,q* variables in a *stationary coordinate system* is

$$\underline{f}_{qds}^{s} = \frac{2}{3}[f_{as} + \underline{a} f_{bs} + \underline{a}^2 f_{cs}] \tag{2.9–2}$$

Figure 2.9 illustrates the concept of a complex vector in terms of the stator cur-

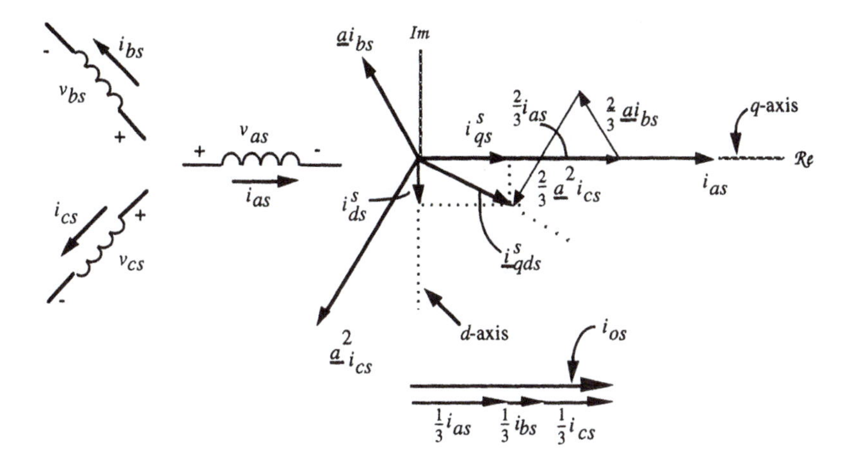

Figure 2.9 Transformation from three phase stator to complex vector voltage

rents [i.e. *f* replace by *i* in eqn (2.9–2)]. In this graphical interpretation, the current i_{as} becomes a vector in the direction of the *as* coil, $\underline{a}i_{bs}$ a vector in the direction of the *bs* coil, etc. The projections of the complex vector $\underline{i}_{qds}^{s}$ on the real and negative imaginary axes are the equivalent *q* and *d* axis two phase currents. For the case illustrated in the figure there is also a zero sequence current as indicated.

A pure rotational transformation, for example from stationary *d,q* axes to what is typically called the *synchronously rotating* axes, would be simply expressed as

$$\underline{f}_{qds}^{e} = \underline{f}_{qds}^{s} e^{-j\theta_e} \tag{2.9–3}$$

where $\theta_e = \omega_e t$ when ω_e is a constant. The symbol *s* is used to denote a *stationary* set of axes while the subscript *e* denotes axes which are rotating *electrically* with the applied voltage vector. Equation (2.9–3) has the graphical interpretation shown in Figure 2.10. At first sight the reverse rotation of the

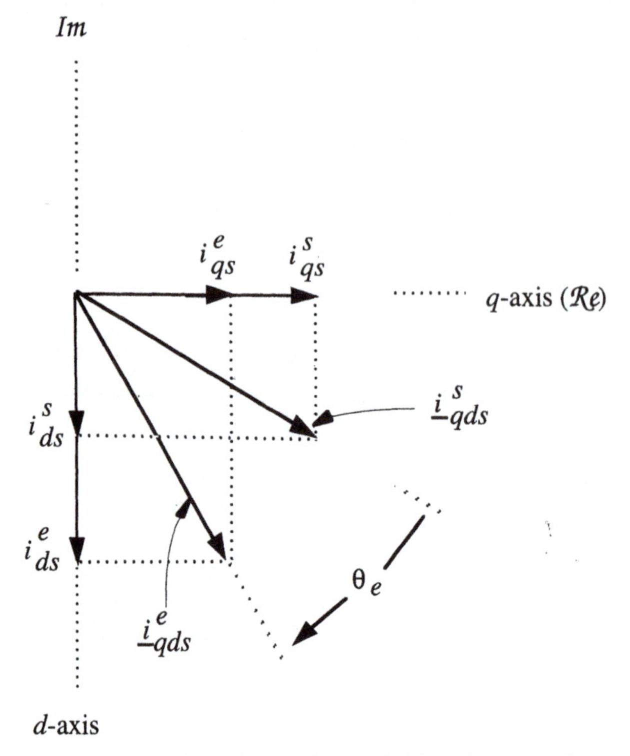

Figure 2.10 Transformation from *d,q* variables in a stationary axes to rotating *d,q* variables, (pure rotation)

complex vector $\underline{i}^e_{qds}$ may seem incorrect since a synchronous frame is usually visualized as rotating with the synchronously rotating machine quantities. However, the rotation in Figure 2.10 is correct since balanced, positive sequence current excitation results in

$$i_{as} = I_s \cos \omega_e t \tag{2.9–4}$$

$$i_{bs} = I_s \cos\left(\omega_e t - \frac{2\pi}{3}\right) \tag{2.9–5}$$

$$i_{cs} = I_s \cos\left(\omega_e t + \frac{2\pi}{3}\right) \tag{2.9–6}$$

which, when expressed as a complex vector yields

$$\underline{i}_{qds}^{s} = \frac{2}{3}\frac{I_s}{2}[e^{j\omega_e t} + e^{-j\omega_e t} + \underline{a}(e^{j(\omega_e t - 2\pi/3)} + e^{-j(\omega_e t - 2\pi/3)})$$
$$+ \underline{a}^2(e^{j(\omega_e t + 2\pi/3)} + e^{-j(\omega_e t + 2\pi/3)})] \tag{2.9–7}$$

$$= \frac{I_s}{3}[3e^{j\omega_e t} + e^{-j\omega_e t}(1 + \underline{a} + \underline{a}^2)]$$
$$= I_s e^{j\omega_e t} \tag{2.9–8}$$

Thus, the complex vector $\underline{i}_{qds}^{s}$ itself rotates at synchronous speed in the positive direction. When transformed to a synchronous reference frame using eqn (2.9–3), the resulting complex vector $\underline{i}_{qds}^{e}$ is stationary because it rotates backward at synchronous speed relative to $\underline{i}_{qds}^{s}$.

It is convenient to generalize the original result of eqn (2.9–1) to handle either rotor or stator quantities by defining the general phase transformation

$$\underline{f}_{qdx}^{x} = \frac{2}{3}[f_{ax} + \underline{a}f_{bx} + \underline{a}^2 f_{cx}] \tag{2.9–9}$$

where x can be replaced with s or r to treat stator or rotor quantities. The general rotation transformation is best stated as

$$\underline{f}_{qdx} = \underline{f}_{qdx}^{s} e^{-j\theta} \tag{2.9–10}$$

and interpreted as the transformation from a stator referred quantity to a freely rotating quantity. The advantage of this interpretation is that all angles are measured from the axis of phase a and hence the angle θ is well defined in all cases. For example, the transformation of any stator referred quantity to axes rotating with the rotor is given by

$$\underline{f}_{qdx}^{r} = \underline{f}_{qdx}^{s} e^{-j\theta_r} \tag{2.9–11}$$

For a rotor quantity, this equation becomes

$$\underline{f}_{qdr}^{r} = \underline{f}_{qdr}^{s} e^{-j\theta_r} \tag{2.9–12}$$

The initial transformation of a rotor quantity derived from eqn (2.9–9) to the stator can be handled by inverting eqn (2.9–12) to obtain

$$\underline{f}_{qdr}^{s} = \underline{f}_{qdr}^{r} e^{j\theta_r} = \frac{2}{3}e^{j\theta_r}[f_{ar} + \underline{a}f_{br} + \underline{a}^2 f_{cr}] \tag{2.9–13}$$

While the proper sign on the exponential term in eqn (2.9–13) can be obtained by careful physical reasoning, the approach enforced by using eqn (2.9–11) as the basic rotation transformation makes it automatic and error free.

The inverse of any rotation transformation is readily obtained by simply transposing the exponential term. The inverse of the phase transformation is more difficult. The inverse can be obtained by using the $\underline{a}$ operator in rectangular form

$$\underline{f}_{qdx}^{x} = \frac{2}{3}\left[f_{ax} + \left(-\frac{1}{2} + j\frac{\sqrt{3}}{2}\right)f_{bx} + \left(-\frac{1}{2} - j\frac{\sqrt{3}}{2}\right)f_{cx}\right]$$
$$= \frac{2}{3}\left[\frac{3}{2}f_{ax} - \frac{1}{2}(f_{ax} + f_{bx} + f_{cx}) + j\frac{\sqrt{3}}{2}(f_{bx} - f_{cx})\right] \tag{2.9–14}$$

By taking the real part and using the definition of the zero sequence from eqn (2.8–3)

$$f_{ax} = Re\,[\underline{f}_{qdx}^{x}] + f_{0x} \tag{2.9–15}$$

The corresponding results for f_{bx} and f_{cx} are obtained in a similar fashion and are

$$f_{bx} = Re\,[\underline{a}^2\underline{f}_{qdx}^{x}] + f_{0x} \tag{2.9–16}$$

$$f_{cx} = Re\,[\underline{a}\underline{f}_{qdx}^{x}] + f_{0x} \tag{2.9–17}$$

If the angle of the complex vector quantity is ϕ

$$\underline{f}_{qdx}^{x} = |f|e^{j\phi} \tag{2.9–18}$$

the inverse relations can be written as

$$f_{ax} = |f|\cos\phi + f_{0x} \tag{2.9–19}$$

$$f_{bx} = |f|\cos\left(\frac{2\pi}{3} - \phi\right) + f_{0x} \tag{2.9–20}$$

$$f_{cx} = |f|\cos\left(\frac{2\pi}{3} + \phi\right) + f_{0x} \tag{2.9–21}$$

The diagram in Figure 2.11a illustrates the angles involved in these equations. This diagram also shows that these same angles can be defined by introducing a three phase reference system and measuring from $\underline{f}_{qdx}^{x}$ to these reference axes instead of rotating $\underline{f}_{qdx}^{x}$. This leads to the graphical interpreta-

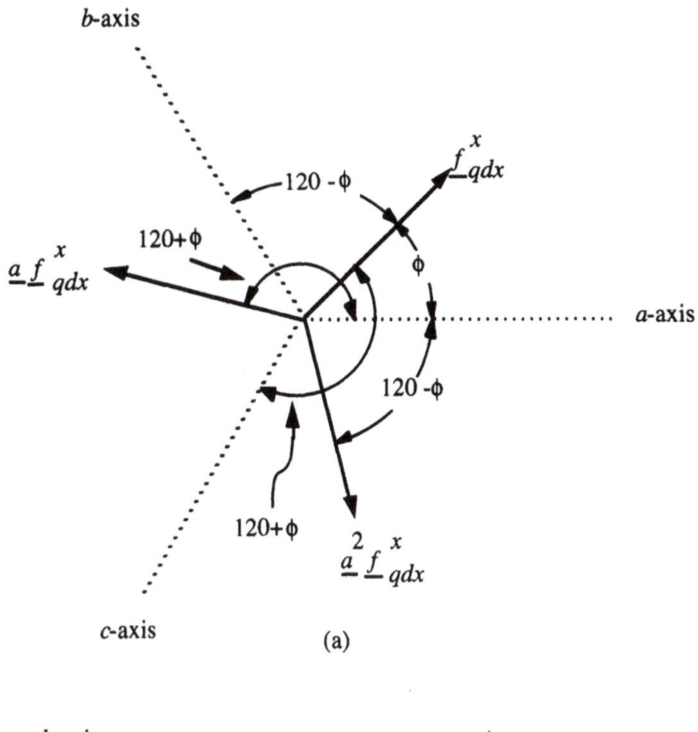

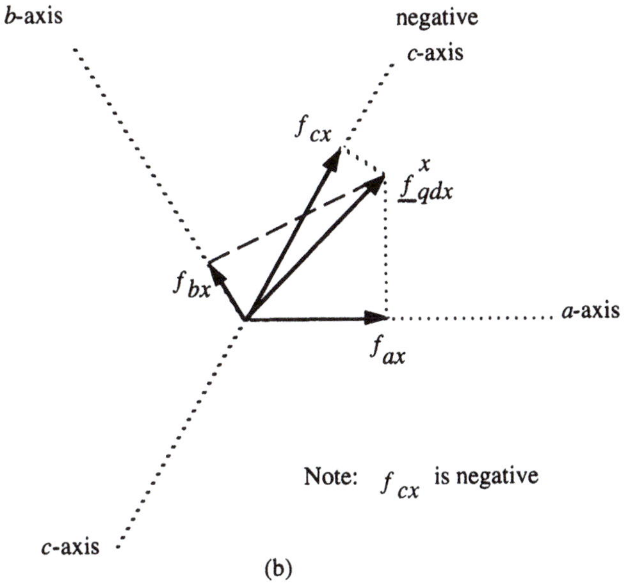

Figure 2.11 Graphical inverse of complex vector quantity

tion of the inverse illustrated in Figure 2.11b. Here the components of the inverse are obtained by simply taking the projections of $\underline{f}^{x}_{qdx}$ on a three phase reference system with phase a oriented along the real axis. This graphical inversion process is very useful in many applications. The zero sequence, if any, must be added to the portion coming from $\underline{f}^{x}_{qdx}$.

2.10 Power Flow in the *d,q* Equivalent Circuit

The power flow into the induction motor in terms of our newly defined $d,q,0$ variables must now be examined. The power flowing into an n phase induction machine is clearly the instantaneous product of the voltages across and currents through the n phases of the machine. In our case the machine is assumed to have three stator and three rotor phases. In order to express the power into the machine we can begin by noting that if '†' denotes the complex conjugate,

$$
\begin{aligned}
\underline{v}_{abcs}\underline{i}^{\dagger}_{abcs} &= \frac{2}{3}(v_{as} + \underline{a}v_{bs} + \underline{a}^2 v_{cs})\frac{2}{3}(i_{as} + \underline{a}^2 i_{bs} + \underline{a}\, i_{cs}) \\
&= \frac{4}{9}[v_{as}i_{as} + \underline{a}(v_{bs}i_{as} + v_{as}i_{cs}) + \underline{a}^2(v_{as}i_{bs} + v_{cs}i_{as} + v_{bs}i_{cs}) \\
&\quad + \underline{a}^3(v_{bs}i_{bs} + v_{cs}i_{cs}) + \underline{a}^4(v_{cs}i_{bs})\ ] \\
&= \frac{4}{9}[v_{as}i_{as} + v_{bs}i_{bs} + v_{cs}i_{cs} + \underline{a}(v_{bs}i_{as} + v_{as}i_{cs} + v_{cs}i_{bs}) \\
&\quad + \underline{a}^2(v_{as}i_{bs} + v_{cs}i_{as} + v_{bs}i_{cs})\ ]
\end{aligned}
$$

Taking the real part of this expression we find that

$$
\begin{aligned}
Re\,[\underline{v}_{abcs}\underline{i}^{\dagger}_{abcs}] = \frac{4}{9}\{&v_{as}i_{as} + v_{bs}i_{bs} + v_{cs}i_{cs} \\
&-\frac{1}{2}[v_{as}(i_{bs} + i_{cs}) + v_{bs}(i_{as} + i_{cs}) + v_{cs}(i_{as} + i_{bs})]\ \}
\end{aligned}
\tag{2.10–1}
$$

If the three currents sum to zero, this expression becomes, finally

$$
Re\,[\underline{v}_{abcs}\underline{i}^{\dagger}_{abcs}] = \frac{2}{3}\{v_{as}i_{as} + v_{bs}i_{bs} + v_{cs}i_{cs}\} \tag{2.10–2}
$$

Thus, assuming a three wire machine without a neutral return, the power into the machine can be written as

$$P_e = \frac{3}{2}\{Re[\underline{v}_{abcs}\underline{i}^{\dagger}_{abcs}] + Re[\underline{v}'_{abcr}\underline{i}^{\dagger\prime}_{abcr}]\} \qquad (2.10\text{–}3)$$

Note that we have immediately used the stator referred rotor variables in eqn (2.10–3) since it is readily seen from Section 2.7 that the rotor voltage/current product is independent of the turns ratio N_s/N_r. Transforming now to *d,q* variables in a rotating axis coordinate system we have from eqn (2.8–5), for the stator variables

$$\underline{f}_{qds} = e^{-j\theta}\underline{f}_{abcs} \qquad (2.10\text{–}4)$$

or, inversely,

$$\underline{f}_{abcs} = e^{j\theta}\underline{f}_{qds} \qquad (2.10\text{–}5)$$

and, from eqn (2.8–6), for the rotor variables,

$$\underline{f}'_{abcr} = e^{j(\theta-\theta_r)}\underline{f}'_{qdr} \qquad (2.10\text{–}6)$$

Thus in terms of *d,q* variables eqn (2.10–3) is written as

$$P_e = \frac{3}{2}Re[(e^{j\theta}\underline{v}_{qds})(e^{-j\theta}\underline{i}^{\dagger}_{qds}) + (e^{j(\theta-\theta_r)}\underline{v}'_{qdr})(e^{-j(\theta-\theta_r)}\underline{i}'^{\dagger}_{qdr})] \qquad (2.10\text{–}7)$$

or simply

$$P_e = \frac{3}{2}Re[\underline{v}_{qds}\underline{i}^{\dagger}_{qds} + \underline{v}'_{qdr}\underline{i}'^{\dagger}_{qdr}] \qquad (2.10\text{–}8)$$

In scalar form

$$P_e = \frac{3}{2}(v_{ds}i_{ds} + v_{qs}i_{qs} + v'_{dr}i'_{dr} + v'_{qr}i'_{qr}) \qquad (2.10\text{–}9)$$

Note that the input power is independent of the reference frame velocity $\omega = d\theta/dt$ as would be expected.

It is interesting to observe that the power into the *d,q* circuits is calculated in a slightly different fashion to that in the actual circuits of the machine due to the appearance of the 3/2 factor. This is a direct result of the 2/3 factor that we selected as the proportionality constant between the *a,b,c* and *d,q* variables, eqn (2.9–1). It is sometimes stated that power is not conserved with this choice of scale factor although it clearly is if we include the 3/2 factor in the *d,q* form of the power equation. Other transformations are possible in which the power, as calculated from the transformed variables, is the same as in the actual phase variable system. In this case the factor $\sqrt{2/3}$ is introduced rather than $2/3$. We will have more to say about the significance of this factor in Section 2.12. Since both types of systems are in widespread use the reader should be alerted to this fact when reading the literature.

2.11 Example – Stator *a,b,c and* Rotor *d,q* Model (Stator Referred)

In problems in which there are switching constraints imposed directly on the stator currents and voltages it is sometimes difficult to transform the constraints to *d,q* variables. For such cases it is often easier to carry out the analysis directly in the three phase variables. However, to avoid the many trigonometric coupling terms in the differential equations, it is possible to represent the rotor by a stator referred *d,q* model while leaving the stator variables in their original form. Derive such an *a,b,c/d,q* stator referred model (with the *qs*–axis aligned with the *as*–axis) using the basic concepts presented in deriving the *d,q/d,q* stator referred model. For generality, assume the zero sequence stator quantities are not zero.

Solution – Starting with the *d,q/d,q* model, [eqn (2.8–26)], implicitly assuming use of a turns ratio transformation (primed rotor variables) at this point,

$$\begin{bmatrix} v^s_{ds} \\ v^s_{qs} \\ v^s_{dr} \\ v^s_{qr} \end{bmatrix} = \begin{bmatrix} r_s + L_s p & -\omega L_s & L_m p & -\omega L_m \\ \omega L_s & r_s + L_s p & \omega L_m & L_m p \\ L_m p & (\omega - \omega_r) L_m & r_r + L_r p & -(\omega - \omega_r) L_r \\ (\omega - \omega_r) L_m & L_m p & (\omega - \omega_r) L_r & r_r + L_r p \end{bmatrix} \begin{bmatrix} i^s_{ds} \\ i^s_{qs} \\ i^s_{dr} \\ i^s_{qr} \end{bmatrix}$$

In a stationary reference frame, $\omega = 0$ and assuming q and a axes are aligned,

$$v_{as} = v^s_{qs} + v_{0s} = (r_s + L_s p)\, i^s_{qs} + L_m p\, i^s_{qr} + v_{0s}$$

But, $i^s_{qs} = i_{as} - i_{0s}$

Hence, $v_{as} = (r_s + L_s p)\, i_{as} + L_m p\, i^s_{qr} + v_{0s} - (r_s + L_s p)\, i_{0s}$

However, $v_{0s} = [r_s + (L_s - L_m)\, p]\, i_{0s}$

which implies that

$$v_{0s} - (r_s + L_s p)\, i_{0s} = -L_m p\, i_{0s}$$

Since $$i_{0s} = \frac{1}{3}(i_{as} + i_{bs} + i_{cs})$$

then $$v_{as} = (r_s + L_s p)\, i_{as} + L_m p i^s_{qr} - \frac{L_m}{3} p\,(i_{as} + i_{bs} + i_{cs})$$

$$v_{as} = \left[r_s + \left(L_s - \frac{L_m}{3}\right)p\right] i_{as} - \frac{L_m}{3} p i_{bs} - \frac{L_m}{3} p i_{cs} + L_m p i^s_{qr}$$

Phase *b*:

$$v_{bs} - v_{0s} = (r_s + L_s p)\, Re(a^2 \underline{i}^s_{qds}) + L_m p Re(a^2 \underline{i}^s_{qdr})$$

$$v_{bs} - v_{0s} = (r_s + L_s p)\,(i_{bs} - i_{0s}) + L_m p\left(-\frac{1}{2} i^s_{qr} - \frac{\sqrt{3}}{2} i^s_{dr}\right)$$

The same substitution for $v_{0s} - (r_s + L_s p)i_{0s}$ yields

$$v_{bs} = \left[r_s + \left(L_s - \frac{L_m}{3}\right)p\right] i_{bs} - \frac{L_m}{3} p i_{as} - \frac{L_m}{3} p i_{cs} - \frac{L_m}{2} p\, i^s_{qr} - \frac{\sqrt{3}}{2} L_m p i^s_{dr}$$

and for *c* phase by symmetry

$$v_{cs} = \left[r_s + \left(L_s - \frac{L_m}{3}\right)p\right] i_{cs} - \frac{L_m}{3} p i_{as} - \frac{L_m}{3} p i_{bs} - \frac{L_m}{2} p\, i^s_{qr} + \frac{\sqrt{3}}{2} L_m p i^s_{dr}$$

For the rotor,

$$\underline{v}^s_{qdr} = (r_r + L_r(p - j\omega_r))\, \underline{i}^s_{qdr} + L_m(p - j\omega_r)\, \underline{i}^s_{qds}$$

Since, $\underline{i}^s_{qds} = \frac{2}{3}(i_{as} + a i_{bs} + a^2 i_{cs})$ and $\underline{i}^s_{qds} = i^s_{qs} - j i^s_{ds}$

$$\underline{i}^s_{qds} = \frac{2}{3} i_{as} - \frac{1}{3} i_{bs} - \frac{1}{3} i_{cs} + j\frac{1}{\sqrt{3}}(i_{bs} - i_{cs})$$

substituting,

$$\underline{v}^s_{qdr} = [r_r + L_r(p - j\omega_r)]\, \underline{i}^s_{qdr} + L_m(p - j\omega_r)\left[\frac{2}{3} i_{as} - \frac{1}{3} i_{bs} - \frac{1}{3} i_{cs} + j\frac{1}{\sqrt{3}}(i_{bs} - i_{cs})\right]$$

The real part is:

$$v_{qr}^s = (r_r + L_r p)\, i_{qr}^s - \omega_r L_r i_{dr}^s + L_m p\left(\frac{2}{3}i_{as} - \frac{1}{3}i_{bs} - \frac{1}{3}i_{cs}\right) + \frac{\omega_r L_m}{\sqrt{3}}(i_{bs} - i_{cs})$$

The imaginary part corresponds to:

$$v_{dr}^s = (r_r + L_r p)\, i_{dr}^s + \omega_r L_r i_{qr}^s + L_m p \frac{1}{\sqrt{3}}(i_{cs} - i_{bs}) + \omega_r L_m\left(\frac{2}{3}i_{as} - \frac{1}{3}i_{bs} - \frac{1}{3}i_{cs}\right.$$

Solution Summary
Stator *a,b,c* and Rotor *d,q* Model

$$\begin{bmatrix} v_{as} \\ v_{bs} \\ v_{cs} \\ v_{qr}^s \\ v_{dr}^s \end{bmatrix} = \begin{bmatrix} r_s + \left(L_s - \frac{1}{3}L_m\right)p & -\frac{1}{3}L_m p & -\frac{1}{3}L_m p & L_m p & 0 \\ -\frac{1}{3}L_m p & r_s + \left(L_s - \frac{1}{3}L_m\right)p & -\frac{1}{3}L_m p & -\frac{1}{2}L_m p & -\frac{\sqrt{3}}{2}L_m p \\ -\frac{1}{3}L_m p & -\frac{1}{3}L_m p & r_s + \left(L_s - \frac{1}{3}L_m\right)p & -\frac{1}{2}L_m p & \frac{\sqrt{3}}{2}L_m p \\ \frac{2}{3}L_m p & -\frac{L_m}{3}(p - \sqrt{3}\omega_r) & -\frac{L_m}{3}(p + \sqrt{3}\omega_r) & r_r + L_r p & -\omega_r L_r \\ \frac{2}{3}L_m \omega_r & -\frac{L_m}{3}(\omega_r + \sqrt{3}p) & -\frac{L_m}{3}(\omega_r - \sqrt{3}p) & \omega_r L_r & r_r + L_r p \end{bmatrix} \begin{bmatrix} i_{as} \\ i_{bs} \\ i_{cs} \\ i_{qr}^s \\ i_{dr}^s \end{bmatrix}$$

(2.11–1)

Note that since $L_{0s} = L_s - L_m = L_{ls}$, and $L_m = \frac{3}{2}L_{ms}$

then $\quad L_s - \frac{1}{3}L_m = \frac{2}{3}L_m + L_{ls} = L_{ms} + L_{ls}$.

2.12 The Electromagnetic Torque

Up to this point we have treated the induction machine as simply an interesting type of coupled magnetic circuit and the electromechanical properties of this device so far has been ignored. It is now time to consider where the energy goes that passes into the machine via the stator and rotor terminals. Upon inserting eqns (2.8–10) and (2.8–11) into the d,q expression for the input power, eqn (2.10–8), we can write that

$$P_e = \frac{3}{2} Re\{ r_s \underline{i}_{qds} + (L_{ls} + L_m)\, p\underline{i}_{qds} + L_m\, p\underline{i}'_{qdr} \tag{2.12–1}$$

$$+ j\omega [(L_{ls} + L_m)\, \underline{i}_{qds} + L_m \underline{i}'_{qdr}] \} \underline{i}^{\dagger}_{qds}$$

$$+ \frac{3}{2} Re\{ r'_r\, \underline{i}'_{qdr} + (L'_{lr} + L_m)\, p\underline{i}'_{qdr} + L_m\, p\underline{i}_{qds}$$

$$+ j(\omega - \omega_r) [(L'_{lr} + L_m)\, \underline{i}'_{qdr} + L_m \underline{i}_{qds}] \} \underline{i}'^{\dagger}_{qdr}$$

This expression in the right hand side can be arranged in the form

Power lost in Conductors — Time Rate of Change of Stored Energy

$$P_e = \overbrace{\frac{3}{2} r_s |\underline{i}_{qds}|^2 + \frac{3}{2} r'_r |\underline{i}'_{qdr}|^2}^{\text{Power lost in Conductors}} + \overbrace{\frac{3}{2} p \left[\frac{L_{ls}}{2} |\underline{i}_{qds}|^2 + \frac{L'_{lr}}{2} |\underline{i}'_{qdr}|^2 + L_m |\underline{i}_{qds} + \underline{i}'_{qdr}|^2 \right]}^{\text{Time Rate of Change of Stored Energy}}$$

$$\left. \begin{aligned} &+ \frac{3}{2} Re\{ j\,\omega [(L_{ls} + L_m) |\underline{i}_{qds}|^2 + L_m \underline{i}'_{qdr}\, \underline{i}^{\dagger}_{qds}] \\ &+ j(\omega - \omega_r) [(L'_{lr} + L_m) |\underline{i}'_{qdr}|] + L_m \underline{i}_{qds} \underline{i}'^{\dagger}_{qdr} \} \end{aligned} \right\} \text{Energy Conversion Term}$$

Hence, the electrical power into the terminals of the machine can be segregated into three terms. The first term clearly accounts for the power dissipated in the stator and rotor resistances. The second term corresponds to the time rate of change of the magnetic energy stored in the inductances of the machine. Since the remaining power must be going somewhere, it is natural to suspect that the third term on the second and third lines account for energy conversion, that is the power being converted from electrical to mechanical form. Let us call this term P_{em}, the electromechanical output power.

Examining the four terms of the energy conversion term itself, it is apparent that the first and third quantities in this expression have only imaginary components. Hence, the energy conversion is accounted for by the second and fourth term. We can write that,

$$P_{em} = \frac{3}{2}Re\{j\omega L_m \underline{i}'_{qdr}\underline{i}^{\dagger}_{qds} + j(\omega - \omega_r) L_m \underline{i}_{qds}\underline{i}'^{\dagger}_{qdr}\}$$
$$= \frac{3}{2}Re\{j\omega L_m (\underline{i}'_{qdr}\underline{i}^{\dagger}_{qds} + \underline{i}_{qds}\underline{i}'^{\dagger}_{qdr}) - j\omega_r L_m \underline{i}_{qds}\underline{i}'^{\dagger}_{qdr}\} \tag{2.12–2}$$

It can be easily shown that if $\underline{a}$ and $\underline{b}$ are complex numbers then

$$\underline{a}\underline{b}^{\dagger} + \underline{a}^{\dagger}\underline{b} = 2(a_r b_r + a_i b_i)$$

That is, it has no imaginary part. It therefore follows that the first term of eqn (2.12–2) has no real part. The output electromechanical power reduces to the last term in this expression,

$$P_{em} = -\frac{3}{2}Re\{j\omega_r L_m \underline{i}_{qds}\underline{i}'^{\dagger}_{qdr}\} \tag{2.12–3}$$

or, alternatively,

$$P_{em} = \frac{3}{2}Im\{\omega_r L_m \underline{i}_{qds}\underline{i}'^{\dagger}_{qdr}\} \tag{2.12–4}$$

which can be easily determined to be equivalent to

$$P_{em} = \frac{3}{2}\omega_r L_m (i_{qs}i'_{dr} - i_{ds}i'_{qr}) \tag{2.12–5}$$

The electrical angular velocity of our equivalent two pole machine, ω_r, is related to the actual mechanical speed by the pole pairs. That is,

$$\omega_r = \frac{P}{2}\omega_{rm} \tag{2.12–6}$$

where P is the number of poles and ω_{rm} is the mechanical speed in radians per second. Since rotational mechanical power is defined as the product of speed times torque, it is apparent that the electromagnetic torque impressed on the shaft of the machine can be expressed as

$$T_e = \frac{3}{2}\frac{P}{2}L_m Im\{\underline{i}_{qds}\underline{i}'^{\dagger}_{qdr}\} \tag{2.12–7}$$

$$= \frac{3}{2}\frac{P}{2}L_m(i_{qs}i'_{dr} - i_{ds}i'_{qr}) \qquad (2.12\text{–}8)$$

It can be shown, however, that the expression for electromagnetic torque has many equivalent forms. For example, the stator flux linkage is related to the stator and rotor currents in complex form as

$$\underline{\lambda}_{qds} = (L_{ls} + L_m)\,\underline{i}_{qds} + L_m\,\underline{i}'_{qdr} \qquad (2.12\text{–}9)$$

Solving for $L_m\,\underline{i}'_{qdr}$ and inserting in eqn (2.12–7) we find that

$$T_e = \frac{3}{2}\frac{P}{2}Im\,[-\,(L_{ls} + L_m)\,\underline{i}_{qds}\,\underline{i}^{\dagger}_{qds} + \underline{i}_{qds}\underline{\lambda}^{\dagger}_{qds}] \qquad (2.12\text{–}10)$$

Since the first term has no imaginary part, this expression reduces to

$$T_e = \frac{3}{2}\frac{P}{2}Im\,(\underline{i}_{qds}\cdot\underline{\lambda}^{\dagger}_{qds}) \qquad (2.12\text{–}11)$$

or, in scalar form

$$T_e = \frac{3}{2}\frac{P}{2}(\lambda_{ds}i_{qs} - \lambda_{qs}i_{ds}) \qquad (2.12\text{–}12)$$

Alternatively, the rotor flux linkages are expressed as

$$\lambda'_{qdr} = (L'_{lr} + L_m)\,\underline{i}'_{qdr} + L_m\underline{i}_{qds} \qquad (2.12\text{–}13)$$

Upon solving for $L_m\,\underline{i}_{qds}$ and inserting in eqn (2.12–7) results in

$$T_e = \frac{3}{2}\frac{P}{2}Im\,(\,\underline{\lambda}'_{qdr}\,\underline{i}'^{\dagger}_{qdr}) \qquad (2.12\text{–}14)$$

Hence, torque production in the machine can be visualized either as the rotor flux linkage vector interacting with the rotor current vector or, alternatively, the stator flux linkage vector crossed with the stator current vector.

If we define the *air gap* or *magnetizing* flux linkages as

$$\underline{\lambda}_{qdm} = L_m(\underline{i}_{qds} + \underline{i}'_{qdr}) \qquad (2.12\text{–}15)$$

solve for $L_m\underline{i}'_{qdr}$, and substitute into eqn (2.12–7) we obtain

$$T_e = \frac{3}{2}\frac{P}{2}Im\,(\underline{i}_{qds}\underline{\lambda}^{\dagger}_{qdm}) \qquad (2.12\text{–}16)$$

Another important expression for electromagnetic torque can be obtained if solve eqn (2.12–13) for rotor current rather than stator current, whereupon

$$\underline{i}'_{qdr} = \frac{1}{L'_{lr} + L_m}\lambda'_{qdr} - \frac{L_m}{L'_{lr} + L_m}\underline{i}_{qds} \tag{2.12–17}$$

Substituting this result into eqn (2.12–7) yields

$$T_e = \frac{3}{2}\frac{P}{2}\frac{L_m}{L'_r} Im\,(\underline{i}_{qds}\lambda'^{\dagger}_{qdr}\,) \tag{2.12–18}$$

where $L_r = L'_{lr} + L_m$. Hence, torque production in an induction machine may also be viewed as the interaction of rotor flux and stator current. In future chapters we will learn that this particular form of the torque equation is extremely useful for purposes of torque control.

Still other expressions are possible. For example, the torque can be expressed as the cross product of stator and rotor flux linkage, i.e.

$$T_e = \frac{3}{2}\frac{P}{2}\frac{L_m}{\sigma L_s L_r} Im\,(\underline{\lambda}_{qds}\lambda'^{\dagger}_{qdr}\,) \tag{2.12–19}$$

where $\sigma = 1 - L_m^2/L_r L_s$. A table summarizing these relationships is given below. Figure 2.12 shows the stator, mutual and rotor flux linkages as well as the stator and rotor currents for a typical operating condition. We have dropped the use of subscripts *d,q* since complex vectors on the *d,q* plane are already implied. The *d*–axis is aligned along the rotor flux vector in order to illustrate those vector cross products involving $\underline{\lambda}_{qdr}$. It is apparent from the construction of Figure 2.12 that the flux linkage vectors produce the same *d*–axis flux com-

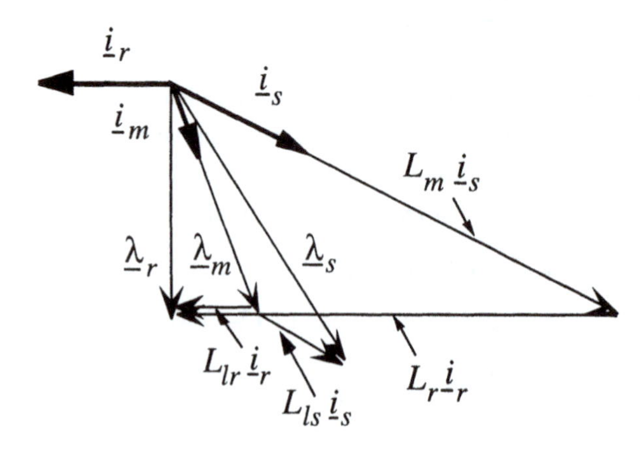

Figure 2.12 Illustrating torque production by vectors on the *d,q* plane

ponent and therefore produce the same electromagnetic torque.In addition to the equations describing the electrical behavior of the machines, additional equations are, of course, necessary to describe its electromechanical behavior. In its simplest form, the electromechanical behavior is described by

$$T_e = J\frac{d\omega_{rm}}{dt} + T_l \tag{2.12–20}$$

where ω_{rm} is the mechanical angular velocity of the rotor in rad/s and T_l is the load torque which may itself be described by additional differential or algebraic equations. Since the actual mechanical speed does not appear in the electrical equations, eqn (2.8–26), it is common to write eqn (2.12–20) as

$$T_e = \frac{2J}{P}\frac{d\omega_r}{dt} + T_l \tag{2.12–21}$$

where $\omega_r = 2\omega_{rm}/P$.

TABLE 2.1 Complex Vector Expressions for Electromagnetic Torque. ‡ denotes variable used to calculate torque

	$\underline{i}_{qds}$	$\underline{i}'_{qdr}$	$\underline{\lambda}_{qds}$	$\underline{\lambda}'_{qdr}$	$\underline{\lambda}_{qdm}$	Torque Expression
1	‡	‡				$T_e = \frac{3}{2}\frac{P}{2}L_m Im\{\underline{i}'^{\dagger}_{qdr}\underline{i}_{qds}\}$
2	‡				‡	$T_e = \frac{3}{2}\frac{P}{2} Im\{\underline{\lambda}^{\dagger}_{qdm}\underline{i}_{qds}\}$
3	‡		‡			$T_e = \frac{3}{2}\frac{P}{2} Im\{\underline{\lambda}^{\dagger}_{qds}\underline{i}_{qds}\}$
4	‡			‡		$T_e = \frac{3}{2}\frac{P}{2}\frac{L_m}{L'_r} Im\{\underline{i}_{qds}\underline{\lambda}'^{\dagger}_{qdr}\}$
5		‡			‡	$T_e = \frac{3}{2}\frac{P}{2} Im\{\underline{i}'^{\dagger}_{qdr}\underline{\lambda}_{qdm}\}$
6		‡		‡		$T_e = \frac{3}{2}\frac{P}{2} Im\{\underline{i}'^{\dagger}_{qdr}\underline{\lambda}'_{qdr}\}$
7		‡	‡			$T_e = \frac{3}{2}\frac{P}{2}\frac{L_m}{L_s} Im\{\underline{i}'^{\dagger}_{qdr}\underline{\lambda}_{qds}\}$
8			‡	‡		$T_e = \frac{3}{2}\frac{P}{2}\frac{L_m}{\sigma L_s L_r} Im\{\underline{\lambda}'^{\dagger}_{qdr}\underline{\lambda}_{qds}\}$

2.13 Analysis of Induction Motor Starting Performance Using *d,q*,0 Variables

As an example of the use of *d,q*,0 variables we will first investigate the starting behavior of a typical 5 horsepower (3.73 kW), 3 phase, 3 wire, 230 V., 4 pole squirrel cage induction machine with parameters:

$r_s = 0.531\ \Omega$ $\quad r_r' = 0.408\ \Omega$ $\quad J = 0.1$ kg–m^2

$L_{ls} = L_{lr}' = 2.52$ mH $\quad L_m = 84.7$ mH

The term *three wire* is generally used to denote that machine has no neutral return path. We will assume that the machine is suddenly connected to a balanced three phase 60 Hz, sinusoidal supply having rated line to line voltage of 230 V rms. The effect of source impedance in series with the supply is neglected. The voltages across the phases are then defined as:

$$v_{as} = \sqrt{\frac{2}{3}}230\cos(377t) \tag{2.13–1}$$

$$v_{bs} = \sqrt{\frac{2}{3}}230\cos(377t - 2\pi/3) \tag{2.13–2}$$

$$v_{cs} = \sqrt{\frac{2}{3}}230\cos(377t + 2\pi/3) \tag{2.13–3}$$

In the first example, Figure 2.13, the machine is simulated directly in terms of phase variables. When the voltage is suddenly applied to the stator windings the current rises rapidly to a large value, usually termed the *inrush current*. The electromagnetic torque develops a unidirectional component which acts to accelerate the rotor of the machine. However, the torque also contains a large pulsating component which decays as time progresses. When the speed approaches rated speed (synchronous speed), the currents in both the stator and rotor windings begin to decrease. Note that while the frequency of the stator current is fixed, the frequency of the rotor currents change continuously as the motor accelerates since the voltage induced in the rotor is always proportional to the relative angular velocity between the synchronously rotating stator MMF and that of the rotor circuits.

Since the motor is assumed as unloaded and second order effects such as windage and friction have been neglected, the simulated speed reaches its synchronous speed. At this point the stator current decreases to the no–load value needed to excite the machine, i.e. magnetizing current. Since the rotor is rotat-

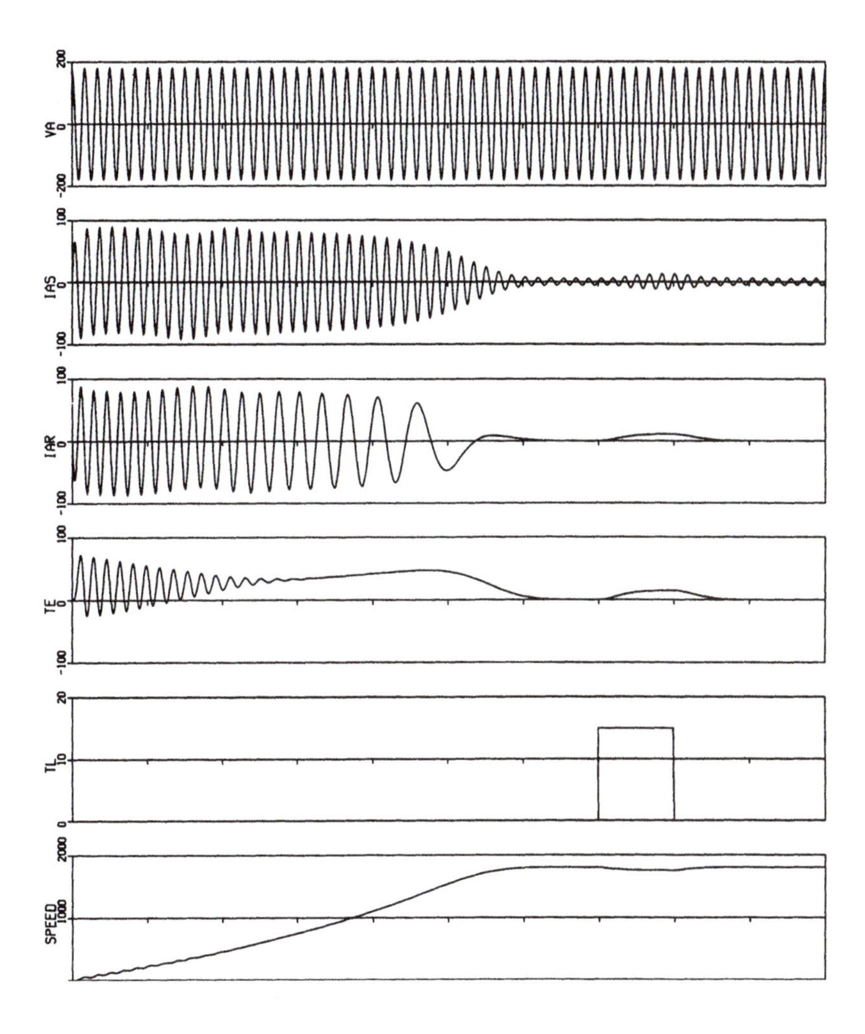

Figure 2.13 Starting performance of 220 V, 5 hp squirrel cage induction motor with a balanced sinusoidal supply showing physical variables. After reaching rated speed the motor is loaded to 0.83 times rated torque. Traces from top to bottom: v_{as}, phase a line to neutral voltage, i_{as}, line current of stator phase as, i'_{ar}, phase current of rotor phase ar (referred to stator turns), T_e, electromagnetic torque, T_l, load torque, rotor speed in RPM; time axis 0.1 s/div.

ing synchronously with the stator MMF, the rotor current becomes zero. In practice, of course, the rotor would actually reach a speed slightly below synchronous speed defined by a no–load slip. For purposes of analysis the windage and friction can be considered as an external effect which can easily be modeled if desired. Some time after achieving synchronous speed, the motor model is loaded with a step torque equivalent to 0.83 times rated torque. The motor responds to the increased torque by slowing down, whereupon the torque increases to an equal (and oppositely directed) value to support the load torque. The stator current increases in amplitude and changes phase with respect to the stator voltage resulting in power passing from the electrical to the mechanical terminals. The steady state currents for both the no–load and the loaded condition can, of course, be correlated with results obtained by conventional phasor analysis.

The pulsating torque observed upon switching on the stator voltages can be explained by considering the transient currents which flow in any inductive circuit. In general, sudden switching of the voltage onto the terminals of the machine results not only in ac current of the same frequency as the supply but also in dc transient components which die out with roughly the rotor open circuit time constant. This stator current component results in a stationary (non–rotating) component of MMF which is represented on the *d,q* plane as a stationary, non–rotating stator current vector. The flux created by this non–rotating MMF acts to induce currents in the rotor which rotate backward at rotor speed (frequency) with respect to the rotor. These induced rotor currents produce a rotor MMF which becomes stationary with respect to the stator MMF (represented as a non–rotating rotor current vector in the *d,q* plane) producing a negative torque which acts to brake the machine. In addition, the rotor MMF also interacts with the synchronously rotating stator MMF (represented as a synchronously rotating stator current complex vector in the *d,q* plane). Since torque is produced by the product of stator and rotor current, eqn (2.12–7), the torque produced by the interaction of these two components oscillates essentially at line frequency (60 Hz). It should be noted that in practice this interpretation is very nearly but not exactly correct due to the fact that the rotor and stator time constant are not completely independent. The rotor current vector (or rotor MMF), in fact, rotates very slowly with respect to the stator current vector (stator MMF) resulting in a 'beating' which can be observed superimposed on the stator currents as well as in the electromagnetic torque.

A plot of the starting of the induction machine is repeated in Figure 2.14. In this case the motor is represented in stationary axis $d,q,0$ variables ($\omega = 0$). Since the motor is connected without a neutral return, the neutral axis current is zero and is not shown in Figure 2.14. The voltages applied to the equivalent circuit, Figure 2.8, are from eqn (2.9-2)

$$\begin{aligned} v_{ds}^s &= -\frac{2}{3} Im\,[v_{as} + \underline{a} v_{bs} + \underline{a}^2 v_{cs}] \\ &= -\frac{2}{3}\frac{\sqrt{3}}{2}(v_{bs} - v_{cs}) \\ &= \sqrt{\frac{2}{3}} \cdot \frac{\sqrt{3}}{2}\left[\sqrt{\frac{2}{3}}220\cos(377t + 2\pi/3) - \sqrt{\frac{2}{3}}220\cos(377t - 2\pi/3)\right] \\ &= -\sqrt{\frac{2}{3}}230\sin(377t) \end{aligned} \tag{2.13–4}$$

$$\begin{aligned} v_{qs}^s &= \frac{2}{3} Re\,[v_{as} + \underline{a} v_{bs} + \underline{a}^2 v_{cs}] \\ &= \frac{2}{3}\left[\sqrt{\frac{2}{3}}220\cos(377t) - \frac{1}{2}\sqrt{\frac{2}{3}}220\cos(377t - 2\pi/3)\right. \\ &\quad \left. - \frac{1}{2}\sqrt{\frac{2}{3}}\cos(377t + 2\pi/3)\right] \\ &= \sqrt{\frac{2}{3}}220\cos(377t) \end{aligned} \tag{2.13–5}$$

Also, the zero sequence component is

$$\begin{aligned} v_{0s} &= \frac{1}{3}\left[\sqrt{\frac{2}{3}}220\cos(377t) + \sqrt{\frac{2}{3}}220\cos(377t - 2\pi/3)\right. \\ &\quad \left. + \sqrt{\frac{2}{3}}220\cos(377t + 2\pi/3)\right] = 0 \end{aligned} \tag{2.13–6}$$

Since the current i_{0s}^s is obviously zero it is not plotted in Figure 2.14. Observe that the behavior of the d,q stator currents is nearly identical to those obtained in terms of phase variables. The rotor currents on the other hand appear quite different. Since both the stator and rotor circuits are represented on stationary d,q axes the rotor currents as well as the stator currents oscillate at 60 Hz. Note that the torque and speed remain identical to Figure 2.13 since they must be invariant under the $d,q,0$ transformation.

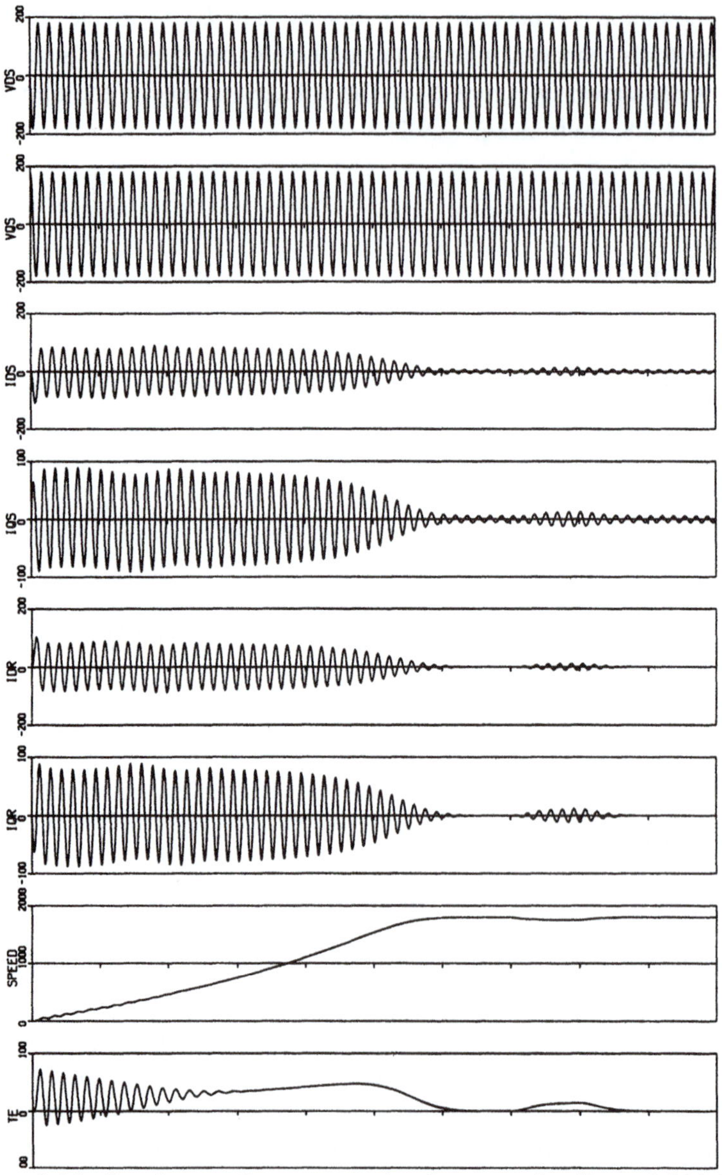

Figure 2.14 Starting performance and subsequent loading of 220 V, 5 hp squirrel cage induction motor with a balanced 60 Hz sinusoidal supply. Motor is modelled in stationary *d,q* axes. (Stationary reference frame). Traces from top to bottom – v^s_{ds} , v^s_{qs} , i^s_{ds} , i^s_{qs} , i'^s_{dr} , i'^s_{qr} , rotor speed in RPM, T_e.; time axis 0.1 s/div.

It is important to note the close similarity of the q–axis stator current i_{qs}^s with the current in phase as, that is i_{as}. In particular, recall from eqn (2.9–2) that

$$i_{qs}^s = \frac{2}{3}\left(i_{as} - \frac{1}{2}i_{bs} - \frac{1}{2}i_{cs}\right) \tag{2.13–7}$$

If in addition,

$$i_{as} + i_{bs} + i_{cs} = 0 \tag{2.13–8}$$

then

$$i_{qs}^s = i_{as} \tag{2.13–9}$$

This is to be expected since we can note from eqns (2.13-1) and (2.13-4) that the voltage applied to the q axis circuit is also equal to the as phase voltage. In essence this feature is the reason why the factor "2/3" was chosen in the initial definition of d,q variables, eqn (2.9–1). Although this introduces a 3/2 factor in the torque and power equations, it has the advantage of making it easier to interpret variables such as voltage and current. For consistency with the work of previous researchers we will employ this so called *scaled transformation* henceforth throughout this book.

In Figure 2.15 the same motor starting problem as studied in Figure 2.13 and Figure 2.14 has been solved in a rotating $d,q,0$ coordinate system. In this case the reference frame angular speed has been set equal to the rotor speed, i.e. a *rotor reference frame*. Note that all of the d,q variables now assume a frequency equal to the rotor slip frequency since the reference frame is now slipping with respect to the synchronously rotating magnetic field. Since the torque and speed are invariant under the d,q transformation they are again identical to Figure 2.13.

In general, the angular velocity of the rotating axes can be freely chosen. In the study of electrical machines, however, it is frequently useful to "attach" the d,q axes to the vector describing the instantaneous spatial position of the motor terminal voltage, current or flux or even to some other vector variable. In most practical cases, the voltage is considered as the independent variable, (i.e. the 'input') so that the angular velocity of this vector is known in advance. It is then a simple matter to attach the reference frame axes to the voltage vector by simply setting ω equal to the angular velocity of this vector. When the phase

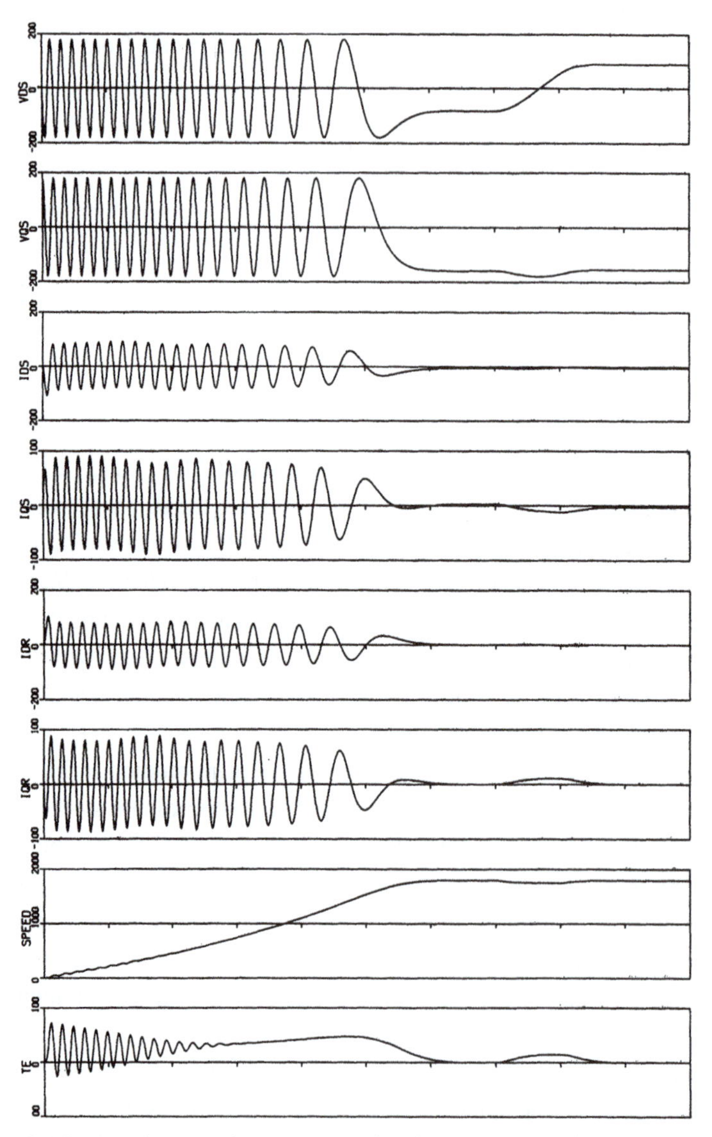

Figure 2.15 Starting performance and subsequent loading of 220 V, 5 hp squirrel cage induction motor with a balanced sinusoidal supply. Motor is modelled in d,q axes rotating at rotor speed. (Rotor reference frame). Traces from top to bottom – v^r_{ds}, v^r_{qs}, i^r_{ds}, i^r_{qs}, i'^r_{dr}, i'^r_{qr}, rotor speed in RPM, torque T_e.; time axis 0.1 s/div.

voltages are balanced and sinusoidal the angular velocity of vector is simply equal to 2π times the frequency of the sine wave. Hence, in our case, if we take

$$\omega = 377 \quad rad/s \tag{2.13–10}$$

and let

$$\theta(0) = 0 \tag{2.13–11}$$

then the q–axis is fixed to the voltage vector itself. It is not difficult to show that the voltages to be applied to the equivalent circuit now become, from eqn (2.9–3),

$$v_{ds}^{sv} = 0 \tag{2.13–12}$$

$$v_{qs}^{sv} = \sqrt{\frac{2}{3}}220\ V \tag{2.13–13}$$

$$v_{0s}^{sv} = 0 \tag{2.13–14}$$

where the superscript sv refers to the 'stator voltage' reference frame.

The solution for the starting problem, this time investigated in a reference frame synchronized with the rotating stator voltage vector is given in Figure 2.16. Note that since the input is now 'dc', all variables take the form of slowly varying dc variables. In effect, the dc offset and 60 Hz ac components observed in Figure 2.13 and Figure 2.14 have interchanged roles. That is, the dc current offset due to sudden switching–on of the source voltages appears now as a decaying 60 Hz ac transient while the ac inrush current obtained during starting now becomes a slowly varying dc component. The electromagnetic torque and speed are, of course, again invariant under this transformation. It is important to note that while the d–axis voltage is zero, the corresponding current is substantial since the d– and q–axes circuits are coupled via the dependant voltage generators of Figure 2.8.

In Figure 2.17 the starting behavior of the same motor is computed one final time. In this case the reference frame has been attached to the rotor flux vector rather than the stator voltage vector. Since the rotor flux is not explicitly known throughout the solution period, the angular velocity of the reference frame which is needed to 'attach' the d,q axes to the flux vector must be solved implicitly. Since the angular velocity of the reference frame is known only implicitly, the stator voltages can now only be expressed in the form

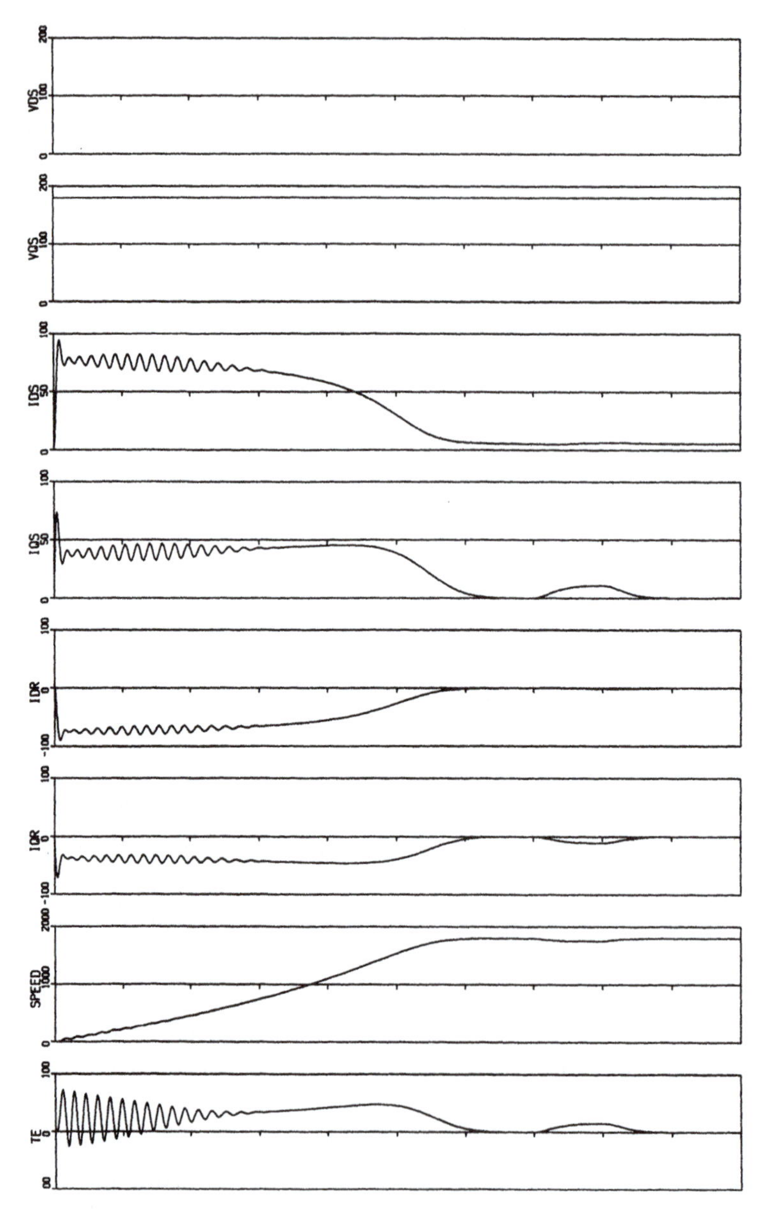

Figure 2.16 Starting performance and subsequent loading of 5 hp, 220 V squirrel cage induction motor with a balanced sinusoidal supply. Motor is modelled in *d,q* axes synchronously rotating with the applied voltage vector. (Synchronous reference frame). Traces from the top: v^{sv}_{ds}, v^{sv}_{qs}, i^{sv}_{ds}, i^{sv}_{qs}, i'^{sv}_{dr}, i'^{sv}_{qr}, rotor speed in RPM, electromagnetic torque T_e; time axis 0.1 s/div.

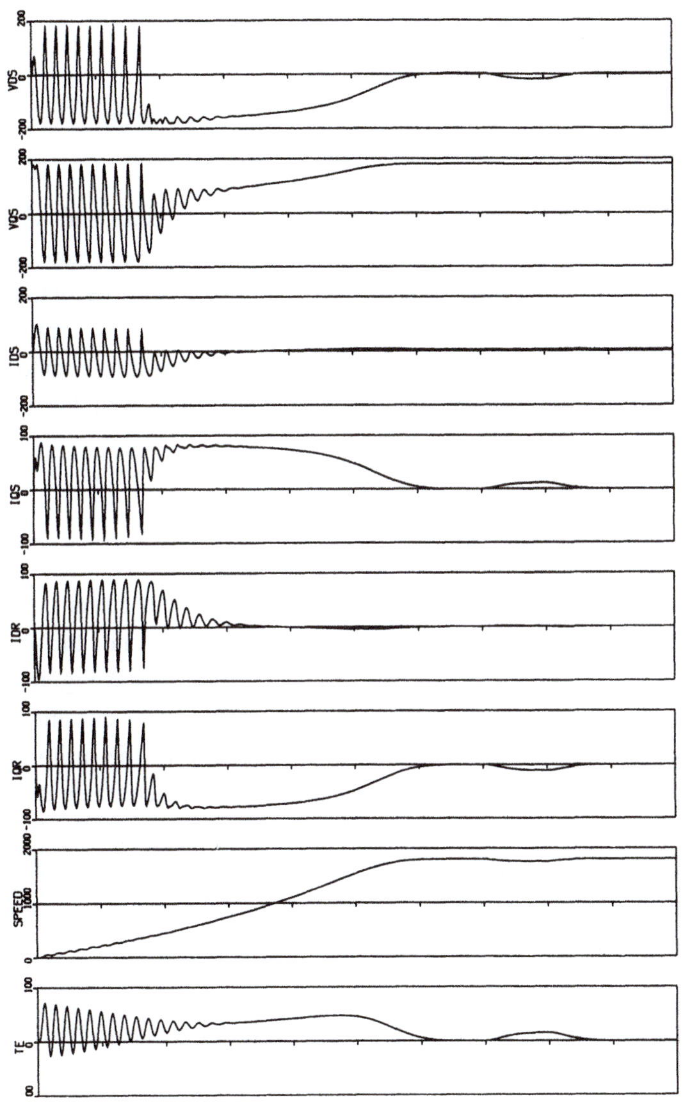

Figure 2.17 Starting performance and subsequent loading of 220 V, 5 hp squirrel cage induction motor with a balanced sinusoidal supply. Motor is modelled in *d,q* axes rotating synchronously with the rotor flux. (Rotor flux reference frame). Traces from the top: v_{ds}^{rf} , v_{qs}^{rf} , i_{ds}^{rf} , i_{qs}^{rf} , i'^{rf}_{dr} , i'^{rf}_{qr} , rotor speed in RPM, electromagnetic torque T_e; time axis 0.1 s/div.

$$v_{ds}^{rf} = -\sqrt{\frac{2}{3}}220\sin(377t - \theta_{rf}) \tag{2.13–15}$$

$$v_{qs}^{rf} = \sqrt{\frac{2}{3}}220\cos(377t - \theta_{rf}) \tag{2.13–16}$$

$$v_{0s}^{rf} = 0 \tag{2.13–17}$$

where θ_{rf} is the instantaneous position of the rotor flux vector and where the superscript *rf* has been used to denote the fact that the *d,q* axes rotate with the 'rotor flux'. The necessary information to determine θ_{rf} can be obtained by means of feedback regulation [4]. In this case, the *d*–axis is continuously aligned with the rotor flux vector so that the *q*–axis rotor flux component is always identically zero. From eqn (2.12–18), the electromagnetic torque equation reduces to

$$T_e = \frac{3}{2}\frac{P}{2}\frac{L_m}{L_r}\lambda_{dr}^{rf} i_{qs}^{rf} \tag{2.13–18}$$

Note that in this case the *q*–axis component of the stator current is always at right angles to the rotor flux and therefore always corresponds to the torque producing component of stator current. While the rotor flux in Figure 2.17 is not constant, the similarity of the traces for the *q*–axis stator current and the electromagnetic torque is apparent. Since we can identify the orthogonal or torque producing component of stator current in this reference frame, the determination of the instantaneous position of the rotor flux is essentially the key to motor torque control. A discussion of the various means by which this is accomplished together with their advantages and limitations is the subject of much of the remainder of this book.

2.14 Extension of *d,q,0* Theory to Analysis of Salient Pole Synchronous Machines

The extensive family of synchronous type machines are another important type of machine for motor drive applications. For the remainder of the chapter we will examine *d,q* theory as related to a salient pole synchronous machine including the effects of non–uniform air gap and non–symmetric rotor windings typical of such a machine. We will then examine how these equations can be modified to model the behavior of other synchronous type machines.

Since the air gap of a salient pole synchronous machine varies along the inner circumference of the stator, the machine is no longer symmetrical in the sense of an induction machine. In particular, the stator self inductances represented by eqn (2.4–9) for the induction machine, now vary with rotor position. Referring to Figure 2.18 it is clear that the self inductance of any winding must

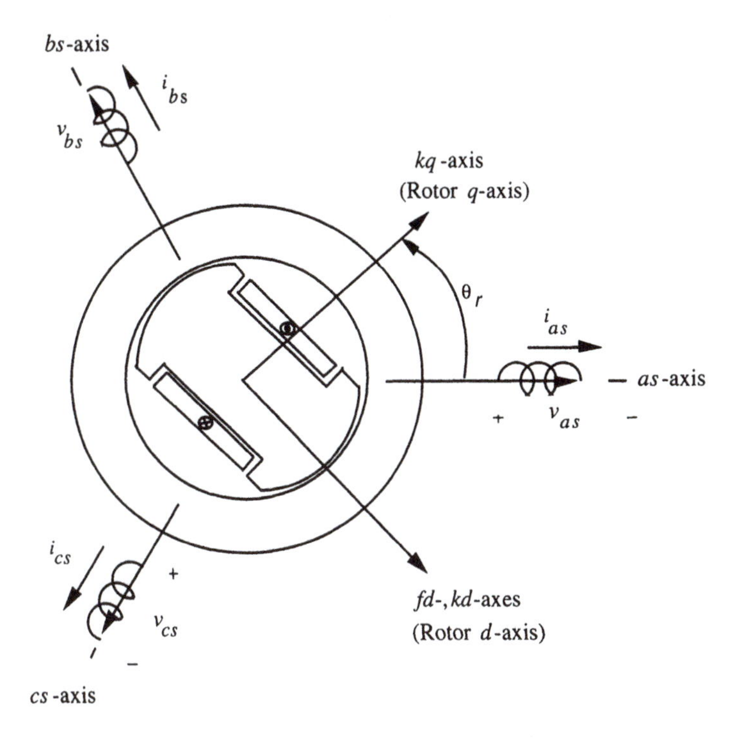

Figure 2.18 Magnetic axes of a salient pole synchronous machine

now pulsate once each time the rotor moves one pole pitch. Neglecting higher order harmonics, the stator self inductance has a second harmonic component in addition to the constant component represented by eqn (2.4–5). For example, the self inductance of phase *as* is

$$L_{as,\,as} = L_{ls} + L_{0s} - L_{2s}\cos 2\theta_r \qquad (2.14\text{–}1)$$

where L_{ls} again represents the leakage inductance of the phase. Using the same approach as Section 2.2, modified for a machine with salient poles, it can be shown that

$$L_{0s} = \mu_0 r l N_s^2 \left(\frac{\pi}{8}\right)\left(\frac{1}{g_{min}} + \frac{1}{g_{max}}\right) \tag{2.14–2}$$

and

$$L_{2s} = \mu_0 r l N_s^2 \left(\frac{\pi}{8}\right)\left(\frac{1}{g_{min}} - \frac{1}{g_{max}}\right) \tag{2.14–3}$$

Inductances for the other phases are found in a similar manner. They are

$$L_{bs,\,bs} = L_{ls} + L_{0s} - L_{2s}\cos(2\theta_r + 2\pi/3) \tag{2.14–4}$$

$$L_{cs,\,cs} = L_{ls} + L_{0s} - L_{2s}\cos(2\theta_r - 2\pi/3) \tag{2.14–5}$$

The mutual inductances between stator phases are

$$L_{as,\,bs} = L_{bs,\,as} = -\frac{1}{2}L_{0s} - L_{2s}\cos(2\theta_r - 2\pi/3) \tag{2.14–6}$$

$$L_{as,\,cs} = L_{cs,\,as} = -\frac{1}{2}L_{0s} - L_{2s}\cos(2\theta_r + 2\pi/3) \tag{2.14–7}$$

$$L_{bs,\,cs} = L_{cs,\,bs} = -\frac{1}{2}L_{0s} - L_{2s}\cos 2\theta_r \tag{2.14–8}$$

The inductances corresponding to flux linking the field winding with the three stator phases can be written as

$$L_{as,\,fd} = L_{fd,\,as} = L_{sfd}\cos\theta_r \tag{2.14–9}$$

$$L_{bs,\,fd} = L_{fd,\,bs} = L_{sfd}\cos(\theta_r - 2\pi/3) \tag{2.14–10}$$

$$L_{cs,\,fd} = L_{fd,\,cs} = L_{sfd}\cos(\theta_r + 2\pi/3) \tag{2.14–11}$$

where

$$L_{sfd} = \mu_0 r l N_s N_{fd}\left(\frac{\pi}{4}\right)\frac{1}{g_{min}} \tag{2.14–12}$$

In like manner, the inductance corresponding to flux linking the *d*–axis damper winding and each of the three stator windings is

$$L_{as,\,kd} = L_{kd,\,as} = L_{skd}\cos\theta_r \tag{2.14–13}$$

$$L_{bs,kd} = L_{kd,bs} = L_{skd}\cos(\theta_r - 2\pi/3) \quad (2.14\text{–}14)$$

$$L_{cs,kd} = L_{kd,cs} = L_{skd}\cos(\theta_r + 2\pi/3) \quad (2.14\text{–}15)$$

where

$$L_{skd} = \mu_0 r l N_s N_{kd}\left(\frac{\pi}{4}\right)\frac{1}{g_{min}} \quad (2.14\text{–}16)$$

Finally, the inductances corresponding to fluxes which link the q–axis damper and the stator windings are

$$L_{as,kq} = L_{kq,as} = -L_{skq}\sin\theta_r \quad (2.14\text{–}17)$$

$$L_{bs,kq} = L_{kq,bs} = -L_{skq}\sin(\theta_r - 2\pi/3) \quad (2.14\text{–}18)$$

$$L_{cs,kq} = L_{kq,cs} = -L_{skq}\sin(\theta_r + 2\pi/3) \quad (2.14\text{–}19)$$

where

$$L_{skq} = \mu_0 r l N_s N_{kq}\left(\frac{\pi}{4}\right)\frac{1}{g_{max}} \quad (2.14\text{–}20)$$

These inductances can be assembled in a matrix equation to represent all of the fluxes linking the three stator windings. The voltages across the three stator phases of the synchronous machine can be written

$$\begin{bmatrix} v_{as} \\ v_{bs} \\ v_{cs} \end{bmatrix} = r_s \begin{bmatrix} i_{as} \\ i_{bs} \\ i_{cs} \end{bmatrix} + p \begin{bmatrix} \lambda_{as} \\ \lambda_{bs} \\ \lambda_{cs} \end{bmatrix} \quad (2.14\text{–}21)$$

where

$$\begin{bmatrix}\lambda_{as}\\ \lambda_{bs}\\ \lambda_{cs}\end{bmatrix} = \begin{bmatrix} L_{ls}+L_{0s}-L_{2s}\cos 2\theta_r & -\frac{1}{2}L_{0s}-L_{2s}\cos(2\theta_r-2\pi/3) & -\frac{1}{2}L_{0s}-L_{2s}\cos(2\theta_r+2\pi/3) \\ -\frac{1}{2}L_{0s}-L_{2s}\cos(2\theta_r-2\pi/3) & L_{ls}+L_{0s}-L_{2s}\cos(2\theta_r+2\pi/3) & -\frac{1}{2}L_{0s}-L_{2s}\cos 2\theta_r \\ -\frac{1}{2}L_{0s}-L_{2s}\cos(2\theta_r+2\pi/3) & -\frac{1}{2}L_{0s}-L_{2s}\cos 2\theta_r & L_{ls}+L_{0s}-L_{2s}\cos(2\theta_r-2\pi/3) \end{bmatrix} \cdot \begin{bmatrix} i_{as}\\ i_{bs}\\ i_{cs}\end{bmatrix}$$

$$+\begin{bmatrix} L_{sfd}\cos\theta_r & L_{skd}\cos\theta_r & -L_{skq}\sin\theta_r \\ L_{sfd}\cos(\theta_r-2\pi/3) & L_{skd}\cos(\theta_r-2\pi/3) & -L_{skq}\sin(\theta_r-2\pi/3) \\ L_{sfd}\cos(\theta_r+2\pi/3) & L_{skd}\cos(\theta_r+2\pi/3) & -L_{skq}\sin(\theta_r+2\pi/3)\end{bmatrix} \cdot \begin{bmatrix} i_{fd}\\ i_{kd}\\ i_{kq}\end{bmatrix} \quad (2.14\text{–}22)$$

The voltages across the field, d–axis damper, and q–axis damper windings are

$$v_{fd} = r_{fd} i_{fd} + p\lambda_{fd} \quad (2.14\text{–}23)$$

$$v_{kd} = r_{kd} i_{kd} + p\lambda_{kd} \quad (2.14\text{–}24)$$

$$v_{kq} = r_{kq} i_{kq} + p\lambda_{kq} \quad (2.14\text{–}25)$$

where,

$$\lambda_{fd} = (L_{lfd}+L_{mfd}) i_{fd} + L_{fkd} i_{kd}' + L_{sfd}[i_{as}\cos\theta_r + i_{bs}\cos(\theta_r-2\pi/3) + i_{cs}\cos(\theta_r+2\pi/3)] \quad (2.14\text{–}26)$$

$$\lambda_{kd} = (L_{lkd}+L_{mkd}) i_{kd} + L_{fkd} i_{fd} + L_{skd}[i_{as}\cos\theta_r + i_{bs}\cos(\theta_r-2\pi/3) + i_{cs}\cos(\theta_r+2\pi/3)] \quad (2.14\text{–}27)$$

$$\lambda_{kq} = (L_{lkq} + L_{mkq})\, i_{kq} - L_{skq}\, [i_{as} \sin\theta_r + i_{bs} \sin(\theta_r - 2\pi/3) + i_{cs} \sin(\theta_r + 2\pi/3)] \qquad (2.14\text{–}28)$$

In these equations, it can be shown that [5]

$$L_{mfd} = \mu_0 r l N_f^2 \left(\frac{\pi}{4}\right) \frac{1}{g_{min}} \qquad (2.14\text{–}29)$$

$$L_{kfd} = \mu_0 r l N_f N_{kd} \left(\frac{\pi}{4}\right) \frac{1}{g_{min}} \qquad (2.14\text{–}30)$$

$$L_{mkd} = \mu_0 r l N_{kd}^2 \left(\frac{\pi}{4}\right) \frac{1}{g_{min}} \qquad (2.14\text{–}31)$$

$$L_{mkq} = \mu_0 r l N_{kq}^2 \left(\frac{\pi}{4}\right) \frac{1}{g_{max}} \qquad (2.14\text{–}32)$$

The quantities L_{lfd}, L_{lkd}, L_{lkq} are the leakage inductances of the field, d–axis damper and q–axis damper windings respectively.

We can write the flux linkage equation for the stator windings in the form

$$\begin{bmatrix} \lambda_{as} \\ \lambda_{bs} \\ \lambda_{cs} \end{bmatrix} = \begin{bmatrix} L_{ls} + L_{0s} & -\frac{1}{2}L_{0s} & -\frac{1}{2}L_{0s} \\ -\frac{1}{2}L_{0s} & L_{ls} + L_{0s} & -\frac{1}{2}L_{0s} \\ -\frac{1}{2}L_{0s} & -\frac{1}{2}L_{0s} & L_{ls} + L_{0s} \end{bmatrix} \cdot \begin{bmatrix} i_{as} \\ i_{bs} \\ i_{cs} \end{bmatrix}$$

$$- \begin{bmatrix} L_{2s}\cos 2\theta_r & L_{2s}\cos(2\theta_r - 2\pi/3) & L_{2s}\cos(2\theta_r + 2\pi/3) \\ L_{2s}\cos(2\theta_r - 2\pi/3) & L_{2s}\cos(2\theta_r + 2\pi/3) & L_{2s}\cos 2\theta_r \\ L_{2s}\cos(2\theta_r + 2\pi/3) & L_{2s}\cos 2\theta_r & L_{2s}\cos(2\theta_r - 2\pi/3) \end{bmatrix} \cdot \begin{bmatrix} i_{as} \\ i_{bs} \\ i_{cs} \end{bmatrix}$$

$$+ \begin{bmatrix} L_{sfd}\cos\theta_r & L_{skd}\cos\theta_r & -L_{skq}\sin\theta_r \\ L_{sfd}\cos(\theta_r - 2\pi/3) & L_{skd}\cos(\theta_r - 2\pi/3) & -L_{skq}\sin(\theta_r - 2\pi/3) \\ L_{sfd}\cos(\theta_r + 2\pi/3) & L_{skd}\cos(\theta_r + 2\pi/3) & -L_{skq}\sin(\theta_r + 2\pi/3) \end{bmatrix} \cdot \begin{bmatrix} i_{fd} \\ i_{kd} \\ i_{kq} \end{bmatrix}$$

(2.14–33)

This equation can be written in complex form as

$$\begin{bmatrix}\lambda_{as}\\ \lambda_{bs}\\ \lambda_{cs}\end{bmatrix} = \begin{bmatrix} L_{ls}+L_{0s} & -\frac{1}{2}L_{0s} & -\frac{1}{2}L_{0s} \\ -\frac{1}{2}L_{0s} & L_{ls}+L_{0s} & -\frac{1}{2}L_{0s} \\ -\frac{1}{2}L_{0s} & -\frac{1}{2}L_{0s} & L_{ls}+L_{0s}\end{bmatrix}\cdot\begin{bmatrix}i_{as}\\ i_{bs}\\ i_{cs}\end{bmatrix}$$

$$-\frac{L_{2s}}{2}\left\{\begin{bmatrix} e^{2j\theta_r} & \underline{a}^2e^{2j\theta_r} & \underline{a}e^{2j\theta_r} \\ \underline{a}^2e^{2j\theta_r} & \underline{a}e^{2j\theta_r} & e^{2j\theta_r} \\ \underline{a}e^{2j\theta_r} & e^{2j\theta_r} & \underline{a}^2e^{2j\theta_r}\end{bmatrix} + \begin{bmatrix} e^{-2j\theta_r} & \underline{a}e^{-2j\theta_r} & \underline{a}^2e^{-2j\theta_r} \\ \underline{a}e^{-2j\theta_r} & \underline{a}^2e^{-2j\theta_r} & e^{-2j\theta_r} \\ \underline{a}^2e^{-2j\theta_r} & e^{-2j\theta_r} & \underline{a}e^{-2j\theta_r}\end{bmatrix}\right\}\cdot\begin{bmatrix}i_{as}\\ i_{bs}\\ i_{cs}\end{bmatrix}$$

$$+\frac{L_{sfd}}{2}\left\{\begin{bmatrix} e^{j\theta_r} \\ \underline{a}^2e^{j\theta_r} \\ \underline{a}e^{j\theta_r}\end{bmatrix} + \begin{bmatrix} e^{-j\theta_r} \\ \underline{a}e^{-j\theta_r} \\ \underline{a}^2e^{-j\theta_r}\end{bmatrix}\right\} i_{fd} + \frac{L_{skd}}{2}\left\{\begin{bmatrix} e^{j\theta_r} \\ \underline{a}^2e^{j\theta_r} \\ \underline{a}e^{j\theta_r}\end{bmatrix} + \begin{bmatrix} e^{-j\theta_r} \\ \underline{a}e^{-j\theta_r} \\ \underline{a}^2e^{-j\theta_r}\end{bmatrix}\right\} i_{kd}$$

$$-\frac{L_{skq}}{2j}\left\{\begin{bmatrix} e^{j\theta_r} \\ \underline{a}^2e^{j\theta_r} \\ \underline{a}e^{j\theta_r}\end{bmatrix} - \begin{bmatrix} e^{-j\theta_r} \\ \underline{a}e^{-j\theta_r} \\ \underline{a}^2e^{-j\theta_r}\end{bmatrix}\right\} i_{kq} \qquad (2.14\text{–}34)$$

Multiplying the second row of this equation by $\underline{a}$ and the third row by $\underline{a}^2$ and adding the result to the first row, one obtains after some simplification,

$$\lambda_{as} + \underline{a}\lambda_{bs} + \underline{a}^2\lambda_{cs} = \left(L_{ls} + \frac{3}{2}L_{0s}\right)(i_{as} + \underline{a}i_{bs} + \underline{a}^2 i_{cs})$$

$$-\frac{3}{2}L_{2s}\,(i_{as} + \underline{a}^2 i_{bs} + \underline{a}i_{cs})e^{j2\theta_r}$$

$$+\frac{3}{2}L_{sfd}i_{fd}e^{j\theta_r} + \frac{3}{2}L_{skd}i_{kd}e^{j\theta_r} - \frac{3}{2}L_{skq}i_{kq}e^{j\left(\theta_r - \frac{\pi}{2}\right)}$$

$$(2.14\text{–}35)$$

Using the basic definitions for the complex vector and its conjugate, eqns (2.5-13) and (2.5–14), this expression becomes finally,

$$\underline{\lambda}_{abcs} = \left(L_{ls} + \frac{3}{2}L_{0s}\right)\underline{i}_{abcs} - \frac{3}{2}L_{2s}\,\underline{i}^{\dagger}_{abcs}\,e^{j2\theta_r} + \frac{3}{2}L_{sfd}i_{fd}e^{j\theta_r} + \frac{3}{2}L_{skd}i_{kd}e^{j\theta_r} - \frac{3}{2}L_{skq}i_{kq}e^{j\left(\theta_r - \frac{\pi}{2}\right)} \tag{2.14–36}$$

It is not difficult to show that the companion differential equation expressing the stator voltages is the same as for the induction motor. It can again be written in complex space vector form as

$$\underline{v}_{abcs} = r_s\underline{i}_{abcs} + p\underline{\lambda}_{abcs} \tag{2.14–37}$$

The space vector concept has again permitted us to express the machine equations in a remarkably simple form. In the case of induction machine we were able to transform the complex vector equations to a freely rotating axis. This was accomplished by essentially multiplying the stator flux linkage equation by $e^{-j\theta}$. However, in this case the necessary symmetry does not exist. Simplification, however, is still possible if we let $\theta = \theta_r$. That is, if we fix the reference frame to the rotor of the machine (*rotor reference frame*), then in this case eqns (2.14–36) and (2.14–37) can be written,

$$\underline{v}_{abcs}e^{-j\theta_r} = r_s\,\underline{i}_{abcs}e^{-j\theta_r} + e^{-j\theta_r}p\underline{\lambda}_{abcs} \tag{2.14–38}$$

$$\underline{\lambda}_{abcs}e^{-j\theta_r} = \left(L_{ls} + \frac{3}{2}L_{0s}\right)\underline{i}_{abcs}e^{-j\theta_r} - \frac{3}{2}L_{2s}\underline{i}^{\dagger}_{abcs}\,e^{j\theta_r} + \frac{3}{2}L_{sfd}i_{fd} + \frac{3}{2}L_{skd}i_{kd} - \frac{3}{2}L_{skq}i_{kq}e^{-j\frac{\pi}{2}} \tag{2.14–39}$$

Multiplying through by 2/3, using the chain rule of differentiation and defining

$$\underline{v}^r_{qds} = v^r_{qs} - jv^r_{ds} = \frac{2}{3}\underline{v}_{abcs}e^{-j\theta_r} \tag{2.14–40}$$

$$\underline{\lambda}^r_{qds} = \lambda^r_{qs} - j\lambda^r_{ds} = \frac{2}{3}\underline{\lambda}_{abcs}e^{-j\theta_r} \tag{2.14–41}$$

$$\underline{i}^r_{qds} = i^r_{qs} - ji^r_{ds} = \frac{2}{3}\underline{i}_{abcs}e^{-j\theta_r} \tag{2.14–42}$$

we arrive at

$$\underline{v}^r_{qds} = r_s \underline{i}^r_{qds} + p\underline{\lambda}^r_{qds} + j\omega_r \underline{\lambda}^r_{qds} \tag{2.14–43}$$

where

$$\underline{\lambda}^r_{qds} = \left(L_{ls} + \frac{3}{2}L_{0s}\right)\underline{i}^r_{qds} - \frac{3}{2}L_{2s}(\underline{i}^r_{qds})^\dagger + L_{sfd}i_{fd} + L_{skd}i_{kd} + jL_{skq}i_{kq} \tag{2.14–44}$$

Note the use of the superscript r to designate the 'rotor' reference frame.

One further simplification is obtained if we again refer the rotor circuits to the stator by the turns ratio. If we define *direct* axis and *quadrature* axis inductance L_{md} and L_{mq} it can be shown that

$$L_{md} = \frac{3}{2}(L_{0s} + L_{2s}) = \frac{3}{2}\frac{N_s}{N_{fd}}L_{sfd} = \frac{3}{2}\frac{N_s}{N_{kd}}L_{skd} \tag{2.14–45}$$

$$L_{mq} = \frac{3}{2}(L_{0s} - L_{2s}) = \frac{3}{2}\frac{N_s}{N_{kq}}L_{skq} \tag{2.14–46}$$

Using the same manipulation as Section 2.7, eqn (2.14–44) becomes

$$\underline{\lambda}^r_{qds} = \left(L_{ls} + \frac{L_{md} + L_{mq}}{2}\right)\underline{i}^r_{qds} + \frac{2}{3}L_{md}(i'_{fd} + i'_{kd}) - \left(\frac{L_{md} - L_{mq}}{2}\right)(\underline{i}^r_{qds})^\dagger + \frac{2}{3}jL_{mq}i'_{kq} \tag{2.14–47}$$

where the prime is used to signify the turns ratio transformation.

It is not difficult to show that the rotor flux linkage equations are also simplified by use of the complex space vector. However, the gain in this case is only marginal because the complex vector cannot be utilized to represent the rotor variables since the rotor is not 'symmetrical'. That is, the inductances and resistances of the d– and q–axis circuits are not the same

$$\lambda_{fd} = (L_{lfd} + L_{mfd})i_{fd} + L_{fkd}i_{kd} + \frac{3L_{sfd}}{4}[\underline{i}^r_{qds} + (\underline{i}^r_{qds})^\dagger] \tag{2.14–48}$$

$$\lambda_{kd} = (L_{lkd} + L_{mkd})\, i_{kd} + L_{fkd} i_{fd} + \frac{3L_{skd}}{4}[\underline{i}^r_{qds} + (\underline{i}^r_{qds})^\dagger] \tag{2.14–49}$$

$$\lambda_{kq} = (L_{lkq} + L_{mkq})\, i_{kq} - j\frac{3L_{skq}}{4}[\underline{i}^r_{qds} - (\underline{i}^r_{qds})^\dagger] \tag{2.14–50}$$

However, since

$$L_{md} = \frac{3}{2}(L_{0s} + L_{2s}) = \frac{3}{2}\frac{N_s^2}{N_{fd}^2} L_{mfd} = \frac{3}{2}\frac{N_s^2}{N_{kd}^2} L_{mkd} = \frac{3}{2}\frac{N_s^2}{N_{fd}N_{kd}} L_{fkd}$$
$$= \frac{3}{2}\frac{N_s}{N_{kd}} L_{skd} = \frac{3}{2}\frac{N_s}{N_{fd}} L_{sfd} \tag{2.14–51}$$

$$L_{mq} = \frac{3}{2}(L_{0s} - L_{2s}) = \frac{3}{2}\frac{N_s^2}{N_{fq}^2} L_{mkq} = \frac{3}{2}\frac{N_s}{N_{kq}} L_{skq} \tag{2.14–52}$$

when eqns (2.14–48) to (2.14–50) are referred to the stator turns, they become,

$$\lambda'_{fd} = L'_{lfd}\, i'_{fd} + L_{md}\{ i'_{fd} + i'_{kd} + \frac{1}{2}[\underline{i}^r_{qds} + (\underline{i}^r_{qds})^\dagger]\} \tag{2.14–53}$$

$$\lambda'_{kd} = L'_{lkd}\, i'_{kd} + L_{md}\{ i'_{kd} + i'_{fd} + \frac{1}{2}[\underline{i}^r_{qds} + (\underline{i}^r_{qds})^\dagger]\} \tag{2.14–54}$$

$$\lambda'_{kq} = L'_{lkq}\, i'_{kq} + L_{mq}\{ i'_{kq} - j\frac{1}{2}[\underline{i}^r_{qds} - (\underline{i}^r_{qds})^\dagger]\} \tag{2.14–55}$$

In this case, however, it has become necessary to define

$$i'_{kd} = \frac{2}{3}\frac{N_{kd}}{N_s} i_{kd} \tag{2.14–56}$$

$$i'_{fd} = \frac{2}{3}\frac{N_{fd}}{N_s} i_{fd} \tag{2.14–57}$$

$$i'_{kq} = \frac{2}{3}\frac{N_{kq}}{N_s} i_{kq} \tag{2.14–58}$$

because of the appearance of the 3/2 factor in the transformed inductances and the absence of the 3/2 factor in eqns (2.14–48) to (2.14–50). Also

$$L'_{lfd} = \frac{3}{2}L_{lfd}\frac{N_s^2}{N_{fd}^2} \tag{2.14–59}$$

$$L'_{lkd} = \frac{3}{2}L_{lkd}\frac{N_s^2}{N_{kd}^2} \tag{2.14–60}$$

$$L'_{lkq} = \frac{3}{2}L_{lkq}\frac{N_s^2}{N_{kq}^2} \tag{2.14–61}$$

The referred flux linkages continue to be defined as

$$\lambda'_{kd} = \frac{N_s}{N_{kd}}\lambda_{kd} \tag{2.14–62}$$

$$\lambda'_{fd} = \frac{N_s}{N_{fd}}\lambda_{fd} \tag{2.14–63}$$

$$\lambda'_{kq} = \frac{N_s}{N_{kq}}\lambda_{kq} \tag{2.14–64}$$

The rotor voltage equations, eqn (2.14–10) to (2.14–25) can also be referred to the stator by the same turns ratio transformation.

When eqns (2.14–23) to (2.14–25), (2.14–43), (2.14–44), and (2.14–48) to (2.14–50) are written in scalar form they become the basis for *Park's Equations*,

$$v_{ds}^r = r_s i_{ds}^r + p\lambda_{ds}^r - \omega_r\lambda_{qs}^r \tag{2.14–65}$$

$$v_{qs}^r = r_s i_{qs}^r + p\lambda_{qs}^r + \omega_r\lambda_{ds}^r \tag{2.14–66}$$

$$v'_{fd} = r'_{fd}\, i'_{fd} + p\lambda'_{fd} \tag{2.14–67}$$

$$v'_{kd} = r'_{kd}\, i'_{kd} + p\lambda'_{kd} \tag{2.14–68}$$

$$v'_{kq} = r'_{kq}\, i'_{kq} + p\lambda'_{kq} \tag{2.14–69}$$

where

$$\lambda_{ds}^r = L_{ls}\, i'_{ds} + L_{md}(i_{ds}^r + i'_{fd} + i'_{kd}) \tag{2.14–70}$$

$$\lambda_{qs}^r = L_{ls} i'_{qs} + L_{mq} (i_{qs}^r + i'_{kq}) \tag{2.14–71}$$

$$\lambda'_{fd} = L'_{lfd} i'_{fd} + L_{md} (i'_{fd} + i'_{kd} + i_{ds}^r) \tag{2.14–72}$$

$$\lambda'_{kd} = L'_{lkd} i'_{kd} + L_{md} (i'_{kd} + i'_{fd} + i_{ds}^r) \tag{2.14–73}$$

$$\lambda'_{kq} = L'_{lkq} i'_{kq} + L_{mq} (i'_{kq} + i_{qs}^r) \tag{2.14–74}$$

Since the necessary decoupling is only obtained in the rotor reference frame, the use of the superscript for the stator *d,q* variables is unnecessary for synchronous machines and the superscript *r* is typically dropped. In addition, it is important to mention that the damper cage is the equivalent of the squirrel cage in an induction machine. Thus, the voltages v'_{kd} and v'_{kd} are identically zero. We have again included these quantities as unknowns only for the purpose of generality.

Note also that because of the use of the 2/3 term in the definition of the primed rotor currents it has been necessary to define the stator referred rotor resistances as $r'_{kd} = \frac{2}{3} r_{kd}$; $r'_{kq} = \frac{2}{3} r_{kq}$; and $r'_{fd} = \frac{2}{3} r_{fd}$. Finally, we have been careful in this chapter to use the prime symbol to indicate when a rotor winding has been referred to the stator by the turns ratio. Henceforth in this book this turns ratio transformation will be tacitly assumed and we will also drop the use of the prime for this purpose.

Using a similar procedure as for the induction machine it is not difficult to show that the power input and torque output equations for the synchronous machine are:

$$P_e = \frac{3}{2} [v_{ds}^r i_{ds}^r + v_{qs}^r i_{qs}^r + v'_{fd} i'_{fd}] \tag{2.14–75}$$

$$T_e = \frac{3}{2} \frac{P}{2} [\lambda_{ds}^r i_{qs}^r - \lambda_{qs}^r i_{ds}^r] \tag{2.14–76}$$

The expression for power assumes that the *d*– and *q*–axis damper circuit voltages are zero.

When the torque equation is expanded by means of eqns (2.14–65) and (2.14–66), we can write

$$T_e = \frac{3}{2} \frac{P}{2} [(L_{ds} - L_{qs}) i_{qs}^r i_{ds}^r + L_{md} i'_{fd} i_{qs}^r + L_{md} i'_{kd} i_{qs}^r - L_{mq} i'_{kq} i_{ds}^r] \tag{2.14–77}$$

The first term in this expression is called the *saliency torque* which exists when the machine has protruding rotor poles. The second term is called the *excitation torque* or *reaction torque* which appears due to the excitation of the field winding with *dc* current (which consequently 'reacts' with the stator current to produce torque). The last two terms are recognized as the *induction motor torque* which we previously obtained for the case of the induction motor. Since this component serves to dampen oscillations in the power grid, this component is also given the name *damping torque*.

Figure 2.19 shows a computer trace of the acceleration of a synchronous machine when modelled in the rotor reference frame. The parameters of this machine are as follows:

$$P_{rated} = 20\ kVA \quad Poles = 4 \quad V_{ll(rms)} = 230\ \text{V} \quad f = 60\ \text{Hz}$$

$$r_s = 0.1\ \Omega \quad r_{fd} = 0.016\ \Omega \quad r_{kd} = 0.17\ \Omega \quad r_{kq} = 0.17\ \Omega$$

$$L_{md} = 4.1\ mH \quad L_{mq} = 2.0\ mH \quad L_{ds} = 4.89\ mH \quad L_{qs} = 2.79\ mH$$

$$L_{fd} = 4.48\ mH \quad L_{kd} = 4.39\ mH \quad L_{kq} = 2.91\ mH \quad J = 0.2\ N-m-s^2$$

Note that the machine takes considerably longer to accelerate than the equivalent induction machine due to the relatively massive poles and weak induction motor torque. While torque pulsations at line frequency are again apparent, oscillations also appear at twice line frequency which is a result of the saliency of the machine. Also note the slight dip in the average torque near half speed. The effect, called the *Görges Effect* after Hans Görges, occurs because of the difference in the impedance of the *d*– and *q*–axis rotor circuits. Because of this asymmetry a forward and backward wave is set up on the rotor at slip frequency. While the forward wave rotates in synchronism with the stator MMF, the backward wave rotates backwards with respect to the rotor at slip frequency or with respect to the stator at $1 - 2f_{slip}$. Below half speed the wave is rotating backward with respect to the stator windings. However, when the rotor reaches half speed the wave begins to rotate forward with respect to the stator. At the point just above half speed the backward wave exerts a strong negative torque which could in severe case result in continuous operation near half speed, perhaps resulting in failure. Observe finally that when the machine is excited as it reaches rated speed, the machine begins to oscillate strongly. Such oscillations or 'swinging' are characteristic of machines operating from a fixed frequency supply. We will learn later in this book that such problems can

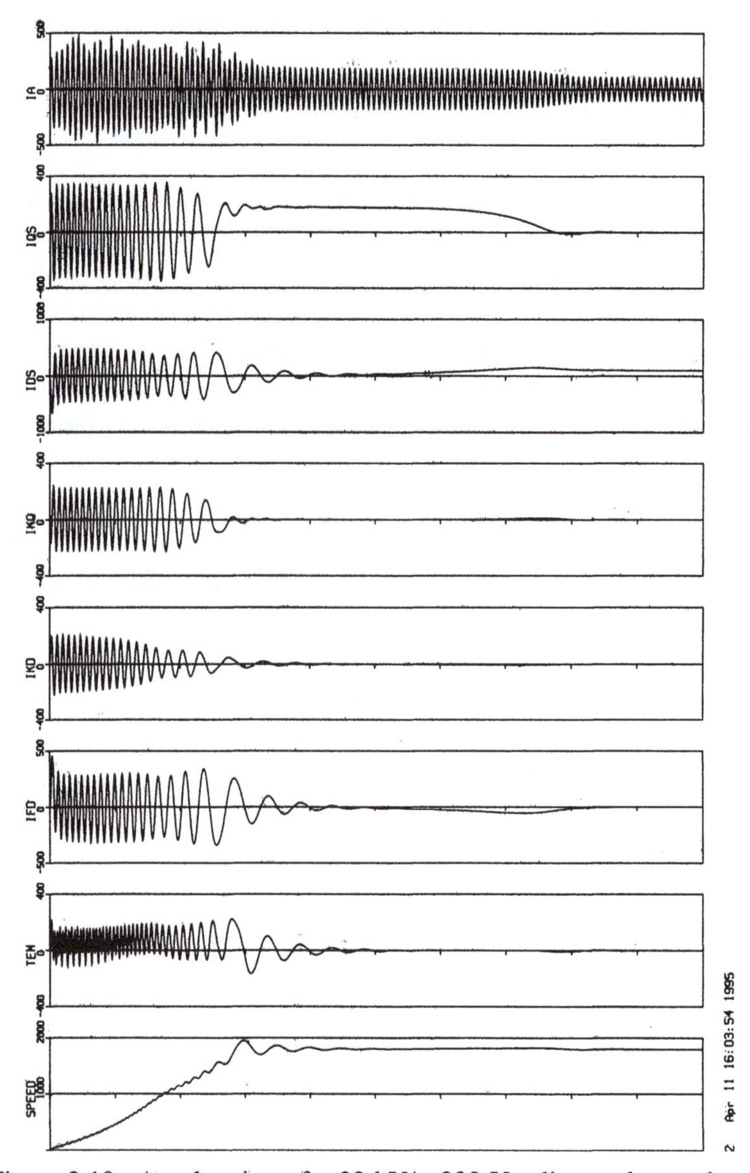

Figure 2.19 Acceleration of a 20 kVA, 230 V salient pole synchronous machine from rest with a shorted field winding. Traces from the top: i_{as}, i^r_{qs}, i^r_{ds}, i_{kq}, i_{kd}, i_{fd}, T_{em}, rotor speed in rpm; time axis – 0.1 s/div.

be completely overcome when operating from a solid state converter by suitable control. An understanding of the source of such problems and the solutions via control are the topic of the remainder of this book.

2.15 Extension of *d,q*,0 Theory to Analysis of Permanent Magnet Motors

Permanent magnet machines form an increasingly important class of high performance ac motor drives. Fortunately, we need not repeat the analysis for this machine since it is simply a special case of a salient pole synchronous machine. In general, many types of permanent magnet machines exist which can be represented by different constraints on Park's Equations. The most general type is obtained by recognizing that permanent magnet excitation is simply equivalent to a constant field current in Park's Equations. Defining the constant

$$\Lambda'_{mf} = L'_{md}\, i'_{fd} \tag{2.15–1}$$

Park's Equations for a permanent magnet machine equipped with a starting cage become

$$v^r_{ds} = r_s i^r_{ds} + p\lambda^r_{ds} - \omega_r \lambda^r_{qs} \tag{2.15–2}$$

$$v^r_{qs} = r_s i^r_{qs} + p\lambda^r_{qs} + \omega_r \lambda^r_{ds} \tag{2.15–3}$$

$$v'_{kd} = r'_{kd}\, i'_{kd} + p\lambda'_{kd} \tag{2.15–4}$$

$$v'_{kq} = r'_{kq}\, i'_{kq} + p\lambda'_{kq} \tag{2.15–5}$$

where

$$\lambda^r_{ds} = L_{ls}\, i'_{ds} + L_{md}\,(i^r_{ds} + i'_{kd}) + \Lambda'_{mf} \tag{2.15–6}$$

$$\lambda^r_{qs} = L_{ls}\, i'_{qs} + L_{mq}\,(i^r_{qs} + i'_{kq}) \tag{2.15–7}$$

$$\lambda'_{kd} = L'_{lkd}\, i'_{kd} + L_{md}\,(i'_{kd} + i^r_{ds}) + \Lambda'_{mf} \tag{2.15–8}$$

$$\lambda'_{kq} = L'_{lkq}\, i'_{kq} + L_{mq}\,(i'_{kq} + i^r_{qs}) \tag{2.15–9}$$

Figure 2.20 shows a line start acceleration for permanent magnet machine having a starting cage. The parameters of this machine are as follows:

$P_{rated} = 5\ hp$ $\quad V_{ll(rms)} = 230\ V$ $\quad f = 60\ hz$ $\quad Poles = 2$

$r_s = 0.32\ \Omega$ $\quad r_{kd} = 0.99\ \Omega$ $\quad r_{kq} = 2.0\ \Omega$

$L_{md} = 23\ mH$ $\quad L_{mq} = 50\ mH$ $\quad L_{ds} = 26.2\ mH$ $\quad L_{qs} = 53.2\ mH$

$L_{kd} = 29.4\ mH$ $\quad L_{kq} = 56.4\ mH$ $\quad J = 0.01\ N-m-s^2$

Note the marked increase in the fundamental frequency torque pulsation which occurs due to the permanent magnet excitation. Since the permanent magnet field cannot be 'turned off' the machine experiences a relatively strong braking torque. In particular, both the excitation torque, saliency torque and the induction motor torque are separately plotted in Figure 2.20. Note that while the induction motor torque is positive as expected, the saliency torque contributes no net average torque while the excitation torque is actually negative. If the magnetic field of the magnet is sufficiently strong or, if the cage is not robust, the machine could actually 'hang up' at a subsynchronous speed. In any case, the acceleration of the machine is prolonged which causes detrimental heating of the rotor which is a potentially serious problem because of the temperature sensitivity of the magnets.

References

[1] A.S. Langsdorf, "Theory of Alternating Current Machinery", McGraw–Hill Book Co., New York, 1955.

[2] W.V. Lyon, "Transient Analysis of Alternating Current Machinery", John Wiley & Sons, Inc., New York, 1954.

[3] P.C. Krause and C.H. Thomas, "Simulation of Symmetrical Induction Machinery", IEEE Trans. on Power Apparatus and Systems, Vol. 84, Nov. 1965, pp. 1038–1053.

[4] Y.K. He and T.A. Lipo, "Computer Simulation of an Induction Machine with Spatially Dependent Saturation", IEEE Trans. on Power Apparatus and Systems, Vol. PAS–103, No. 4, April 1984, pp. 707–714.

[5] T.A. Lipo, "Analysis of Synchronous Machines", Course Notes for ECE 511, University of Wisconsin, 1994.

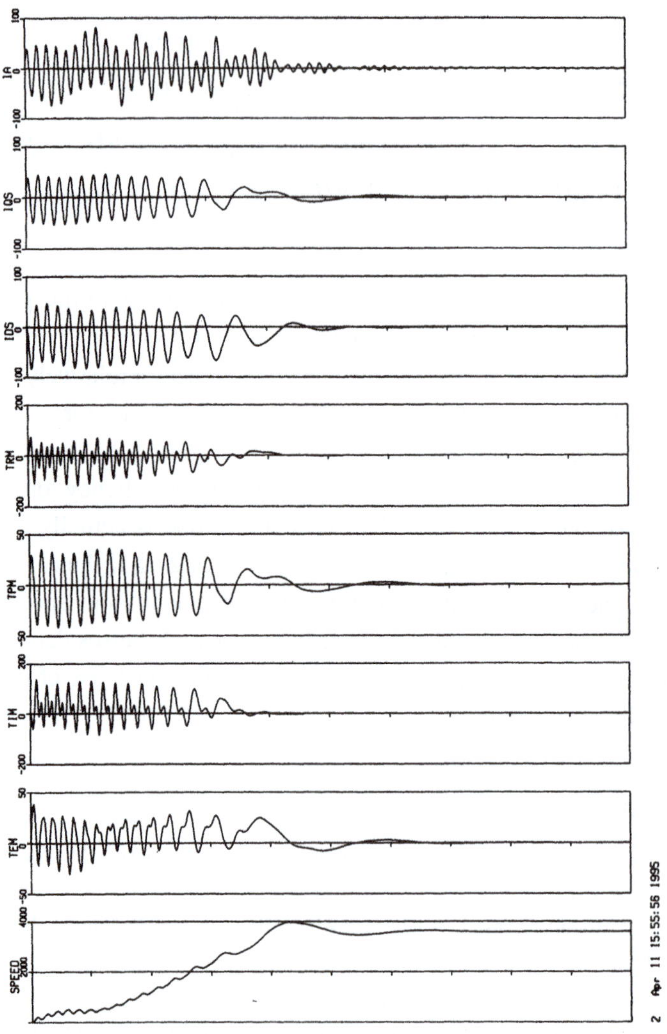

Figure 2.20 Acceleration of 5 HP, 230 V. permanent magnet machine started from rest. Traces from top to bottom: i^r_{qs}, q–axis stator current (A.), i^r_{ds}, d–axis stator current (A.), T_{rm}, reluctance component of torque (Nt–m), T_{pm}, PM excitation component of torque (Nt–m), T_{im}, induction component of torque (Nt–m), $T_{em} = T_{rm}+T_{pm}+T_{im}$, total electromagnetic torque (Nt–m), rotor speed (RPM); time axis 0.1 s/div.

Problems

Problem 2-1 Reciprocity of Winding Inductances

Verify eqn 2.2-25 by direct application of the general expression for flux linkage.

Problem 2-2 Complex Variable Coil Variable Equations

Verify eqn 2.6-14 by direct combination of the scalar equations describing the rotor circuits.

Problem 2-3 Complex Variable *d,q* Equations

Derive the rotor circuit *d,q* equation given in eqn 2.8-11.

Problem 2-4 Power in *d,q* Model

Derive the equation for the total three phase power in the stator windings of the uniform air gap machine model in terms of *d,q* variables including the zero sequence quantities.

Problem 2-5 Source Voltages

Calculate the *d,q* voltages in stator, rotor and synchronous frames for a balanced three phase, 460 volt, 60 hz system. Take the *a-b* line voltage as the reference quantity (i.e. zero phase cosine wave) and align the phase '*a*' voltage with the *q*-axis for the stator and rotor frames. Put all of the voltage in the *q*-axis for the synchronous frame. Express the results in real variables and in complex form.

Problem 2-6 Source Voltages-Unbalanced Supply

Repeat 2-5 for a 460 Volt, 60 Hz supply with $v_{ab} = v_{ac}$ and $v_{bc} = 0$ (put all of the positive sequence in the *q*-axis for the synchronous frame).

Problem 2-7 Alternate Forms For Torque

Derive the torque equation in terms of the quantities listed below starting from the basic form

$$T_e = \left(\frac{3}{2}\right)\left(\frac{P}{2}\right)L_m [i_{qs}i_{dr} - i_{ds}i_{qr}] = \left(\frac{3}{2}\right)\left(\frac{P}{2}\right)L_m Im [\underline{i}^{\dagger}_{qdr}\underline{i}_{qds}]$$

a) airgap flux and rotor current,

b) airgap flux and stator current,

c) stator flux and rotor current,

d) stator flux and stator current,

e) rotor flux and rotor current,

f) rotor flux and stator current,

g) rotor flux and stator flux.

Problem 2–8 Torque Expressed in Phase Variables

Working backwards from eqn (2.12-12) derive an expression for the electromagnetic torque in terms stator 'phase variable' flux linkages and currents, i.e. in terms of i_{as}, i_{bs}, i_{cs}, λ_{as}, λ_{bs}, and λ_{cs}. Can you provide a physical (or space vector based) interpretation of your result?

Problem 2-9 Balanced Capacitor Circuits in Expressed in a Rotating Reference Frame

Three balanced (equal) capacitors are wye connected across a three phase line. Determine the equations and corresponding equivalent circuit expressing the circuit behavior of these capacitors in a freely rotating *d,q* reference frame.

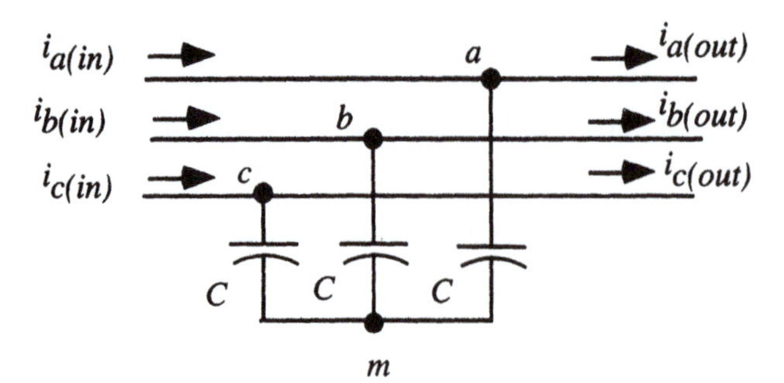

Problem 2-10 Incorporating Iron Loss Resistance Into the *d,q,0* equivalent circuit.

Determine the *d,q,0* rotating reference frame equivalent circuit of the balanced *r—L* circuit shown below. If r_m and L_m represent the iron loss resistance and magnetizing inductance of an induction machine respectively, construct the $d,q,0$ equivalent circuit for an induction machine with iron loss.

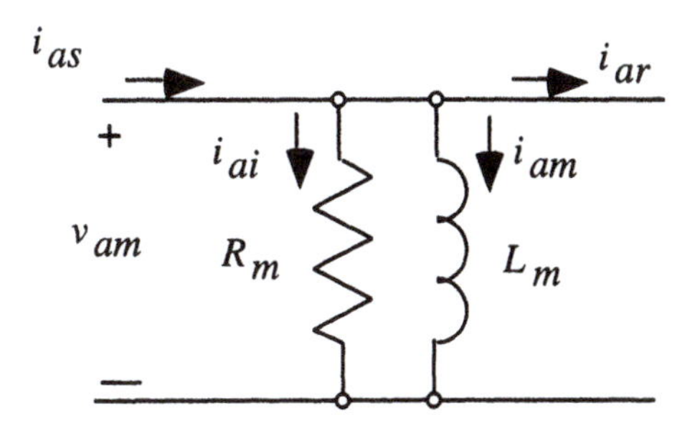

Note: Same circuit for phases b and c.

3 d,q Models for Solid State Power Converters

3.1 Introduction

For analysis of motor drives, it is sometimes convenient to characterize the inverter in the same framework as used for the induction motor, that is, in terms of a $d,q,0$ components. As will be shown, the d,q orthogonal reference system also yields converter models which are useful apart from their convenience in combined machine–converter modeling. This is particularly evident when a complex vector d,q representation is employed and the value of such models becomes apparent in later discussion of, for example, current regulated inverters.

3.2 d,q Model for Voltage Source Inverter

The d,q model for a voltage source inverter (VSI) system is obtained by simply applying the d,q transformation developed earlier in Chapter 2 to the inverter constraint equations on a mode by mode basis. Recall that the d,q transformation from phase variables to d,q axes fixed in the stator is written in scalar form as

$$f_{qs}^{s} = \frac{2}{3}f_{as} - \frac{1}{3}f_{bs} - \frac{1}{3}f_{cs} \tag{3.2–1}$$

$$f_{ds}^{s} = \frac{1}{\sqrt{3}}f_{cs} - \frac{1}{\sqrt{3}}f_{bs} \tag{3.2–2}$$

$$f_{0s}^{s} = \frac{1}{3}(f_{as} + f_{bs} + f_{cs}) \tag{3.2–3}$$

where the symbol f represents voltage v, current i or flux linkage λ.

To illustrate the development of the d,q equations which describe the six step voltage source inverter, consider first the connection 612 in Figure 1.4. The voltages across the machine are then,

$$v_{ab} = -v_{ca} = v_i \tag{3.2–4}$$

$$v_{bc} = 0 \tag{3.2–5}$$

The current in phase *as* is

$$i_{as} = i_i \tag{3.2–6}$$

where i_i is the instantaneous current supplied to the inverter from the dc link and is defined in Figure 1.7.

To apply the *d,q* transformation, the voltage relations must be converted to phase voltages which is readily accomplished by observing that the absence of a neutral connection forces the constraint that,

$$i_{as} + i_{bs} + i_{cs} = 0 \tag{3.2–7}$$

For any balanced load (for example, a wye connected induction machine) the line to neutral voltages are then constrained such that

$$v_{as} + v_{bs} + v_{cs} = 0 \tag{3.2–8}$$

The phase voltages become, from eqns (3.2–4) and (3.2–5),

$$v_{as} = \frac{1}{3}(v_{ab} - v_{ca}) = \frac{2}{3}v_i \tag{3.2–9}$$

$$v_{bs} = \frac{1}{3}(v_{bc} - v_{ab}) = -\frac{1}{3}v_i \tag{3.2–10}$$

$$v_{cs} = \frac{1}{3}(v_{ca} - v_{bc}) = -\frac{1}{3}v_i \tag{3.2–11}$$

Application of the *d,q* transformation then yields,

$$v_{qs}^s = \frac{2}{3}v_i \tag{3.2–12}$$

$$v_{ds}^s = 0 \tag{3.2–13}$$

$$v_{0s}^s = 0 \tag{3.2–14}$$

Hence, modelling of the inverter during this time interval can be accomplished by applying 2/3 the inverter dc link voltage to the *q*–axis equivalent circuit of the induction machine while applying zero volts (i.e. shorting) the *d*– and 0–axis circuits.

Applying the d,q transformation to the current constraints of eqns (3.2–6) and (3.2–7) yields,

$$i_{qs}^{s} = i_{i} \tag{3.2–15}$$

$$i_{ds}^{s} = \frac{1}{\sqrt{3}}(i_{cs} - i_{bs}) \tag{3.2–16}$$

$$i_{0s}^{s} = 0 \tag{3.2–17}$$

Note that although zero current flows in the 0–component equivalent circuit, short circuit current will clearly flow in the d–axis circuit due to the speed voltage which appear in the $d,q,0$ equivalent circuit, see Figure 2.8. Equation (3.2–16) cannot be reduced beyond its basic definition since the current which flows in this portion of the equivalent circuit is defined simply by induction machine d–axis circuit conditions. The current i_{ds}^{s} clearly does not contribute to the inverter current i_i.

If the same type of analysis is carried out for the remaining five modes illustrated in Figure 1.4, the results summarized in Figure 3.1 are obtained for the d,q variables. Since v_{0s}^{s} and i_{0s}^{s} are zero in all cases, they are not shown. The d,q relations for the VSI can be conveniently described by defining two *switching functions* to express the constraint equations in Figure 3.1. The two switching functions are illustrated in Figure 3.2 and permit writing the VSI d,q constraint equations as

$$v_{qs}^{s} = \frac{2}{\pi} v_i g_{qs}^{s} \tag{3.2–18}$$

$$v_{ds}^{s} = \frac{2}{\pi} v_i g_{ds}^{s} \tag{3.2–19}$$

$$\frac{\pi}{3} i_i = i_{qs}^{s} g_{qs}^{s} + i_{ds}^{s} g_{ds}^{s} \tag{3.2–20}$$

These expressions relate the instantaneous inverter input quantities v_i and i_i to the instantaneous d,q output quantities. The functions g_{qs}^{s} and g_{ds}^{s} are independent time functions which express the inverter switching operations. Note that g_{qs}^{s} is related to the ideal phase voltage of the VSI and that g_{ds}^{s} is related to the ideal line to line voltage. The choice of the quantities $\pi/3$ and $\pi\sqrt{3}/6$ as the amplitude of g_{qs}^{s} and g_{ds}^{s} is arbitrary and have been chosen such that the

1	$-\frac{\pi}{6} < \omega_e t < \frac{\pi}{6}$	+ i_i a, v_i, − b c	$v_{qs}^s = \frac{2}{3}v_i$ $v_{ds}^s = 0$ $i_i = i_{qs}^s$	$\underline{v}_{qds}^s = \frac{2}{3}v_i e^{j0}$
2	$\frac{\pi}{6} < \omega_e t < \frac{\pi}{2}$	+ b a, v_i, − i_i c	$v_{qs}^s = \frac{v_i}{3}$ $v_{ds}^s = -\frac{v_i}{\sqrt{3}}$ $i_i = \frac{1}{2}i_{qs}^s - \frac{\sqrt{3}}{2}i_{ds}^s$	$\underline{v}_{qds}^s = \frac{2}{3}v_i e^{j\frac{\pi}{3}}$
3	$\frac{\pi}{2} < \omega_e t < \frac{5\pi}{6}$	+ i_i b, v_i, − c a	$v_{qs}^s = -\frac{v_i}{3}$ $v_{ds}^s = -\frac{v_i}{\sqrt{3}}$ $i_i = -\frac{1}{2}i_{qs}^s - \frac{\sqrt{3}}{2}i_{ds}^s$	$\underline{v}_{qds}^s = \frac{2}{3}v_i e^{j\frac{2\pi}{3}}$
4	$\frac{5\pi}{6} < \omega_e t < \frac{7\pi}{6}$	+ c b, v_i, − i_i a	$v_{qs}^s = -\frac{2}{3}v_i$ $v_{ds}^s = 0$ $i_i = -i_{qs}^s$	$\underline{v}_{qds}^s = \frac{2}{3}v_i e^{j\pi}$
5	$\frac{7\pi}{6} < \omega_e t < \frac{3\pi}{2}$	+ i_i c, v_i, − a b	$v_{qs}^s = -\frac{v_i}{3}$ $v_{ds}^s = \frac{v_i}{\sqrt{3}}$ $i_i = -\frac{1}{2}i_{qs}^s + \frac{\sqrt{3}}{2}i_{ds}^s$	$\underline{v}_{qds}^s = \frac{2}{3}v_i e^{j\frac{4\pi}{3}}$
6	$\frac{3\pi}{2} < \omega_e t < \frac{11\pi}{6}$	+ a c, v_i, − i_i b	$v_{qs}^s = \frac{v_i}{3}$ $v_{ds}^s = \frac{v_i}{\sqrt{3}}$ $i_i = \frac{1}{2}i_{qs}^s + \frac{\sqrt{3}}{2}i_{ds}^s$	$\underline{v}_{qds}^s = \frac{2}{3}v_i e^{j\frac{5\pi}{3}}$

Figure 3.1 *d,q* equations for the six modes of a VSI

fundamental component coefficient in the Fourier series for the two functions is unity. These Fourier series are,

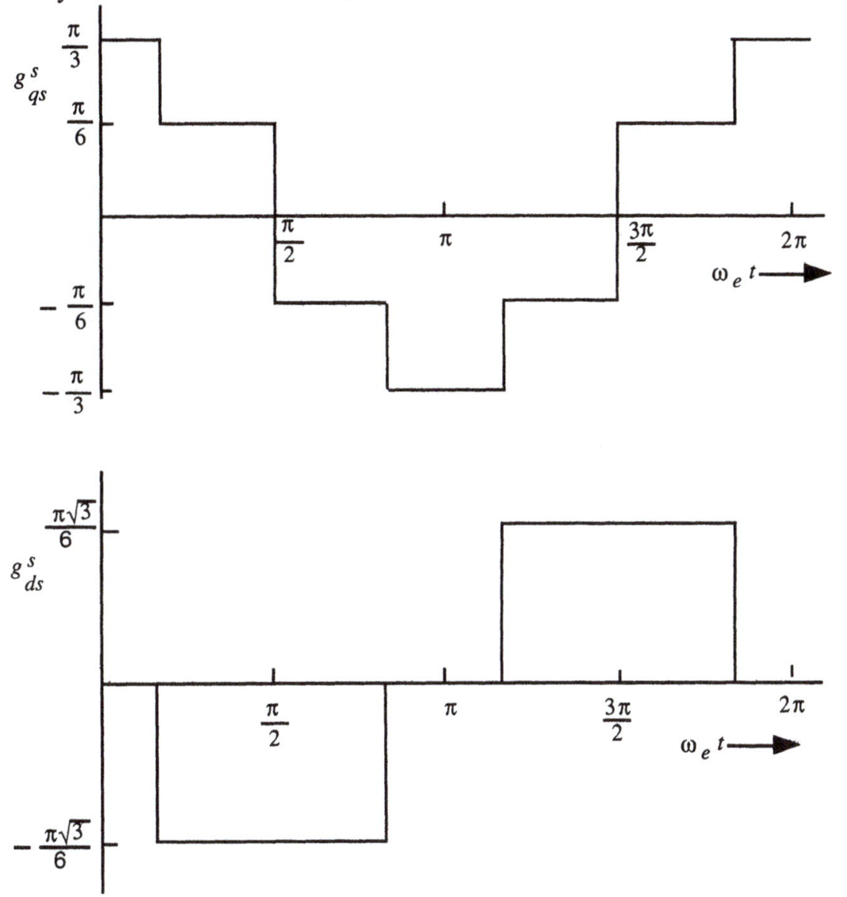

Figure 3.2 The VSI switching functions g^s_{qs} and g^s_{ds}

$$g^s_{qs} = \cos\theta_e + \frac{1}{5}\cos 5\theta_e - \frac{1}{7}\cos 7\theta_e - \ldots \tag{3.2–21}$$

$$g^s_{ds} = -\sin\theta_e + \frac{1}{5}\sin 5\theta_e + \frac{1}{7}\sin 7\theta_e - \ldots \tag{3.2–22}$$

The complex vector form of these *d,q* equations for the VSI take on very simple forms and are very useful in visualizing and manipulating the equations. Combining eqns (3.2–18) and (3.2–19) yields the complex *d,q* equations for the VSI voltages,

$$\underline{v}_{qds}^{s} = v_{qs}^{s} - jv_{ds}^{s} = \frac{2}{\pi}v_i(g_{qs}^{s} - jg_{ds}^{s}) = \frac{2}{\pi}v_i\underline{g}_{qds}^{s} \tag{3.2–23}$$

From the Fourier series for g_{qs}^{s} and g_{ds}^{s}, the complex function $\underline{g}_{qds}^{s}$ is given by

$$\underline{g}_{qds}^{s} = e^{j\omega_e t} + \frac{1}{5}e^{-j5\omega_e t} - \frac{1}{7}e^{j7\omega_e t} - \ldots \tag{3.2–24}$$

and is thus seen to represent a series of rotating complex vectors moving in alternate directions at the speed ω_e and multiples of ω_e. The complex vector form of the current constraint equation in eqn (3.2–20) is easily shown to be

$$\frac{\pi}{3}i_i = Re\,[\underline{i}_{qds}^{s}(\underline{g}_{qds}^{s})^{\dagger}] \tag{3.2–25}$$

An alternative and very useful form of the *d,q* complex vector voltage equation can be obtained by writing out the complex vector voltage expression for each of the modes illustrated in Figure 3.1. Thus, in mode 1

$$\underline{v}_{qds}^{s} = \frac{2}{3}v_i e^{j0} \qquad \text{(mode 1)} \tag{3.2–26}$$

and in mode 2

$$\underline{v}_{qds}^{s} = \frac{2}{3}v_i\left(-\frac{1}{2} + j\frac{\sqrt{3}}{2}\right) = \frac{2}{3}v_i e^{j\pi/3} \qquad \text{(mode 2)} \tag{3.2–27}$$

Repeating the process for all six modes shows that in each mode the complex vector voltage has a constant amplitude of $\frac{2}{3}v_i$ and a constant phase angle equal to $(k-1)\frac{\pi}{3}$ where k is the mode number. The complex vector output voltage of a VSI thus consists of six equi–spaced complex vectors (60° apart), each having an amplitude of $\left(\frac{2}{3}\right)v_i$ which is, of course, constant if v_i is constant. Figure 3.3 illustrates these complex vectors. A simple analytical expression for the complex vector voltage is

$$\underline{v}_{qds}^{s} = \frac{2}{3}v_i e^{j(k-1)\pi/3} \qquad k = 1,2,\ldots,6 \tag{3.2–28}$$

This six vector concept is very useful in visualizing inverter operation and in explaining various inverter properties.

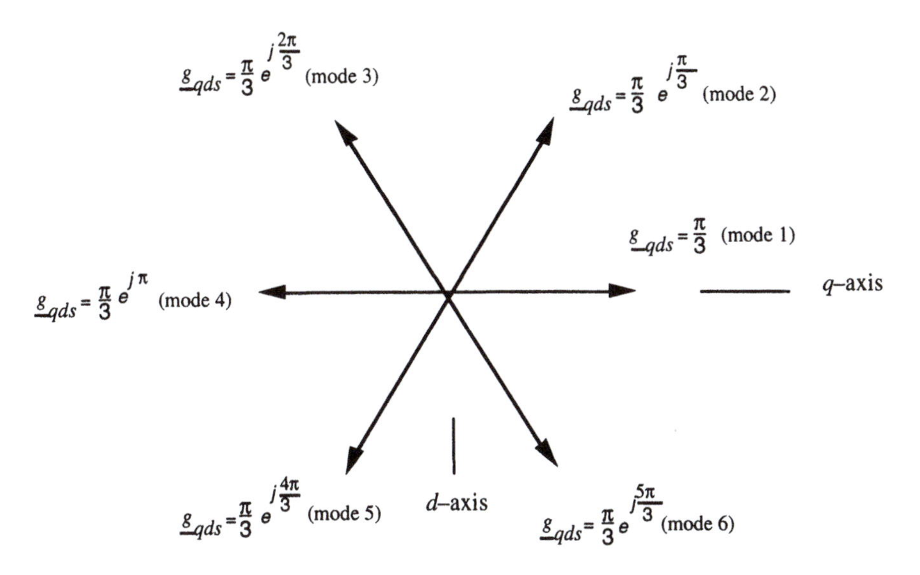

Figure 3.3 The six voltage vectors characterizing VSI operation

3.3 *d,q* Model for PWM Operation

In a PWM inverter the various modulation techniques will alter the switching functions g_{qs}^{s} and g_{ds}^{s} such that each type of PWM has its own unique switching functions. Although it is a straight forward process to establish the mode constraints and the associated switching functions, the results become very involved and are seldom actually required for analytical purposes. In most cases the PWM inverter can be modelled in terms of its fundamental component with harmonics treated as a separate issue. The fundamental component model will be presented in a later section.

There is one important aspect of PWM operation which distinguishes it from modulated VSI operation. This fundamental difference is the existence of two additional zero voltage states as illustrated in the switching mode diagram of Figure 3.4. In addition to the six possible voltage complex vectors associated with the VSI (Figure 3.3), there are two zero voltage states associated with having all three of the positive pole switches on or all three of the negative pole switches on. Pulse width modulation is thus often considered as an eight state operation described by the equation

$$\underline{v}^s_{qds} = \frac{2}{3}v_i e^{j(k-1)\pi/3} \qquad k = 1,2,\ldots 6$$
$$= 0 \qquad k = 7,8 \tag{3.3–1}$$

The zero states are important in PWM operation and play significant roles, for example, in the operation of current regulated PWM inverters.

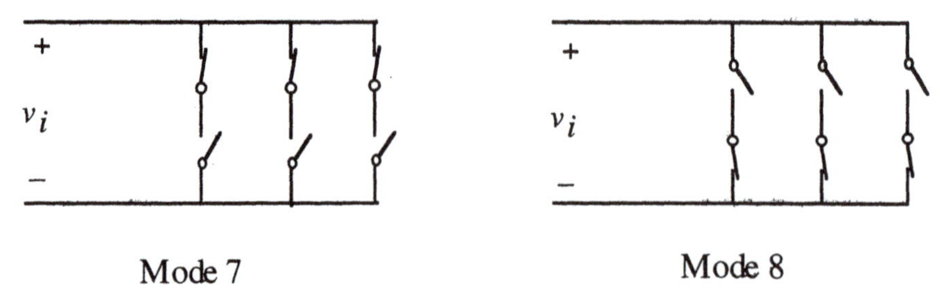

Mode 7 Mode 8

Figure 3.4 Switch conditions corresponding to the zero voltage vector state for a PWM inverter

3.4 *d,q* Model for CSI System

The duality of VSI and CSI systems implies that switching function models, which are the direct duals of those of the VSI, must exist for the CSI. In general, the exact dual of a VSI feeding a Y connected load is a CSI feeding a Δ connected load. However, since it is more convenient to always consider Y connected loads, the CSI model presented here is not the exact dual of the VSI. There are, however, very close dual relationships between the analysis and the results for the two types of systems.

The *d,q* equations for each mode are obtained by the same process as for the VSI and are illustrated in Figure 3.5. The scalar equations can again be written in terms of CSI inverter based switching functions *h*, such that

$$i^s_{qs} = \frac{2\sqrt{3}}{\pi}h^s_{qs}i_i \tag{3.4–1}$$

$$i^s_{ds} = \frac{2\sqrt{3}}{\pi}h^s_{ds}i_i \tag{3.4–2}$$

$$v_i = \frac{3\sqrt{3}}{\pi}(v^s_{qs}h^s_{qs} + v^s_{ds}h^s_{ds}) \tag{3.4–3}$$

Mode	Time Interval	Connection	dq Equations	Complex Vector Current
1	$0 < \omega_e t < \frac{\pi}{3}$	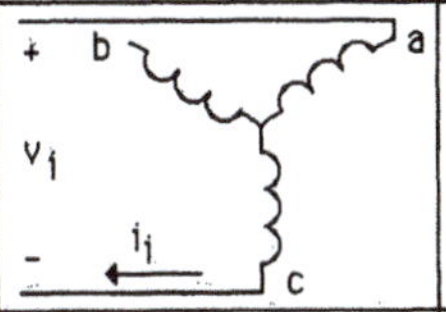 + b, a; v_i; − c; i_i	$i_{qs}^s = i_i$ $i_{ds}^s = -\frac{1}{\sqrt{3}} i_i$ $v_i = \frac{3}{2} v_{qs}^s - \frac{\sqrt{3}}{2} v_{ds}^s$	$\underline{i}_{qds} = \frac{2}{\sqrt{3}} i_i \epsilon^{j\frac{\pi}{6}}$
2	$\frac{\pi}{3} < \omega_e t < \frac{2\pi}{3}$	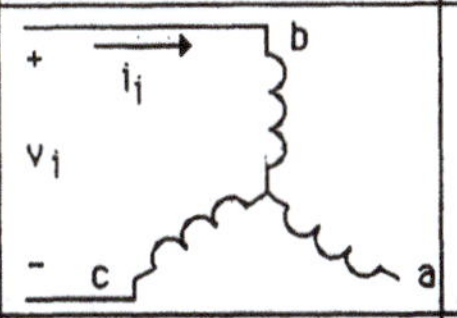 + b; i_i; v_i; − c, a	$i_{qs}^s = 0$ $i_{ds}^s = -\frac{2}{\sqrt{3}} i_i$ $v_i = -\sqrt{3}\, v_{ds}^s$	$\underline{i}_{qds} = \frac{2}{\sqrt{3}} i_i \epsilon^{j\frac{\pi}{2}}$
3	$\frac{2\pi}{3} < \omega_e t < \pi$	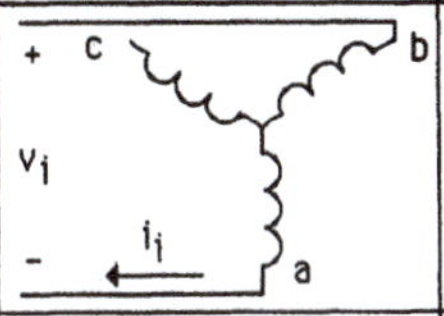 + c, b; v_i; − a; i_i	$i_{qs}^s = -i_i$ $i_{ds}^s = -\frac{1}{\sqrt{3}} i_i$ $v_i = -\frac{3}{2} v_{qs}^s - \frac{\sqrt{3}}{2} v_{ds}^s$	$\underline{i}_{qds} = \frac{2}{\sqrt{3}} i_i \epsilon^{j\frac{5\pi}{6}}$
4	$\pi < \omega_e t < \frac{4\pi}{3}$	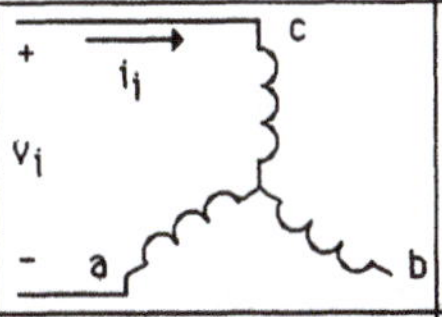 + c; i_i; v_i; − a, b	$i_{qs}^s = -i_i$ $i_{ds}^s = \frac{1}{\sqrt{3}} i_i$ $v_i = -\frac{3}{2} v_{qs}^s + \frac{\sqrt{3}}{2} v_{ds}^s$	$\underline{i}_{qds} = \frac{2}{\sqrt{3}} i_i \epsilon^{j\frac{7\pi}{6}}$
5	$\frac{4\pi}{3} < \omega_e t < \frac{5\pi}{3}$	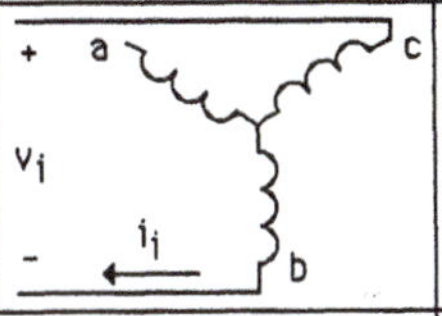 + a, c; v_i; − b; i_i	$i_{qs}^s = 0$ $i_{ds}^s = \frac{2}{\sqrt{3}} i_i$ $v_i = \sqrt{3}\, v_{ds}^s$	$\underline{i}_{qds} = \frac{2}{\sqrt{3}} i_i \epsilon^{j\frac{3\pi}{2}}$
6	$\frac{5\pi}{3} < \omega_e t < 2\pi$	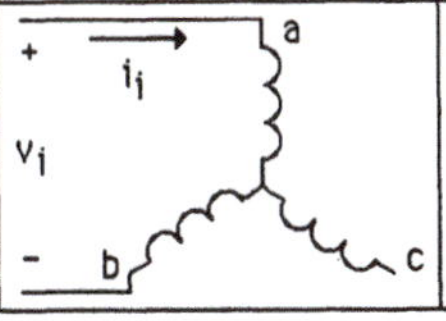 + a; i_i; v_i; − b, c	$i_{qs}^s = i_i$ $i_{ds}^s = \frac{1}{\sqrt{3}} i_i$ $v_i = \frac{3}{2} v_{qs}^s + \frac{\sqrt{3}}{2} v_{ds}^s$	$\underline{i}_{qds} = \frac{2}{\sqrt{3}} i_i \epsilon^{j\frac{11\pi}{6}}$

Figure 3.5 *d,q* equations for the six modes of CSI

where the CSI switching functions h_{qs}^s and h_{ds}^s are shown in Figure 3.6. The Fourier series for the switching functions are,

$$h_{qs}^s = \cos\omega_e t - \frac{1}{5}\cos 5\omega_e t + \frac{1}{7}\cos 7\omega_e t - \ldots \qquad (3.4\text{–}4)$$

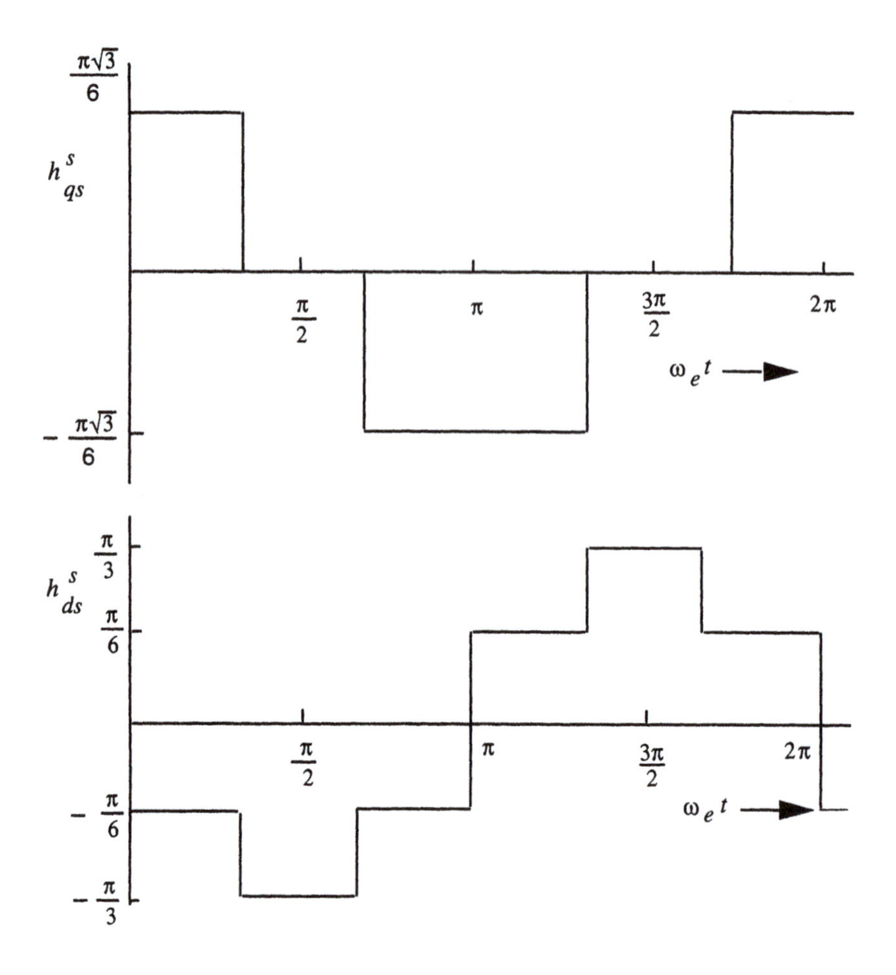

Figure 3.6 The CSI switching functions h^s_{qs} and h^s_{ds}

$$h^s_{ds} = \sin\omega_e t - \frac{1}{5}\sin 5\omega_e t + \frac{1}{7}\sin 7\omega_e t - \ldots \tag{3.4–5}$$

In complex form these equations become

$$\underline{i}^s_{qds} = \frac{2\sqrt{3}}{\pi} i_i (h^s_{qs} - jh^s_{ds}) = \frac{2\sqrt{3}}{\pi} i_i \underline{h}^s_{qds} \tag{3.4–6}$$

$$v_i = \frac{3\sqrt{3}}{\pi} Re\,[\underline{v}^s_{qds} \underline{h}^{s\dagger}_{qds}] \tag{3.4–7}$$

where

$$\underline{h}^{s}_{qds} = e^{j\omega_e t} - \frac{1}{5}e^{-j5\omega_e t} + \frac{1}{7}e^{j7\omega_e t} - \ldots \tag{3.4–8}$$

Note the close relationship between the switching functions for the VSI and the CSI. In particular, note that in $\underline{h}^{s}_{qds}$ only the signs of the harmonic terms are altered. An alternate expression for the complex vector current can also be written based again on the fact that the complex vector current is constant in each mode and shifts by 60° at each mode transition. The expression is

$$\underline{i}^{s}_{qds} = \frac{2}{\sqrt{3}}i_i e^{j[\pi/6 + (k-1)(\pi/3)]} \qquad k = 1,2,\ldots,6 \tag{3.4–9}$$

as can be easily verified by comparison with the last column in Figure 3.5. The six vectors corresponding to the switching of a CSI is shown in Figure 3.7.

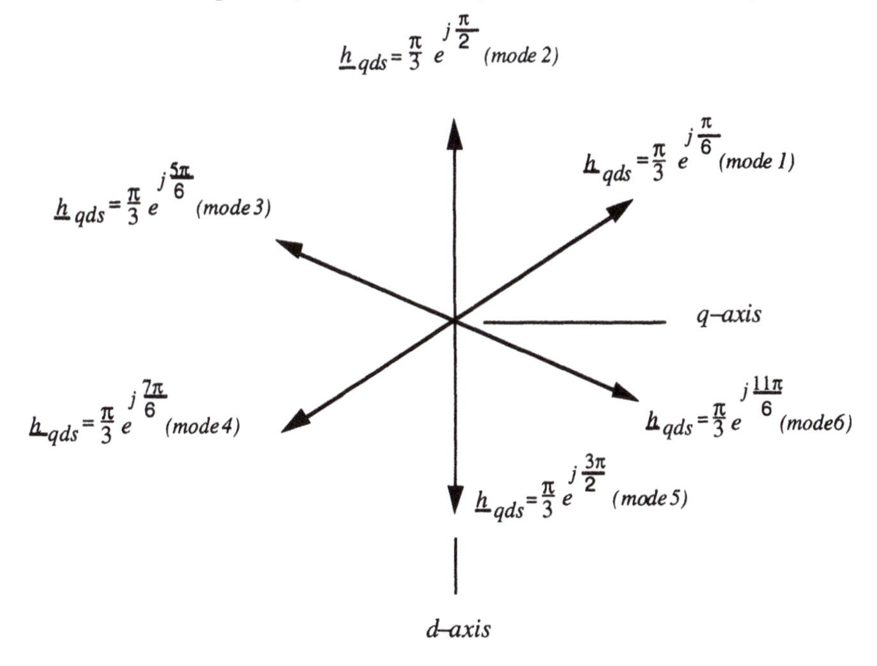

Figure 3.7 The six current vectors characterizing CSI operation

It should be noted in passing that the idealized *d,q* model of a CSI as presented above is a less accurate representation (for inductive loads) than the corresponding model for a VSI. The inherent commutation delay associated with switching the current is much more significant than the commutation time associated with switching the voltage in a VSI. In most systems the CSI output currents will be decidedly trapezoidal instead of rectangular in waveform, especially at frequencies at the high end of the inverter operating frequency range.The errors associated with using the idealized CSI model are usually quite minimal. Corrections for the commutation phase shift and for the change in fundamental component can be incorporated where necessary. Note, also, that the accuracy question would be reversed if the load were capacitive; another dual property of VSI and CSI systems.

3.5 Inverter *d,q* Models in a Synchronous Reference Frame

Although the stationary *d,q* reference frame is appropriate for many purposes, a synchronously rotating reference frame offers some advantages. It is especially useful in simulation and in the development of small signal models and transfer functions.

The most direct development is to simply transform the stationary frame models of the preceding sections to the synchronous frame using the basic transformation equations from Chapter 2. Repeating, the pertinent transformation is

$$f_{qs}^{e} = f_{qs}^{s}\cos\theta_e - f_{ds}^{s}\sin\theta_e \tag{3.5–1}$$

$$f_{ds}^{e} = f_{qs}^{s}\sin\theta_e + f_{ds}^{s}\cos\theta_e \tag{3.5–2}$$

where θ_e is the electrical angle corresponding to the inverter operating frequency ω_e

$$\theta_e = \int_0^t \omega_e dt + \theta_{e0} \tag{3.5–3}$$

The manipulations are more readily handled using the complex vector form of the transformation equations,

$$\underline{i}_{qds}^{e} = \underline{i}_{qds}^{s} e^{-j\theta_e} \tag{3.5–4}$$

For the VSI, applying the transformation to eqns (3.2–23) and (3.2–25) respectively,

$$
\begin{aligned}
\underline{v}^e_{qds} &= \underline{v}^s_{qds} e^{-j\theta_e} \\
&= \frac{2}{\pi} v_i \underline{g}^s_{qds} e^{-j\theta_e} \\
&= \frac{2}{\pi} v_i \underline{g}^e_{qds}
\end{aligned}
\tag{3.5–5}
$$

$$
\begin{aligned}
\frac{\pi}{3} i_i &= Re\,[\underline{i}^e_{qds} e^{j\theta_e} (\underline{g}^e_{qds})^\dagger] \\
&= Re\,[\underline{i}^e_{qds} \underline{g}^{e\dagger}_{qds}]
\end{aligned}
\tag{3.5–6}
$$

where the new complex switching function $\underline{g}^e_{qds}$ is defined as

$$
\begin{aligned}
\underline{g}^e_{qds} &= \underline{g}^s_{qds} e^{-j\theta_e} \\
&= g^e_{qs} - j g^e_{ds}
\end{aligned}
\tag{3.5–7}
$$

Written explicitly in terms of their components, eqns (3.5–5) and (3.5–6) are therefore

$$
v^e_{qs} = \frac{2}{\pi} v_i g^e_{qs} \tag{3.5–8}
$$

$$
v^e_{ds} = \frac{2}{\pi} v_i g^e_{ds} \tag{3.5–9}
$$

$$
\frac{\pi}{3} i_i = i^e_{qs} g^e_{qs} + i^e_{ds} g^e_{ds} \tag{3.5–10}
$$

Using the expression for $\underline{g}^s_{qds}$ given in eqn (3.2–24), the complex Fourier series for $\underline{g}^e_{qds}$ is

$$
\underline{g}^e_{qds} = e^{j(\omega_e t - \theta_e)} + \frac{1}{5} e^{-j(5\omega_e t + \theta_e)} - \frac{1}{7} e^{j(7\omega_e t - \theta_e)} - \ldots \tag{3.5–11}
$$

Expanding and separating into two components yields the scalar switching functions

$$
g^e_{qs} = 1 + \frac{2}{35} \cos 6\omega_e t - \frac{2}{143} \cos 12\omega_e t + \ldots \tag{3.5–12}
$$

$$g_{ds}^{e} = \frac{12}{35}\sin 6\omega_e t - \frac{24}{143}\sin 12\omega_e t + \ldots \tag{3.5–13}$$

In the above two equations we have assumed that the rotating *d,q* axes have been synchronized with the fundamental frequency of the inverter output voltage, i.e. $\theta_e = \omega_e t$. The time functions which generate these series are illustrated in Figure 3.8.

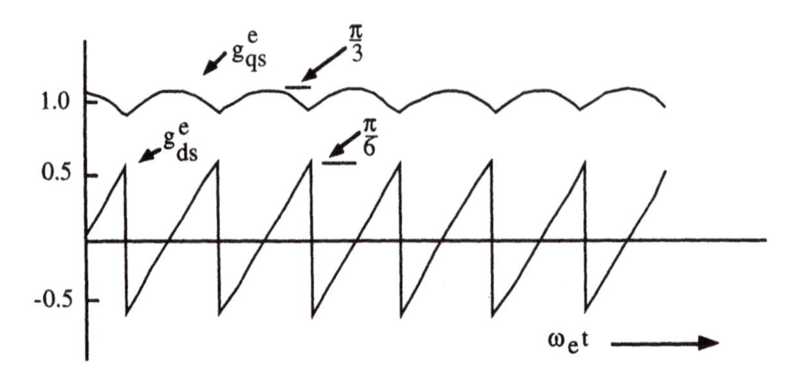

Figure 3.8 The synchronous frame VSI switching functions

The choice of phase reference in these equations ($\theta_{e0} = 0$) has placed all of the effect of the fundamental frequency (dc in the synchronous frame) in the *q*–axis. The harmonics are also shifted in frequency by ω_e such that only a sixth harmonic ripple (plus multiples of the sixth) now exists in the inverter voltages and currents.

By the same method the synchronous frame CSI equations become

$$\begin{aligned} \underline{i}_{qds}^{e} &= \underline{i}_{qds}^{s} e^{-j\theta_e} \\ &= \frac{2\sqrt{3}}{\pi} i_i \underline{h}_{qds}^{s} e^{-j\theta_e} \\ &= \frac{2\sqrt{3}}{\pi} i_i \underline{h}_{qds}^{e} \end{aligned} \tag{3.5–14}$$

$$v_i = \frac{3\sqrt{3}}{\pi} Re\,[\underline{v}_{qds}^{e} \underline{h}_{qds}^{e\dagger}] \tag{3.5–15}$$

where

$$\underline{h}^e_{qds} = e^{j(\omega_e t - \theta_e)} - \frac{1}{5}e^{-j(5\omega_e t + \theta_e)} + \frac{1}{7}e^{j(7\omega_e t - \theta_e)} \qquad (3.5\text{–}16)$$

and the scalar switching functions are

$$h^e_{qs} = 1 - \frac{2}{35}\cos 6\omega_e t - \frac{2}{143}\cos 12\omega_e t + \ldots \qquad (3.5\text{–}17)$$

$$h^e_{ds} = -\frac{12}{35}\sin 6\omega_e t - \frac{24}{143}\sin 12\omega_e t + \ldots \qquad (3.5\text{–}18)$$

The scalar equations corresponding to eqns (3.5–14) and (3.5–15) follow directly. These functions are illustrated in Figure 3.9. Note that the functions

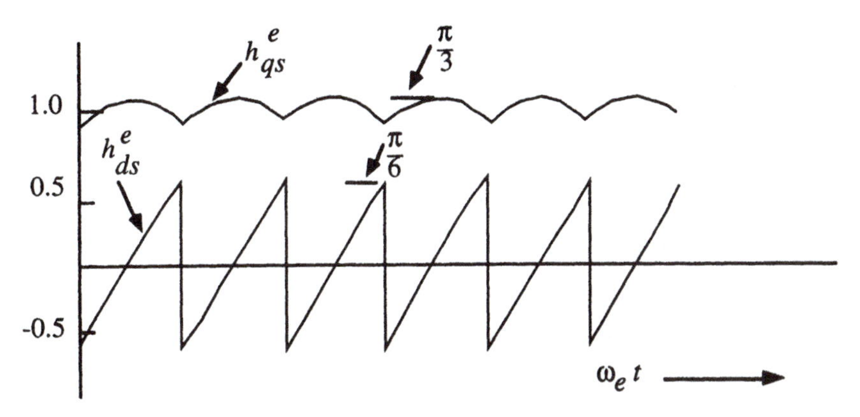

Figure 3.9 The synchronous frame CSI switching functions

are the same as the switching functions for the VSI except for the fact that the initial condition has been chosen differently for the CSI case.

3.6 Examples of Inverter–Induction Motor Models

To illustrate the application of the inverter *d,q* models of the preceding sections, the overall *d,q* model of a voltage stiff inverter and an induction motor is presented here. All that is required is to combine the inverter models developed above with the motor models previously developed. Figure 3.10 shows the stationary *d,q* model of a VSI driving an induction machine. The inverter current i_i is represented by two current sources dependent on the two motor *d,q* currents i^s_{qs} and i^s_{ds} . This circuit represents a complete dynamic model in which the inverter voltage v_i can be any independent time function and the inverter

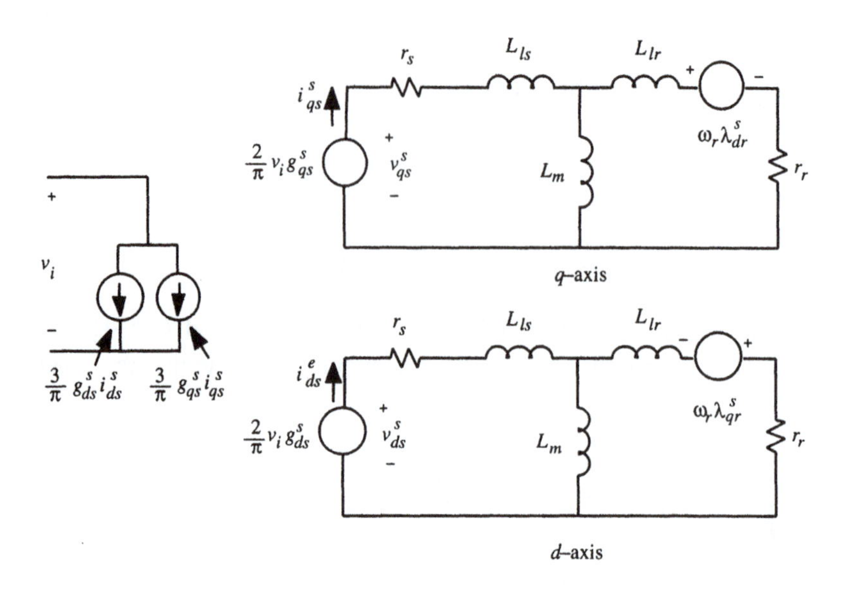

Figure 3.10 Stationary frame *d,q* model of a VSI driven induction machine

switching is completely modelled including all harmonics. The model of the machine and the switching functions can also be represented in the synchronous frame. The equivalent circuit becomes that of Figure 3.11. Similar equivalent circuits for the motor driven from a current stiff inverter supply are shown in Figure 3.12 and Figure 3.13.

For many purposes, only the fundamental component of the Fourier series representing the output waveforms of an inverter need be retained. In electric machine analysis this is a common approximation since it is easily shown that the harmonics result in very little average torque. Thus, in cases where only the basic overall electromechanical response is of interest, the system can be modelled with sufficient accuracy by simply neglecting the harmonics caused by inverter excitation. Examination of Figure 3.8 and eqns (3.5–12) and (3.5–13) indicate that if we wish to neglect harmonics, the inverter can be modelled in the synchronous reference frame by simply neglecting the effects of the 6th harmonic and multiples of the 6th. The switching functions become simply,

$$g_{qs}^e \cong 1 \tag{3.6–1}$$

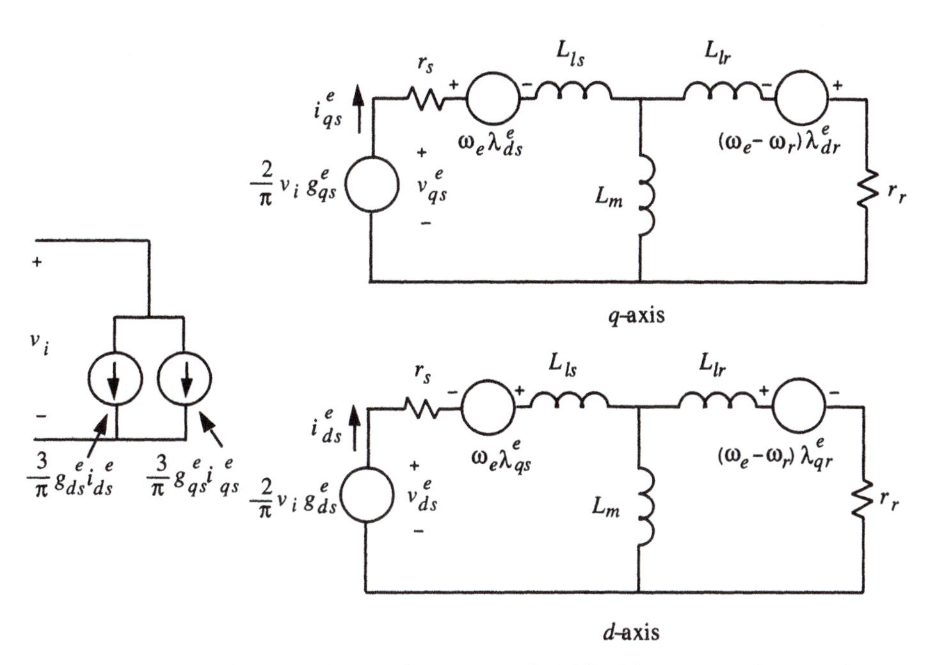

Figure 3.11 Synchronous frame d,q model of a VSI driven induction machine

$$g_{ds}^e \cong 0 \tag{3.6–2}$$

The simplified equivalent circuit is illustrated in Figure 3.14. If we take note of the fact that

$$i_i \cong \frac{3}{\pi} i_{qs}^e \text{ and } v_{qs}^e \cong \frac{2}{\pi} v_i \tag{3.6–3}$$

the need for the current source models to represent the inverter current disappear entirely. In a similar manner the harmonics can be neglected when modelling a CSI driven machine. In this case, from eqns (3.5–17) and (3.5–18),

$$h_{qs}^e \cong 1 \tag{3.6–4}$$

$$h_{ds}^e \cong 0 \tag{3.6–5}$$

The corresponding simplified equivalent circuit is shown in Figure 3.15.

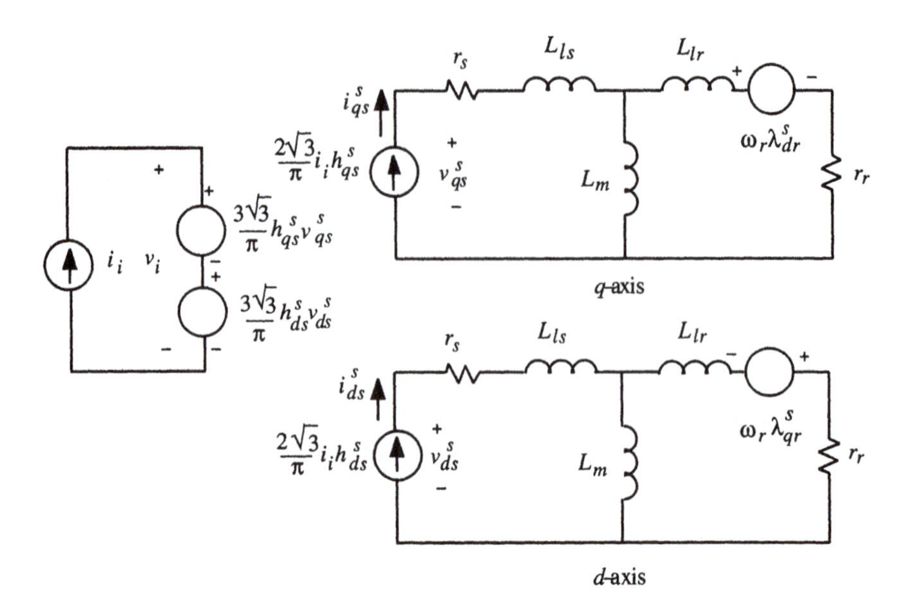

Figure 3.12 Stationary frame *d,q* model of a CSI driven induction machine

The models represented by Figure 3.10 and Figure 3.11 can be used to obtain the transient response of a VSI driven induction motor. Figure 3.16 and Figure 3.17 show computer traces of the acceleration of a VSI driven induction machine in both the stationary and synchronous reference frames. Note that the representations of v_{qs}, i_{qs} and i_{qr}' are quite different in the two simulations, whereas the remaining quantities are all identical.

Figure 3.18 portrays behavior of the system when the effects of the inverter harmonics are neglected. It is apparent that these harmonics have little effect on the basic electromechanical response of the system. It also should be noted that these simulations assume that the dc link voltage is ideal, i.e. a constant. In practice, this voltage can fluctuate substantially, producing, in reality, an oscillatory or even unstable operating condition [1],[2]. In such cases the dc link components must be modelled in more detail.

Figure 3.19 shows the synchronous reference frame representation of the inverter and induction motor which also incorporates the effect of the dc link filter. For simplicity, the rectifier input, which may in fact be a single phase or three phase rectifier, remains modelled as a voltage source in which the current

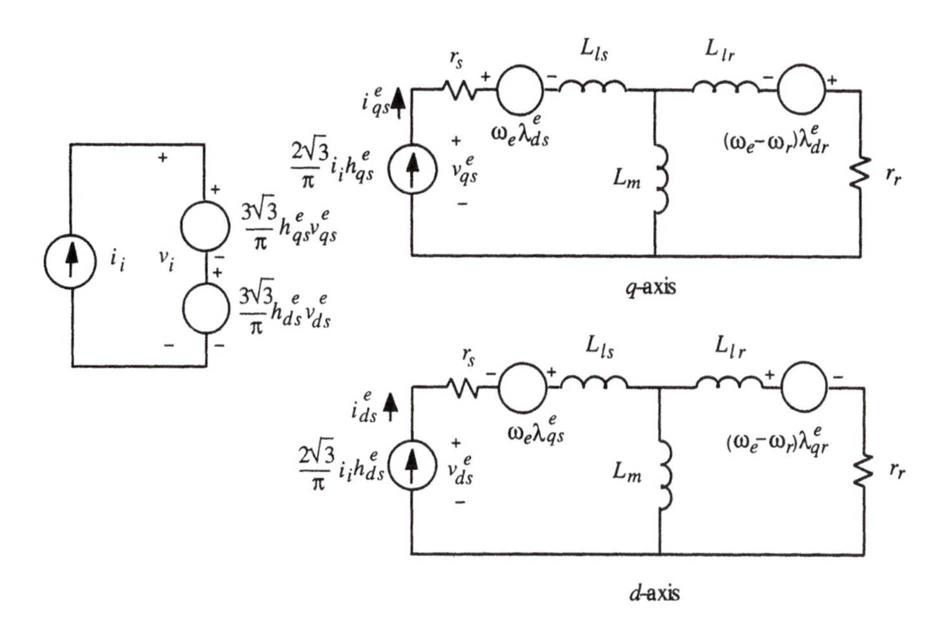

Figure 3.13 Synchronous frame d,q model of a CSI driven induction machine

cannot reverse. This effect is modelled with a simple diode. The equations of the L–C filter are

$$v_r = v_i + (r_{lf} + pL_f)\, i_r \qquad i_r > 0 \qquad (3.6\text{–}6)$$

$$v_r = v_i \qquad i_r = 0 \qquad (3.6\text{–}7)$$

$$v_i = \frac{1}{p}\frac{1}{C_f}(i_r - i_i) \qquad (3.6\text{–}8)$$

where p again denotes the time derivative operator d/dt and $1/p$ represents the integral with respect to time.

Observing the proportionality constants in eqns (3.5–8) to (3.5–10), these equations can be rewritten as

$$\frac{2}{\pi}v_r = \frac{2}{\pi}v_i + \frac{6}{\pi^2}(r_{lf} + pL_f)\left(\frac{\pi}{3}i_r\right) \qquad i_r > 0 \qquad (3.6\text{–}9)$$

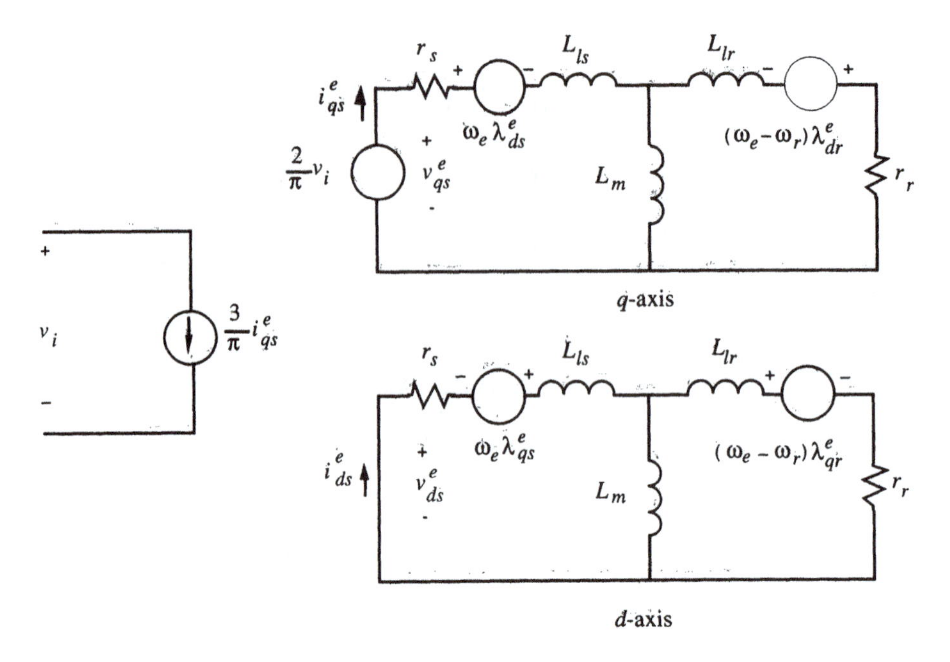

Figure 3.14 Simplified synchronous frame *d,q* model of VSI driven induction machine

$$\frac{2}{\pi}v_r = \frac{2}{\pi}v_i \qquad i_r = 0 \tag{3.6–10}$$

$$\frac{2}{\pi}v_i = \frac{1}{p}\left(\frac{6}{\pi^2}\frac{1}{C_f}\right)\left(\frac{\pi}{3}i_r - \frac{\pi}{3}i_i\right) \tag{3.6–11}$$

These equations combine with eqns (3.5–8) to (3.5–10) and the induction motor equations in the synchronous frame to form Figure 3.20. Note that now the filter model has been conveniently coupled directly to the *q*–axis machine model and the inverter harmonics appear as a series voltage source equal to $(2/\pi)\,v_i\,(g^e_{qs} - 1)$. When the effects of the inverter harmonics are neglected, the two voltage dependant sources in Figure 3.20 now become short circuits and the equivalent circuit of Figure 3.20 results. In this case primed variables have been defined to represent the dc link variables which can now be considered as referred to the stator in much the same manner as the turns ratio transformation used to refer the rotor variables to the stator in Section 2.7.

To show the effect of the filter on the acceleration of an induction machine from rest, Figure 3.21 can be compared with Figure 3.16. In this computer run,

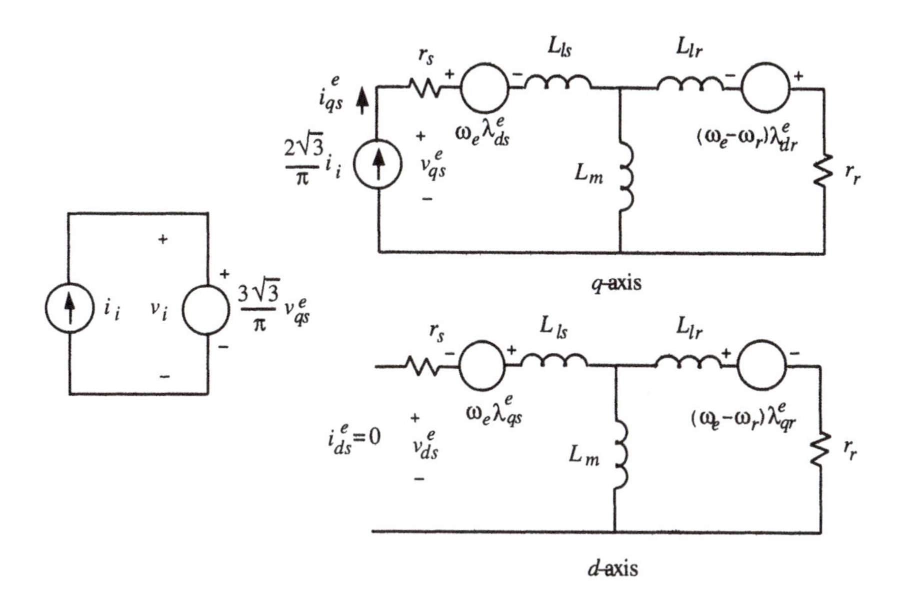

Figure 3.15 Simplified synchronous frame d,q model of CSI driven induction machine

the dc link capacitor is assumed to be completely charged before commands are given to the inverter to supply voltage to the motor. The results are nearly the same as before indicating that the effects of the input supply impedance have essentially been decoupled from the output by the action of the large filter capacitor, a useful attribute from the point of view of ease of control.

It should be mentioned that the condition studied corresponds to a fixed inverter frequency of 30 Hz. In practice, the motor is rarely, if ever, started under such conditions since the resulting inrush currents to the motor would destroy the inverter switches. In most practical cases, open loop general purpose drives such as the type modelled here are normally started with the inverter set at the permissible minimum frequency, usually only a few hertz. The frequency command is then ramped up to ensure that the motor continues to operate with a slip frequency of only a few hertz as it accelerates up to the desired speed.

The equivalent circuits for a CSI fed induction machine including the effect of the dc link inductor can be developed in much the same manner. Once again, scaled variables can be introduced on the dc side to correspond to the inverter

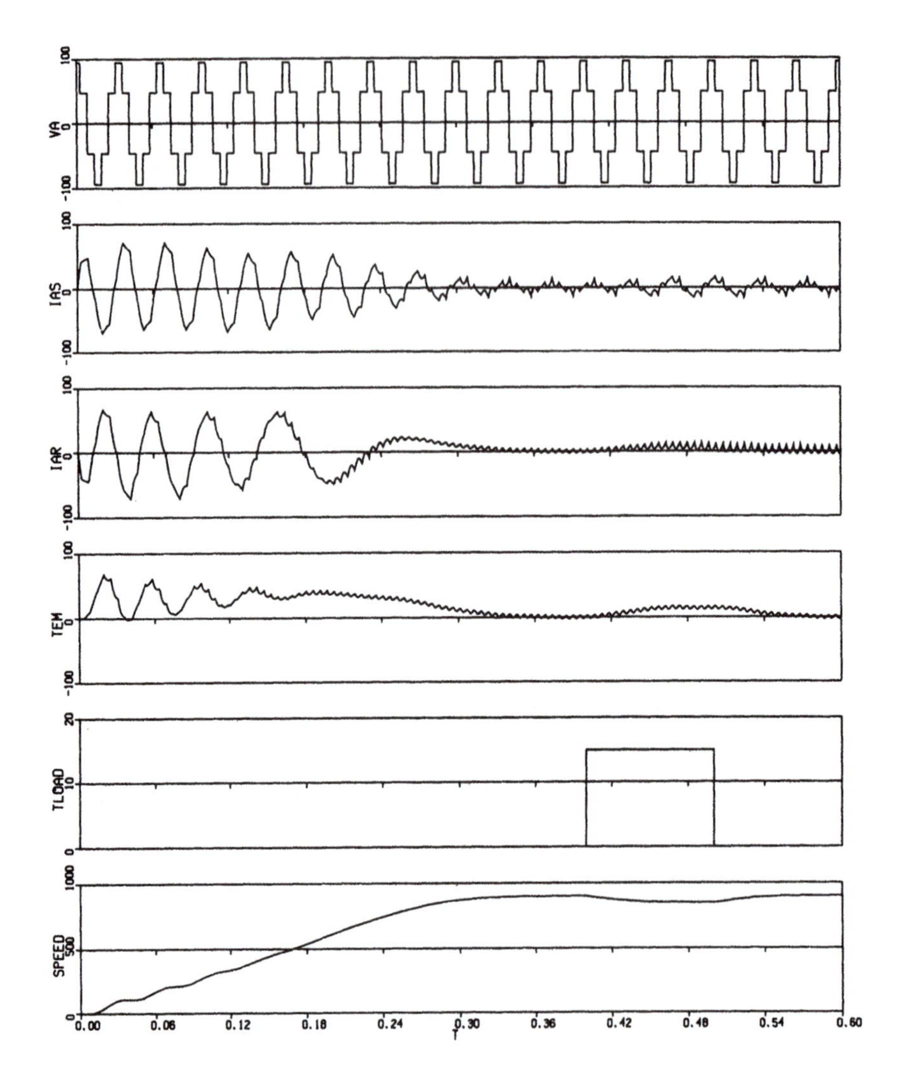

Figure 3.16 Simulation of an induction machine accelerating from rest using a VSI supply with the system represented in the stationary reference frame. The inverter frequency is fixed at 30 Hz. VA = $v_{as} = v_{qs}^{s}$ (V), IAS = $i_{as} = i_{qs}^{s}$ (A), IAR = $i_{ar}' = i_{qr}^{r}{}'$ (A), TEM = T_e (N–m), TLOAD = Load Torque (N–m), SPEED = $\omega_{mech} = 2\omega_r/P$ (rad/s)

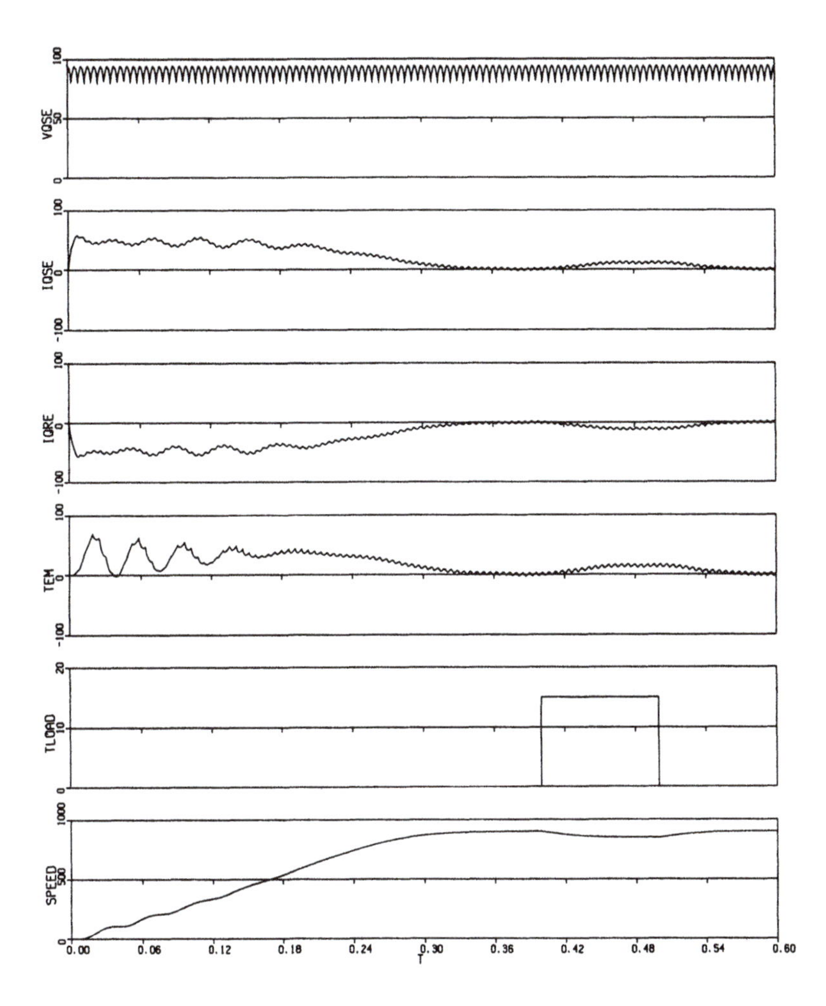

Figure 3.17 Simulation of an induction machine accelerating from rest with a VSI supply with the system represented in the synchronous reference frame. The inverter frequency is fixed at 30 Hz. VQSE = v_{qs}^{e} (V), IDSE = i_{ds}^{e} (A), IQRE = $i_{dr}^{e'}$ (A), TEM = T_e, (N–m),TLOAD = T_{load} (N–m), SPEED = $(2\omega_r)/P$ rad/s

switching relations, eqns (3.5–14) and (3.5–15). Figure 3.22 shows the resulting equivalent circuit. When the harmonics impressed on the motor are neglected, eqns (3.6–4) and (3.6–5) apply and the equivalent circuit reduces to that of Figure 3.23. Note that whereas the VSI is modelled by connecting the scaled voltage dc bus to the q–axis and shorting the d–axis, Figure 3.20, the CSI simplified circuit is obtained by connecting the scaled dc current bus to the q–axis and open circuiting the d–axis.

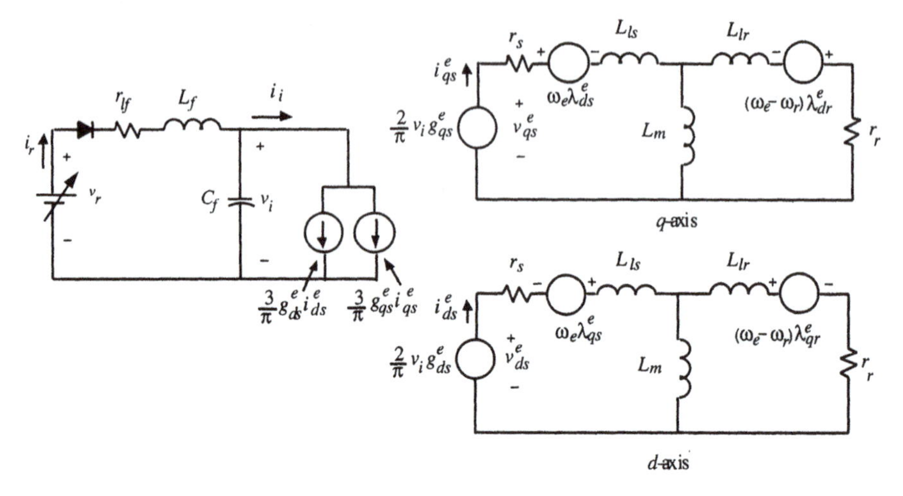

Figure 3.19 Synchronous frame d,q model of VSI driven induction machine including the effects of the dc link filter

3.7 Fundamental Component Approximation for Steady State Operation

It has already been noted that when the drive torque characteristics are the primary consideration or when only the approximate overall characteristics are desired, neglecting all harmonics and representing only the fundamental component behavior is quite acceptable. Incorporating the harmonics (or carrying out time domain solutions) is justified only in cases where waveforms, torque pulsations, efficiencies, or other internal details of operation are the desired results. The simplification resulting from neglecting harmonics is very significant and is the basis for the steady state inverter equivalent circuits presented in this section.

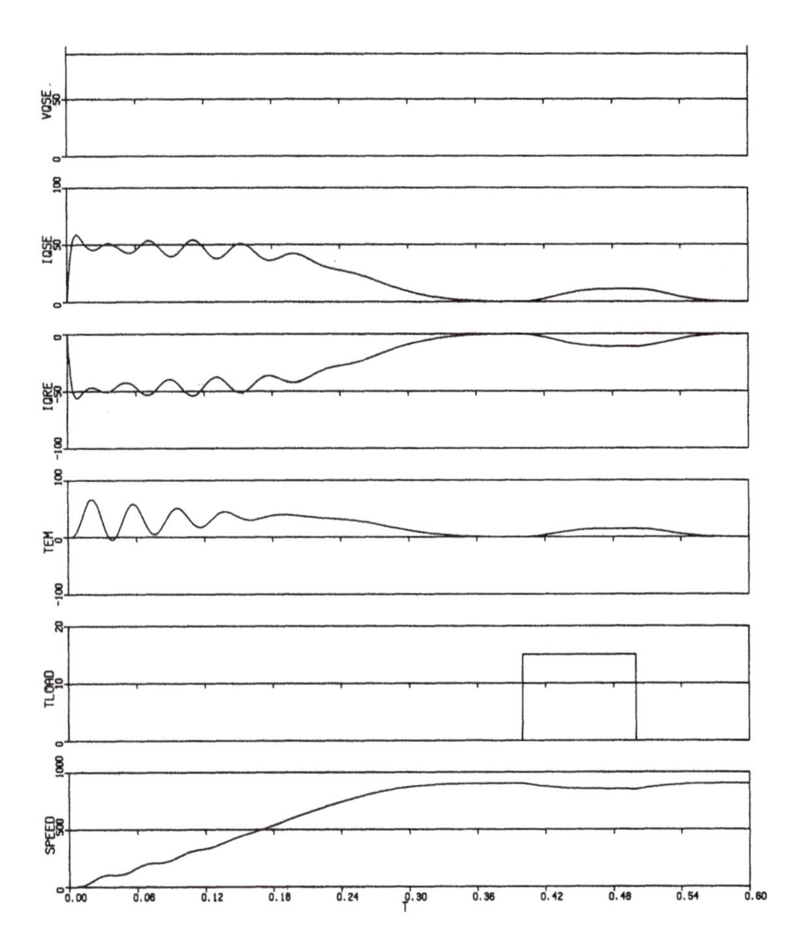

Figure 3.18 Simulation of an induction machine accelerating from rest with a VSI supply with the system represented in the synchronous reference frame neglecting the effect of the inverter harmonics. The inverter frequency is fixed at 30 Hz. VDSE = v_{ds}^{e} (V), VQSE = v_{qs}^{e} (V), IDSE = i_{ds}^{e} (A), IQSE = i_{qs}^{e} (A), IQRE = $i_{qr}^{e\,\prime}$ (A), TEM = T_e (N–m), SPEED = $(2\omega_r)/P$ rad/s

The steady state output waveforms for six step VSI and CSI inverters are repeated in Figure 3.24 with the fundamental component shown as shaded curves. Although the phase shift of the waveform on a per phase basis is arbitrary, a typical phase and line voltage can be described by reference to Figure

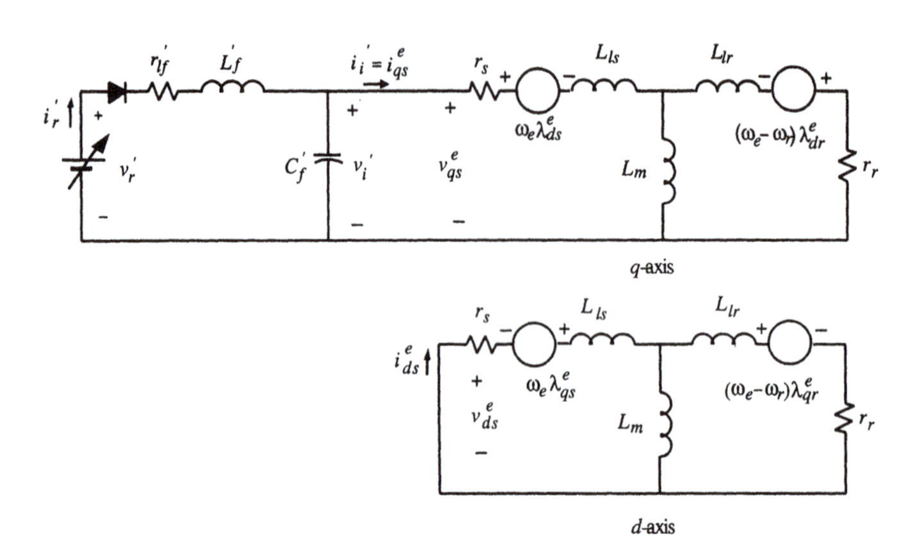

Figure 3.20 Simplified synchronous frame d,q model incorporating the effects of the dc link but neglecting the effect of the inverter harmonics

3.24. Considering phase voltage v_{as} and line voltage $-v_{bc}$ as typical, the Fourier series representing these waveforms for the VSI are:

$$v_{line}(= v_{bc}) = \frac{2\sqrt{3}}{\pi}V_i\left[\cos\theta - \frac{1}{5}\cos 5\theta - \frac{1}{7}\cos 7\theta + \frac{1}{11}\cos 11\theta + \ldots\right] \quad (3.7\text{–}1)$$

$$v_{phase}(= v_{as}) = \frac{2}{\pi}V_i\left[\sin\theta + \frac{1}{5}\sin 5\theta - \frac{1}{7}\sin 7\theta - \frac{1}{11}\sin 11\theta + \ldots\right] \quad (3.7\text{–}2)$$

where $\theta = \omega_e t + \phi$ is determined by the inverter switching rate and the time zero reference. The corresponding results for the CSI are obtained by replacing v with V and i with I (to denote steady state). Thus, from Figure 1.15, using the instant of switching from interval 61 to 12 as the $t = 0$ reference,

$$i_{line}(= i_{as}) = \frac{2\sqrt{3}}{\pi}I_i\left[\cos\theta - \frac{1}{5}\cos 5\theta - \frac{1}{7}\cos 7\theta + \frac{1}{11}\cos 11\theta + \ldots\right] \quad (3.7\text{–}3)$$

$$i_{phase}(= i_{bc}) = \frac{2}{\pi}I_i\left[\sin\theta + \frac{1}{5}\sin 5\theta - \frac{1}{7}\sin 7\theta - \frac{1}{11}\sin 11\theta + \ldots\right] \quad (3.7\text{–}4)$$

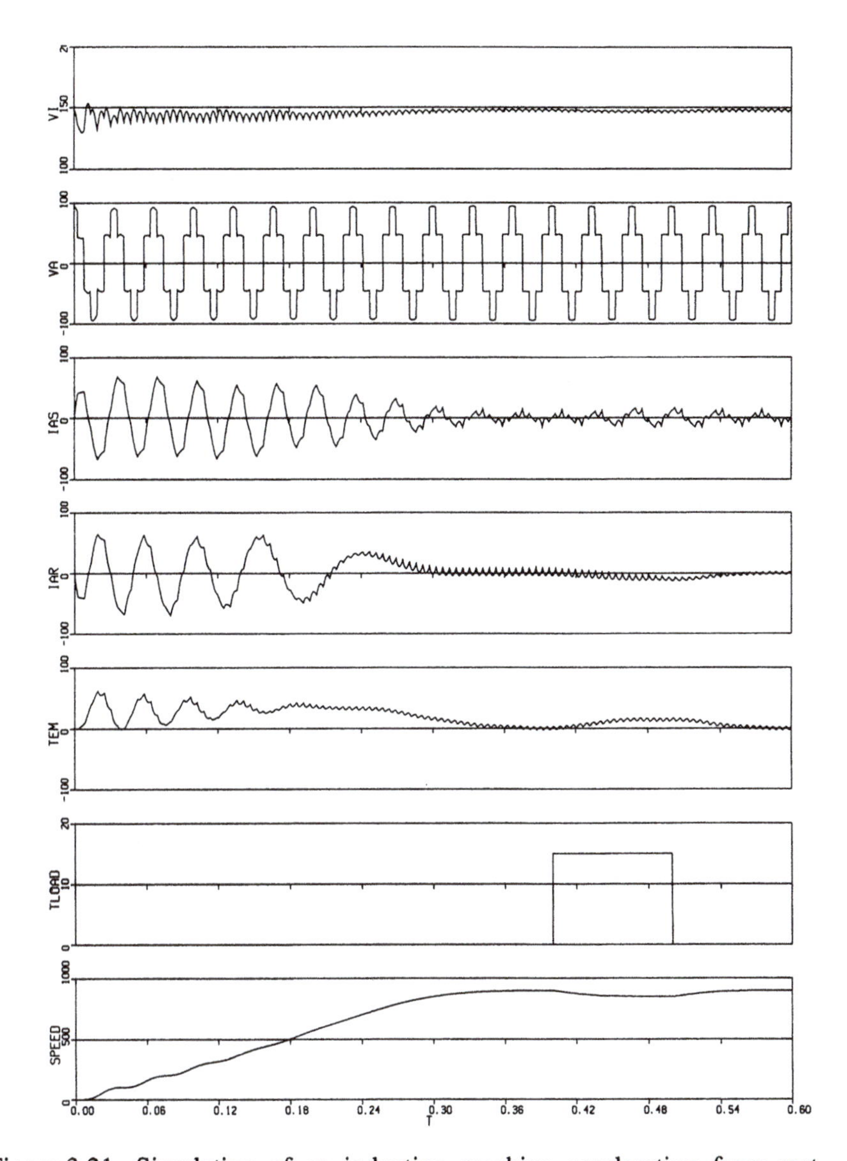

Figure 3.21 Simulation of an induction machine accelerating from rest with an voltage source inverter supply. Simulation incorporates the effect of the dc link filter. The inverter frequency is fixed at 30 Hz. VI = V_i (V), VA = $v_{as} = v_{qs}^{s}$ (V), IA = $i_{as} = i_{qs}^{s}$ (A), IAR = i_{ar}' (A), TEM = T_e (N–m), TLOAD = T_{load} (N–m), SPEED = $(2\omega_r)/P$ (rad/s)

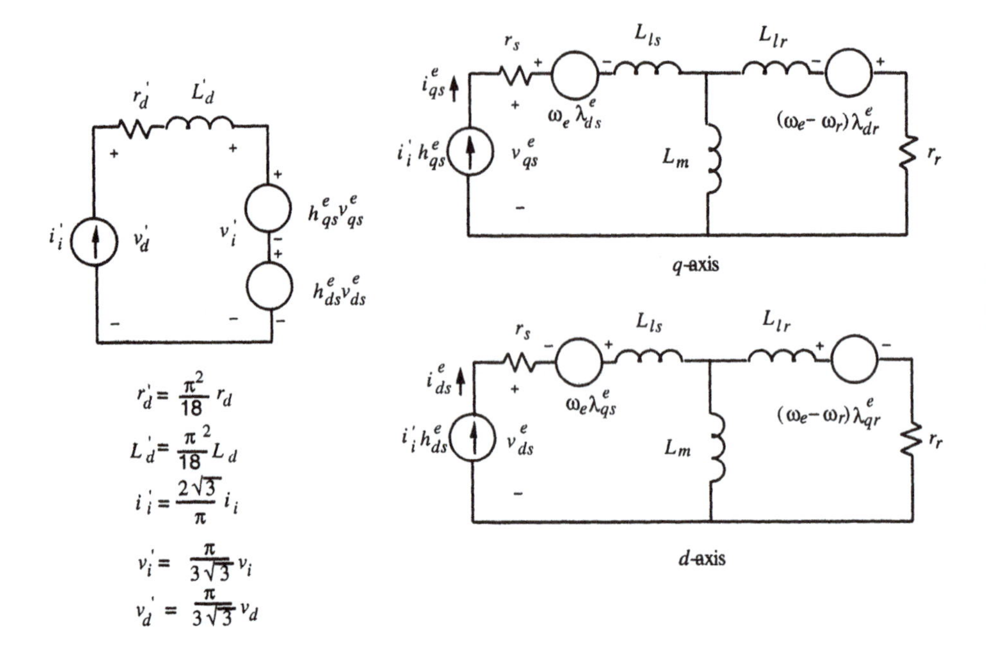

Figure 3.22 Synchronous frame *d,q* equivalent circuit of CSI fed induction motor incorporating the effect of the dc link inductor. Link variables referred to the induction motor stator

The output voltage waveforms for the PWM–VSI inverter described in Section 1.7 are also shown in Figure 3.24. In general the Fourier series describing this type of waveform can be written in the form

$$v_{phase} = \frac{2}{\pi} V_i \left[a_1 \sin\theta_1 + \frac{a_5}{5}\sin\theta_5 - \frac{a_7}{7}\sin\theta_7 + \frac{a_{11}}{11}\sin\theta_{11} + \ldots \right] \qquad (3.7\text{–}5)$$

where the *PWM coefficients* a_k and phase shifts θ_k are dependent on the specific type of PWM modulation employed. In addition to the odd harmonics shown in eqn (3.7–5) some PWM algorithms may even contain non–*triplen* even harmonics (even harmonics not a multiple of three). For the simple case illustrated in Figure 3.24, the values of a_k can be readily written as a function of the length of the short circuit submode. Note that as defined, a_1 is always equal to or less than one.

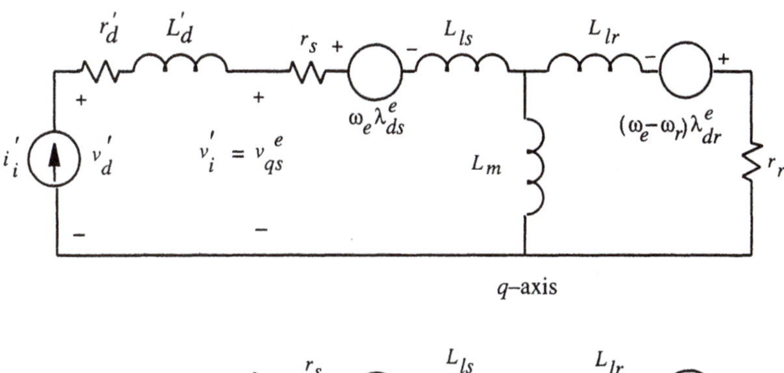

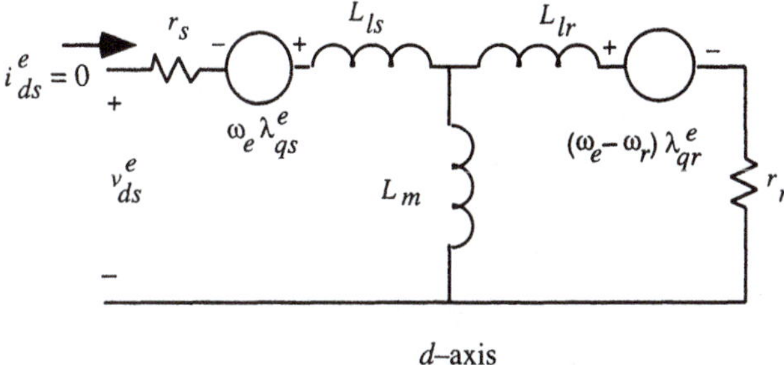

Figure 3.23 Simplified synchronous frame d,q model of CSI fed induction motor incorporating the effects of the dc link but neglecting the effect of the inverter harmonics

Whereas in transient analysis the peak values are of importance, it is conventional in steady state analysis to use rms quantities. Retaining only the fundamental term yields the following relations between the rms line voltage V_{line} and phase voltage V_{phase} and the inverter dc voltage V_i in a VSI.

$$V_{line,\,rms} = \frac{\sqrt{6}}{\pi} V_i \tag{3.7–6}$$

$$V_{phase,\,rms} = \frac{\sqrt{2}}{\pi} V_i \tag{3.7–7}$$

Similar relations for the CSI relating the rms line and phase currents I_{line} and I_{phase} (Δ connection) to the inverter dc current I_i can also be written

$$I_{line,\,rms} = \frac{\sqrt{6}}{\pi} I_i \quad \text{(a)} \tag{3.7–8}$$

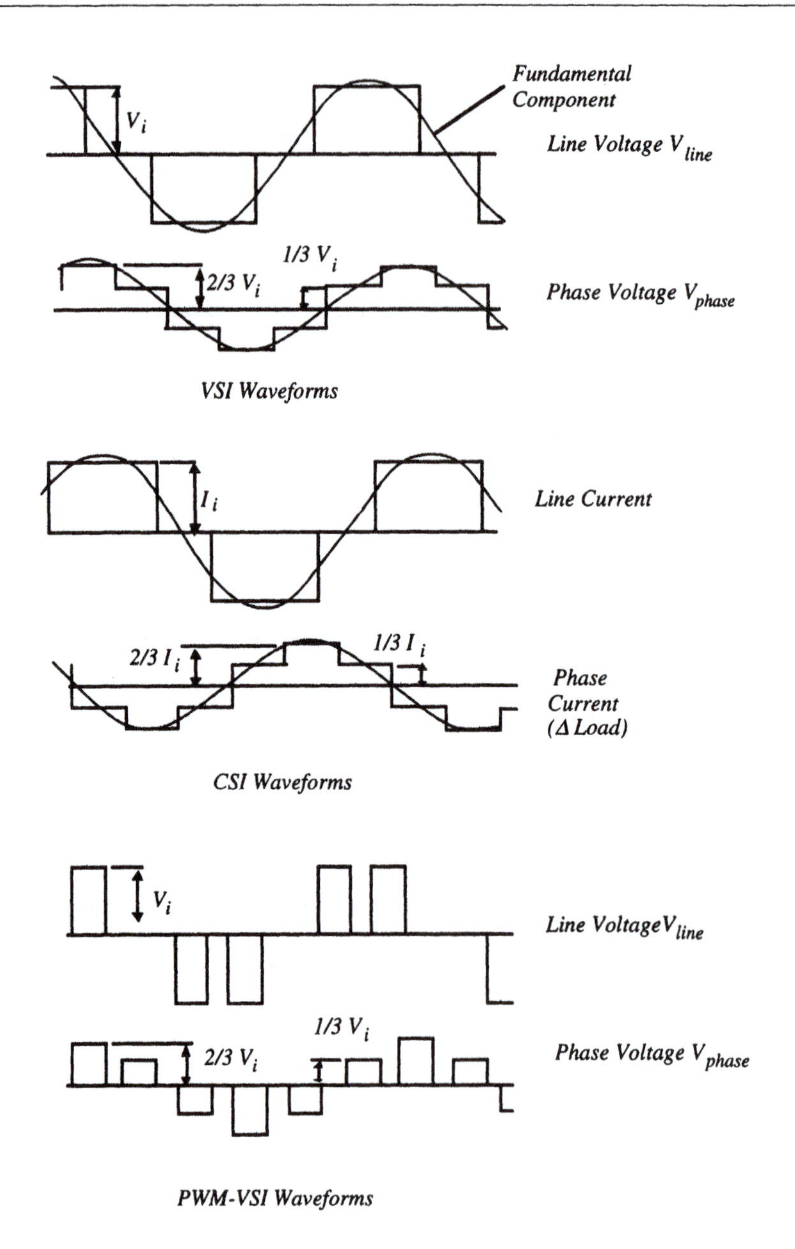

Figure 3.24 Output waveforms for basic inverters with constant dc input

$$I_{phase,\, rms} = \frac{\sqrt{2}}{\pi} I_i \tag{3.7–9}$$

For the PWM–VSI, the relations are similar to those for the VSI in eqn (3.7–6) except that the PWM ratio a_1 is present as a multiplier as shown in eqn (3.7–10).

$$V_{line,\, rms} = \frac{\sqrt{6}}{\pi} a_1 V_i \tag{3.7–10}$$

$$V_{phase,\, rms} = \frac{\sqrt{2}}{\pi} a_1 V_i \tag{3.7–11}$$

The relations given in eqns (3.7–6) to (3.7–10) represent what can be described as the basic or primary switching constraints of the three types of inverters under consideration. They describe the dc to ac fundamental component transfer relation for the directly controlled quantity for each type of inverter; voltage in the VSI, current in the CSI. There remains the question of the transfer relation for the other (auxiliary) variable; current in the VSI and voltage in the CSI. This auxiliary relation is not determined by the inverter alone since it is clear that the switching constraints do not completely specify the auxiliary variable. Because of the degree of freedom in the auxiliary variable switching constraints, the nature of the load has an important impact on the transfer relation. A detailed analysis of the auxiliary variable transfer relation is presented in [3]. For our purposes it is much simpler to consider the overall input–output balance and to determine the auxiliary variable transfer relation on this basis. The procedure is illustrated for the six step VSI; the other cases are handled in exactly the same way and only the results are presented.

For the VSI, the input–output voltage relation is known, eqn (3.7–6), but there remains the problem of determining the input–output current relation. If we assume that the inverter behaves as an ideal dc to sine wave converter (i.e., assume the total dc input power is converted to fundamental frequency power in the output) we can write for a Y–connected load

$$V_i I_i = 3 V_{phase,\, rms} I_{line,\, rms} \cos\phi, \quad \phi = \text{power factor angle} \tag{3.7–12}$$

which when combined with eqn (3.7–6)(b), yields the following input–output current relation

$$I_i = \frac{3\sqrt{2}}{\pi} I_{line,\ rms} \cos\phi \tag{3.7–13}$$

The interpretation of this result is that only the in–phase or power component of the output current is reflected to the input of the inverter and hence, by implication, that the quadrature or reactive component circulates in the inverter in the form of the short circuit current shown in the conduction mode diagram of Figure 1.4.

The corresponding result for the CSI, obtained with the same type of power balance relation, is

$$V_i = \frac{3\sqrt{6}}{\pi} V_{phase,\ rms} \cos\phi \tag{3.7–14}$$

In this relation, and for the remainder of this discussion we assume Y–connected loads on both the VSI and CSI.

For the PWM–VSI, the result corresponding to eqn (3.7–13) becomes

$$I_i = \frac{3\sqrt{2}}{\pi} a_1 I_{line,\ rms} \cos\phi \tag{3.7–15}$$

where the only change is the addition of the PWM ratio a_1 as a multiplying factor in the equation.

The input–output transfer relations given by eqns (3.7–6) and (3.7–13) for the VSI, eqns (3.7–8) and (3.7–14) for the CSI, and eqns (3.7–10) and (3.7–15) for the PWM–VSI can be conveniently expressed in equivalent circuit form if all quantities of interest are referred to the ac side of each inverter. Again using the VSI to illustrate the method, observe that

1) the ac side equivalent of the inverter input voltage is given, on a per phase basis, by eqn (3.7–7)

$$V_{phase,\ rms} = \frac{\sqrt{2}}{\pi} V_i \tag{3.7–16}$$

2) the current constraint relation, eqn (3.7–13), can be rewritten as

$$I_{line,\ rms} \cos\phi = \frac{\pi}{3\sqrt{2}} I_i \tag{3.7–17}$$

which expresses the ac side equivalent of the inverter input current I_i in terms of the *in–phase component* of the load current. A simple equivalent circuit which is described by eqns (3.7–16) and (3.7–17) is shown in Figure 3.25.

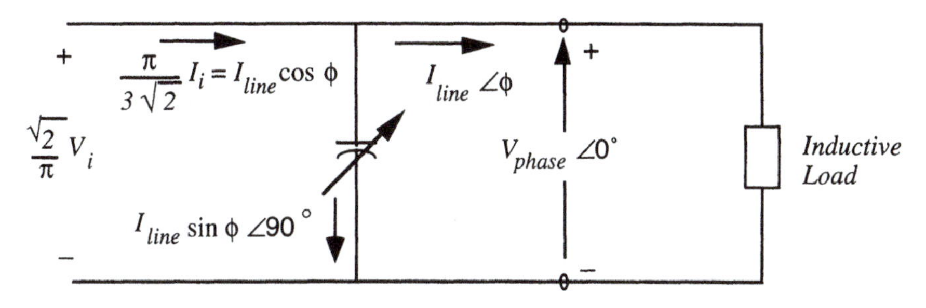

Figure 3.25 Partial per phase fundamental component equivalent circuit of three phase six step VSI

At the right hand end of the circuit are the inverter output terminals connected to the load. The rms phasor voltage and current at the load are $V_{phase}\angle 0° = (\sqrt{2}/\pi)\, V_i$ and $I_{line}\angle\theta$. Connected in parallel is a variable reactive element (a capacitor for an inductive load) which cancels the reactive part of the load current. This element represents the effect of the inverter short circuit currents and is inherently continuously adjusted to be in parallel resonance with the load impedance in the steady state. To the left of this element, the current is the inverter input current referred to the ac side of the inverter which is given by the expression $(\pi/3\sqrt{2})\, I_i$ (eqn (3.7–17)). The input and output voltage are the same and equal to the phase voltage $V_{phase,\ rms}$ given by $(\sqrt{2}/\pi)\, V_i$, eqn (3.7–16). The inverter functions as a source of the reactive power required by the load, continuously adjusting its supply of reactive power (adjusting the value of the shunt reactive element) to be exactly equal to the reactive demand of the load.

A more complete equivalent circuit including the resistance r_{lf} in the dc link filter inductance supplying the inverter input terminals can be constructed as shown in Figure 3.26. The actual voltage drop is $r_{lf}\, I_i$, but this must be referred to the ac side by multiplying by the voltage transfer ratio $\sqrt{2}/\pi$ to yield $(\sqrt{2}/\pi)\, r_{lf} I_i$. Since the ac side inverter input current is $(\pi/3\sqrt{2})\, I_i$, the referred value of r_{lf} becomes

$$\frac{(\sqrt{2}/\pi)\, r_{lf} I_i}{(\pi/3\sqrt{2})\, I_i} = \frac{6}{\pi^2} r_{lf} \tag{3.7–18}$$

The input voltage to the dc link V_r, can also be referred to the ac side by multiplying by $\sqrt{2}/\pi$ to complete the diagram. The power balance equation for Figure 3.26

$$\left[\frac{\sqrt{2}}{\pi}V_r\right]\left[\frac{\pi}{3\sqrt{2}}I_i\right] - \left[\frac{\pi}{3\sqrt{2}}I_i\right]^2\left[\frac{6}{\pi^2}r_{lf}\right] = V_{phase,\,rms}I_{line,\,rms}\cos\phi \tag{3.7–19}$$

$$\frac{1}{3}[V_rI_i] - \frac{1}{3}[I_i^2r_{lf}] = P_{phase}$$

verifies that the circuit correctly represents the per phase power relations of the inverter.

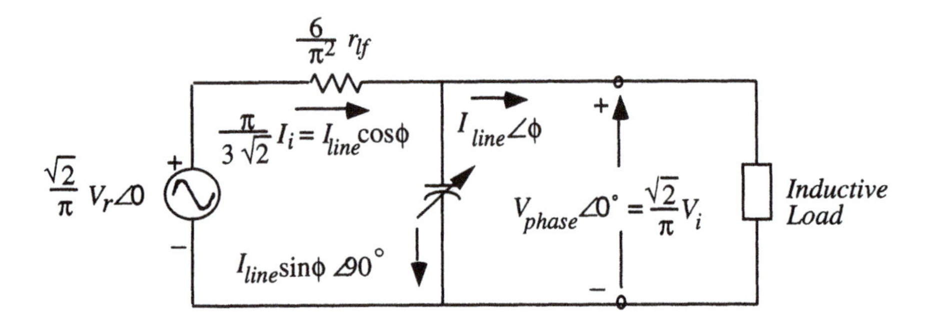

Figure 3.26 Per phase fundamental component equivalent circuit of three phase six step VSI

Figure 3.27 presents the corresponding equivalent circuit for the CSI. In this case the reactive element is in series with the load current to represent the reactive voltage developed as a result of the open circuit inherent in the CSI switching modes. As for the VSI, the reflected effect of dc link resistance is included in the circuit for those cases where the source feeding the CSI is a voltage source. If the input to the CSI can be represented as a current source, the resistance and equivalent ac side voltage source can be omitted and replaced by the current source.

For the PWM–VSI inverter, comparison of eqn (3.7–6)(b) with (3.7–10)(b) and eqn (3.7–13) with (3.7–15) reveals that the influence of the PWM control is to introduce the PWM ratio a_1 in exactly the form of the turns ratio of a transformer. This suggests the modification of the VSI equivalent circuit of Figure 3.27 to produce the PWM VSI equivalent circuit shown in Figure 3.28.

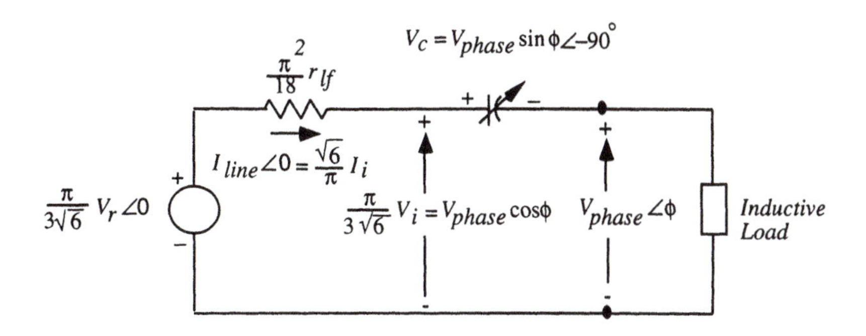

Figure 3.27 Per phase fundamental component equivalent circuit of three phase six step CSI

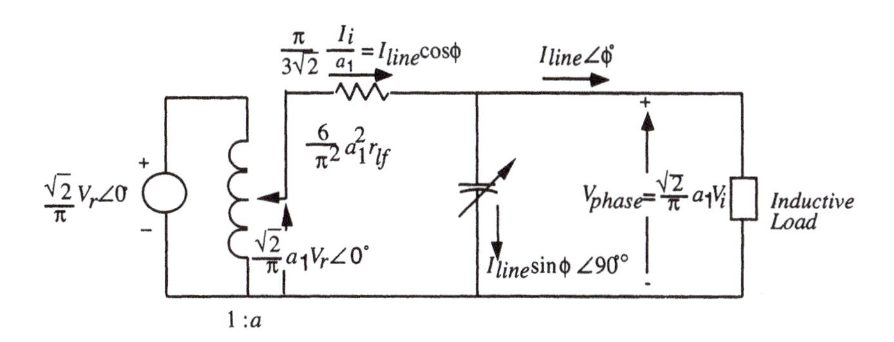

Figure 3.28 Per phase fundamental component equivalent circuit of the three phase PWM–VSI inverter

The variable ratio 1: a_1 transformer is a very compact and convenient means of representing the effect of PWM–control.

The circuits of Figure 3.26, Figure 3.27 and Figure 3.28 can be used to analyze the steady state fundamental frequency behavior of VSI and CSI inverters feeding any type of balanced linear load. The variable reactive elements in the circuits are the fundamental frequency representations of the inherent ability of the inverters to supply the reactive power required by the load. Since the loads are assumed to be balance three phase loads, all that is really required is a means of transferring the stored energy from phase to phase. For inductive loads this transfer mechanism is provided by the short circuit paths in the VSI and by the commutation of current from phase to phase in the CSI. For capacitive loads the transfer occurs at commutation in the VSI and in the open circuited phase in the CSI. Thus the accuracy of the representation depicted in the circuits is better for the natural cases of inductive loads on the VSI or capacitive loads on the CSI, but can be applied to the other cases if care is used in interpreting the results and the likely occurrence of commutation spikes associated with the reactive energy transfer is recognized.

3.8 Duality of VSI and CSI Systems

The duality of VSI and CSI systems has been stressed in the preceding sections. In terms of the basic power circuit, the two systems are exact duals if we compare a VSI with a wye connected load with a CSI with a delta connected load. The equivalent circuits of the preceding section do not exhibit exact duality because a wye connected load is assumed in all cases. However, the duality principle is still a valid and important concept. Table 3.1 summarizes the dual properties of these two basic systems.

Table 3.1 Duality of VSI and CSI Systems

VSI	CSI
1) Output is constrained voltage	1) Output is constrained current
2) dc bus dominated by shunt capacitor	2) dc bus dominated by series inductor
3) dc bus current proportional to motor power and hence dependent on motor power factor	3) dc bus voltage proportional to motor power and hence dependent on motor power factor
4) Output contains voltage harmonics varying inversely as harmonic order	4) Output contains current harmonics varying inversely as harmonic order
5) Prefers motors with larger leakage reactance	5) Prefers motors with lower leakage reactance
6) Can handle motors smaller than inverter rating	6) Can handle motors larger than inverter rating

Table 3.1 Duality of VSI and CSI Systems

VSI	CSI
7) dc bus current reverses in regeneration	7) dc bus voltage reverses in regeneration
8) Immune to open circuits	8) Immune to short circuits

References

[1] T.A. Lipo and P.C. Krause, "Stability Analysis of a Rectifier–Inverter Induction Motor Drive", IEEE Trans. on Power Apparatus and Systems, Vol. PAS–88, No. 1, January 1969, pp. 55–66.

[2] T. A. Lipo, P. C. Krause and H. E. Jordan, "Harmonic Torque and Speed Pulsations in a Rectifier–Inverter Induction Motor Drive", IEEE Trans.on Power Apparatus and Systems, Vol. PAS–88, No. 5, May 1969, pp. 579–587.

[3] D.W. Novotny, "Switching Function Representation of Polyphase Invertors", 1975 Industry Applications Society Annual Meeting, Conf. Record, Atlanta GA, Sept. 28–Oct. 2, pp. 823–831.

[4] D.W. Novotny and T. King, "Equivalent Circuit Representation of Current Inverter Driven Synchronous Machines," IEEE PAS, Vol. 100, No 6, June 1981, pp. 2920–2926.

Problems

Problem 3–1 Six Step Waveform (VSI)

For the six step waveform shown below (with each step of duration π/3 radians)

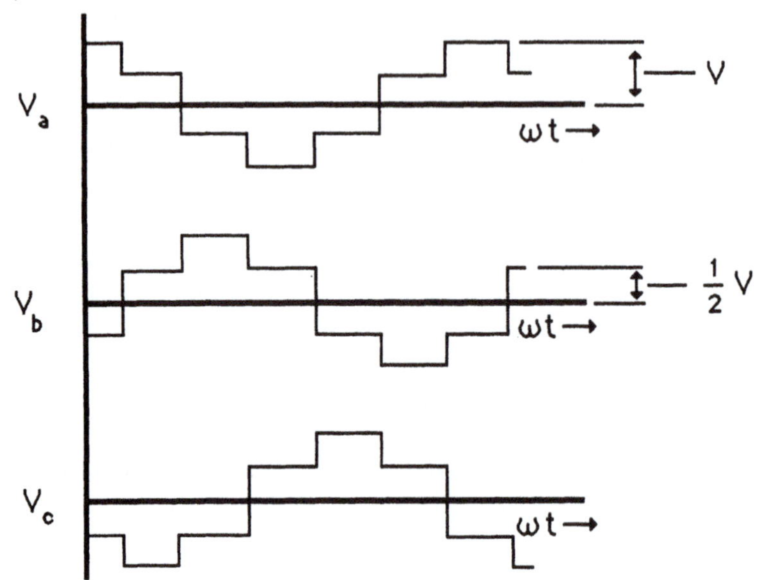

Find:

a) v_{qs}^s and v_{ds}^s as waveform sketches.

b) expressions for v_{qs}^s and v_{ds}^s as Fourier series.

c) complex variable Fourier series for $\underline{v}_{qds}$.

d) complex variable Fourier series for $\underline{v}_{qds}$.

Problem 3–2 VSI Switching Functions

Verify that the Fourier series in eqns (3.2–21) and(3.2–22) are correct representations of the switching functions g_{qs}^s and g_{ds}^s shown in Figure 3.2.

Problem 3–3 CSI Switching Functions

Verify that the Fourier series in eqns (3.4–4) and (3.4–5) are correct representations of the switching functions h_{qs}^s and h_{ds}^s shown in Figure 3.6.

Problem 3–4 Synchronous Frame Switching Functions

Derive the Fourier series in eqns (3.5–12) and (3.5–13) from the complex series in eqn (3.5–11).

Problem 3–5 VSI Waveforms

For a VSI operating with a constant voltage input v_i = 100 volts at a constant fundamental frequency of 60 hz, sketch the steady state waveforms of i_{qs}^s, i_{ds}^s and i_i for :

a) a balanced three phase, Y– connected resistive load of 10 Ω/ϕ.

b) a balanced three phase, Y– connected inductive load of L=5 mH/ϕ.

c) a balanced three phase, Y– connected R–L load with r=10 Ω/ϕ and L=5mh/ϕ.

d) a balanced three phase, Y– connected R–C load with r=10 Ω/ϕ and C=50 $\mu f/\phi$ where '/ϕ' denotes 'per phase'.

Problem 3–6 CSI Waveforms

For a CSI operating with a constant current input i_i = 10 amps at a constant fundamental frequency of 60 hz, sketch the steady state waveforms of v_{qs}^s, v_{ds}^s and v_i for :

a) a balanced three phase, Y– connected resistive load of 10 Ω/ϕ.

b) a balanced three phase, Y– connected capacitive load of 50 $\mu f/\phi$

c) a balanced three phase, Y– connected R–L load with R=10 Ω/ϕ and L=5 mh/ϕ.

d) a balanced three phase, Y– connected R–C load with R=10 Ω/ϕ and C=50 $\mu f/\phi$ where '/ϕ' denotes 'per phase'.

Problem 3–7 VSI Harmonic Analysis

For the conditions specified in problem 3–5, find the fundamental and 5th harmonic components of i_{qs}^s and i_{ds}^s and the dc and 6th harmonic components of i_i.

Problem 3–8 CSI Harmonic Analysis

For the conditions specified in problem 3–6, find the fundamentaland 5th harmonic components of v_{qs}^s and v_{ds}^s and the dc and 6th harmonic components of v_i.

Problem 3–9 VSI Synchronous Frame Model

The synchronous frame model of a VSI with harmonics neglected is given by eqn (3.6–1) and (3.6–2). The zero reference for the synchronous frame for this case is chosen such that the q–axis coincides with phase a at t=0. Derive the general model corresponding to these equations where the q–axis is at an angle ϕ_{oe} from the a–axis. Sketch the general form of the equivalent circuit shown in Figure 3.14.

Problem 3–10 VSI Driven Synchronous Machine

Derive a fundamental component equivalent circuit for a VSI driven non–salient pole synchronous machine. Use the circuit to derive the torque–speed characteristic of this type of drive operated with a constant value of torque angle δ (between V and E) and constant I_f .What type of dc machine has a similar characteristic?

Problem 3–11 CSI Driven Synchronous Machine

Show that the fundamental component equivalent circuit for a salient pole synchronous machine driven by a CSI is as shown in the figure below. The angle γ is the internal power factor angle (between E and I) of the machine and R_{eq} =(1/2) $(X_q - X_d)\sin 2\gamma$

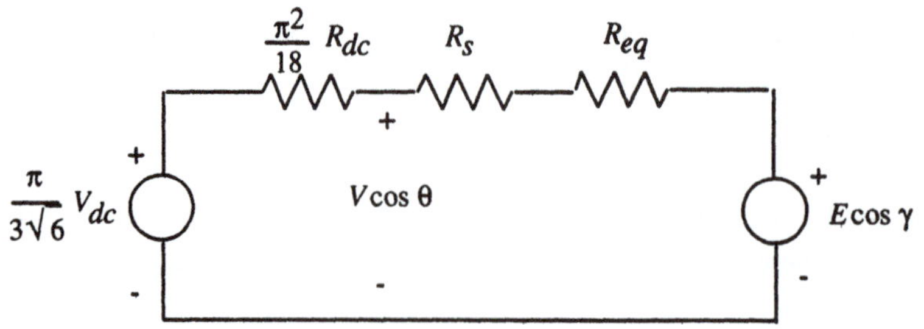

Problem 3–12 VSI Self–Excitation

Consider a VSI driven induction machine in which the dc supply is replaced by a load resistor. This system can be operated as a self excited induction generator with the VSI equivalent capacitor serving as the excitation source (reactive power). Show that for no–load $(R_{dc} \rightarrow \infty)$, the relation $S(\omega_e L_m)^2 \approx -r_1 r_2$ must hold and hence the level of saturation (and output voltage) is determined by the slip s and the machine resistances r_1 and r_2.

Problem 3–13 CSI Constant γ System

Derive the torque–speed characteristic for a system of the type in problem 3–11 operated with constant γ and constant I_f. What type of dc machine has a similar characteristic?

Problem 3–14 CSI Self–Excitation

Find the relation for CSI self–excitation corresponding to the result given in problem 3–12 for a VSI (the dc terminals are short–circuited in the CSI case).

4 Complex Vector Analysis of Induction Machines

4.1 Introduction

The development of the complex vector model of an induction machine presented in Chapter 2 clearly demonstrates the utility of the complex vector concept in simplifying and manipulating the differential equations describing the machine. The resulting model can be interpreted as a set of ordinary, real variable differential equations which can be used to represent an induction machine in specific situations. For each application a set of external constraint relations describing the specific situation is developed and imposed on the general model to obtain the application specific model.

When the external constraints are unsymmetrical (i.e. single phase operation, unequal phase impedances) the analysis is usually best carried out in terms of the real variable d,q differential equations. However, in those situations where the fundamental polyphase symmetry is retained, analysis can be conveniently carried out in complex vector form. The result is that the four electrical d,q equations are combined into two complex vector equations and the fourth order real variable electrical model is reduced to a second order complex variable model [1,2]. This reduction in order permits considerable simplification in analysis. For example, in the case of constant speed operation the system eigenvalues are obtained from a second order complex variable characteristic equation rather than from a fourth order real variable characteristic equation [3,4]. Even in cases where a lack of symmetry requires that the real variable equations be used, it is often conceptually useful to visualize the solution in terms of complex vectors.

The direct use of the complex vector model is especially valuable in the constant speed situation where the electrical d–q equations reduce to linear differential equations with constant coefficients. In this special case it is possible to obtain closed form solutions and the complex vector model provides a greatly simplified means of obtaining these solutions. Equivalent circuits which correctly represent the solutions play a significant role in understanding constant speed operation and these circuits are also readily derived using complex vector modelling.

4.2 Complex Vector Equivalent Circuit

An equivalent circuit representing the arbitrary frame complex vector differential equations developed in Chapter 2 can be constructed if the *d–q* equations (2.8-10) and (2.8-11)are expressed as

$$\underline{v}_{qds} = (r_s + L_s p)\,\underline{i}_{qds} + L_{sr} p \underline{i}_{qdr} + j\omega\underline{\lambda}_{qds} \quad (4.2\text{–}1)$$

$$\underline{v}_{qdr} = L_{sr} p \underline{i}_{qds} + (r_r + L_r p)\,\underline{i}_{qdr} + j(\omega - \omega_r)\,\underline{\lambda}_{qdr} \quad (4.2\text{–}2)$$

where $L_{sr} = (N_r/N_s)\,L_m$ and where the flux linkages are given by

$$\underline{\lambda}_{qds} = L_s \underline{i}_{qds} + L_{sr} \underline{i}_{qdr} \quad (4.2\text{–}3)$$

$$\underline{\lambda}_{qdr} = L_{sr} \underline{i}_{qds} + L_r \underline{i}_{qdr} \quad (4.2\text{–}4)$$

The complex vector equivalent circuit corresponding to these equations is illustrated in Figure 4.1. Note that this circuit is of exactly the same form as the

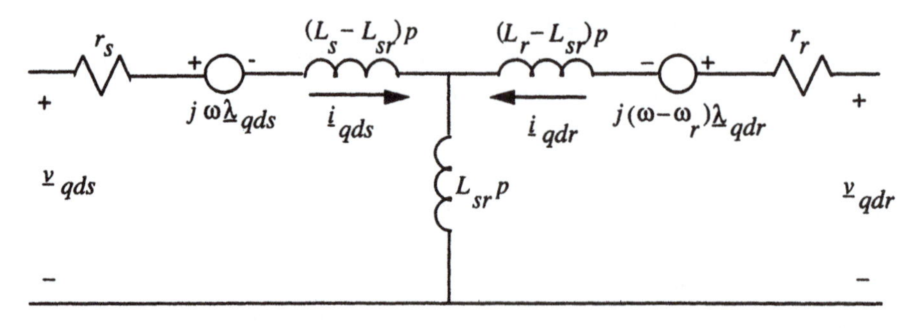

Figure 4.1 Complex vector equivalent circuit

real variable circuits. The speed voltages again occur as dependent generators; in this case both appear as counter emf voltages with the sign reversal between the *d* and *q* circuits handled by the *j* operator. Clearly the two real variable circuits are easily obtained by inspection from the complex vector circuit.

An operational form of equivalent circuit can also be employed to represent the complex vector equations. This form is obtained by multiplying through the rotor equation (4.2–2) by the operator $(p + j\omega) / [p + j(\omega - \omega_r)]$ to obtain (with $\underline{v}_{qdr} = 0$)

$$0 = L_{sr}(p + j\omega)\,\underline{i}_{qds} + \left[\frac{r_r(p + j\omega)}{p + j(\omega - \omega_r)} + L_r(p + j\omega)\right]\underline{i}_{qdr} \tag{4.2–5}$$

The equivalent circuit suggested by this result is shown in Figure 4.2. It has the

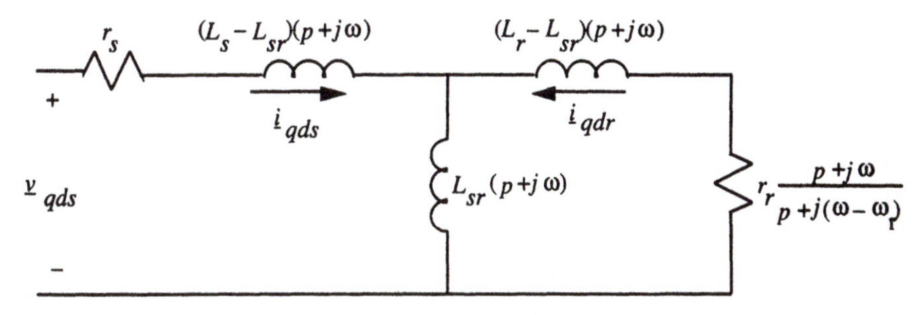

Figure 4.2 Complex vector operational equivalent circuit (zero external rotor voltage)

merit of simplicity and clearly the original complex variable differential equations can be recovered by writing out the circuit equations. It is useful for circuit manipulations and as a simple representation of the dynamic model of a machine with zero applied rotor voltage (induction machine). In steady state analysis it also can be used as a rapid means of obtaining the steady state equivalent circuit for balanced operation and demonstrating several important steady state results.

4.3 Turns Ratio and Modified *d,q* Models

The selection of the 'turns ratio' between the stator and rotor circuits of the *d–q* model can be shown to be essentially completely arbitrary when the machine is operated with zero applied rotor voltage. If we interpret the basic complex vector model as describing a machine in actual 'non–referred' parameters and introduce an arbitrary constant *b*, the basic equations can be written in the form

$$\underline{v}_{qds} = [r_s + L_s(p + j\omega)]\,\underline{i}_{qds} + bL_{sr}(p + j\omega)\frac{\underline{i}_{qdr}}{b} \tag{4.3–1}$$

$$b\underline{v}_{qdr} = bL_{sr}(p + j(\omega - \omega_r))\,\underline{i}_{qds} + b^2[r_r + L_r(p + j(\omega - \omega_r))]\,\frac{\underline{i}_{qdr}}{b} \quad (4.3\text{–}2)$$

These equations, with parameters,

$$r_s,\, L_s,\, b^2 r_r,\, b^2 L_r,\, bL_{sr},$$

and 'referred' variables,

$$\underline{v}_{qds},\, \underline{i}_{qds},\, b\underline{v}_{qdr},\, \underline{i}_{qdr}/b$$

where the 'referral ratio b is any constant we might want to choose (except $b = 0$ or ∞). Note that if we are willing to give up knowledge of the actual value of the rotor current (and voltage), we do not need to know the value of b. Thus, in induction machines where $\underline{v}_{qdr} = 0$, we can utilize this freedom to simplify the model by letting b have a convenient value. For example:

1) Let $L_s = b^2 L_r$ or $b = (L_s/L_r)^{1/2}$. This selection yields equal self inductances and results in a circuit model with equal *leakage* inductances.

2) Let $bL_{sr} = b^2 L_r$ or $b = L_{sr} / L_r$. This choice produces equal mutual and rotor self inductances and results in a circuit model with all 'leakage' located in the stator.

3) Let $L_s = bL_{sr}$ or $b = L_s / L_{sr}$. Selecting equal mutual and stator self inductance results in a model with all 'leakage' located in the rotor.

4) Let $b = N_s / N_r$. Choosing b as the stator to rotor turns ratio results in the "real" leakage inductance model based on the designer's computation of leakage inductances. This is the most commonly used version of the various choices.

The operational equivalent circuits resulting from these various selections of b are illustrated in Figure 4.3. Circuit 4 is the most common selection and the one which has been used in the basic model as given in Chapter 2. When machine parameters are determined from laboratory tests, circuit 1 is often used since there is no way to measure the actual turns ratio unless measurements in the rotor circuit can be made. Circuit 2 is very convenient in systems which control rotor flux, such as in field orientation and in switching transient problems where a very widely used approximate method based on constant rotor flux linkages is applicable. These situations will be examined in more detail in later sections.

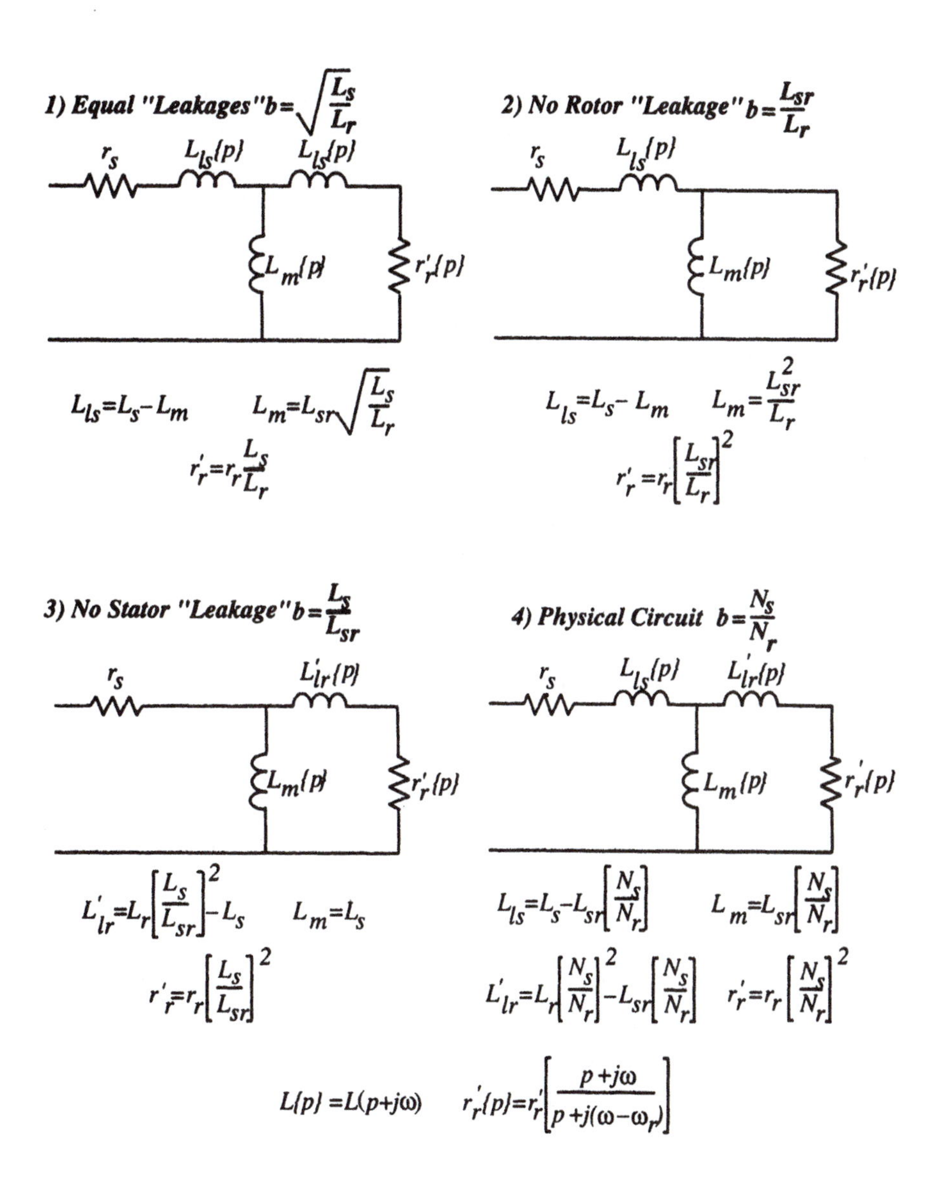

Figure 4.3 Modified *d,q* models as obtained from various selections of the referral ratio *b*

4.4 Steady State Equivalent Circuits

For operation at constant rotor speed, the volt–ampere equations become linear differential equations. In addition, the torque equation, which remains nonlinear, does not have to be solved in combination with the volt–ampere equations and can simply be evaluated as a last step after the currents are known.

For any periodic excitation voltage, an expansion in a Fourier series can be carried out and an analysis on a term by term basis is valid because of the linearity of the equations. The analysis for an arbitrary set of sinusoidal excitation voltages is therefore sufficient to handle any periodic input. The complex vector model will be employed in the analysis both as an illustration of the application of this model and because the algebraic operations are easily carried out in this form.

4.4.1 Sinusoidal Excitation in Complex Vector Form

Consider a set of general three phase voltages of unequal amplitude and arbitrary phase (it is assumed that the zero sequence voltage is zero)

$$\begin{aligned} v_{as} &= V_{as}\cos(\omega_e t + \phi_a) \\ v_{bs} &= V_{bs}\cos(\omega_e t + \phi_b) \\ v_{cs} &= V_{cs}\cos(\omega_e t + \phi_c) \end{aligned} \tag{4.4–1}$$

To facilitate evaluating the complex vector voltage associated with this excitation, each of the voltages is written as the sum of two complex exponentials, for example v_{as} can be written as

$$v_{as} = \frac{1}{2}[V_{as}e^{j(\omega_e t + \phi_a)} + V_{as}e^{-j(\omega_e t + \phi_a)}] \tag{4.4–2}$$

If the exponential term involving V_{as} is combined with the voltage amplitude to form the conventional peak value phasor voltage

$$\tilde{V}_{as} = V_{as}e^{j\phi_a} \tag{4.4–3}$$

the equation for v_{as} can be written as

$$v_{as} = \frac{1}{2}\tilde{V}_{as}e^{j\omega_e t} + \frac{1}{2}\tilde{V}^{\dagger}_{as}e^{-j\omega_e t} \tag{4.4–4}$$

Similar expressions can be written for v_{bs} and v_{cs}.

Substituting these voltage expressions into the defining equation for $\underline{v}^s_{qds}$, eqn 2.9–2, yields

$$\begin{aligned}\underline{v}^s_{qds} &= \frac{2}{3}[v_{as} + \underline{a}v_{bs} + \underline{a}^2 v_{cs}] \\ &= \frac{1}{3}[\tilde{V}_{as} + \underline{a}\tilde{V}_{bs} + \underline{a}^2\tilde{V}_{cs}]\, e^{j\omega_e t} + \frac{1}{3}[\tilde{V}^{\dagger}_{as} + \underline{a}\tilde{V}^{\dagger}_{bs} + \underline{a}^2\tilde{V}^{\dagger}_{cs}]\, e^{-j\omega_e t}\end{aligned} \tag{4.4–5}$$

The bracketed term multiplying $e^{j\omega_e t}$ is the formal definition of the positive sequence voltage of steady state symmetrical component theory

$$\tilde{V}_{sp} = \frac{1}{3}[\tilde{V}_{as} + \underline{a}\tilde{V}_{bs} + \underline{a}^2\tilde{V}_{cs}] \tag{4.4–6}$$

and if the second bracketed term is rewritten using the fact that the conjugate of a product is the product of the conjugates, i.e.

$$\tilde{Z}^{\dagger}_1 \tilde{Z}^{\dagger}_2 = (\tilde{Z}_1 \tilde{Z}_2)^{\dagger} \tag{4.4–7}$$

it becomes the conjugate of the negative sequence voltage of steady state symmetrical components

$$\tilde{V}^{\dagger}_{sn} = \frac{1}{3}[\tilde{V}_{as} + \underline{a}^2\tilde{V}_{bs} + \underline{a}\tilde{V}_{cs}]^{\dagger} \tag{4.4–8}$$

The complex vector voltage can then be written as

$$\underline{v}^s_{qds} = \tilde{V}_{sp} e^{j\omega_e t} + \tilde{V}^{\dagger}_{sn}\, e^{-j\omega_e t} \tag{4.4–9}$$

If the excitation voltages form a balanced three phase set, $\tilde{V}_{sn} = 0$ and the vector voltage reduces to

$$\underline{v}^s_{qds} = \tilde{V}_{sp} e^{j\omega_e t} = V_m e^{j\omega_e t} \tag{4.4–10}$$

where V_m is the peak value of the input voltage and its phase angle ϕ_m is arbitrarily taken as zero.

The stator referred complex vector voltage can be transformed to an arbitrary rotating reference using

$$\underline{v}_{qds} = \underline{v}^s_{qds}\, e^{-j\theta} \tag{4.4–11}$$

which yields

$$\underline{v}_{qds} = \tilde{V}_{sp} e^{j(\omega_e t - \theta)} + \tilde{V}^{\dagger}_{sn} e^{-j(\omega_e t + \theta)} \quad (4.4\text{–}12)$$

Table 4.2 presents the transformed voltage for the three most commonly used reference systems. Note that in a stator reference frame both the positive and negative sequence voltage are at the frequency, ω_e. The difference between the two voltage sets is the phase sequence. In a rotor reference the positive sequence voltage is at slip frequency, $\omega_e - \omega_r$, and the negative sequence at the sum frequency, $\omega_e + \omega_r$. Again the phase sequence is reversed. In a synchronous reference the positive sequence becomes dc and the negative sequence is at twice frequency, $2\omega_e$.

4.4.2 Stator Referred Equivalent Circuits

As a first method of developing the steady state equivalent circuits, the direct approach of applying the excitation constraints of the previous section directly to the complex vector equations will be employed. Using the basic stator referred equations yields (with zero external rotor voltage)

$$\tilde{V}_{sp} e^{j\omega_e t} + \tilde{V}^{\dagger}_{sn} e^{-j\omega_e t} = (r_s + L_s p)\, \underline{i}^s_{qds} + L_m p \underline{i}^s_{qdr} \quad (4.4\text{–}13)$$

$$0 = L_m (p - j\omega_r)\, \underline{i}^s_{qds} + [r_r + L_r (p - j\omega_r)]\, \underline{i}^s_{qdr} \quad (4.4\text{–}14)$$

Since the equations are linear, the two components of the excitation can be handled separately. For the positive sequence, the steady state solution will be of exponential form and will be symbolized by

$$\underline{i}^s_{qdsp} = \tilde{I}_{sp} e^{j\omega_e t}, \qquad \underline{i}^s_{qdrp} = \tilde{I}_{rp} e^{j\omega_e t} \quad (4.4\text{–}15)$$

Substituting these definitions and cancelling the common factor $e^{j\omega_e t}$ yields

$$\tilde{V}_{sp} = (r_s + j\omega_e L_s)\, \tilde{I}_{sp} + j\omega_e L_m \tilde{I}_{rp} \quad (4.4\text{–}16)$$

$$0 = j(\omega_e - \omega_r)\, L_m \tilde{I}_{sp} + [r_r + j(\omega_e - \omega_r)\, L_r]\, \tilde{I}_{rp} \quad (4.4\text{–}17)$$

which can be interpreted as a set of conventional phasor equations. Multiplying the rotor equation by $j\omega_e / j(\omega_e - \omega_r)$ results in

$$0 = j\omega_e L_m \tilde{I}_{sp} + \left[r_r \left(\frac{\omega_e}{\omega_e - \omega_r} \right) + j\omega_e L_r \right] \tilde{I}_{rp} \quad (4.4\text{–}18)$$

Table 4.1 *d,q* voltages for sinusoidal excitation

Positive Sequence	Negative Sequence
$\tilde{V}_{sp} = \frac{1}{3}[\tilde{V}_{as} + \underline{a}\tilde{V}_{bs} + \underline{a}^2\tilde{V}_{cs}]$	$\tilde{V}_{sn} = \frac{1}{3}[\tilde{V}_{as} + \underline{a}^2\tilde{V}_{bs} + \underline{a}\tilde{V}_{cs}]$
$= V_p e^{j\phi_p}$	$= V_n e^{j\phi_n}$

Ref. Type	Reference Frame Speed & Angle $\theta = \omega t + \phi$	Reference Frame Voltage Positive Sequence Peak = V_p Freq = ω_e	Reference Frame Voltage Negative Sequence Peak = V_n Freq = ω_e
Stator	$\omega = 0$ $\phi = 0$	$\underline{v}^s_{qds} = \tilde{V}_{sp} e^{j\omega_e t}$ $v^s_{qs} = V_p \cos(\omega_e t + \phi_p)$ $v^s_{ds} = -V_p \sin(\omega_e t + \phi_p)$	$\underline{v}^s_{qds} = \tilde{V}_{sn}{}^{\dagger} e^{-j\omega_e t}$ $v^s_{qs} = V_n \cos(\omega_e t + \phi_n)$ $v^s_{ds} = V_n \sin(\omega_e t + \phi_n)$
Rotor	$\omega = \omega_r$ $\theta = \omega_r t + \phi$	$\underline{v}^r_{qds} = \tilde{V}_{sp} e^{j[(\omega_e - \omega_r)t - \phi]}$ $v^r_{qs} = V_p \cos[(\omega_e - \omega_r)t + \phi_p - \phi]$ $v^r_{ds} = -V_p \sin[(\omega_e - \omega_r)t + \phi_p - \phi]$	$\underline{v}^r_{qds} = \tilde{V}^{\dagger}_{sn} e^{-j[(\omega_e + \omega_r)t + \phi]}$ $v^r_{qs} = V_n \cos[(\omega_e + \omega_r)t + \phi_n + \phi]$ $v^r_{ds} = V_n \sin[(\omega_e + \omega_r)t + \phi_n + \phi]$
Synch.	$\omega = \omega_e$ $\theta = \omega_e t + \phi$	$\underline{v}^e_{qds} = \tilde{V}_{sp} e^{-j\phi}$ $v^e_{qs} = V_p \cos(\phi_p - \phi)$ $v^e_{ds} = -V_p \sin(\phi_p - \phi)$	$\underline{v}^e_{qds} = \tilde{V}^{\dagger}_{sn} e^{-j(2\omega_e t + \phi)}$ $v^e_{qs} = V_n \cos(2\omega_e t + \phi_n + \phi)$ $v^e_{ds} = V_n \sin(2\omega_e t + \phi_n + \phi)$

which now has the same apparent frequency as the stator equation. These two equations, (4.4–16) and (4.4–18), are the same as the equations of the conventional equivalent circuit of the induction machine as illustrated in Figure 4.4 where the slip is defined as

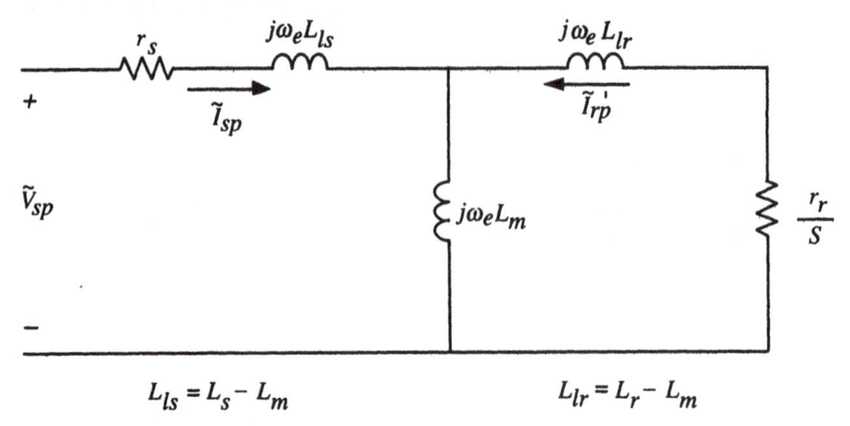

Figure 4.4 Positive sequence equivalent circuit

$$S = \frac{\omega_e - \omega_r}{\omega_e} \tag{4.4–19}$$

Note that eqn (4.4–18) is not valid at $\omega_e = 0$ (dc excitation) since this would mean we have multiplied by zero in obtaining the result. For dc excitation one must return to eqn (4.4–17) and abandon the equivalent circuit model. The rotor current is not zero for dc excitation as is easily seen from eqn (4.4–17).

The interpretation of eqns (4.4–16) and (4.4–17) as conventional phasor equations is quite acceptable but is not the complex vector interpretation. In conventional phasor methods, to obtain the time function solution we multiply by $e^{j\omega_e t}$ and take the real part. In complex vector notation, both the real and imaginary parts have meaning; the real part being the q–axis solution and the (negative of the) imaginary part the d–axis solution. One can, however, make the association with ordinary phasor methods to have a familiar point of reference for the final result.

An exactly parallel development for the negative sequence can be carried out. Defining the negative sequence complex vector components as

$$\underline{i}^s_{qdsn} = \tilde{I}^{\dagger}_{sn}\, e^{-j\omega_e t}; \quad \underline{i}^s_{qdrn} = \tilde{I}^{\dagger}_{rn}\, e^{-j\omega_e t} \tag{4.4–20}$$

and substituting yields

$$\tilde{V}^{\dagger}_{sn} = (r_s - j\omega_e L_s)\,\tilde{I}^{\dagger}_{sn} - j\omega_e L_m \tilde{I}^{\dagger}_{rn} \tag{4.4–21}$$

$$0 = -j(\omega_e + \omega_r)\, L_m \tilde{I}^{\dagger}_{sn} + [r_r - j(\omega_e + \omega_r)\, L_r]\, \tilde{I}^{\dagger}_{rn} \tag{4.4–22}$$

The occurrence of the minus sign on each of the $j\omega_e$ terms is unusual in terms of real variable machine theory. These minus signs occur because of the $e^{-j\omega_e t}$ multiplier in the negative sequence portion of the complex vector voltage. To eliminate these minus signs, the entire set of equations can be replaced by their conjugate set. Clearly there is no loss of information in this operation and the negative sequence voltage $\tilde{V}_{sn}$ now occurs rather than its conjugate. The result is

$$\tilde{V}_{sn} = (r_s + j\omega_e L_s)\,\tilde{I}_{sn} + j\omega_e L_m \tilde{I}_{rn} \tag{4.4–23}$$

$$0 = j(\omega_e + \omega_r)\, L_m \tilde{I}_{sn} + [r_r + j(\omega_e + \omega_r)\, L_r]\, \tilde{I}_{rn} \tag{4.4–24}$$

Multiplying the rotor equation by $j\omega_e / j(\omega_e + \omega_r)$ yields the equations of the conventional negative sequence circuit of Figure 4.5 where

$$\frac{\omega_e + \omega_r}{\omega_e} = 2 - \frac{\omega_e - \omega_r}{\omega_e} = 2 - S \tag{4.4–25}$$

The comment regarding zero frequency operation as for the positive sequence holds here also and again the interpretation for complex vector analysis differs from conventional phasor analysis.

The steady state torque is obtained by direct substitution of the solution for the current expressed in complex vector form in the torque equation given in eqn (2.12–7).

$$\begin{aligned} T_e &= \frac{3}{2}\frac{P}{2} L_m Im\,[\underline{i}^s_{qds}\underline{i}^{s\dagger}_{qdr}] \\ &= \frac{3}{2}\frac{P}{2} L_m Im\,[\,(\tilde{I}_{sp} e^{j\omega_e t} + \tilde{I}^{\dagger}_{sn}\, e^{-j\omega_e t})\,(\tilde{I}^{\dagger}_{rp}\, e^{-j\omega_e t} + \tilde{I}_{rn} e^{j\omega_e t})\,] \end{aligned} \tag{4.4–26}$$

which yields, after expansion, two terms; the average torque

$$T_{eo} = \frac{3}{2}\frac{P}{2}L_m Im[\tilde{I}_{sp}\tilde{I}^{\dagger}_{rp} + \tilde{I}^{\dagger}_{sn}\tilde{I}_{rn}] \tag{4.4–27}$$

and a pulsation torque at $2\omega_e$

$$T_{e2} = \frac{3}{2}\frac{P}{2}L_m Im[\tilde{I}_{sp}\tilde{I}_{rn}e^{j2\omega_e t} + \tilde{I}^{\dagger}_{sn}\tilde{I}^{\dagger}_{rp}e^{-j2\omega_e t}] \tag{4.4–28}$$

These expressions can be converted to the conventional results in terms of rotor currents only by using the relations between rotor and stator sequence currents obtained from the equivalent circuits

$$\tilde{I}_{sp} = \frac{1}{\omega_e L_m}\left[-\omega_e(L_{lr}+L_m) + j\frac{r_r}{S}\right]\tilde{I}_{rp} \tag{4.4–29}$$

$$\tilde{I}^{\dagger}_{sn} = \frac{1}{\omega_e L_m}\left[-\omega_e(L_{lr}+L_m) - j\frac{r_r}{2-S}\right]\tilde{I}^{\dagger}_{rn} \tag{4.4–30}$$

Using these results yields

$$T_{eo} = \frac{3}{2}\frac{P}{2}\frac{1}{\omega_e}\left[|\tilde{I}_{rp}|^2\frac{r_r}{S} - |\tilde{I}_{rn}|^2\frac{r_r}{2-S}\right] \tag{4.4–31}$$

and

$$T_{e2} = \frac{3}{2}\frac{P}{2}\frac{1}{\omega_e}\left(\frac{r_r}{S} - \frac{r_r}{2-S}\right)Re[\tilde{I}_{rp}\tilde{I}_{rn}e^{j2\omega_e t}] \tag{4.4–32}$$

which are the standard forms resulting from conventional symmetrical component theory (except that rms values are normally used and one of the factors of 1/2 therefore does not occur in the leading factor).

As an alternative approach to obtaining the equivalent circuits in Figure 4.4 and Figure 4.5, the operational circuit of Figure 4.2 can be employed. With $\underline{v}^s_{qds}$ known to be an exponential function of frequency ω_e, the steady state model is obtained by recognizing that the p–operator becomes the complex operator $\pm j\omega_e$ depending on the sign of the exponential frequency. Thus, for a stator referred system,

$$\omega = 0 \tag{4.4–33}$$

and for the positive sequence

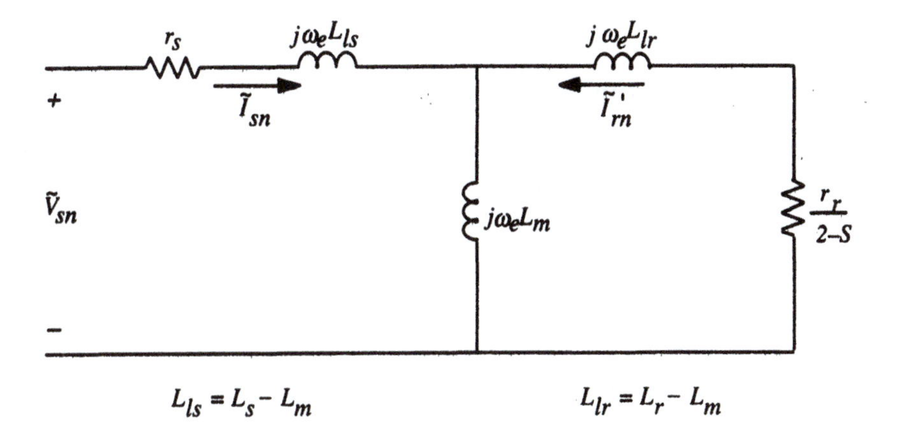

Figure 4.5 Negative sequence equivalent circuit

$$p = j\omega_e \tag{4.4–34}$$

In the operational circuit of Figure 4.2, the operator $p + j\omega$ of each inductor therefore becomes

$$p + j\omega \to j\omega_e \tag{4.4–35}$$

and the operator of the rotor resistance becomes

$$\frac{p + j\omega}{p + j(\omega - \omega_r)} \to \frac{\omega_e}{\omega_e - \omega_r} \tag{4.4–36}$$

which immediately converts the general operational circuit to the positive sequence circuit of Figure 4.4.

By similar arguments, for the negative sequence

$$\omega = 0,\ p = -j\omega_e \tag{4.4–37}$$

and

$$p + j\omega \to -j\omega_e \tag{4.4–38}$$

$$\frac{p + j\omega}{p + j(\omega - \omega_r)} \to \frac{\omega_e}{\omega_e + \omega_r} \tag{4.4–39}$$

which (after conjugating) yields the negative sequence circuit of Figure 4.5.

4.5 Invariance of the Steady State Equivalent Circuit

The steady state equivalent circuits shown in Figure 4.4 and Figure 4.5 were derived in a stator reference system. They are, however, exactly the same in any reference system. This is easily shown since the complex vector voltage in an arbitrary reference is, eqn (4.4–12),

$$\underline{v}_{qds} = \tilde{V}_{sp}\, e^{j(\omega_e t - \theta)} + \tilde{V}_{sn}^{\dagger}\, e^{-j(\omega_e t + \theta)} \tag{4.5–1}$$

Thus, for a frame rotating at ω, the frequency of the positive sequence excitation is $\omega_e - \omega$ and hence in the steady state

$$p = j(\omega_e - \omega) \tag{4.5–2}$$

The operators in the operational equivalent circuit become

$$p + j\omega \rightarrow j(\omega_e - \omega) + j\omega = j\omega_e \tag{4.5–3}$$

and

$$\frac{p + j\omega}{p + j(\omega - \omega_r)} \rightarrow \frac{\omega_e}{\omega_e - \omega_r} \tag{4.5–4}$$

which are the same as previously obtained for a stator reference. Similarly for the negative sequence the frequency is $-(\omega_e + \omega)$ and

$$p = -j(\omega_e + \omega) \tag{4.5–5}$$

and

$$p + j\omega \rightarrow -j\omega_e \tag{4.5–6}$$

$$\frac{p + j\omega}{p + j(\omega - \omega_r)} \rightarrow \frac{\omega_e}{\omega_e + \omega_r} \tag{4.5–7}$$

which again are the same as obtained for a stator reference.

Physically, the reason the circuits are all identical is, of course, that they all represent the same machine. In terms of transformation theory, the change in the excitation frequency in different frames is exactly counterbalanced by the speed voltage arising from the motion of the reference frame so that the frequency dependent inductors are always evaluated at ω_e and the resistance multiplier is always $\omega_e/(\omega_e + \omega_r)$ or $\omega_e/(\omega_e - \omega_r)$. The interpretation of the circuits as conventional phasor circuits is, however, not possible except in the stator frame. For example, in a rotor frame, the excitation frequency is $\omega_e - \omega_r$ (positive sequence) and yet the reactances in the circuit are evaluated at a frequency ω_e. The synchronous frame is especially interesting since here the excitation frequency is zero or dc (positive sequence) and yet we have a complex parameter equivalent circuit. The complex vector interpretation is straightforward. The real parts of the complex quantities are the *q*–axis dc values and the $-j$ parts are the *d*–axis values. Physically, we can think of moving in synchronism with the traveling waves in the machine so all quantities are stationary. Figure 4.6 illustrates a *d,q* axis diagram in the synchronous frame where the time zero reference of the rotation transformation is selected such that the air gap flux is in the *d*–axis. Note that while the diagram is labeled as a synchronous frame diagram, it is really indistinguishable from a stator frame diagram where the complex vectors are interpreted as conventional phasors. Note also that if the time reference is chosen such that the positive sequence voltage has a zero phase angle

$$\tilde{V}_{sp} = |\tilde{V}_{sp}|e^{j0} \tag{4.5–8}$$

then the reference for the rotating transformation would have to be

$$\theta = \omega_e t - \varphi \tag{4.5–9}$$

where φ is the phase angle of $\tilde{V}_{sp}$ in Figure 4.6.

4.6 Modified Equivalent Circuits

The modified *d–q* models introduced in Section 4.3 can be used to obtain modified steady state equivalent circuits [5]. The simplest way to obtain these circuits is by converting the operational circuits shown in Figure 4.3 to steady state circuits by the method used in Section 4.4.2. With this approach, the modified circuits would be interpreted as resulting from *turns ratio* transformations of the basic coupled circuit model.

There is, however, another way to obtain these modified circuits which offers a different and, in many ways, a more useful interpretation of their origin. This alternative approach is in terms of the spatial orientation of the rotating frame *d–q* reference system and its relation to the flux waves in the actual

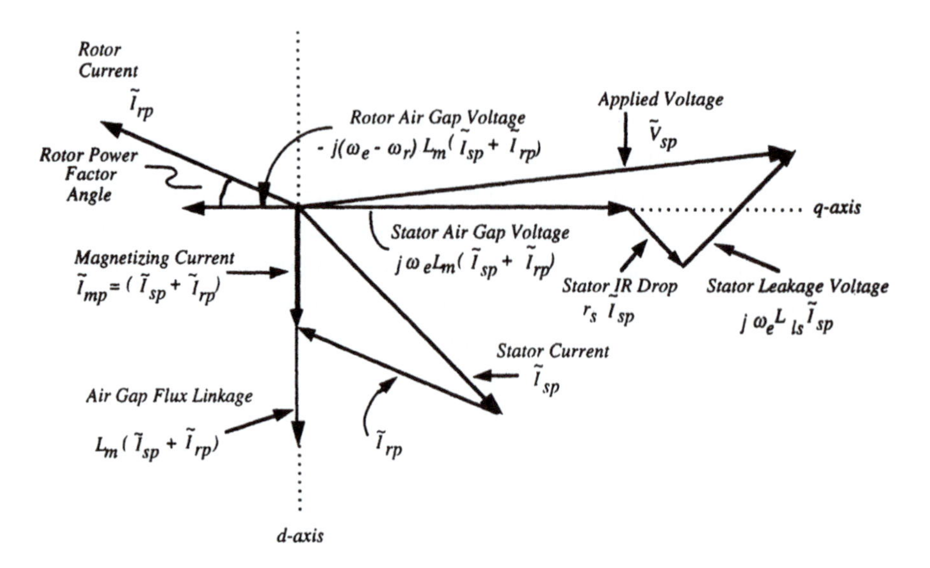

Figure 4.6 Complex vector diagram for positive sequence synchronous reference frame – air gap flux reference, subscript p dentes positive sequence component

machine. The concept is most easily interpreted in a synchronous frame; the resulting circuits are of course valid in any frame.

Consider the positive sequence synchronous frame complex vector diagram shown in Figure 4.7 where the reference system is chosen to coincide with the air gap flux. This diagram is similar to Figure 4.6 except that the rotor and stator fluxes are also shown. From the figure it is clear that for this reference system (the subscript p to indicate positive sequence is omitted)

$$I_{qs} = -I_{qr} \qquad \text{(since } \lambda_{mq} = 0\text{)} \tag{4.6–1}$$

$$I_{ds} + I_{dr} = I_m \tag{4.6–2}$$

The components of the stator flux linkage vector are

$$\lambda_{qs} = L_s I_{qs} + L_m I_{qr} = (L_s - L_m) I_{qs} \tag{4.6–3}$$

$$\lambda_{ds} = L_s I_{ds} + L_m I_{dr} = (L_s - L_m) I_{ds} + L_m (I_{ds} + I_{dr}) \tag{4.6–4}$$

from which the stator flux linkage vector can be written as

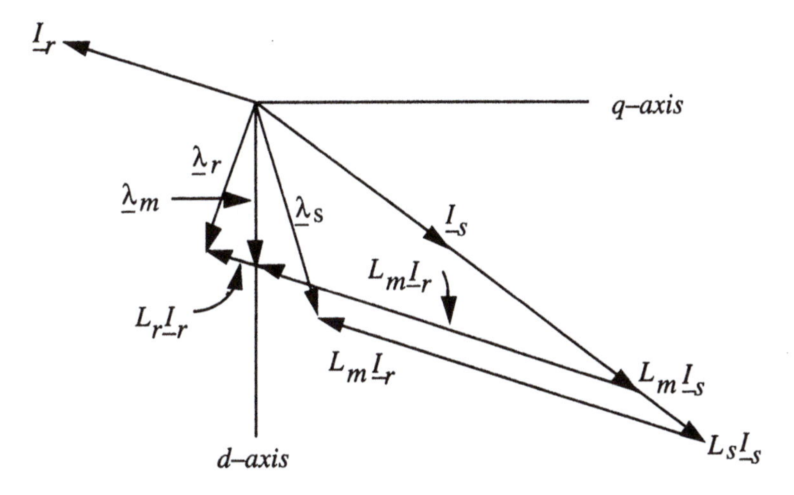

Figure 4.7 Complex vector diagram for positive sequence synchronous reference frame – air gap flux reference

$$\underline{\lambda}_s = (L_s - L_m)\,\underline{I}_s - jL_m I_m \tag{4.6–5}$$

Similarly, the rotor flux vector is

$$\underline{\lambda}_r = (L_r - L_m)\,\underline{I}_r - jL_m I_m \tag{4.6–6}$$

The voltage equations in the steady state are

$$\underline{V}_s = r_s \underline{I}_s + j\omega_e \underline{\lambda}_s \tag{4.6–7}$$

$$0 = \frac{r_r}{S}\underline{I}_r + j\omega_e \underline{\lambda}_r \tag{4.6–8}$$

Using eqns (4.6–5) and (4.6–6), the voltage equations become

$$\underline{V}_s = [r_s + j\omega_e (L_s - L_m)]\,\underline{I}_s + j\omega_e L_m (-jI_m) \tag{4.6–9}$$

$$0 = j\omega_e L_m (-jI_m) + \left[\frac{r_r}{S} + j\omega_e (L_r - L_m)\right]\underline{I}_r \tag{4.6–10}$$

which are clearly the equations of the positive sequence circuit of Figure 4.4.

If we now consider the d,q axes reference shifted such that the d–axis coincides with the rotor flux as illustrated in Figure 4.8, the new conditions for the currents are

$$I_{dr} = 0 \tag{4.6–11}$$

$$L_m I_{qs} + L_r I_{qr} = 0 \qquad (\text{since } \lambda_{qr} = 0) \tag{4.6–12}$$

The stator flux components become

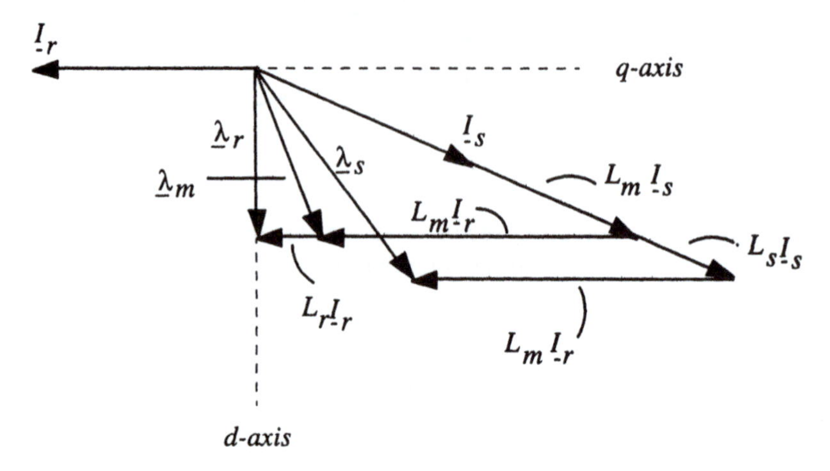

Figure 4.8 Complex vector diagram for positive sequence synchronous reference frame – rotor flux linkage reference

$$\lambda_{qs} = L_s I_{qs} + L_m I_{qr} = \left(L_s - \frac{L_m^2}{L_r}\right) I_{qs} = L_s' I_{qs} \tag{4.6–13}$$

$$\lambda_{ds} = L_s I_{ds} + L_m I_{dr} = \left(L_s - \frac{L_m^2}{L_r}\right) I_{ds} + \frac{L_m^2}{L_r} I_{ds} \tag{4.6–14}$$

which results in the stator flux vector

$$\underline{\lambda}_s = L_s' \underline{I}_s + \frac{L_m^2}{L_r}(-jI_{ds}) \tag{4.6–15}$$

The rotor flux vector is simply

$$\underline{\lambda}_r = -j\lambda_{dr} = L_m(-jI_{ds}) \tag{4.6–16}$$

and the voltage equations become

$$\underline{V}_s = (r_s + j\omega_e L_s')\,\underline{I}_s + j\omega_e \frac{L_m^2}{L_r}(-jI_{ds}) \tag{4.6–17}$$

$$0 = \frac{r_r}{S} I_{qr} + j\omega_e L_m (-jI_{ds}) \tag{4.6–18}$$

Multiplying the rotor voltage equation by L_m / L_r and using eqn (4.6–12) to eliminate I_{qr} yields

$$0 = -\left(\frac{L_m}{L_r}\right)^2 \left(\frac{r_r}{S}\right) I_{qs} + j\omega_e \left(\frac{L_m^2}{L_r}\right)(-jI_{ds}) \tag{4.6–19}$$

Equations (4.6–17) and (4.6–19) describe the equivalent circuit shown in Figure 4.9 which is the same as would be obtained from Figure 4.3 (2) using a 'turns ratio' of L_m / L_r

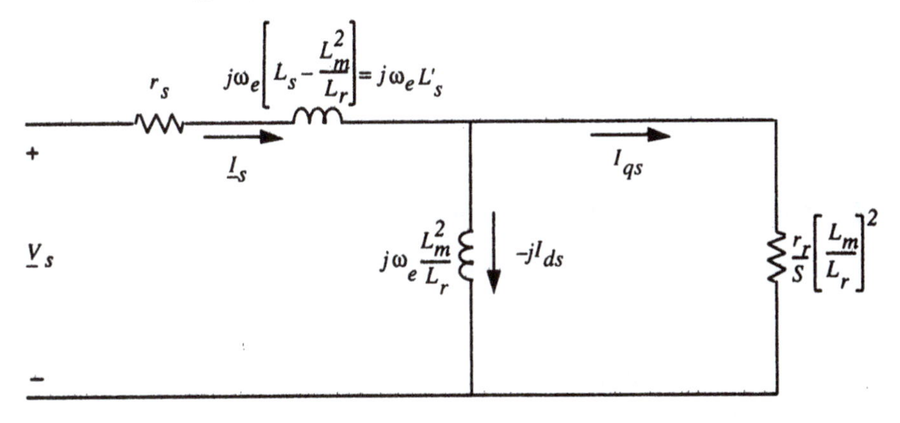

Figure 4.9 Complex vector equivalent circuit – rotor flux reference

This circuit, which places all of the 'leakage' on the stator side, is often called a rotor flux based circuit since it contains a magnetizing branch which depicts the total magnetizing current producing the rotor flux. Although the circuit can be viewed as resulting from a particular choice of 'turns ratio', the derivation from a rotor flux oriented, synchronous frame dc model is much more useful. This circuit has an important role in developing the concept of field oriented control and will be employed in later chapters for this purpose.

If a stator flux oriented synchronous frame is employed then, as one might expect, a circuit with all of the 'leakage' referred to the rotor is obtained. The circuit is the same as would be obtained from the operational circuit of Figure 4.3 (3) and is illustrated using d,q complex vector notation in Figure 4.10. The derivation is left as an exercise for the reader.

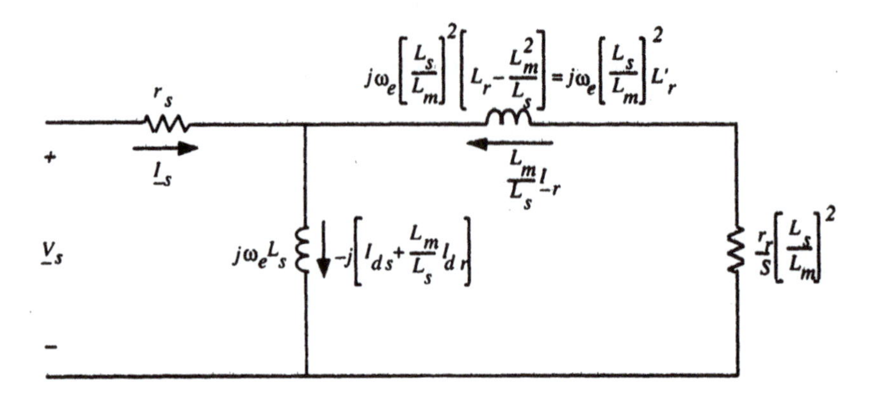

Figure 4.10 Complex vector equivalent circuit – stator flux reference

4.7 Electrical Transients at Constant Speed

There are many important practical situations where the purely electrical behavior of a machine is of concern and over the time period of interest the speed of the machine can be considered to be constant. This situation can arise because the electrical transients are simply very fast compared to the slow mechanical transients which occur when the system mechanical inertia is large, or it may be that we are only concerned with very short time periods during which the speed can be assumed constant. In any case, the constant speed constraint yields linear electrical differential equations with constant coefficients for which general solutions are known. The electrical torque, although a nonlinear function of the electrical variables, can be evaluated after the solution for the electrical quantities is completed. The following sections are restricted to the case of uniform air gap machines where the external rotor voltage is zero (cage induction machines).

4.7.1 Complex Vector Operational Impedances

The complex vector electrical equations in an arbitrary rotating reference are (eqns (4.3–1) and (4.3–2) with $b = 1$)

$$\underline{v}_{qds} = [r_s + L_s(p + j\omega)]\,\underline{i}_{qds} + L_m(p + j\omega)\,\underline{i}_{qdr} \tag{4.7–1}$$

$$0 = L_m(p + j(\omega - \omega_r))\,\underline{i}_{qds} + [r_r + L_r(p + j(\omega - \omega_r))]\,\underline{i}_{qdr} \tag{4.7–2}$$

To shorten the notation so we can focus on the basic nature of the system, the complex operational impedances defined below are introduced:

$$\begin{aligned} \underline{Z}_{ss} &= r_s + L_s(p + j\omega) & \underline{Z}_{sr} &= L_m(p + j\omega) \\ \underline{Z}_{rs} &= L_m(p + j(\omega - \omega_r)) & \underline{Z}_{rr} &= r_r + L_r(p + j(\omega - \omega_r)) \end{aligned} \tag{4.7–3}$$

The electrical equations can then be written as

$$\begin{aligned} \underline{v}_{qds} &= \underline{Z}_{ss}\underline{i}_{qds} + \underline{Z}_{sr}\underline{i}_{qdr} \\ 0 &= \underline{Z}_{rs}\underline{i}_{qds} + \underline{Z}_{rr}\underline{i}_{qdr} \end{aligned} \tag{4.7–4}$$

Note that the complex operational impedances always contain the derivative operator p but that the frequencies present depend upon the selected reference system. Thus, in a stator reference ($\omega = 0$) the impedances are

$$\begin{aligned} \underline{Z}^s_{ss} &= r_s + L_s p & \underline{Z}^s_{sr} &= L_m p \\ \underline{Z}^s_{rs} &= L_m(p - j\omega_r) & \underline{Z}^s_{rr} &= r_r + L_r(p - j\omega_r) \end{aligned} \tag{4.7–5}$$

in which the stator impedances are real and the only frequency present is the rotor speed. In a rotor reference ($\omega = \omega_r$), the situation is reversed

$$\begin{aligned} \underline{Z}^r_{ss} &= r_s + L_s(p + j\omega_r) & \underline{Z}^r_{sr} &= L_m(p + j\omega_r) \\ \underline{Z}^r_{rs} &= L_m p & \underline{Z}^r_{rr} &= r_r + L_r p \end{aligned} \tag{4.7–6}$$

but again only the rotor speed is present. Finally, in a synchronous reference where $\omega = \omega_e$ all impedances are complex

$$\begin{aligned} \underline{Z}^e_{ss} &= r_s + L_s(p + j\omega_e) & \underline{Z}^e_{sr} &= L_m(p + j\omega_e) \\ \underline{Z}^e_{rs} &= L_m(p + j(\omega_e - \omega_r)) & \underline{Z}^e_{rr} &= r_r + L_r(p + j(\omega_e - \omega_r)) \end{aligned} \tag{4.7–7}$$

and the stator excitation frequency occurs in the stator impedances and the rotor electrical frequency (slip frequency) occurs in the rotor impedances

4.7.2 Example – Sudden Application of Stator Current Source Excitation

As an example of a constant speed switching transient, consider the sudden application of controlled positive sequence stator currents to an induction machine turning at a speed ω_r. The excitation will be taken as

$$\underline{i}^s_{qds} = \begin{cases} 0 & t < 0 \\ \tilde{I}_{sp} e^{j\omega_e t} & t \geq 0 \end{cases} \tag{4.7–8}$$

Since the only unknown current is $\underline{i}_{qdr}$ only the rotor equation

$$0 = \underline{Z}_{rs} \underline{i}_{qds} + \underline{Z}_{rr} \underline{i}_{qdr} \tag{4.7–9}$$

must be solved. To illustrate the differences between solutions in the various reference frames, the analysis will be carried out in a general frame moving at a speed ω. In this general frame, the stator excitation becomes (all the following analysis is for $t \geq 0$)

$$\underline{i}_{qds} = \tilde{I}_{sp} e^{j(\omega_e - \omega)t} \tag{4.7–10}$$

and the rotor differential equation becomes

$$\underline{Z}_{rr} \underline{i}_{qdr} = -\underline{Z}_{rs} \tilde{I}_{sp} e^{j(\omega_e - \omega)t} \tag{4.7–11}$$

where

$$\underline{Z}_{rr} = r_r + L_r [p + j(\omega - \omega_r)] \tag{4.7–12}$$

The complete solution to this equation is

$$\underline{i}_{qdr} = \underline{C} e^{\underline{\lambda} t} + \tilde{I}_{rp} e^{j(\omega_e - \omega)t} \tag{4.7–13}$$

wherein $\underline{C}$ is a complex constant depending on initial conditions, $\underline{\lambda}$ is the complex eigenvalue of the (first order complex) system and $\tilde{I}_{rp}$ is the steady state phasor solution for the rotor current. The steady state solution can be obtained from the equivalent circuits of the previous section or directly from eqn (4.7–11) by setting $p = j(\omega_e - \omega)$

$$\tilde{I}_{rp} = \frac{-j(\omega_e - \omega_r) L_m}{r_r + j(\omega_e - \omega_r) L_r} \tilde{I}_{sp} \tag{4.7–14}$$

The eigenvalue $\underline{\lambda}$ is formally obtained from the characteristic equation obtained by substituting $\underline{C}e^{\underline{\lambda}t}$ into the differential equation with the source set to zero (the reduced equation) which yields

$$r_r\underline{C}e^{\underline{\lambda}t} + L_r(\underline{\lambda} + j(\omega - \omega_r))\underline{C}e^{\underline{\lambda}t} = 0 \tag{4.7–15}$$

or

$$r_r + L_r(\underline{\lambda} + j(\omega - \omega_r)) = 0 \tag{4.7–16}$$

Note that this formalism is equivalent to setting $p = \underline{\lambda}$ in the differential equation with the source removed. Equation (4.7–16) yields the eigenvalue as

$$\underline{\lambda} = -\frac{r_r}{L_r} - j(\omega - \omega_r) = -\frac{1}{\tau_r} - j(\omega - \omega_r) \tag{4.7–17}$$

so that the complete solution can now be written as

$$\underline{i}_{qdr} = \underline{C}e^{-t/\tau_r}e^{-j(\omega - \omega_r)t} + \tilde{I}_{rp}e^{j(\omega_e - \omega)t} \tag{4.7–18}$$

To evaluate the constant $\underline{C}$, the initial value of $\underline{i}_{qdr}$ must be known. This initial condition can be found from the knowledge that the rotor flux linkage must be continuous and hence

$$\underline{\lambda}_{qdr}(0) = L_m\tilde{I}_{sp} + L_r\underline{i}_{qdr}(0) \tag{4.7–19}$$

or

$$\underline{i}_{qdr}(0) = \frac{\underline{\lambda}_{qdr}(0)}{L_r} - \frac{L_m}{L_r}\tilde{I}_{sp} \tag{4.7–20}$$

Combining this result with eqn (4.7–18) evaluated at $t = 0$ yields

$$\begin{aligned} \underline{C} &= \frac{\underline{\lambda}_{qdr}(0)}{L_r} - \frac{L_m}{L_r}\tilde{I}_{sp} - \tilde{I}_{rp} \\ &= \frac{\underline{\lambda}_{qdr}(0)}{L_r} - \frac{1}{L_r}(L_m\tilde{I}_{sp} + L_r\tilde{I}_{rp}) \end{aligned} \tag{4.7–21}$$

This expression can be simplified by observing that the final steady state value of the rotor flux linkage is

$$\underline{\lambda}_{qdr} = (L_m\tilde{I}_{sp} + L_r\tilde{I}_{rp})\, e^{j(\omega_e - \omega)t} \tag{4.7–22}$$

and defining

$$\underline{\lambda}_{qdr}(\infty) = L_m\tilde{I}_{sp} + L_r\tilde{I}_{rp} \tag{4.7–23}$$

allows writing the final solution as

$$\underline{i}_{qdr} = \frac{1}{L_r}(\underline{\lambda}_{qdr}(0) - \underline{\lambda}_{qdr}(\infty))\, e^{-\frac{t}{\tau_r}} e^{-j(\omega - \omega_r)t} + \tilde{I}_{rp} e^{j(\omega_e - \omega)t} \tag{4.7–24}$$

The torque can be evaluated from the basic torque equation

$$T_e = \frac{3}{2}\frac{P}{2} L_m Im[\underline{i}_{qds}\underline{i}^{\dagger}_{qdr}] \tag{4.7–25}$$

and is easily shown to consist of the expected steady state component, see eqn (4.4–31)

$$T_{eo} = \frac{3}{2}\frac{P}{2} L_m Im[\tilde{I}_{sp}\tilde{I}^{\dagger}_{rp}] = \frac{3}{2}\frac{P}{2}\frac{1}{\omega_e}|\tilde{I}_{rp}|^2 \frac{r_r}{S} \tag{4.7–26}$$

and a transient component given by

$$T_{et} = \frac{3}{2}\frac{P}{2}\frac{L_m}{L_r} Im[\tilde{I}_{sp}(\underline{\lambda}^{\dagger}_{qdr}(0) - \underline{\lambda}^{\dagger}_{qdr}(\infty))\, e^{-t/\tau_r} e^{j(\omega_e - \omega_r)t}] \tag{4.7–27}$$

which varies at slip frequency and decays with the rotor open circuit time constant τ_r. Note that the reference frame speed does not occur in the torque expression; the torque is not a transformed quantity and must clearly be the same in any reference frame.

Returning to the expression for the rotor current, it is clear that the frequencies present in the current do depend on the reference frame. There are two frequencies in eqn (4.7–24); one from the applied excitation, $\omega_e - \omega$, and one from the eigenvalue, $\omega - \omega_r$. Both depend on the choice of reference frame. The "natural" or real frame for the rotor current is clearly a rotor frame (i.e. $\omega = \omega_r$). Viewed from the rotor, the excitation frequency is slip frequency, $\omega_e - \omega_r$, and the eigenvalue frequency is zero. This implies that the transient component of the rotor current is simply a decaying dc component which is 'trapped' at the instant t = 0 when the excitation is applied. Note that the characteristic equation given in eqn (4.7–16) yields the eigenfrequency as zero in a rotor reference as it should.

Viewed in a stator frame ($\omega = 0$) the excitation frequency is ω_e and the eigenfrequency is the rotor frequency ω_r. The stator voltage will contain a voltage component at this frequency during the transient period. In a synchronous frame, the excitation frequency becomes dc and the eigenfrequency is slip frequency, $\omega_e - \omega_r$. In all reference frames the amplitudes of the components (coefficients of the complex exponential time functions) are the same. Graphically, the lengths of the rotating vectors are preserved, only the frequency or speed of rotation changes as the reference frame is changed.

The solution for the current is of interest also from the point of view that it is possible to execute the switching operation such that the transient component is zero. The necessary condition is that the initial rotor flux linkage be the same as the final value (both magnitude and phase). This is, in fact, the scheme employed in a field oriented controller as will be shown in later chapters. Note that with the transient component zero, the torque responds instantaneously to the stator excitation current. This is the major goal of field oriented control – to obtain a torque response which follows a torque command instantaneously. This simple example actually illustrates a basic requirement for this type of response: that the rotor flux linkage must remain constant in amplitude and phase.

As a final step in the solution, the d and q real variable expressions for the rotor current will be obtained. From the general form given in eqn (4.7–24), the real variable expressions are:

$$\begin{aligned} i_{qr} = {} & \frac{1}{L_r}\left|\underline{\lambda}_{qdr}(0)\right| e^{-t/\tau_r}\cos\left[(\omega-\omega_r)t-\phi_{ro}\right] \\ & -\frac{1}{L_r}\left|\underline{\lambda}_{qdr}(\infty)\right| e^{-t/\tau_r}\cos\left[(\omega-\omega_r)t-\phi_{rf}\right] \\ & -\left|\tilde{I}_{rp}\right|\cos\left[(\omega_e-\omega)t+\phi_{rp}\right] \end{aligned} \tag{4.7–28}$$

$$\begin{aligned} i_{dr} = {} & \frac{1}{L_r}\left|\underline{\lambda}_{qdr}(0)\right| e^{-t/\tau_r}\sin\left[(\omega-\omega_r)t-\phi_{ro}\right] \\ & -\frac{1}{L_r}\left|\underline{\lambda}_{qdr}(\infty)\right| e^{-t/\tau_r}\sin\left[(\omega-\omega_r)t-\phi_{rf}\right] \\ & -\left|\tilde{I}_{rp}\right|\sin\left[(\omega_e-\omega)t+\phi_{rp}\right] \end{aligned} \tag{4.7–29}$$

where ϕ_{ro}, ϕ_{rf} and ϕ_{rp} are the phase angles of the complex quantities $\underline{\lambda}_{qdr}(0)$, $\underline{\lambda}_{qdr}(\infty)$ and I_{rp} respectively.

In the special case where $\underline{\lambda}_{qdr}(0)$ is zero, $\underline{\lambda}_{qdr}(\infty)$ can be expressed in terms of machine parameters by using eqn (4.7–23) and (4.7–14), and the complex current for this case can be written as

$$\underline{i}_{qdr} = -\frac{\tilde{I}_{sp} L_m}{r_r + j(\omega_e - \omega_r) L_r}\left[\frac{r_r}{L_r}\varepsilon^{-t/\tau_r} e^{-j(\omega - \omega_r)t} + j(\omega_e - \omega_r) e^{j(\omega_e - \omega)t}\right] \tag{4.7–30}$$

which involves only the excitation current, rotor speed and machine parameters. If the rotor impedance is defined as

$$Z_r e^{j\phi_r} = r_r + j(\omega_e - \omega_r) L_r \tag{4.7–31}$$

this expression becomes

$$\underline{i}_{qdr} = -\frac{\tilde{I}_{sp} L_m}{Z_r}\left[\frac{r_r}{L_r} e^{-t/\tau_r} e^{-j((\omega - \omega_r)t + \phi_r)} + j(\omega_e - \omega_r) e^{j((\omega_e - \omega)t - \phi_r)}\right] \tag{4.7–32}$$

from which the d and q components can be written as

$$i_{qr} = -\frac{|\tilde{I}_{sp}| L_m}{Z_r}\left[\frac{r_r}{L_r} e^{-t/\tau_r} \cos((\omega - \omega_r)t + \phi_r + \phi_p) \right. \tag{4.7–33}$$
$$\left. - (\omega_e - \omega_r) \sin((\omega_e - \omega)t - \phi_r + \phi_p)\right]$$

$$i_{dr} = -\frac{|\tilde{I}_{sp}| L_m}{Z_r}\left[\frac{r_r}{L_r} e^{-t/\tau_r} \sin((\omega - \omega_r)t + \phi_r + \phi_p) \right.$$
$$\left. - (\omega_e - \omega_r) \cos((\omega_e - \omega)t - \phi_r + \phi_p)\right] \tag{4.7–34}$$

where ϕ_p is the angle of $\tilde{I}_{sp}$.

The real variable results illustrate that in real variable analysis the single complex eigenvalue which occurs in complex vector analysis is augmented by its conjugate value to yield a complex conjugate pair of eigenvalues. This complex conjugate pair produces the damped sine and cosine terms which occur in the real variable results. A mathematical interpretation of this simplification is that the complex vector model is only one half of the total model and the complete complex vector model also involves the conjugate complex vector $(f_q + jf_d)$. In this case there are four complex equations just as there are four real equations. Because of the machine symmetry the two sets of two equations are conjugates of each other and only one set need be retained; the other set can be obtained by conjugation.

4.7.3 General Solution for Switching Transients

In general, the solution for switching transients involves the simultaneous solution of the stator and rotor differential equations. The example in the previous section only required the rotor equation because the stator current was a known function. The basic approach to the solution is the same, however, and consists of finding the steady state and transient parts of the solution and evaluating the unknown coefficients from initial conditions. The steady state solution is again available from the steady state equivalent circuits or by direct solution of the appropriate steady state form of the machine equations.

The system characteristic equation is obtained from the machine equations with the source removed

$$0 = \underline{Z}_{ss}\underline{i}_{qds} + \underline{Z}_{sr}\underline{i}_{qdr} \tag{4.7–35}$$

$$0 = \underline{Z}_{rs}\underline{i}_{qds} + \underline{Z}_{rr}\underline{i}_{qdr} \tag{4.7–36}$$

again by substituting for $\underline{i}_{qds}$ and $\underline{i}_{qdr}$ an arbitrary complex exponential solution. The result of this formalism is the characteristic equation

$$\underline{Z}_{ss}\underline{Z}_{rr} - \underline{Z}_{sr}\underline{Z}_{rs} = 0 \tag{4.7–37}$$

in which, as we have shown, the operator p is replaced by the unknown eigenvalues symbolized by $\underline{\lambda}$. Once the eigenvalues are known, the solution proceeds as in the example in Section 4.7.2.

4.8 The Constant Speed Eigenvalues

The evaluation of the roots of eqn (4.7–37) can be carried out in a general reference frame rotating at a speed ω to obtain the general reference frame values of the system eigenvalues [3,4]. Carrying out the multiplication indicated in eqn (4.7–37), with $p = \underline{\lambda}$, the following quadratic equation with complex coefficients is obtained

$$\sigma L_s L_r \underline{\lambda}^2 + [r_s L_r + r_r L_s + j\sigma L_s L_r (2\omega - \omega_r)]\underline{\lambda} \\ + r_s r_r - \sigma L_s L_r \omega(\omega - \omega_r) + j[r_s L_r(\omega - \omega_r) + r_r L_s \omega] = 0 \tag{4.8–1}$$

where the *coupling factor* σ is

$$\sigma = 1 - \frac{L_m^2}{L_s L_r} \tag{4.8–2}$$

Solving for the two complex roots of this quadratic yields, after some algebraic manipulation, the roots can be written in either of the following two equivalent forms,

$$\underline{\lambda}_1, \underline{\lambda}_2 = -\frac{1}{2\sigma\tau_r}(1+\alpha) + j\left(\frac{\omega_r}{2} - \omega\right) \\ \pm \frac{1}{2\sigma\tau_r}\sqrt{(1+\alpha)^2 - 4\sigma\alpha - (\omega_r\sigma\tau_r)^2 + j2(\alpha - 1)\omega_r\sigma\tau_r} \tag{4.8–3}$$

or

$$\underline{\lambda}_1, \underline{\lambda}_2 = -\frac{1}{2}\left(\frac{1}{\tau'_s} + \frac{1}{\tau'_r}\right) + j\left(\frac{\omega_r}{2} - \omega\right) \\ \pm \frac{1}{2}\sqrt{\left(\frac{1}{\tau'_s} + \frac{1}{\tau'_r}\right)^2 - \frac{4\sigma}{\tau'_s\tau'_r} - \omega_r^2 + j2\omega_r\left(\frac{1}{\tau'_s} - \frac{1}{\tau'_r}\right)} \tag{4.8–4}$$

where

$$\tau_r = \frac{L_r}{r_r} \quad \text{(rotor open circuit time constant)} \tag{4.8–5}$$

$$\tau_s = \frac{L_s}{r_s} \quad \text{(stator open circuit time constant)} \tag{4.8–6}$$

$$\tau'_r = \sigma\tau_r \quad \text{(rotor short circuit time constant)} \tag{4.8–7}$$

$$\tau'_s = \sigma\tau_s \quad \text{(stator short circuit time constant)} \tag{4.8–8}$$

$$\alpha = \frac{\tau_r}{\tau_s} = \frac{r_s L_r}{r_r L_s} \tag{4.8–9}$$

The other two eigenvalues of the complete real variable system are the conjugates of the two given by eqn (4.8–3) or (4.8–4).

As can be expected on the basis of transformation theory, the speed of the reference system enters as a simple linear shift in frequency. Thus, if the eigenvalues are known in a stationary reference, they can be transformed to any other reference system by simply subtracting ω from the frequency of the eigenvalues (i.e., the same as for excitation frequency). A plot of the migration of the eigenvalues in the stationary reference frame ($\omega = 0$) as a function of rotor speed is shown in Figure 4.11. The parameters of the machine used for the study are for a 5 horsepower four pole machine and are given as r_s = 0.44Ω, r_r = 0.708 Ω, $L_s = L_r$ = 0.069 mH, and L_m= 0.0668 mH.

It is interesting to observe that in a stationary reference the eigenvalues are always complex except for zero speed where both are real. The induction machine, therefore, always has complex roots (damped sinusoidal oscillations in the transient response) except at zero speed. There is, in fact, one condition for a machine with $\alpha = 1$ for which the eigenvalues occur as a double complex pair at the rotor speed which makes the square root terms in eqn (4.8–3) equal to zero. In a reference system rotating at one half this speed the machine is also characterized by real roots (four equal real roots in fact) but this is a single special exception to the general rule of complex conjugate roots. Since all of the motor parameters are involved in determination of both of the pairs of eigenvalues, it can be recalled that in the stationary reference frame, a speed voltage proportional to rotor speed appears in the rotor circuit, see Figure 2.8 or Figure 4.1. Hence, one expects that the pole which migrates as a function of rotor speed corresponds to the effects of the rotor circuit. Consequently, these poles are termed the *rotor poles*. On the other hand, in the stationary frame, speed voltages do not appear in the stator circuit so that the poles remain near the real axis and form what are termed the *stator poles*.

When the eigenvalues are calculated for the synchronous reference frame the plot of Figure 4.12 results. In this case we have set $\omega = 377$ since this corresponds to the usual sixty hertz operation. Note that since the reference frame ω appears only as a single linear term in eqn (4.8–3), the roots have simply undergone a spectral shift. Upon analyzing our machine in the synchronous frame we would now expect to see damped sinusoidal oscillations of the cur-

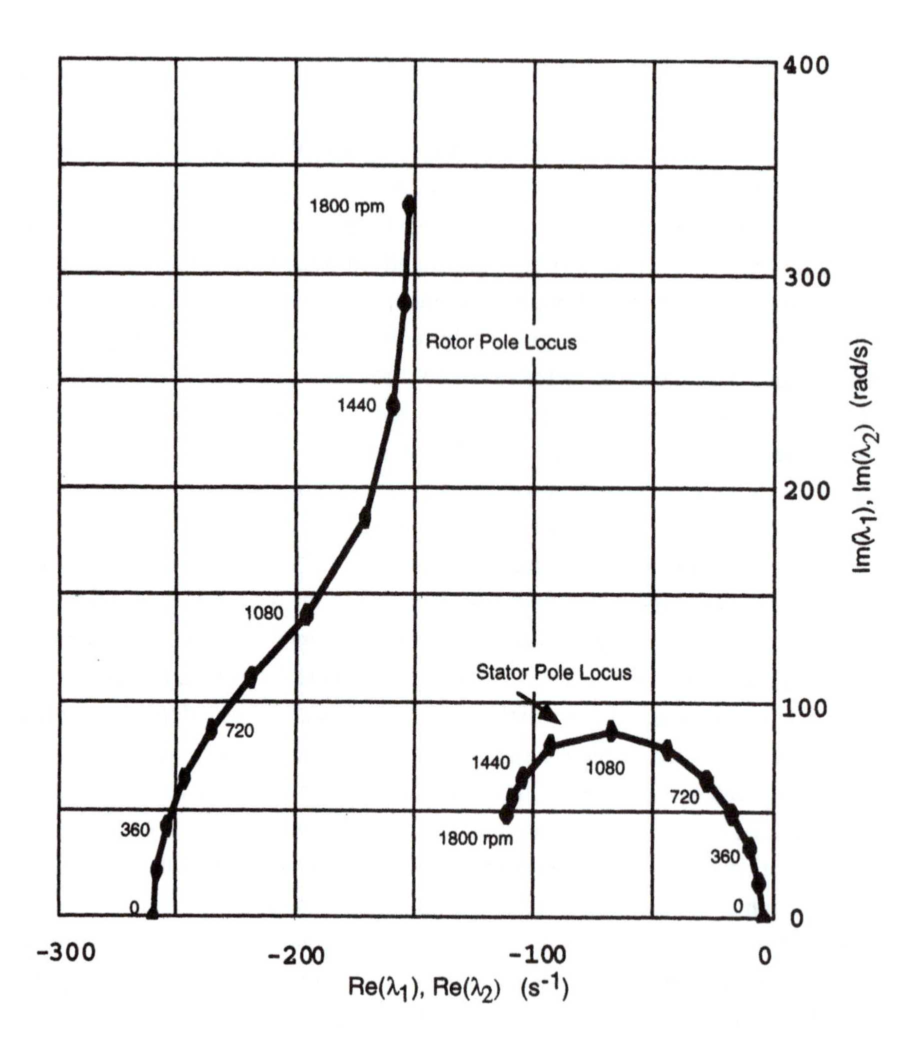

Figure 4.11 Eigenvalue vs. speed loci in the stationary reference frame

rents at sixty hertz when the rotor speed is zero. This tendency can be observed in the transient solution of Figure 2.16. However, it is important to caution that the eigenvalue approach assumes constant speed operation whereas Figure 2.16 clearly pertains to variable speed. Nonetheless, these observations remain roughly true when the acceleration of the rotor is not too rapid.

In Figure 4.13 the eigenvalues are plotted in the rotor reference frame. In this case the stator poles undergo a wide variation in frequency while the fre-

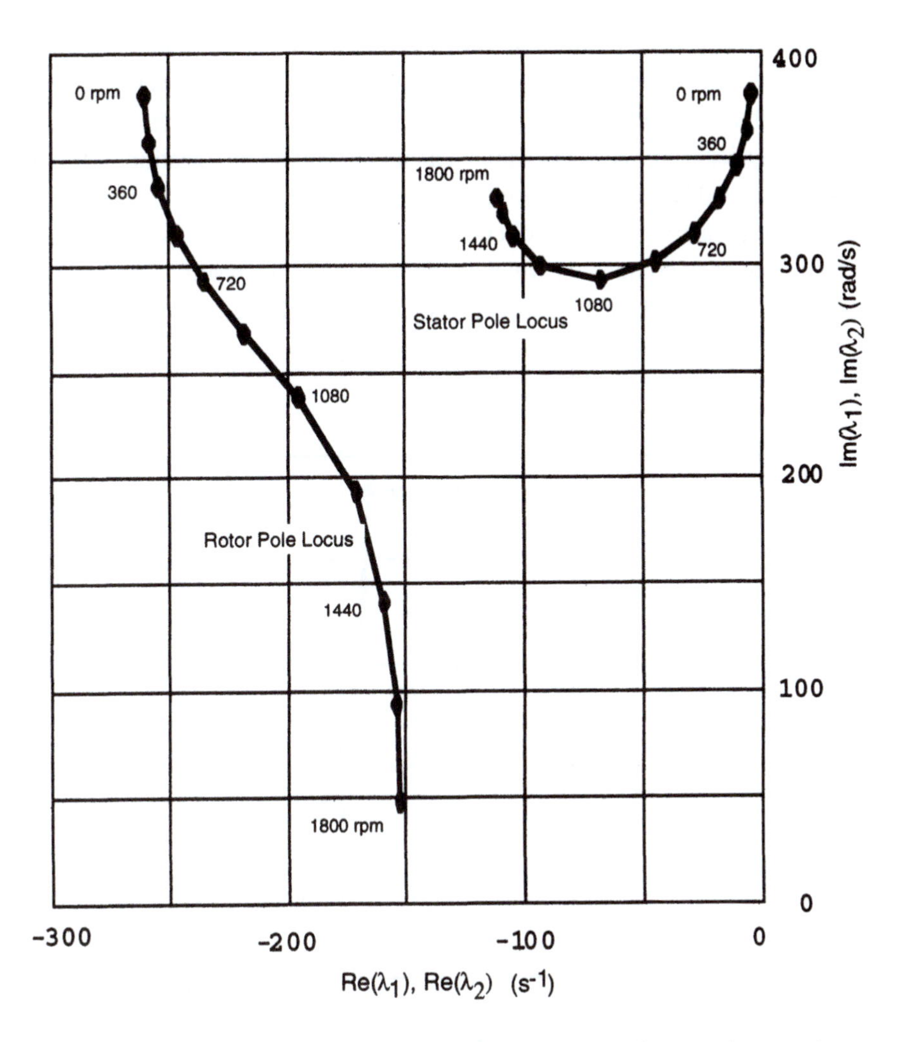

Figure 4.12 Eigenvalue vs. speed loci in the synchronously rotating reference frame

quency of the rotor poles remain roughly constant. Examination of Figure 4.11 and Figure 4.13 are both needed to reveal the physical nature of the stator and rotor poles. In general, the effects of stator and rotor poles both show up in the stator current transients. However, a complete eigenvector analysis would reveal that the stator poles of Figure 4.11 show the *predominant* nature of the physical stator currents while effect of the rotor poles on the stator current is relatively weak. Figure 4.11 also shows the nature of the *equivalent* rotor currents which depend primarily on the rotor poles. However, it is important to note that these rotor currents are not the physical currents which are, in fact, rotating with the rotor. On the other hand, Figure 4.13 represents the pole positions in the rotor reference frame and therefore shows the predominant nature of the physical rotor currents which depend heavily on the location of the rotor poles and weakly on the location of the equivalent stator poles. In each case the pole dominating the physical circuit is located near the real axis and has only a small frequency component which is fundamentally the result of the rotor rotation (or, equivalently stator rotation with respect to the rotor).

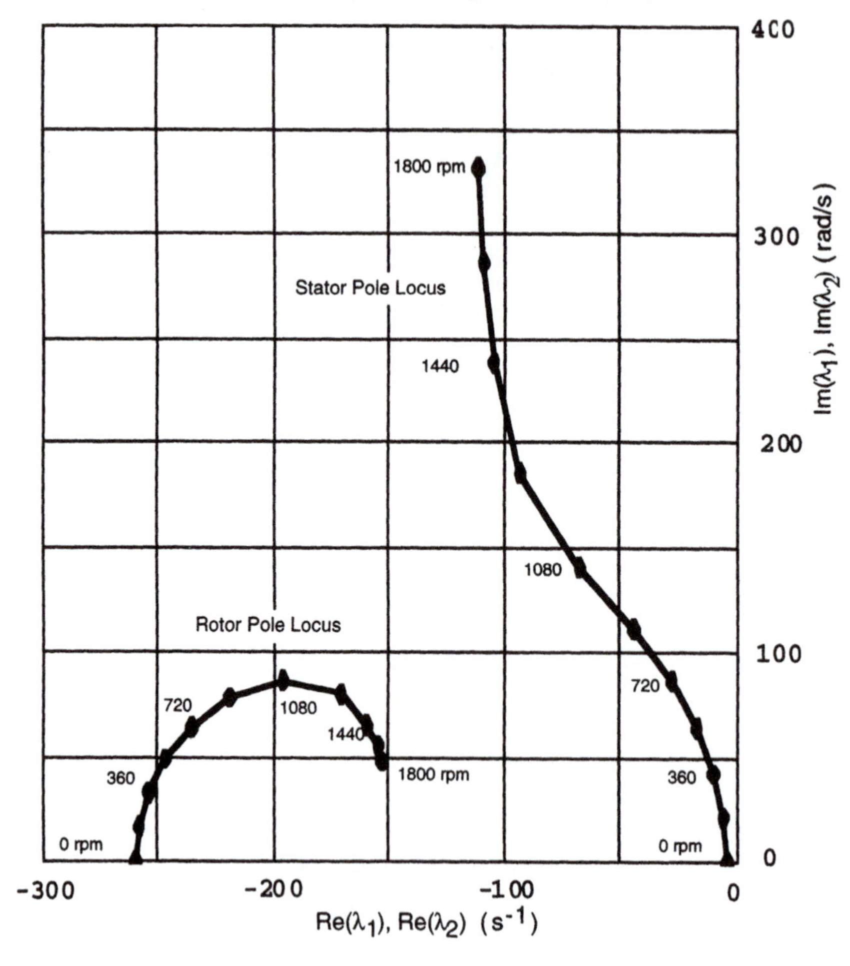

Figure 4.13 Eigenvalue vs. speed loci in the rotor reference frame

Two important properties of eqn (4.8–3) can be identified by examining its form [6]. Specifically if the two eigenvectors $\underline{\lambda}_1$ and $\underline{\lambda}_2$ are added, one obtains:

$$Re\,[\underline{\lambda}_1 + \underline{\lambda}_2] = -\frac{1}{\sigma\tau_r}(1+\alpha) = -\left(\frac{1}{\tau'_s} + \frac{1}{\tau'_r}\right) \qquad (4.8\text{–}10)$$

Therefore the sum of the two eigenvalues add to a constant value which is independent of rotor speed (or reference frame). Since the stator and rotor transient time constants can be readily determined from machine parameters, this fact can be used either as a check on computations or as a means of calculating the second eigenvalue, given the first. A sketch of the real parts of the eigenvalues as a function of speed are plotted in Figure 4.14.

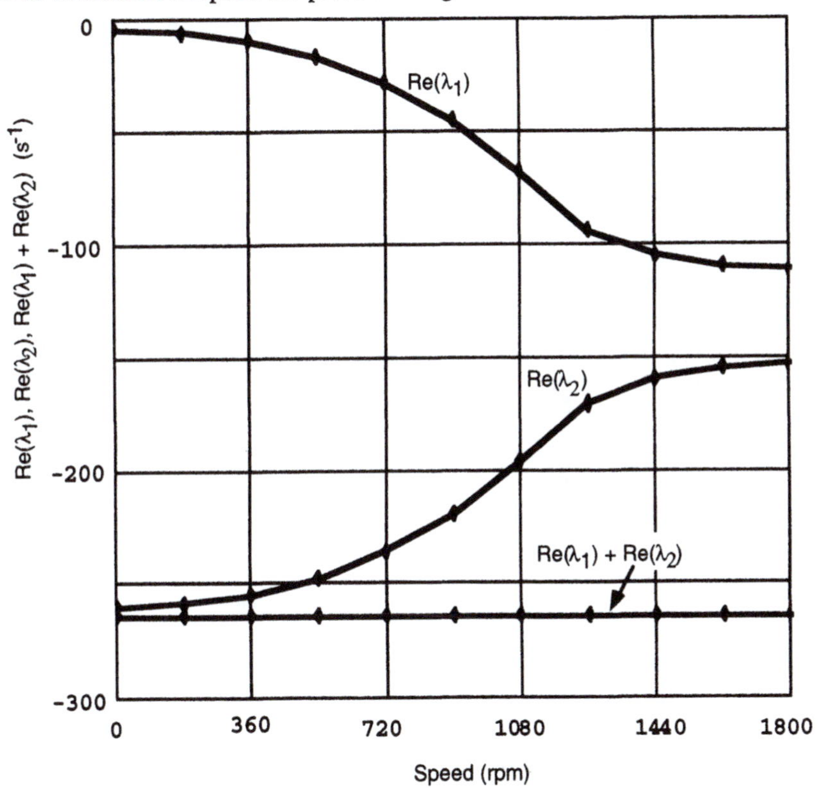

Figure 4.14 Real parts of $\underline{\lambda}_1$ and $\underline{\lambda}_2$ and their sum vs. rotor speed

For the imaginary part one obtains:

$$Im\,[\underline{\lambda}_1 + \underline{\lambda}_2] = \omega_r - 2\omega \qquad (4.8\text{–}11)$$

Given the imaginary part of one of the eigenvalues, the imaginary part of the second can also be obtained. Note that when the reference frame speed is set to zero or to rotor speed the result is $+\omega_r$ and $-\omega_r$, respectively, indicating the symmetrical nature of the reference frame on the location of the stator and rotor poles. The result is essentially the same since the sum of the conjugates would simply reverse the signs of the answer. A plot of the imaginary parts of the eigenvalues and their sum as speed changes is shown in Figure 4.15.

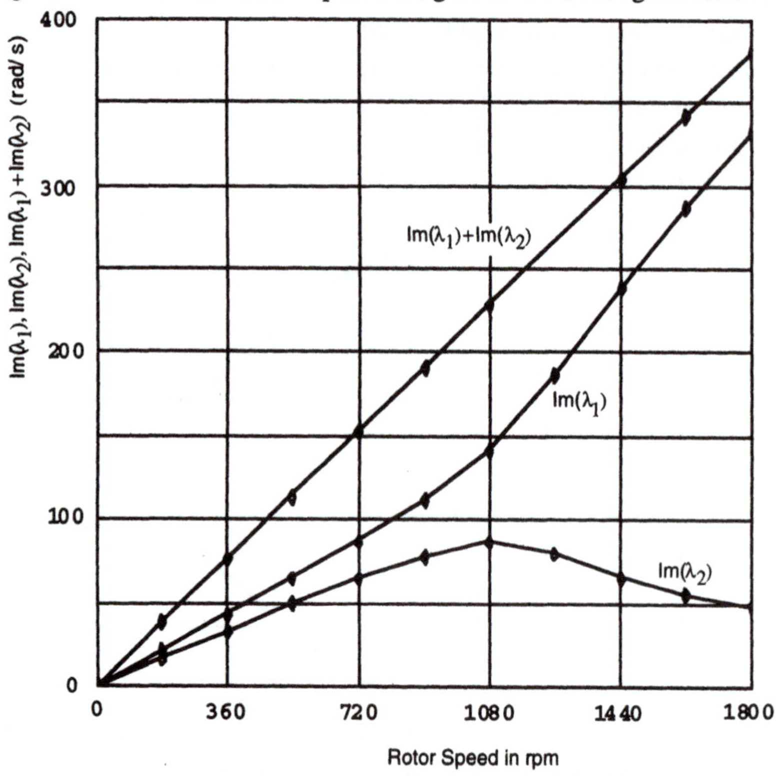

Figure 4.15 Imaginary parts of λ_1 and λ_2 and their sum vs. rotor speed in the rotor reference frame

There are several special or limiting cases for the eigenvalues which are often useful as approximations in theoretical work or as illustrations of typical behavior. These include:

1) $\alpha = 1$ and $\omega_r \sigma \tau_r \gg 1$

$$\underline{\lambda}_1 = -\frac{1}{\tau'_r} - j\omega \qquad \underline{\lambda}_2 = -\frac{1}{\tau'_r} + j(\omega_r - \omega) \tag{4.8–12}$$

Since most conventional machines have α close to one and the inequality is satisfied at all but very low speeds, this is often a useful approximation for real machines. Note that α is a fundamental parameter in the sense that it is unaffected by the turns ratio transformations discussed in Section 4.3. An α close to one implies nearly equal rotor and stator I^2R losses.

2) $\qquad r_s = 0$

$$\underline{\lambda}_1 = 0 - j\omega \qquad \underline{\lambda}_2 = -\frac{1}{\tau'_r} + j(\omega_r - \omega) \tag{4.8–13}$$

This limiting case implies no stator damping and as a result one of the eigenvalues (i.e. the stator pole) has a zero real part. It can often be used as an approximation for cases where only a short period of time is of interest. It also has value as a limiting case in stability analysis.

3) $\qquad r_s = 0, \ r_r = 0$

$$\underline{\lambda}_1 = -j\omega, \ \underline{\lambda}_2 = j(\omega_r - \omega) \tag{4.8–14}$$

This is another limiting case where all damping is neglected. It is often used as an approximation to the transient behavior for a very short period following a disturbance. This approximation is directly related to the constant rotor flux linkage transient model discussed in the next section.

4) $\qquad \sigma = 0$

$$\underline{\lambda}_1 = 0, \quad \underline{\lambda}_2 = -\frac{\alpha}{(1+\alpha)\tau_r} + j\left(\frac{\alpha\omega_r}{1+\alpha} - \omega\right) \tag{4.8–15}$$

The limiting case of $\sigma = 0$ implies perfect rotor to stator coupling or $L_s L_r = L_m^2$ and results in one eigenvalue at the origin. It is essentially never a good approximation to a real machine but does serve as a useful theoretical result in stability analysis.

5) $\qquad \sigma \ll 1 \qquad 1 + \alpha \gg \omega_r \sigma \tau_r$

$$\lambda_1 = 0 + j\left(\frac{\omega_r}{2} - \omega\right), \tag{4.8–16}$$

$$\lambda_2 = -\frac{1+\alpha}{\sigma\tau_r} + j\left(\frac{\omega_r}{2} - \omega\right) = -\left(\frac{1}{\tau_r'} + \frac{1}{\tau_s'}\right) + j\left(\frac{\omega_r}{2} - \omega\right) \tag{4.8–17}$$

This is sometimes used as a low speed approximation for the eigenvalues in real machines. Since $\sigma << 1$ is essentially always satisfied except in very small machines, this approximation is nearly always applicable below some particular speed.

Table 4.3 indicates typical ranges for the machine parameters involved in determining the eigenvalues. The largest range is for the open circuit time con-

Table 4.3 Ranges of Machine Parameters for Normal (NEMA –B) 4–Pole Machines

Parameter	Range	Comments
σ	0.2 – 0.04	Smaller in large or high efficiency machines
α	0.5 – 2.0	Generally of the order of 1.5 with no significant size influence
$\sigma\tau_r = \tau_r'$	4 – 200 ms	Larger in large or high efficiency machines
τ_r	0.03 – 5.0 s	Larger in large or high efficiency machines

stant τ_r which varies over a 100:1 range from small to large machines. The least variable parameter is the time constant ratio α which perhaps varies over a 3:1 range with no relation to machine size. Table 4.4 also gives several sets of typical values of machine parameters for different size machines. The corresponding eigenvalues at $\omega_r = 377$ rad/sec are also given in Table 4.4. Considerable variations from these typical values are possible for any specific machine;

the values given are only to indicate order of magnitude. The trends with size are, however, well established and have a sound theoretical basis.

Table 4.4 Typical Values of Machine Parameters and Eigenvalues

Power (hp)	σ	α	$\sigma\tau_r$ (ms)	τ_r (s)	Eigenvalues at ω_r=377 (rad./sec.)
1	0.15	1.5	5	0.035	–105+j121 –370+j256
10	0.075	1.5	15	0.200	–65+j17 –101+j360
100	0.05	1.5	50	1.0	–20+j1.5 –30+j375
1000	0.04	1.5	150	4.0	–6.3+j1.5 –9.5+j377

4.8.1 Example – Sudden Short Circuit of Unloaded Induction Machine

As an illustration of a transient solution using the full set of eigenvalues, consider a sudden short circuit of an induction machine initially operating in the steady state at no load. Assuming the system inertia maintains the speed constant, the transient stator current following the short circuit will be

$$\underline{i}^s_{qds} = \underline{C}_1 e^{\underline{\lambda}_1 t} + \underline{C}_2 e^{\underline{\lambda}_2 t} \qquad (4.8\text{–}18)$$

where $\underline{\lambda}_1$ and $\underline{\lambda}_2$ are the eigenvalues corresponding to the rotor speed. The constants $\underline{C}_1$ and $\underline{C}_2$ are found from the initial current and initial current derivative

$$\underline{i}^s_{qds}(0) = \underline{C}_1 + \underline{C}_2 \qquad (4.8\text{–}19)$$

$$p\underline{i}^s_{qds}(0) = \underline{\lambda}_1 \underline{C}_1 + \underline{\lambda}_2 \underline{C}_2 \qquad (4.8\text{–}20)$$

from which

$$\underline{C}_1 = \frac{\underline{\lambda}_2 \underline{i}^s_{qds}(0) - p\underline{i}^s_{qds}(0)}{\underline{\lambda}_2 - \underline{\lambda}_1} \qquad (4.8\text{–}21)$$

$$\underline{C}_2 = \frac{p\underline{i}^s_{qds}(0) - \underline{\lambda}_1 \underline{i}^s_{qds}(0)}{\underline{\lambda}_2 - \underline{\lambda}_1} \tag{4.8–22}$$

Prior to the short circuit, the motor currents will be

$$\underline{i}^s_{qds} = \underline{I}^s_{qds} e^{j\omega_e t} \tag{4.8–23}$$

$$\underline{i}^s_{qdr} = 0 \tag{4.8–24}$$

so the initial values are

$$\underline{i}^s_{qds}(0) = \underline{I}^s_{qds} \tag{4.8–25}$$

$$\underline{i}^s_{qdr}(0) = 0 \tag{4.8–26}$$

To maintain constant flux linkage at $t = 0$ requires that these currents be continuous at the switching instant. The derivative of the stator current at $t = 0$, $p\underline{i}^s_{qds}(0)$ can be found from the machine equations at $t = 0$

$$0 = (r_s + L_s p)\,\underline{i}^s_{qds}(0) + L_m p\underline{i}^s_{qdr}(0) \tag{4.8–27}$$

$$0 = L_m(p - j\omega_r)\,\underline{i}^s_{qds}(0) + L_r p\underline{i}^s_{qdr}(0) \tag{4.8–28}$$

or

$$L_s p\underline{i}^s_{qds}(0) + L_m p\underline{i}^s_{qdr}(0) = -r_s \underline{I}^s_{qds} \tag{4.8–29}$$

$$L_m p\underline{i}^s_{qds}(0) + L_r p\underline{i}^s_{qdr}(0) = j\omega_r L_m \underline{I}^s_{qds} \tag{4.8–30}$$

from which

$$\begin{aligned} p\underline{i}^s_{qds}(0) &= \frac{-L_r r_s - j\omega_r L_m^2}{\sigma L_s L_r}\underline{I}^s_{qds} \\ &= -\left(\frac{1}{\sigma\tau_s} + j\frac{1-\sigma}{\sigma}\omega_r\right)\underline{I}^s_{qds} \end{aligned} \tag{4.8–31}$$

Substituting these results, eqns (4.8–25) and (4.8–31), into the expressions for the constants $\underline{C}_1$ and $\underline{C}_2$ yields

$$\underline{C}_1 = \frac{\underline{\lambda}_2 + \dfrac{1}{\sigma\tau_s} + j\dfrac{1-\sigma}{\sigma}\omega_r}{\underline{\lambda}_2 - \underline{\lambda}_1} \underline{I}^s_{qds} \quad (4.8\text{–}32)$$

$$\underline{C}_2 = \frac{\underline{\lambda}_1 + \dfrac{1}{\sigma\tau_s} + j\dfrac{1-\sigma}{\sigma}\omega_r}{\underline{\lambda}_1 - \underline{\lambda}_2} \underline{I}^s_{qds} \quad (4.8\text{–}33)$$

which completes the solution.

If we assume the speed is high enough so the approximation for the eigenvalues given in eqn (4.8–12) applies,

$$\underline{\lambda}_1 = -\frac{1}{\sigma\tau_r} + j\omega_r, \quad \underline{\lambda}_2 = -\frac{1}{\sigma\tau_r}, \text{ and } \alpha = 1 \quad (4.8\text{–}34)$$

and the expressions for the constants $\underline{C}_1$ and $\underline{C}_2$ reduce to

$$\underline{C}_1 = -\frac{1-\sigma}{\sigma}\underline{I}^s_{qds}, \quad \underline{C}_2 = \frac{1}{\sigma}\underline{I}^s_{qds} \quad (4.8\text{–}35)$$

The short circuit current is then given by

$$\underline{i}^s_{qds} = \left[\frac{\sigma-1}{\sigma}e^{j\omega_r t} + \frac{1}{\sigma}\right]\underline{I}^s_{qds} e^{-t/\sigma\tau_r} \quad (4.8\text{–}36)$$

which consists of a decaying dc component of magnitude

$$\text{dc component magnitude} = \frac{\left|\underline{I}^s_{qds}\right|}{\sigma} \quad (4.8\text{–}37)$$

and a rotor frequency component with a magnitude

$$\text{ac component magnitude} = \frac{1-\sigma}{\sigma}\left|\underline{I}^s_{qds}\right| \quad (4.8\text{–}38)$$

Since σ can be as small as 0.03 in a large machine, these short circuit currents can be as large as 30 times the no load steady state current. The no load current is given approximately by

$$\underline{I}^s_{qds} = \frac{\underline{V}^s_{qds}}{j\omega_e L_s} \quad (4.8\text{–}39)$$

so the short circuit current components are approximately

$$\approx \frac{|\underline{V}^s_{qds}|}{\omega_e \sigma L_s} \text{ and } \frac{|\underline{V}^s_{qds}|}{\omega_e \sigma L_s}(1-\sigma) \tag{4.8–40}$$

each of which is also approximately equal to the steady state locked rotor current. At the instant when the ac and dc components add in one of the machine phases, the peak current is roughly twice the locked rotor current. The actual peak value is

$$\text{peak short circuit current} = \frac{2-\sigma}{\sigma}\frac{|\underline{V}_{qds}|}{\omega_e L_s} \tag{4.8–41}$$

Both components of the short circuit current decay with the transient time constant $\sigma\tau_r$.

4.9 Transient Equivalent Circuits (Constant Rotor Flux Linkage)

There are many instances in which the behavior of a machine for a very short period following an electrical disturbance is of interest. Examples include evaluating worst case fault currents or peak transient torque and modelling the sub–cycle behavior of PWM inverters or other fast switching controllers applied to machines. In such cases it is possible to ignore the damping of the transient solutions and greatly reduce the complexity of the analysis. This is equivalent to neglecting the resistances of the machine and results in the simple eigenvalue expressions given in eqn (4.8–14).

4.9.1 Constant Rotor Flux Linkage Model

A very useful conceptual model can be obtained for this situation by starting from the flux linkage form of the machine equations

$$\underline{v}_{qds} = r_s \underline{i}_{qds} + (p + j\omega)\underline{\lambda}_{qds} \tag{4.9–1}$$

$$0 = r_r \underline{i}_{qdr} + [p + j(\omega - \omega_r)]\underline{\lambda}_{qdr} \tag{4.9–2}$$

With the rotor IR drop neglected, eqn (4.9–2) becomes

$$[p + j(\omega - \omega_r)]\underline{\lambda}_{qdr} = 0 \tag{4.9–3}$$

or in a rotor reference

$$p\underline{\lambda}_{qdr}^{r} = 0 \tag{4.9–4}$$

from which the solution for the rotor flux is

$$\underline{\lambda}_{qdr}^{r} = \underline{\lambda}_{qdr}^{r}(0) = \text{constant} = \underline{\lambda}_{qdro}^{r} \tag{4.9–5}$$

Transforming to a general frame moving at speed ω yields

$$\underline{\lambda}_{qdr} = \underline{\lambda}_{qdro}^{r} e^{-j(\omega - \omega_r)t} \tag{4.9–6}$$

which simply indicates that the trapped constant rotor flux generates sinusoidally varying rotor flux linkages in any other frame of reference.

4.9.2 Transient Equivalent Circuit

To utilize the constant rotor flux linkage expression in eqn (4.9–6) in the stator equation and obtain a circuit model, we employ the flux linkage equations

$$\underline{\lambda}_{qds} = L_s \underline{i}_{qds} + L_m \underline{i}_{qdr} \tag{4.9–7}$$

$$\underline{\lambda}_{qdr} = L_m \underline{i}_{qds} + L_r \underline{i}_{qdr} \tag{4.9–8}$$

Eliminating the rotor current in these equations yields

$$\underline{\lambda}_{qds} = \left(L_s - \frac{L_m^2}{L_r}\right)\underline{i}_{qds} + \frac{L_m}{L_r}\underline{\lambda}_{qdr} \tag{4.9–9}$$

but

$$L_s - \frac{L_m^2}{L_r} = L_s\left(1 - \frac{L_m^2}{L_s L_r}\right) = \sigma L_s = L'_s \tag{4.9–10}$$

where L'_s is the stator transient inductance which was originally introduced in Section 4.6 in connection with turns ratio considerations. Substituting these results in the stator equation, eqn (4.9–1), yields

$$\underline{v}_{qds} = [r_s + L'_s(p + j\omega)]\,\underline{i}_{qds} + (p + j\omega)\frac{L_m}{L_r}\underline{\lambda}_{qdr} \tag{4.9–11}$$

and finally using eqn (4.9–6)

$$\underline{v}_{qds} = [r_s + L'_s(p + j\omega)]\,\underline{i}_{qds} + j\left(\omega_r \frac{L_m}{L_r}\underline{\lambda}^r_{qdro}\right)e^{-j(\omega-\omega_r)t} \tag{4.9–12}$$

If we define the *voltage behind transient reactance*, $\underline{E}'_{qd}$ to be

$$\underline{E}'_{qd} = j\omega_r \frac{L_m}{L_r}\underline{\lambda}^r_{qdro} \tag{4.9–13}$$

Equation (4.9–12) becomes

$$\underline{v}_{qds} = [r_s + L'_s(p + j\omega)]\,\underline{i}_{qds} + \underline{E}'_{qd}\,e^{-j(\omega-\omega_r)t} \tag{4.9–14}$$

which describes the *transient equivalent circuit* shown in Figure 4.16. The most commonly used circuit is the stator referred circuit shown in Figure 4.17, obtained by setting $\omega = 0$ in the general circuit. From this circuit it is clear that in a stator reference, the voltage behind transient reactance $\underline{E}'_{qd}$ is a balanced sinusoidal voltage oscillating at rotor frequency with respect to the stator d,q frame. Figure 4.18 illustrates the evolution of the transient equivalent circuit from the complex vector equivalent circuit of Figure 4.1.

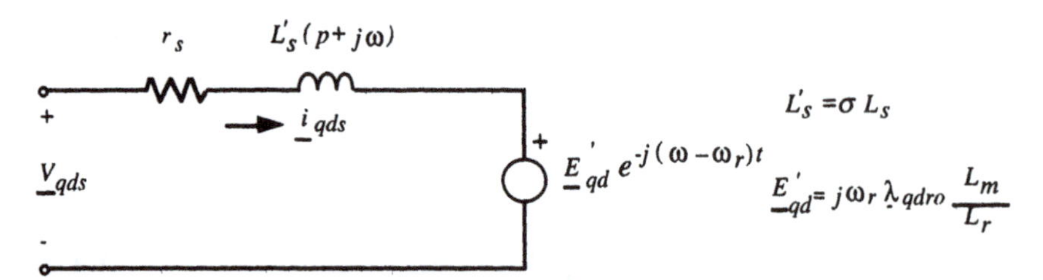

Figure 4.16 Transient equivalent circuit in reference frame rotating at speed ω

4.9.3 Evaluation of Voltage Behind Transient Reactance

The voltage $\underline{E}'_{qd}$ is proportional to the rotor flux linkage which exists at the instant the transient is initiated. Assuming the machine is in the steady state prior to the initiation of the transient, the steady state rotor flux just prior to the transient can be used to find $\underline{E}'_{qd}$. In the steady state stator referred circuit of Figure 4.19, the voltage across the rotor resistor is given by

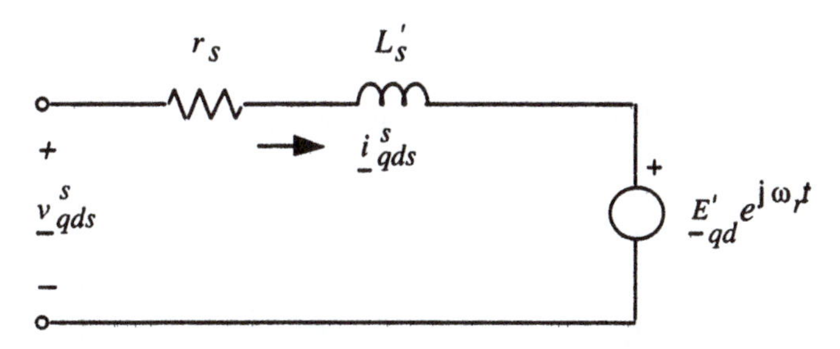

Figure 4.17 Stator referred equivalent circuit

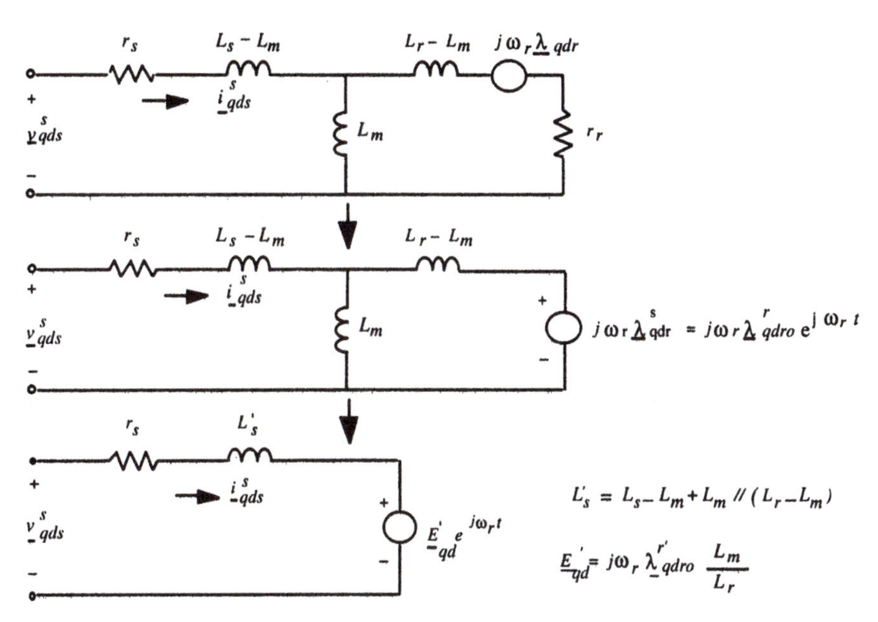

Figure 4.18 Evolution of transient circuit from equivalent circuit. The symbol '//' denotes the parallel combination of the two inductances.

$$\underline{E}_r = j\omega_e (L_m \underline{I}^s_{qds} + L_r \underline{I}^s_{qdr}) = j\omega_e \underline{\lambda}^s_{qdr} \tag{4.9–15}$$

The flux linkage $\underline{\lambda}^s_{qdr}$ will become the trapped flux $\underline{\lambda}^r_{qdro}$ and hence

$$\underline{E}'_{qd} = j\omega_r \underline{\lambda}^r_{qdro} \frac{L_m}{L_r} = \frac{\omega_r}{\omega_e} \frac{L_m}{L_r} \underline{E}_{ro} \tag{4.9–16}$$

or in terms of the slip S

$$\underline{E}'_{qd} = \frac{L_m}{L_r}(1-S)\,\underline{E}_{ro} \qquad (4.9\text{–}17)$$

where $\underline{E}_{ro}$ is the value of $\underline{E}_r$ at the switching instant.

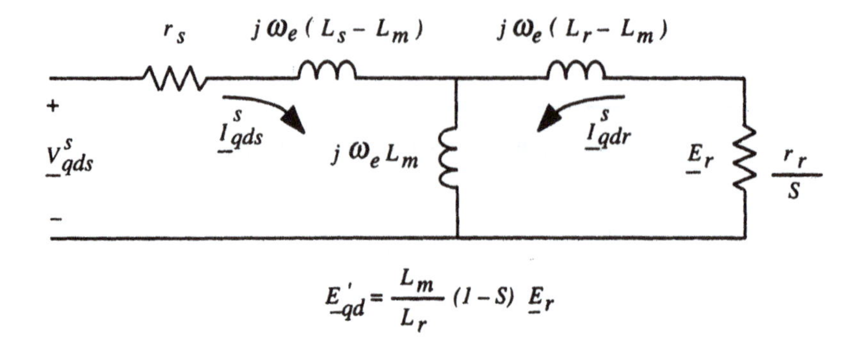

Figure 4.19 Evaluation of $\underline{E}'_{qd}$ from steady state circuit prior to initiation of transient

4.9.4 Example – Sudden Short Circuit of Unloaded Induction Machine

A similar example as was treated in Section 4.8.1 by eigenvalue methods will now be examined using the transient equivalent circuit.

The first step is to evaluate the transient voltage $\underline{E}'_{qd}$ for the condition just prior to the short circuit. Since the machine is operating at no load the slip is zero and the steady state circuit takes the form shown in Figure 4.20The voltage $\underline{E}_r$ is clearly (neglecting r_s)

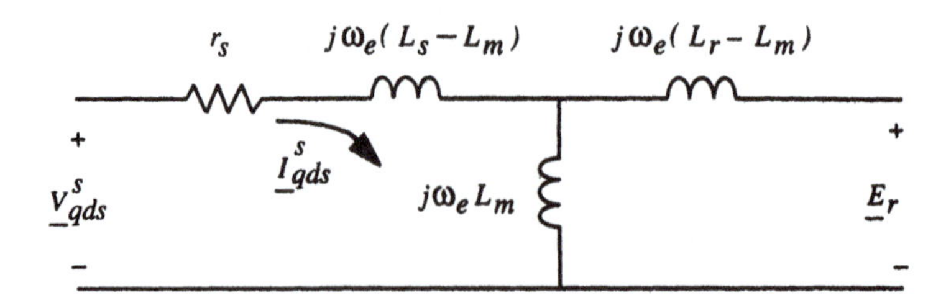

Figure 4.20 Steady state equivalent circuit at $S = 0$ for example problem

$$\underline{E}_{ro} = j\omega_e L_m \underline{I}^s_{qds} = \underline{V}^s_{qds} \frac{L_m}{L_s} \tag{4.9–18}$$

and from eqn (4.9–17), the voltage $\underline{E}_{qd}'$ is

$$\underline{E}'_{qd} = \frac{L_m^2}{L_s L_r} \underline{V}^s_{qds} = (1-\sigma)\, \underline{V}^s_{qds} \tag{4.9–19}$$

The transient equivalent circuit is as shown in Figure 4.21. Subsequent to

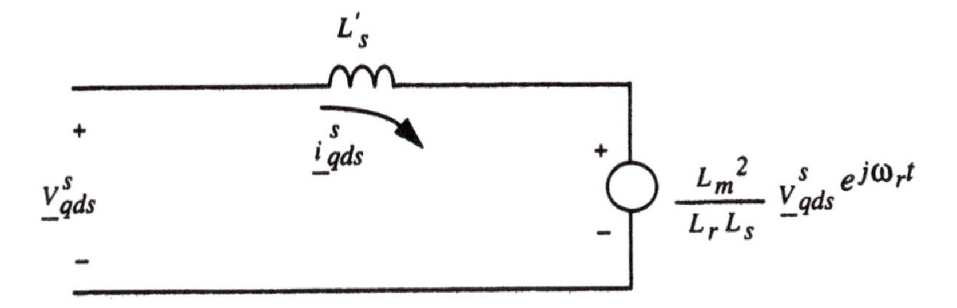

Figure 4.21 Transient equivalent circuit for example short circuit problem

the short circuit, the differential equation describing the short circuit current is

$$0 = L'_s\, p\underline{i}^s_{qds} + (1-\sigma)\, \underline{V}^s_{qds} e^{j\omega_r t} \tag{4.9–20}$$

from which

$$\underline{i}^s_{qds}(0) = -\frac{1}{p}(1-\sigma) \frac{\underline{V}^s_{qds}}{L'_s} e^{j\omega_r t} \tag{4.9–21}$$

$$= -(1-\sigma) \frac{\underline{V}^s_{qds}}{j\omega_r L'_s} e^{j\omega_r t} + \underline{C}$$

To evaluate the integration constant, the initial value of the stator current

$$\underline{i}^s_{qds}(0) = \frac{\underline{V}^s_{qds}}{j\omega_e L_s} = \frac{\underline{V}^s_{qds}}{j\omega_r L_s} \tag{4.9–22}$$

is used with the result

$$\underline{C} = \underline{V}^s_{qds}\left(\frac{1}{j\omega_r L_s} + \frac{(1-\sigma)}{j\omega_r L_s'}\right)$$

$$= \frac{\underline{V}^s_{qds}}{j\omega_r L_s}\left(1 + \frac{(1-\sigma)}{\sigma}\right) \tag{4.9–23}$$

$$= \frac{\underline{V}^s_{qds}}{j\omega_r L_s}\left(\frac{1}{\sigma}\right)$$

and the total solution is

$$\underline{i}^s_{qds} = \frac{\underline{V}^s_{qds}}{j\omega_r L_s}\left[\frac{\sigma-1}{\sigma}e^{j\omega_r t} + \frac{1}{\sigma}\right] \tag{4.9–24}$$

which is the same as eqn (4.8–36) except for the exponential decay term.

4.9.5 Comments on the Transient Equivalent Circuit

Although in the development of the transient equivalent circuit model the stator resistance can be retained as was done in Section 4.9.2, the time constant associated with this resistance is not a correct description of the time behavior of the system. Since the rotor resistance was neglected in the development of the circuit, the time dependence of the rotor flux is neglected. As has been shown in the section on constant speed eigenvalues, there is a strong interaction between the rotor and stator damping and to ignore one and retain the other is not generally valid. If one is very long compared to the other (i.e. if α is very large or very small) there is some validity in neglecting the smaller time constant. Since, in general, the two time constants τ_r and τ_s are comparable, there is no valid reason to retain only one or the other. Thus, the transient equivalent circuit should be viewed as a means of evaluating the undamped behavior of the machine and the stator resistance should be neglected. It is left in the model as developed in Section 4.9.3 only as an indication that at the initiation of a transient, the difference between the terminal voltage and the transient voltage does appear across the stator resistance and the transient inductance.

References

[1] W.V. Lyon, "Transient Analysis of Alternating Current Machinery", John Wiley and Sons, Inc., New York, 1954.

[2] K.P. Kovacs, "Symmetrical Components in Polyphase Machines", (in German), Druck von Rosch and Co., Bern, 1962.

[3] D.W. Novotny, and J. H. Wouterse, "Induction Machine Transfer Functions and Dynamic Response by Means of Complex Time Variables," IEEE PAS, Vol. 95, No. 4 July/Aug. 1976, pp. 1125–1333.

[4] R. Stern and D.W. Novotny, "A Simplified Approach to Induction Machine Dynamic Response," IEEE PAS, Vol. 97, No. 4, July/Aug. 1978, pp. 1430–1439.

[5] R. DeDoncker and D.W. Novotny, "The Universal Field Oriented Controller," IEEE–IAS Trans., Vol. 30, No. 1, January/February 1994, pp. 92–100.

[6] B.T. Ooi and T.H. Barton, "The Modal Approach to Induction Motor Dynamics, Part I and II" Papers C72–045–8, C 72 046–6, IEEE Winter Meeting, New York, Jan. 30–Feb. 4, 1972.

Problems

Problem 4–1 DC Excitation of Induction Machine

Consider a three phase induction motor in which stator phases b and c are connected to a dc current source of amplitude I_{dc}. Phase a is left disconnected. Use a stationary reference system with $\phi = 0$ and find:

a) the stator d and q currents;

b) the stator d and q steady state voltages if the rotor speed is constant at ω_{ro} (rad/sec);

c) the rotor currents for the condition of part (b);

d) the torque as a function of the dc current, the rotor speed and the machine parameters.

Problem 4–2 DC Excitation of Induction Machine

Repeat problem 4–1 using a rotor reference.

Problem 4–3 AC Current Source Excitation of Induction Machine

Repeat problem 4–1 if the current source is an ac supply of amplitude I_m and frequency ω_e. Discuss the advantage (or disadvantage) of a synchronous reference for this situation.

Problem 4–4 Single Phase Excitation of Induction Machine

For a three phase induction machine operated with phase a open and a voltage v_{bc} suppled to terminals b–c:

a) derive a d, q, 0 stator referred differential equation model which explicitly includes v_{bc} and i_b variables. Give both the voltage–current equations and the torque equation;

b) for an input voltage $v_{bc} = V\cos\omega_e t$ and a constant rotor speed ω_{ro}, derive the steady state stator referred phasor equations and the torque equation;

c) show that these equations can be represented by the (symmetrical component) equivalent circuit shown below and express $\tilde{V}_x$ and $\tilde{I}_x$ in terms of the phasors representing v_{bc} and i_b;

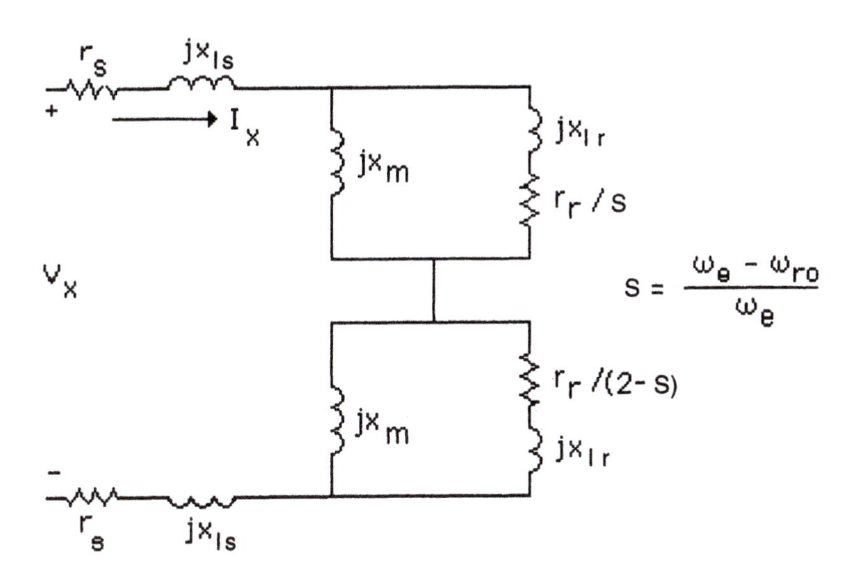

Problem 4–5 Torque Pulsation Caused by Unbalanced Excitation

For a 100 hp, 460 volt, 60 Hz, 4 pole, three phase induction machine with equivalent circuit parameters (in per unit):

$$r_s = 0.010 \qquad x_{ls} = 0.10 \qquad x_m = 3.0$$

$$r_r = 0.015 \qquad x_{lr} = 0.10$$

operated at rated voltage and frequency and a slip of $S = 0.0175$ under the single phase condition of problem 4–4, find (assume the inertia is large enough to maintain constant rotor speed):

a) the phasor current $\tilde{I}_b$ (take $\tilde{V}_{bc}$ as the reference quantity, i.e. zero phase angle);

b) the positive and negative sequence average torque and the total average torque;

c) the peak value of the second harmonic pulsating torque.

Problem 4–6 Induction Machine Constant Speed Switching Transient

The 100 hp, 460 volt, 60 Hz, 4 pole, three phase induction machine of problem 4–5 is operating at full load ($S = 0.0175$) at rated V and f. A Y–connected set of 10.0 pu resistors is in parallel with the stator. At $t = 0$ the source supplying the motor and resistors is disconnected.

a) Find the values of r_s and r_r (in ohms) as well as L_r, L_s and L_m (in henries) for this machine;

b) Find the constant speed eigenvalues which characterize the switching transient following removal of the source;

c) Find the initial values of the complex vectors representing the rotor and stator fluxes in a stator reference;

d) Find the time domain expression for the stator voltage v^s_{qds} (stator reference) following the switching operation assuming the speed remains constant.

Approximately how long will it take for this voltage to drop to 10% of rated voltage?

Problem 4–7 Transient Current and Torque by 'Plug' Reversal

For the 100 hp, 460 volt, 60 Hz, 4 pole, three phase induction machine of problems 4–5 and 4–6 operated at rated voltage and frequency and no load (assume the inertia is large enough to maintain constant rotor speed):

a) use the constant rotor flux linkage transient model to calculate the transient current immediately following the instantaneous reversal of two line wires (plugging) if the switching occurs at a voltage maximum;

b) find the maximum instantaneous transient torque resulting from the switching operation.

Problem 4–8 VSI–Induction Machine Harmonic Analysis

For a VSI driving a 100 hp induction machine with pu parameters $r_s = 0.015$, $x_{ls} = 0.10$, $x_m = 2.0$, $r_r = 0.020$, $x_{lr} = 0.10$ with a constant input voltage V_i such that fundamental output voltage is 1.0 pu, find the fundamental and 5–th harmonic components of i_{qs} and i_{ds} and the dc and 6th harmonic components of i_i. Take the fundamental frequency equal to 1.0 pu and the slip equal to rated slip ($S_R = 0.0247$). Locate all of the fundamental voltage in the q–axis.

5 Principles of Vector Control and Field Orientation

5.1 Introduction

In this chapter the basic concepts of torque control and field orientation are introduced based on steady state considerations for both synchronous and induction machines. A basic set of three requirements which enable direct control of machine torque production is developed and used to illustrate the type of controller needed in both types of ac machines. These controllers are generally referred to as *vector controllers* because they control both the amplitude and phase of the ac excitation. The vector control of currents and voltages results in control of the spatial orientation of the electromagnetic fields in the machine and has led to the term *field orientation* [1,2,3]. Usually, this term is reserved for controllers which maintain a 90° spatial orientation between critical field components and hence we will adopt the term *field angle control* or simply *angle control* for systems which depart from the 90° orientation. The dynamic performance of the systems outlined in this chapter is considered in later chapters.

5.2 DC Machine Torque Control

Before proceeding with the development of the principles of vector control and field orientation, a brief review of torque control in dc machines will be presented. The close parallel between the physical torque control principles in a dc machine and in ac vector controllers will be emphasized in later sections.

A dc machine consists of a stationary field structure utilizing a stationary dc excited winding or permanent magnets and a rotating armature winding supplied through a commutator and brushes. This basic structure is schematically illustrated in Figure 5.1 along with the resulting orientation of the armature MMF and the field flux. The action of the commutator is to reverse the direction of the armature winding currents as the coils pass the brush position such that the armature current distribution is fixed in space no matter what

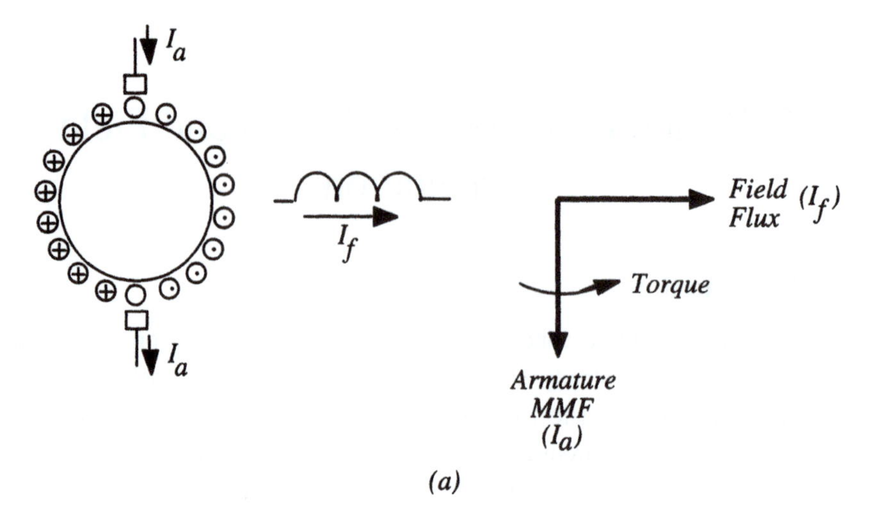

(a)

Orientation of Armature MMF and Field Flux

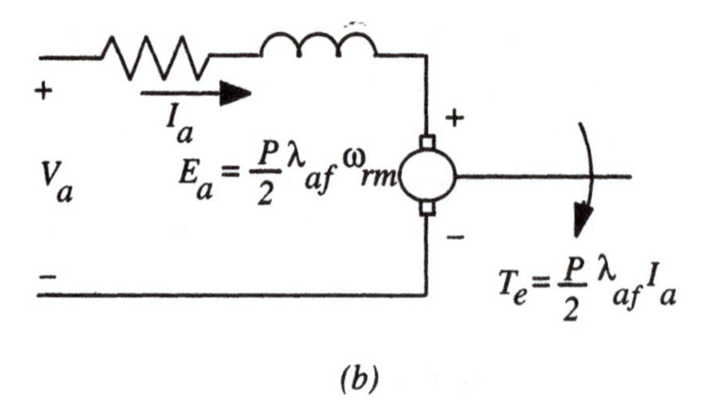

(b)

Equivalent Circuit

Torque = Constant x Field Flux x Armature Current

$$T_e = \frac{P}{2}\frac{L_{af}}{L_f} \times \lambda_f \times I_a$$

Figure 5.1 DC machine model

rotor speed exists. As shown in the figure, the field flux and armature MMF are maintained in a mutually perpendicular orientation independent of rotor speed. The result of this orthogonality is that the field flux is unaffected by the armature current except for second order, nonlinear effects.

The electromagnetic interaction between the field flux and the armature MMF results in two basic outputs: an induced voltage proportional to rotor speed,

$$E_a = \frac{P}{2}\lambda_{af}\omega_{rm} \tag{5.2–1}$$

and an electromagnetic torque proportional to the armature current

$$T_e = \frac{P}{2}\lambda_{af}I_a \tag{5.2–2}$$

where ω_{rm} is the rotor speed in mechanical radians per second, P is the number of poles and λ_{af} is the flux produced by the field current which links the armature winding. (The constants in the two equations are unit dependent but are identically equal to $P/2$ in SI units). The flux linking the armature is related to the total field flux linkages λ_f by the expression

$$\lambda_{af} = \frac{L_{af}}{L_{lf}+L_{af}}\lambda_f = \frac{L_{af}}{L_f}\lambda_f \tag{5.2–3}$$

L_{af}, L_{lf}, and L_f are the mutual inductance between field and armature windings, the field leakage inductance and the field self inductance respectively. The torque can then be written in the alternative form

$$T_e = \frac{P}{2}\frac{L_{af}}{L_f}\lambda_f I_a \tag{5.2–4}$$

Inclusion of the armature resistance (and inductance for transient currents) completes the model as illustrated in Figure 5.1. It is important to note that the simplicity of the model is strongly dependent on the mutually perpendicular orientation of the flux and MMF. If this orthogonality were disturbed (by shifting the brushes for example) two major complications occur:

1) the field flux is no longer independent of the armature current since there will be an MMF component in the field axis, and
2) the voltage and torque relations in eqns (5.2–1) and (5.2–2) will be

modified by addition of an angle dependent function (sine of the angle between axes to a first approximation).

Adjustable speed operation is normally attained by operating with a fixed field flux and a varying the armature voltage. The no load speed is then determined as the speed where the induced voltage is essentially equal to the applied voltage. Torque production requires armature current and results in an armature *IR* drop. This *IR* drop must come at the expense of a smaller induced voltage which necessitates a small speed reduction. Thus, every armature voltage has an associated torque–speed curve and the actual speed of operation is determined by the armature voltage and the torque required to supply the load. The situation is analogous to variable frequency, constant volts/hertz operation of an induction machine. In both cases the torque, current and speed are determined by the load with the torque–speed characteristic set by the excitation.

Adjustable torque operation is readily attainable in the dc machine by the simple action of controlling the armature current instead of the voltage. As expressed in eqn (5.2–4), with a constant value of field flux, the torque is directly proportional to armature current. Thus, the torque can be adjusted as accurately and as rapidly as the armature current can be adjusted and controlled. In practice this is accomplished by using a feedback current regulator and a power electronic supply. If both positive and negative values of torque (and speed) are desired, the supply must be capable of bidirectional current (and voltage). The armature voltage required to supply the controlled current is always the sum of the speed dependent induced voltage and the *IR* and *L(dI/dt)* voltages of the armature circuit. Very rapid current changes thus require large transient armature voltages to overcome the inductive effects and operation at high speed requires a large average voltage to supply the speed dependent induced voltage. However, the torque will exactly follow the current and to the extent that the current regulator follows its reference input, the torque will be proportional to the current reference to the regulator. Figure 5.2 illustrates the elements of a typical torque controlled dc motor drive.

5.3 Requirements for Torque Control

Following the concepts outlined for the dc machine, the requirements for torque control are:

1) an independently controlled armature current to overcome the effects of armature winding resistance, leakage inductance and induced voltage;

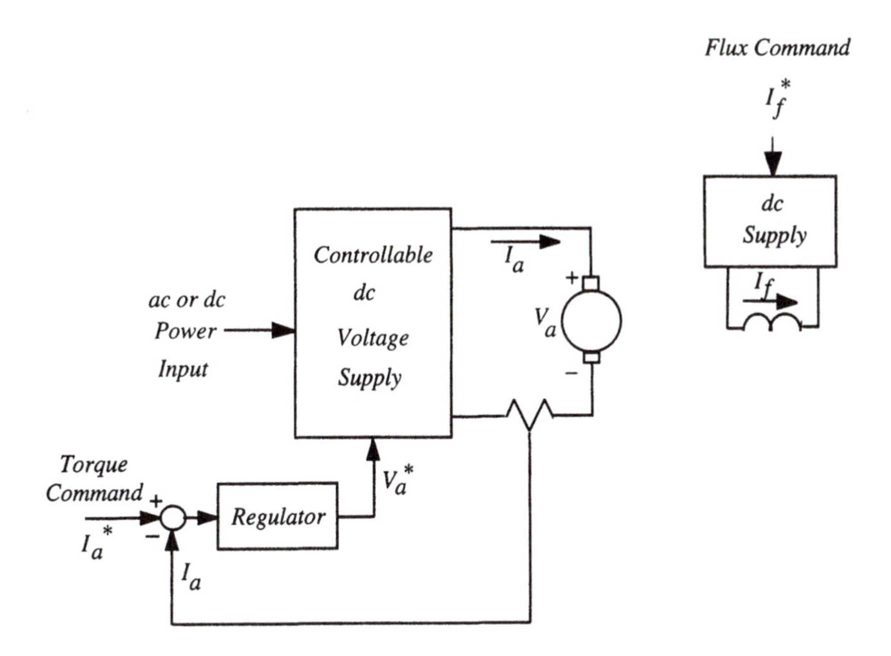

Figure 5.2 Elements of a torque controlled separately excited dc motor drive

2) an independently controlled or constant value of the field flux;

3) an independently controlled orthogonal spatial angle between the flux axis and the MMF axis to avoid interaction of the MMF and the flux.

If all three of these requirements are met at every instant of time, the torque will instantaneously follow the current and instantaneous torque control will result. If, as can occur in certain systems, these requirements are only met for steady state conditions, only steady state torque control will result. During the transient period, the torque will not follow the current exactly and this may or may not be satisfactory for the specific application.

In the dc machine, requirements 2 and 3 are assured by the commutator and the separate field excitation system (dc winding or permanent magnet). In ac machines, these two requirements must be achieved by external controls and thus the situation is more complex and somewhat more difficult to understand. For this reason, the development which follows first examines torque control in synchronous motors where the separate field winding and known position of the flux axis on the rotor are easily visualized as similar to the dc machine situation.

5.4 Synchronous Machine Vector Control

The CSI driven synchronous machine is a natural starting point since it combines a number of features which suggest vector control and angle control. These include:

1) the CSI is a current supply which can be controlled in both amplitude and phase,

2) the field winding of the machine is physically available and can be controlled as in the dc machine, and

3) the space position of the rotor dc field is clearly located in space by the position of the rotor.

These electrical and physical features suggest the possibility of controlling the CSI – synchronous machine combination so as to emulate a dc machine. The resulting *brushless dc machine* or *commutatorless dc machine* has been utilized for many years, especially in Japan [4,5].

5.4.1 CSI – Synchronous Machine (Commutatorless DC Machine)

The basic open loop CSI – synchronous machine system is illustrated in Figure 5.3. The CSI can be force commutated or in many high power applications is load commutated by operating the synchronous machine with leading current (overexcited). In either case, the concept of sensing the rotor position to locate the field winding axis and using this rotor position information to control the firing of the SCRs in the inverter and hence control the field angle can be employed. The scheme is illustrated in Figure 5.4 and the analogy to a dc motor is apparent in the sense of producing a fixed space angle between the field winding and the stator winding MMF. Clearly the system is self–synchronous since the frequency of the inverter will always be the shaft frequency.

Steady state performance of the machine for the fundamental component can be treated by the standard equivalent circuit approach as illustrated in Figure 5.5. For simplicity, a non–salient pole machine is considered since saliency does not affect the basic system behavior and only serves to obscure the underlying concepts to be considered here. Saliency effects are considered later when dynamic behavior is presented using d,q modeling.

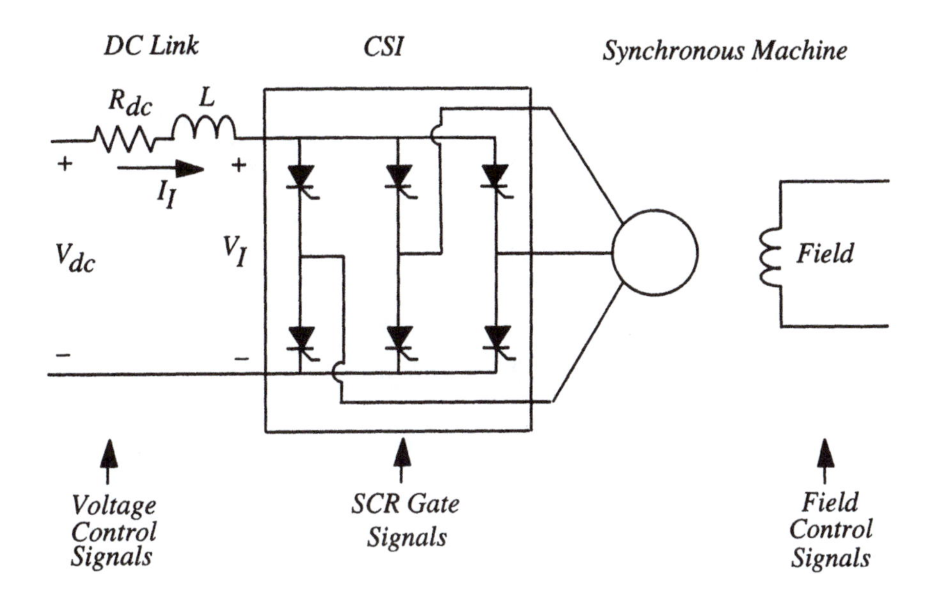

Figure 5.3 CSI–synchronous machine system

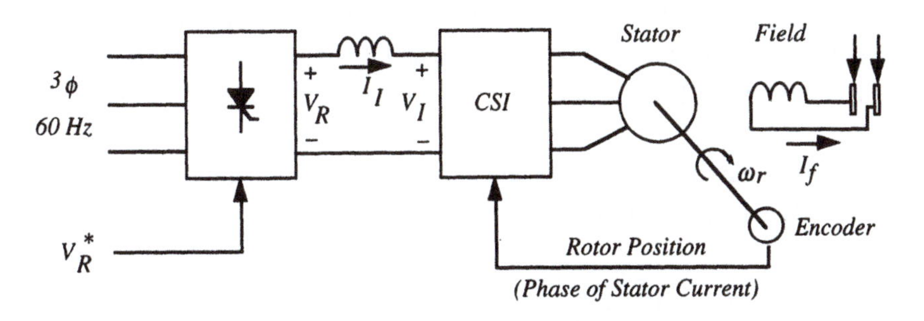

Figure 5.4 Commutatorless dc motor utilizing direct feedback of rotor position to control phase of stator current

In the circuit of Figure 5.5 the voltage E_a is the internal voltage produced by the dc field current and ω_{re} is the speed of the rotor in equivalent electrical radians per second and, under steady state conditions, is equal to the electrical angular frequency ω_e. The amplitude of E_a is proportional to the rotor speed and the flux produced by the field current

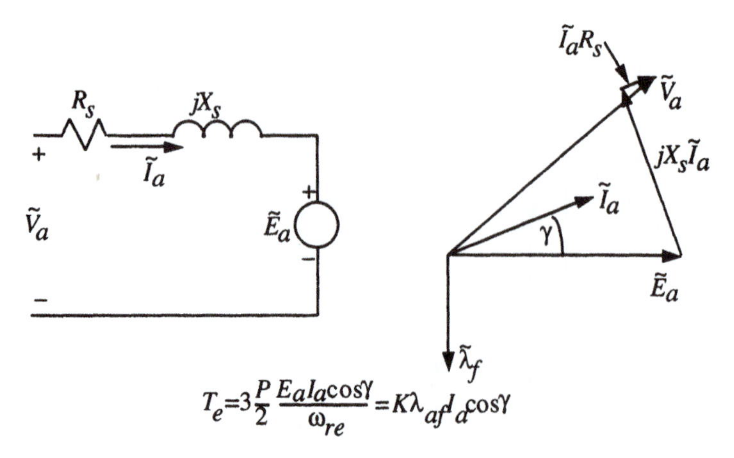

Figure 5.5 Phasor diagram and torque equation for commutatorless dc motor with positive γ

$$E_a = \omega_{re}\lambda_{af} = \frac{P}{2}\omega_{rm}\lambda_{af} \tag{5.4–1}$$

just as for the counter EMF in a dc machine, and the time phase of E_a is directly related to the rotor position. Thus, the rotor position feedback loop can be viewed as directly controlling the angle γ between the phasors of E_a and I_a as shown in Figure 5.5. In this figure, γ is shown as a small positive angle with I_a leading E_a. Note that for this condition, adding in the $I_a X_s$ and $I_a R_s$ voltages yields a terminal voltage which results in a lagging current at the machine terminals.

The machine torque can be evaluated from the power input to the voltage E_a divided by mechanical speed

$$T_e = 3\frac{p}{2}\frac{E_a I_a \cos\gamma}{\omega_{re}} \tag{5.4–2}$$

Substituting for E_a from eqn (5.4–1) yields

$$T_e = 3\frac{P}{2}\lambda_{af} I_a \cos\gamma \tag{5.4–3}$$

which is identical in form to the corresponding equation for a dc machine if γ is zero. Note also that the torque per ampere can be maximized by choosing $\gamma = 0°$. However, if load commutation is desired, a relatively large value of γ is needed to create a leading power factor at the machine terminals as shown in Figure 5.6.

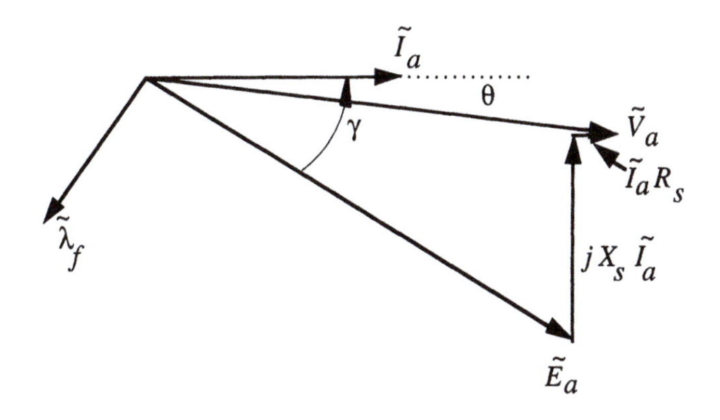

Figure 5.6 Phasor diagram for larger value of γ showing current leading voltage

The analogy of the system to a dc machine can be extended to include the armature counter EMF if we consider the dc side variables of the inverter. Equating the dc power input to the power associated with the fundamental component on the ac side yields

$$V_I I_I = (V_R - I_I R_{dc}) I_I = 3 V_a I_a \cos\theta \tag{5.4–4}$$

As shown in Chapter 3, the inverter current I_I is directly related to the motor current (fundamental component) by

$$I_I = \frac{\pi}{\sqrt{6}} I_a \tag{5.4–5}$$

and from the phasor diagrams in Figure 5.5 or Figure 5.6, the voltage component $V_a \cos\theta$ can be expressed as

$$V_a \cos\theta = E_a \cos\gamma + I_a R_s \tag{5.4–6}$$

Combining these three equations, the following relation between the rectifier output voltage V_R, the armature current I_a and the back EMF $E_a \cos\gamma$ can be obtained

$$V_R \frac{\pi}{3\sqrt{6}} = E_a \cos\gamma + \left(R_s + \frac{\pi^2}{18} R_{dc} \right) I_a \qquad (5.4\text{–}7)$$

This result is illustrated in the form of an equivalent circuit in Figure 5.7.

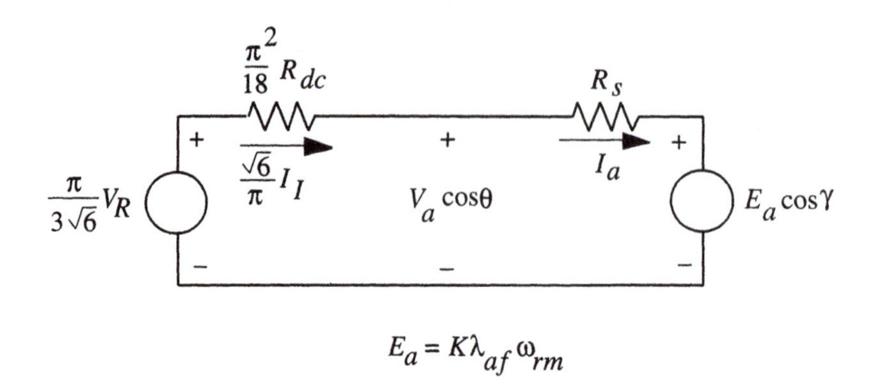

Figure 5.7 Equivalent circuit showing analogy to dc motor if γ = constant

When γ is constant, the circuit is identical to the armature circuit of a dc machine. Note that the system of Figure 5.4 can be operated in an exactly analogous manner as a dc shunt motor; the speed of operation will be essentially set by the rectifier voltage V_R with the system in steady state equilibrium when the counter voltage $E_a \cos\gamma$ plus the IR drops is equal to $(\pi/3\sqrt{6})V_R$. The speed–torque curves will be straight lines just as for the dc machine.

5.4.2 Torque Control and Choice of γ

For direct torque control, eqn (5.4–3) indicates that just as in the d.c. machine the requirement is to be able to control the stator current I_a. As noted previously, the choice of $\gamma = 0°$ is attractive in terms of maximizing the torque per ampere. As will be shown subsequently, this choice is also advantageous in terms of decoupling the transient response of the system. One drawback to choosing $\gamma = 0°$ is the inescapable fact that the terminal power factor will be lagging. This is unacceptable in large hp drives where load commutation is necessary for other reasons and hence a relatively large γ (40°–60°) is often used in these cases. Figure 5.8 summarizes these properties of the angle γ.

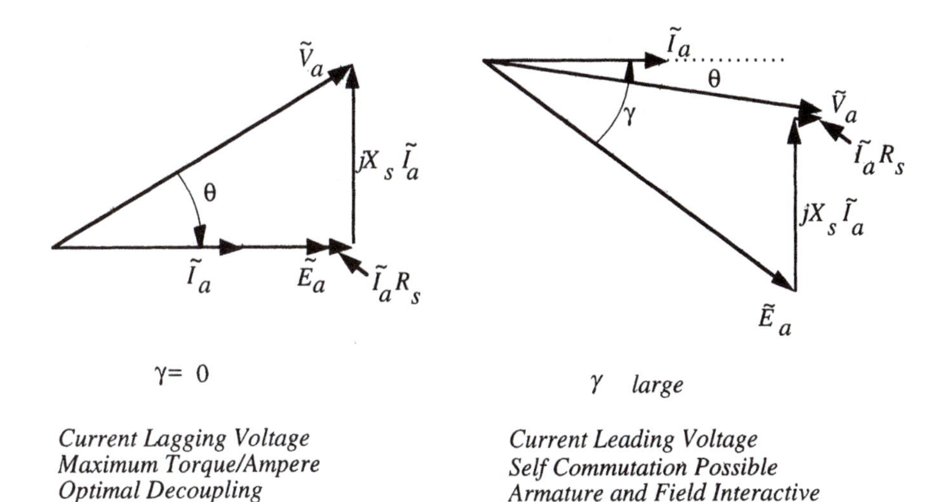

Figure 5.8 Influence of internal angle

5.5 Synchronous Machine Steady State *d,q* Model

To facilitate the transition to the dynamic analysis to be presented in Chapter 6, it is convenient to introduce the steady state *d,q* model of a synchronous machine. The notation employed in the dynamic analysis of Chapter 6 will then correspond to the steady state notation employed in this section.

5.5.1 Steady State Conditions in d,q Variables

To relate the steady state concepts of Section 5.4 to the *d,q* model, it is only necessary to specialize the *d,q* equations to the steady state. The constraints for steady state are:

1) Constant amplitude and phase of stator current – implies $i_{qs} = I_{qs}$ and $i_{ds} = I_{ds}$ are constants,

2) Constant rotor flux linkages – implies having zero currents in the damper windings, $i_{qr} = i_{dr} = 0$,

3) Constant field current – implies $i_{fr} = I_f$.

The basic torque equation from Chapter 2 is

$$T_e = \frac{3}{2}\frac{P}{2}(\lambda_{ds}\, i_{qs} - \lambda_{qs}\, i_{ds}) \tag{5.5–1}$$

or, by substituting for the flux linkages in terms of currents using the flux linkage expressions in Chapter 2,

$$T_e = \frac{3}{2}\frac{P}{2}[L_{md}(i_{fr} + i_{dr})\, i_{qs} - L_{mq}\, i_{qr}\, i_{ds} + (L_{ds} - L_{qs})\, i_{ds}\, i_{qs}] \tag{5.5–2}$$

where we now consider, for generality, a salient pole machine.

Since the damper currents are zero, the steady state torque becomes

$$T_e = \frac{3}{2}\frac{P}{2}[L_{md} I_f I_{qs} + (L_{ds} - L_{qs})\, I_{ds}\, I_{qs}] \tag{5.5–3}$$

which consists of the torque produced by the field winding

$$\textit{Reaction Torque} = \frac{3}{2}\frac{P}{2} L_{md} I_f I_{qs} \tag{5.5–4}$$

and the reluctance torque

$$\textit{Reluctance Torque} = \frac{3}{2}\frac{P}{2}(L_{ds} - L_{qs})\, I_{ds}\, I_{qs} \tag{5.5–5}$$

The steady state stator flux linkages are

$$\lambda_{qs} = L_{qs} I_{qs} \tag{5.5–6}$$

$$\lambda_{ds} = L_{ds} I_{ds} + L_{md}\, I_f \tag{5.5–7}$$

and the steady state stator voltage components are ($p = 0$ in the steady state)

$$V_{qs} = r_s I_{qs} + \omega_e (L_{ds} I_{ds} + L_{md} I_f) \tag{5.5–8}$$

$$V_{ds} = r_s I_{ds} - \omega_e L_{qs} I_{qs} \tag{5.5–9}$$

5.5.2 d,q Variable Vector Diagrams

The voltage equations are illustrated in Figure 5.9 in the *d,q* axes. The voltage component

$$E_a = \omega_e L_{md} I_f \tag{5.5–10}$$

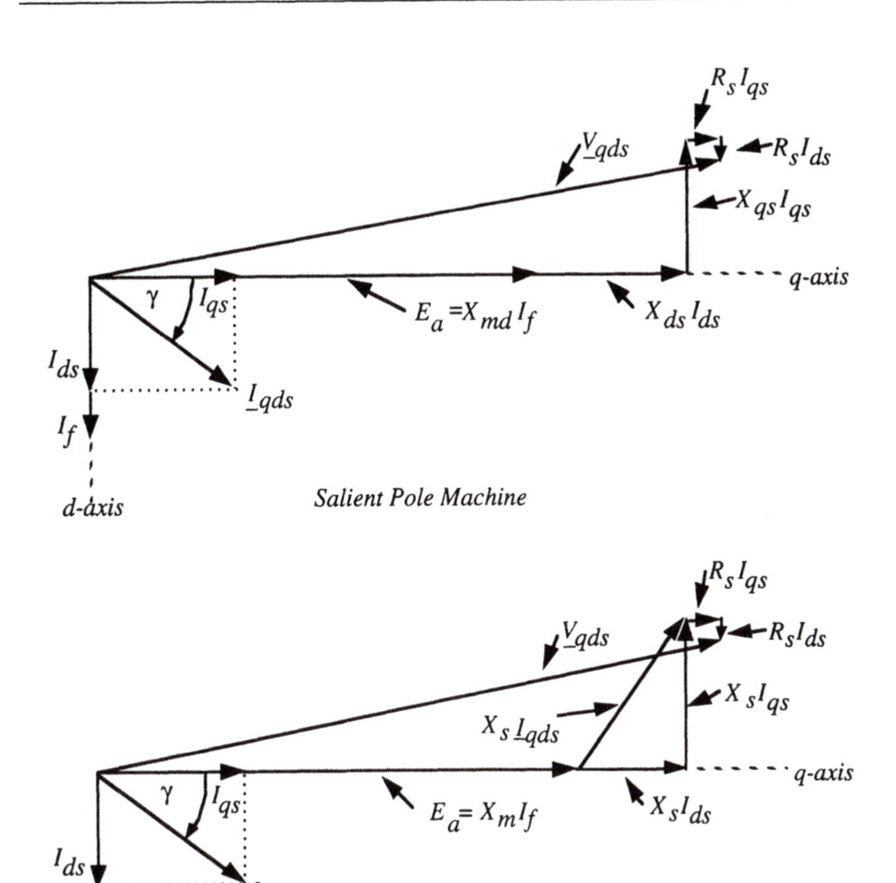

Note: All Reactances Evaluated at $\omega = \omega_e$

Figure 5.9 Steady state vector diagrams showing internal voltage E_a, terminal voltage $\underline{V}_{qds}$ and the internal angle γ

is the induced voltage E_a of steady state theory and the diagram in Figure 5.9 for the non–salient pole machine ($X_{ds} = X_{qs}$, $X_{md} = X_{mq}$) is clearly the same as the phasor diagrams presented in Section 5.4‡.

The diagrams in Figure 5.9 are *d,q* diagrams representing constant *d,q* quantities in the two axes, i.e. the complex vector interpretation of the scalar *d–q* axis quantities. They are, however, identical to the (time) phasor diagrams of conventional steady state theory. Note that the angle γ which appeared as a phase angle in conventional steady state theory is now interpreted as a space angle between the *q*–axis (where E_a is located) and the vector position of the current $\underline{I}_{qds}$. The two interpretations are a result of the shift in reference system; a stationary reference for conventional steady state theory where all electrical quantities vary at stator frequency and a reference system rotating with the rotor where all quantities are dc.

The torque can be rewritten in terms of the angle γ

$$Reaction\ Torque = \frac{3}{2}\frac{P}{2}\frac{1}{\omega_e}E_a I_{qds}\cos\gamma \tag{5.5–11}$$

$$Reluctance\ Torque = \frac{3}{2}\frac{P}{2}(L_{ds} - L_{qs})\, I_{qds}^2 \cos\gamma \sin\gamma \tag{5.5–12}$$

These torque components are the same as found from steady state theory. The amplitude of the vector current $\underline{I}_{qds}$ is clearly the same as the amplitude of the stator (peak) current as can be seen by comparing the phasor diagrams of Section 5.3 with the vector diagrams of Figure 5.9.

An important special case is illustrated in Figure 5.10 where the stator current is all in the *q*–axis ($\gamma = 0$). This is the situation usually referred to as *field–orientation.* Note that for this situation the field current in the d–axis and the stator current in the *q*–axis are 90° apart as is the case in the dc machine. Note also, that because of the absence of *d*–axis stator current there is no reluctance torque and only the *q*–axis reactance is involved in finding the terminal voltage, i.e. there is no direct magnetization or demagnetization of the *d*–axis, only the field winding acts to produce flux in this direction.

‡. E_a is usually expressed as an rms quantity in steady state analysis but represents the peak value when using *d,q* variables.

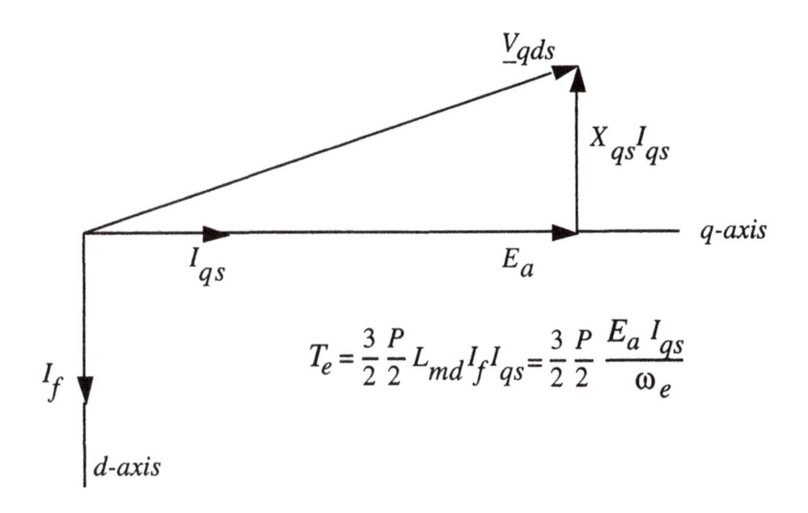

Figure 5.10 Steady state vector diagram for field orientation ($\gamma = 0, I_{ds} = 0$)

5.6 Torque Control Implementations – Synchronous Machines

The implementation of the angle control and field orientation concepts developed in the preceding sections calls for control of the magnitude and phase of the stator current with respect to the location of the field winding axis. In general, this vector control of the stator current must be maintained for both steady state and transient conditions. Several possible implementation schemes are now presented.

5.6.1 Torque Control Using Field Orientation with a CSI

Figure 5.11 suggests a direct implementation of field orientation ($\gamma = 0$) using absolute rotor position sensing and a CSI. With $\gamma = 0$, the stator current is entirely q–axis current and is equivalent to a torque command as illustrated. The rotor position information is directly utilized to set $\gamma = 0$ by controlling the inverter firing times. Because of the commutation delay inherent in the CSI, some form of compensation is necessary to maintain $\gamma = 0$ for different levels of current and different operating frequencies. This compensation is suggested in the figure by the commutation delay compensation block and by the inclusion of a phase regulator block. In principle, a γ^* command could be entered in

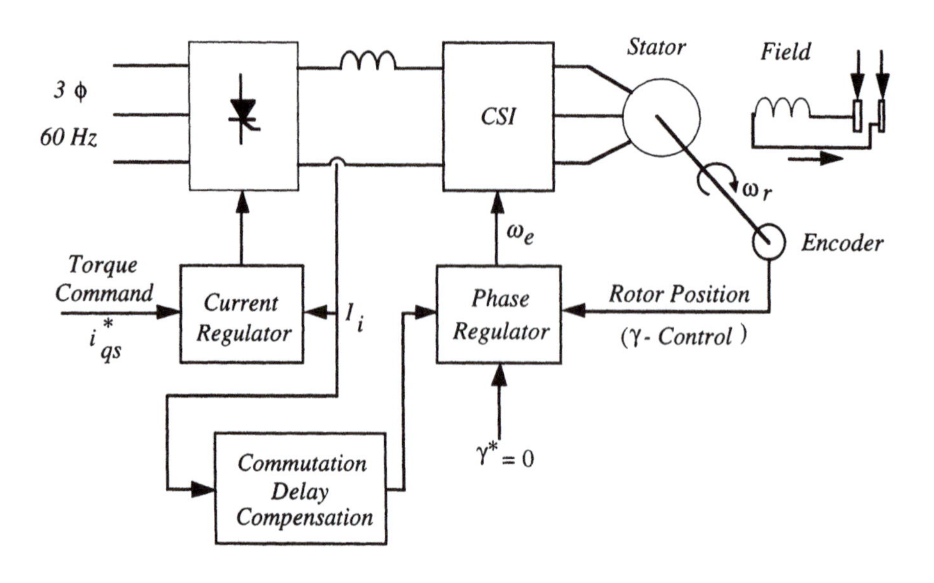

Figure 5.11 Torque control via field orientation using a current regulated CSI

the phase regulator in order to operate at other than $\gamma = 0$. In this case the input current would no longer represent a torque command since it would contain both a *q*–axis and a *d*–axis component. An implementation for such a scheme will be presented after introducing the concept in connection with the current regulated based implementation of angle control in the next section.

5.6.2 Torque Control Using a CRPWM

The regulation of the stator current by means of a fast switching power converter provides a conceptually simple means for implementing torque control with independent *q*–axis and *d*–axis current inputs [6]. Figure 5.12 illustrates the basic system typically called a current regulated pulse width modulated (CRPWM) inverter. In essence, all that is required is to use absolute rotor position information to convert the i_{qs}^* and i_{ds}^* commands in the rotor reference frame to a stator reference. The stator referred currents, at stator frequency, become the current commands for the CRPWM as shown in the figure. Normal field orientation is obtained by simply setting $i_{ds}^* = 0$. Other choices for i_{ds}^* allow for controlling the motor power factor or other performance features.

The rotor to stator transformation indicated in block form in Figure 5.12 is the same basic transformation discussed in Chapter 2. The purpose here is to

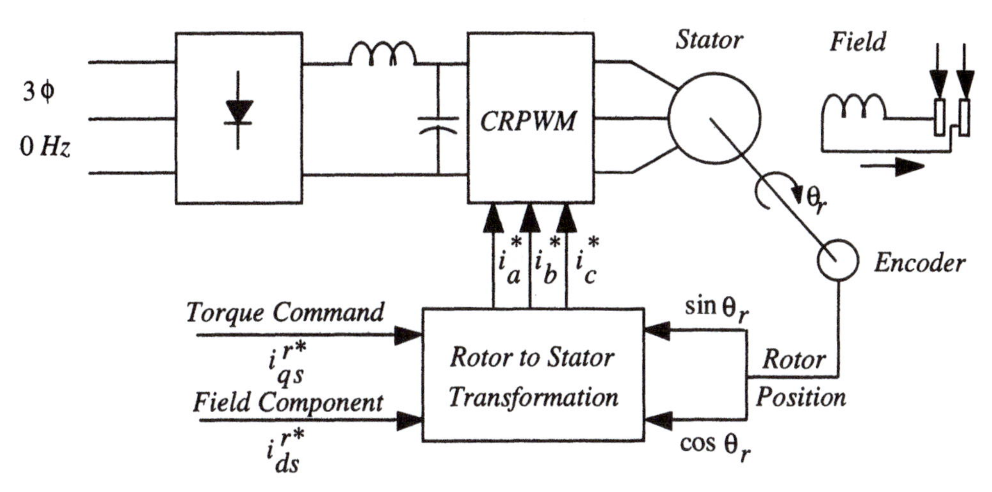

Figure 5.12 Torque control using a CPRWM. Field orientation requires $i_{ds} = 0$, $(\gamma = 0)$

convert the dc signals representing the torque command i_{qs}^* and the field component command i_{ds}^* into ac signals which become the current commands for the CRPWM. In complex vector form the rotation transformation is simply

$$i_{qds}^{s*} = e^{j\theta_r} i_{qds}^{r*} \tag{5.6–1}$$

since the commands are in a rotor reference and must be converted to a stator frame. The two phase stator current commands i_{qs}^{s*} and i_{ds}^{s*} are then converted to three phase current commands i_{as}^{s*}, i_{bs}^{s*} and i_{cs}^{s*} and supplied to the CRPWM. The actual equations implemented in the *rotor to stator transformation* block in Figure 5.12 are then the equivalent of the combined *rotating to stationary transformation* and the *two phase to three phase transformation* resulting in the expressions

$$i_{as}^{s*} = i_{qs}^{r*} \cos\theta_r + i_{ds}^{r*} \sin\theta_r \tag{5.6–2}$$

$$i_{bs}^{s*} = \left(-\frac{1}{2} i_{qs}^{r*} - \frac{\sqrt{3}}{2} i_{ds}^{r*}\right)\cos\theta_r + \left(\frac{\sqrt{3}}{2} i_{qs}^{r*} - \frac{1}{2} i_{ds}^{r*}\right)\sin\theta_r \tag{5.6–3}$$

$$i_{cs}^{s*} = \left(-\frac{1}{2} i_{qs}^{r*} + \frac{\sqrt{3}}{2} i_{ds}^{r*}\right)\cos\theta_r + \left(\frac{\sqrt{3}}{2} i_{qs}^{r*} + \frac{1}{2} i_{ds}^{r*}\right)\sin\theta_r \tag{5.6–4}$$

For field orientation ($\gamma = 0$) the flux command current i^{r*}_{ds} would be zero. In later sections dealing with induction machines, the angle θ_r will be chosen as the position angle of a particular flux vector in the machine but the same equations are applicable with the new choice of angle.

5.6.3 Magnitude – Angle Resolver and Use in CSI Torque Control

The CSI torque control shown in Figure 5.11 can be generalized to the case of independent *q*– and *d*– axis input currents by introducing the concept of a magnitude–angle (or vector) resolver. The concept is illustrated in Figure 5.13 and

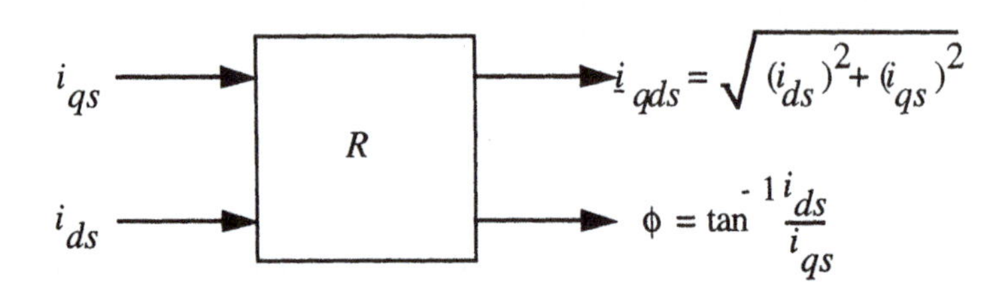

Figure 5.13 *d–q* to magnitude–angle resolver

is easily understood in the context of thinking of the *d,q* set as a vector. From this perspective, the resolver simply expresses two ways to describe a vector; orthogonal components or polar form.

Using a resolver as an input element for the CSI system of Figure 5.11 permits treating the input quantities as i^{r*}_{qs} and i^{r*}_{ds} as in Figure 5.12. The magnitude output of the resolver would become the current command of the CSI (which would no longer be a torque command). The angle output would become a γ command and would be fed to the phase regulator to be combined with the rotor position information. The combined θ_r signal and the ϕ signal from the resolver would set the phase of the current to give the proper position to the current vector $\underline{i}_{qds}$ in the rotor reference frame. With this approach the CSI system can be made fully equivalent to the system of Figure 5.12. Again compensation for the commutation delay would be required to obtain equivalent performance from the two systems. In both cases, setting $i^{r*}_{ds} = 0$ results in field orientation ($\gamma = 0$) which would be the preferred option for torque control and would be the obvious selection unless there is a need to regulate the power factor or some other terminal variable.

5.6.4 Torque Control Requirements – Synchronous Machine

The vector controlled synchronous machine systems of Figure 5.11 and Figure 5.12 can be interpreted in terms of the basic torque control requirements outlined in Section 5.2.

For the field oriented case of $\gamma = 0$, examination of Fig. 5.11 reveals that:

1) the current amplitude regulator provides the controlled current of requirement 1;

2) the field winding is the exact counterpart of the dc machine field winding;

3) the rotor position feedback loop controlling the phase of the stator current provides the field orientation of requirement 3.

This system is readily seen to be a close relative of the dc machine. It is, in fact, a machine with an 'electronic commutator' provided by the rotor position controlled inverter switching times. The physical commutator of a dc machine provides exactly the same function of rotor position controlled current switching. In actuality all three torque control requirements are provided in very much the same way as in the dc machine.

The only conceptual difference between the system of Figure 5.12 and the CSI system of Figure 5.11 is that the separate amplitude and phase controls of the CSI are combined into a single instantaneous current controller and both functions are handled simultaneously. Thus the CRPWM handles both the independent current excitation (requirement 1) and the field orientation (requirement 3) necessary for torque control in a single operation. In this system it is not possible to isolate the electronic commutator and the current control function from each other in the same physical way as in the system of Figure 5.11.

The current reference expressions in eqns (5.6–2) to (5.6–4), however, offer a new viewpoint which provides further insight into the torque control concept. Consider that for any fixed speed ω_r, the rotor angle can be expressed as

$$\theta_r = \omega_r t + \beta \tag{5.6–5}$$

where β is simply a position reference (position at $t = 0$). Using this result, the expressions in eqns (5.6–2) to (5.6–4) clearly indicate the frequency and phase of the current references and demonstrate that the stator currents are always at synchronous frequency and at a fixed phase with respect to the field axis. Note that at zero speed the currents are dc currents and no special problem is encountered. Note also, that when the speed is varying, the frequency (and hence phase) of the currents is also varying so as to maintain the required fixed space angle between the MMF and the field axis.

5.7 Electrical Measurement of the Rotor Field Angle θ_r

There are many situations in which the use of an encoder or resolver on the shaft of the machine is undesirable for cost or reliability reasons. It is possible to avoid the need for direct shaft position information by making electrical measurements from which the rotor position (more specifically, rotor flux position) information can be calculated. Such measurements generally require signal processing to obtain the desired information and the computations require knowledge of the stator parameters of the machine. The presentation of actual measurement schemes and the associated computational requirements are deferred to the next chapter on dynamic analysis.

Figure 5.14 illustrates the nature of an angle control or field orientation ($i_{ds}^{r*} = 0$) system based on electrical determination of rotor field angle. Although the illustration shows a CSI based system, the same techniques are applicable using a CRPWM. Many high power, load commutated inverter drives (LCIs) utilize a measurement and computational scheme like that illustrated in Figure 5.14 and do not employ an encoder or resolver.

5.8 Brushless DC Machines

The principles developed in the preceding sections have been widely used to create machines which possess properties similar to dc machines but do not employ a commutator and brushes. These machines are referred to collectively as brushless dc machines but there is, in fact, a great deal of variation in the details of how they are constructed and controlled [7,8]. While they all share the basic concept of field orientation and most utilize a rotating permanent magnet field and a stationary armature winding, there are major differences in flux distribution, position sensing and current control which result in a variety of performance characteristics.

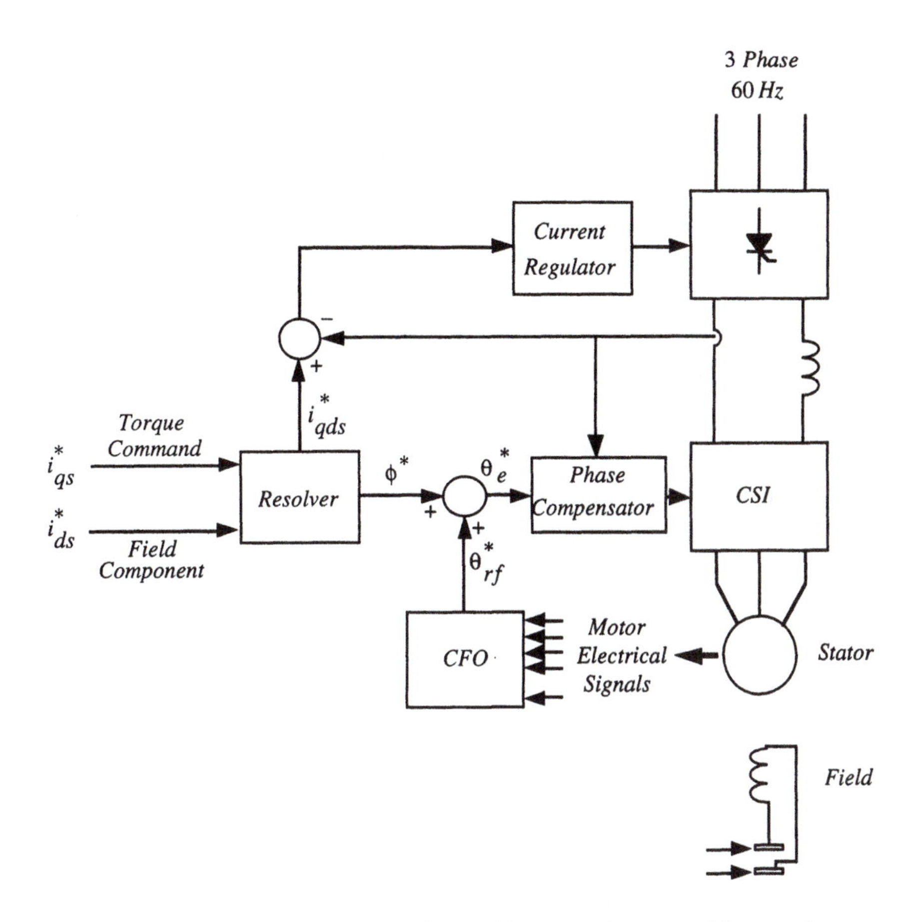

Figure 5.14 Field orientation ($i^*_{ds} = 0$) or angle control in synchronous machine using electrical estimation of rotor flux angle

5.8.1 Sine Wave Machines with Current Control

Permanent magnet machines with distributed windings resulting in sinusoidal back EMF voltages are used to produce high performance brushless dc machines for demanding servo applications [6,8]. These drives typically employ a high resolution encoder or resolver and a CRPWM power supply very much as illustrated in Figure 5.12 and are true field oriented drives. Because of the PM excitation a field component of current typically has little effect and is seldom used except to produce a very moderate increase in speed at reduced torque loads.

With a zero value of the field component of current, the armature MMF and field flux are orthogonal and the torque is directly proportional to current. With a PM field, the cross magnetization caused by this armature current has very little effect on the total flux and hence the linearity of the torque vs. current characteristic is excellent for a wide range of current. Since machines with nearly sine wave back EMF are readily built and with modern CRPWM supplies easily capable of excellent sinusoidal current waveforms, the torque ripple in this class of drives is generally very low. They therefore offer very nearly ideal dc machine characteristics with better torque/inertia and torque ripple properties and wider speed ranges than brush type dc machines.

5.8.2 Trapezoidal Wave Machines with Current Control

In a surface magnet PM machine with concentrated coil windings the waveform of the back EMF is typically quasi–square wave, or more accurately, trapezoidal [8,9]. This results from the nearly constant magnet MMF and the resulting nearly constant flux density along the rotor surface. An essentially constant torque can be obtained by controlling the current to be distributed to the individual phases so as to coincide with the back EMF as shown in Figure 5.15. Note that it is only necessary to keep the 120° current pulse centered on the back EMF wave. Since the 60° intervals of zero current correspond to the regions where the EMF is changing polarity, it is only necessary to have the constant voltage region be approximately 120° in width. The rise time of the current and any asymmetries in the structure create torque ripple.

Control of the machine is much simpler than for sine wave machines since only an indication of the center of the back EMF wave is needed. The most common method is the use of three Hall sensors to provide timing pulses for the three phases. In many machines of this type three timing pulses are used to provide gating information for an inverter which uses 120° switch operation as in a CSI as illustrated in Figure 5.16. A dc link current regulator completes the control.

Machines of this type are field oriented in the sense that the field flux and armature MMF are held to a fixed relative spatial position. However, limitations of the control (current rise time, etc.) can cause variations in this spatial relationships and create torque pulsations and reduced average torque.

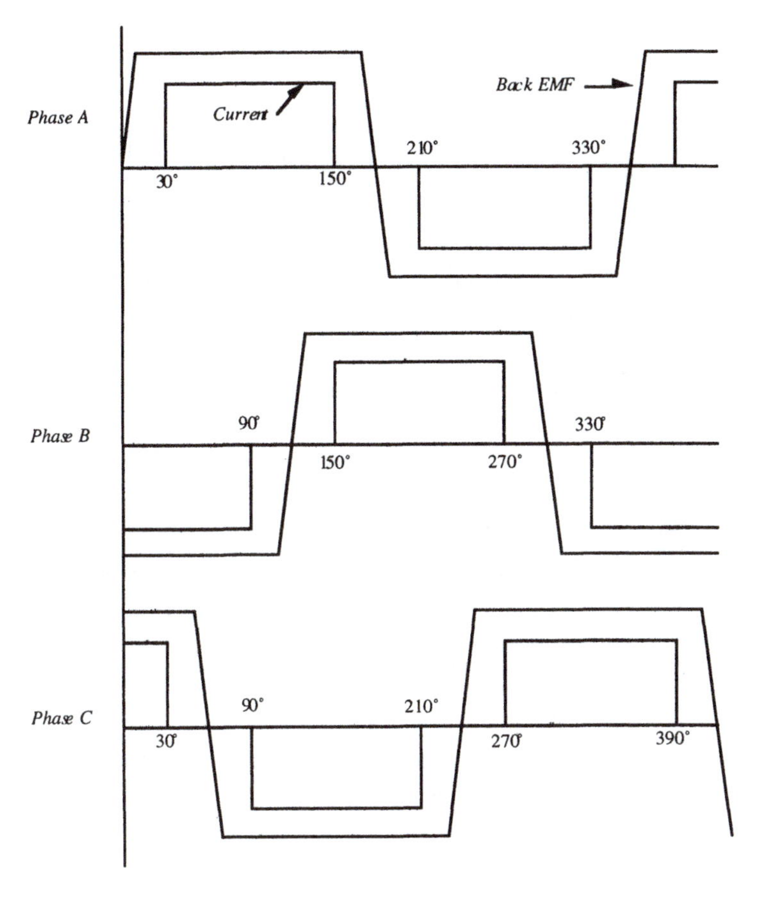

Figure 5.15 Back EMF and current in trapezoidal flux brushless dc machine

5.8.3 Other Brushless Machines

There are many other variants of the fundamental concept of field orientation which are included under the term brushless dc machine. To illustrate, consider the machine in Figure 5.17 where the rotor position information is obtained by sensing the EMF zeros. The inverter in this case is operated with 120° gating so there is always one phase off and the voltage of that phase is the back EMF. In this drive no current loop is used (except for current limit) and the speed is set by regulating the voltage applied to the motor. Drives of this type are generally limited to small sizes.

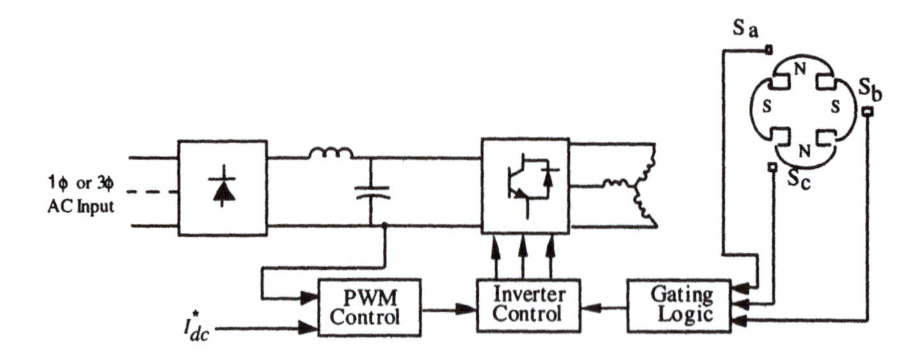

Figure 5.16 Trapezoidal EMF motor used as brushless dc machine

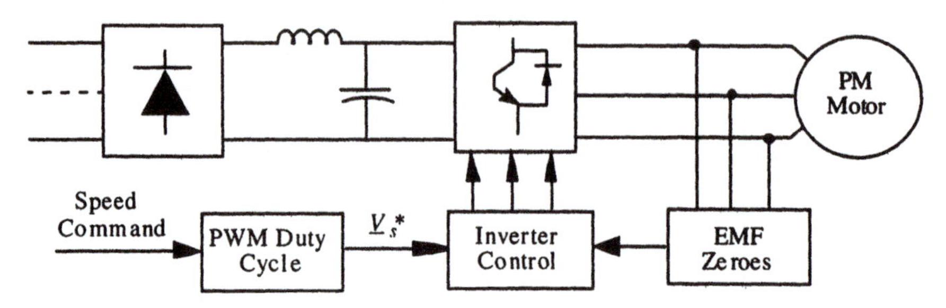

Figure 5.17 Brushless dc machine using voltage sensing to control spatial angle (no current loop)

5.9 Induction Machine Vector Control — Steady State

With synchronous machine vector control as background, the question of induction machine torque control can be approached from the perspective of asking how vector control of induction machine stator current can be employed to directly control torque [1]. The following development utilizes conventional steady state induction machine theory but attempts to parallel the development used in synchronous machines. Transient analysis using *d,q* modeling is treated in the next chapter.

5.9.1 Conventional Equivalent Circuit Considerations

Figure 5.18 is the conventional induction motor equivalent circuit containing two series reactances normally referred to as the stator and rotor leakage reactances. To parallel the synchronous machine model and *d,q* theory, the rotor

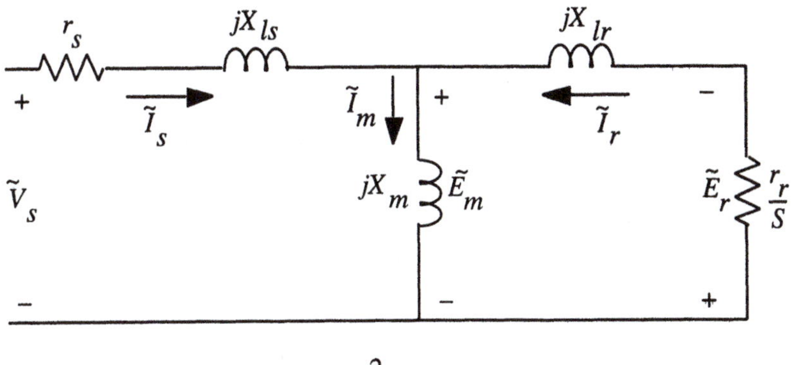

$$T = 3\,\frac{P}{2}\,\frac{I_r^2 r_r}{S\omega_e} = 3\,\frac{P}{2}\,\frac{E_r I_r}{\omega_e}$$

Figure 5.18 Conventional induction motor equivalent circuit showing rotor induced voltage E_r

current I_r is shown in a direction opposing the stator current; this is opposite to the usual steady state convention.

The torque is represented in the equivalent circuit as being proportional to the air gap power which is the power in the resistor r_r/S

$$T_e = 3\frac{P}{2}\frac{I_r^2}{S}\frac{r_r}{\omega_e} \tag{5.9–1}$$

where ω_e is the stator electrical frequency. This equation can be rewritten in terms of the voltage magnitude E_r across the resistor r_r/S to obtain

$$T_e = 3\frac{P}{2}\frac{E_r I_r}{\omega_e} \tag{5.9–2}$$

which is a similar form to the torque expression in a field oriented synchronous machine ($\gamma = 0$). In the synchronous machine, the induced voltage E_a was controlled directly by the field current; hence to parallel the synchronous machine, some means for independently controlling E_r in the induction machine is needed. If this can be accomplished, eqn (5.9–2) implies that torque control can be achieved in a similar manner as the synchronous machine. Note that the phase angle between $\tilde{E}_r$ and $\tilde{I}_r$ is automatically zero, corresponding to the field oriented synchronous machine ($\gamma = 0$), so that this aspect of torque control is automatically achieved in the induction machine (for steady state conditions).

5.9.2 *Modified Equivalent Circuit*

In order to illustrate the similarity in torque production between the dc and induction machines, the conventional equivalent circuit is not well suited to our purposes. It has been shown in Chapter 4 that there are many possible equivalent circuits for steady state operation. The very general model with an arbitrary referral ratio a is shown in Figure 5.19. It can be recalled that the

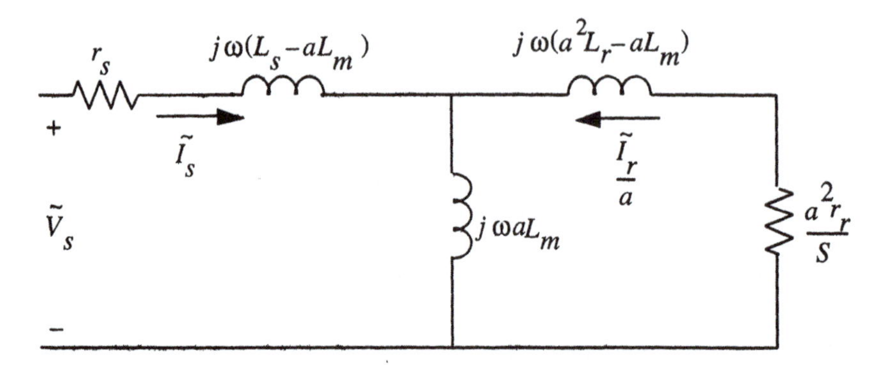

Figure 5.19 General equivalent circuit showing arbitrary value of referral ratio a ($a = N_s/N_r$ yields the conventional circuit of Figure 5.18)

choice of a is completely free (except for $a = 0$) and hence an infinite number of circuits can be obtained from Figure 5.19 by choosing different values of a. The conventional circuit is obtained by choosing a to be the stator to rotor effective turns ratio.

An especially useful form of the circuit for torque control analysis is obtained by choosing the referral ratio such that the series reactance in the rotor branch is zero. Setting this element equal to zero in the general circuit of Figure 5.19 yields the value of the required referral ratio as

$$a = \frac{L_m}{L_r} \tag{5.9–3}$$

With this choice for a the general circuit reduces to the circuit shown in Figure 5.20. Note that the new current in the rotor branch is L_r/L_m times the rotor cur-

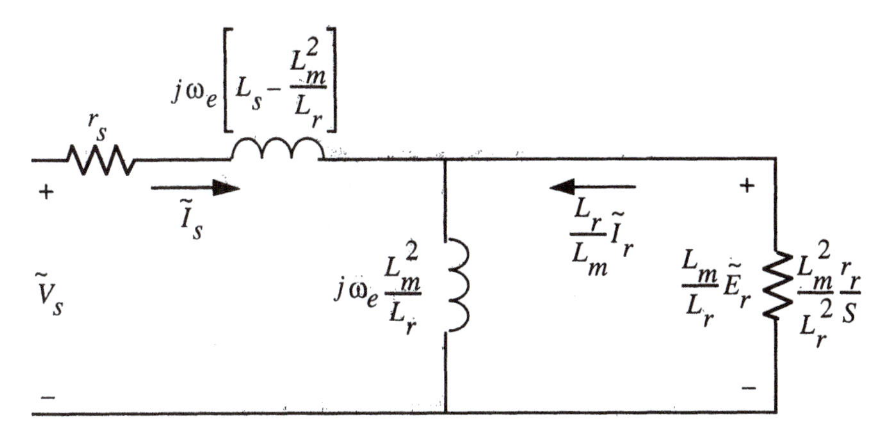

$$L_s - \frac{L_m^2}{L_r} = L_s' = \textit{stator transient inductance}$$

Figure 5.20 Induction motor equivalent circuit without rotor leakage – referral ratio $a = L_m/L_r$

rent in the conventional equivalent circuit and the new rotor voltage is L_m/L_r times the usual rotor voltage. Note also, that the new magnetizing reactance has this same voltage across its terminals and can therefore be directly associated with the flux producing the voltage E_r. In effect, the new circuit represents behavior in terms of rotor flux whereas the conventional circuit emphasizes air gap flux. This is important for torque control because it places in evidence the magnetizing component of current responsible for the rotor flux and E_r rather than that creating the air gap flux and E_m as in the conventional circuit. This same circuit was derived in Chapter 4 by starting from the synchronous frame model with the d–axis fixed to the rotor flux.

The new circuit is redrawn in Figure 5.21 with new labels on the circuit elements and current components. The stator side reactance is identified as the *stator short circuit transient reactance*

$$X_s' = \omega_e L_s' = \omega_e\left(L_s - \frac{L_m^2}{L_r}\right) \tag{5.9–4}$$

which is a well known induction machine transient parameter. More important, the stator current is shown divided into two components; one through the new magnetizing branch called $I_{s\phi}$ and one through the new rotor resistance called

I_{sT}. These are the two components of stator current which respectively control the rotor flux and torque as is demonstrated by the following analysis.

5.9.3 Torque Control in Terms of $I_{s\phi}$ and I_{sT}

The voltage E_r has been identified as the voltage drop across r_r/S. It is therefore equal to the time rate of change of the rotor flux

$$E_r = j\omega_e\lambda_r \tag{5.9–5}$$

From the circuit of Figure 5.21, the current $I_{s\phi}$ is given by

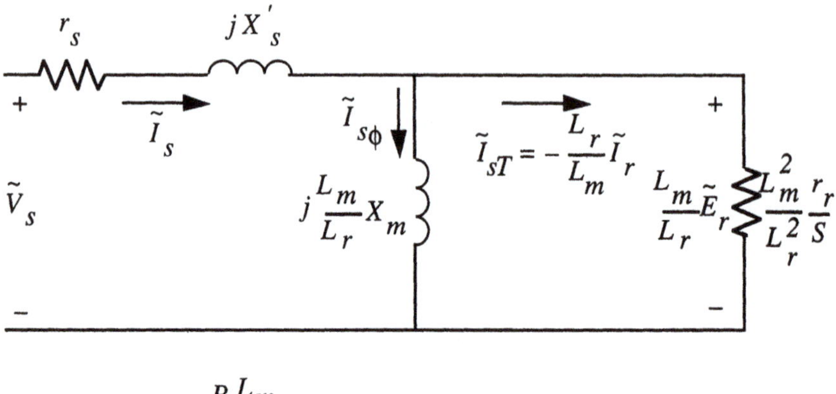

Figure 5.21 Equivalent circuit showing torque component (I_{sT}) and rotor flux component ($I_{s\phi}$) of stator current

$$\tilde{I}_{s\phi} = \frac{\frac{L_m}{L_r}\tilde{E}_r}{j\frac{L_m}{L_r}X_m} = \frac{\tilde{E}_r}{jX_m} = \frac{\tilde{E}_r}{j\omega_e L_m} \tag{5.9–6}$$

Combining eqn (5.9–5) and (5.9–6) yields

$$\tilde{\lambda}_r = L_m\tilde{I}_{s\phi} \tag{5.9–7}$$

which clearly demonstrates that the rotor flux is controlled by $I_{s\phi}$.

The torque component of stator current is immediately identified as

$$\tilde{I}_{sT} = -\frac{L_r}{L_m}\tilde{I}_r \tag{5.9–8}$$

and the torque developed can be expressed using eqns (5.9–2), (5.9–6) and (5.9–8) to form

$$T_e = 3\frac{P}{2}\frac{E_r I_r}{\omega_e} = 3\frac{P}{2}\frac{1}{\omega_e}(\omega_e L_m I_{s\phi})\left(\frac{L_m}{L_r}I_{sT}\right) = 3\frac{P}{2}\frac{L_m^2}{L_r}I_{s\phi}I_{sT} \tag{5.9–9}$$

which demonstrates the desired torque control properties in terms of the current components $I_{s\phi}$ and I_{sT}. Note the similarity to the field oriented synchronous machine (and the *dc* machine) with $I_{s\phi}$ playing the role of field current and I_{sT} the stator (or armature) current. A phasor diagram illustrating the current components is given in Figure 5.22.

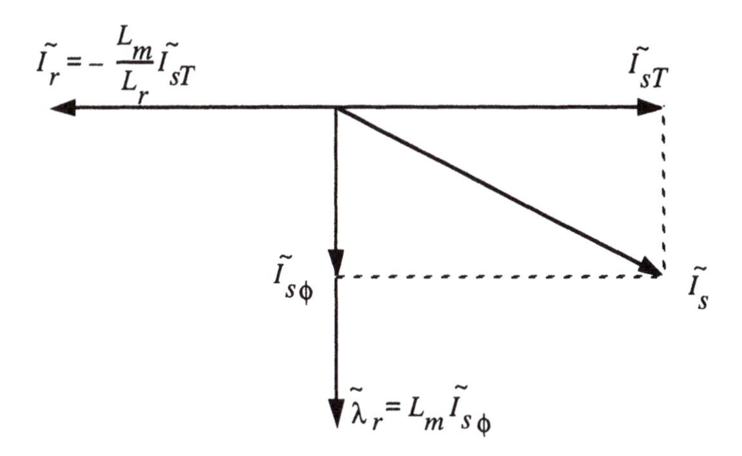

Figure 5.22 Phasor diagram showing torque and flux components of stator current

It is important to note that the circuit of Figure 5.21 contains one more relation involving the torque component of current and E_r. In particular

$$\tilde{I}_{sT} = \frac{\dfrac{L_m}{L_r}\tilde{E}_r}{\dfrac{L_m^2}{L_r^2}\dfrac{r_r}{S}} = \frac{L_r}{L_m}\frac{S\tilde{E}_r}{r_r} \tag{5.9–10}$$

Combining eqns (5.9–6) and (5.9–10) yields a relation between I_{sT} and $I_{s\phi}$

$$\tilde{I}_{sT} = j\frac{L_r}{r_r}S\omega_e\tilde{I}_{s\phi} \tag{5.9–11}$$

which is a consequence of the fact that the voltage across the magnetizing reactance and the voltage across the equivalent rotor resistance are equal. This is an extremely important result since it implies that for torque control in terms of $I_{s\phi}$ and I_{sT} as expressed in eqn (5.9–9), there is also a unique value of slip frequency associated with each set of values of the current components. Rewriting eqn (5.9–11) as a slip expression in terms of current magnitudes

$$S\omega_e = \frac{r_r I_{sT}}{L_r I_{s\phi}} \tag{5.9–12}$$

emphasizes this relationship and the slip variation which must accompany torque control via the current $I_{s\phi}$ and I_{sT}.

This relation is simply an expression of the fact that specifying stator current and slip frequency in an induction machine completely determines the torque. The special feature of the current components $I_{s\phi}$ and I_{sT} is that they specify both rotor flux and torque (and they are orthogonal). Note that once $I_{s\phi}$ and I_{sT} are chosen, eqn (5.9–12) determines the one and only slip frequency value which will yield the proper torque and flux. In fact, this represents a means of steady state control; choose $I_{s\phi}$ and I_{sT} and compute $S\omega_e$ from (5.9–12) to attain the proper operating point. This same concept will be shown to hold for transient conditions and is the basis of *indirect field orientation* [1].

It is also interesting to note that steady state torque control can be achieved in terms of air gap flux and the normal magnetizing current. Conceptually, this involves the conventional equivalent circuit of Figure 5.18 and the resolution of the stator current into components representing I_m and I_r in Figure 5.18. There is an expression like eqn (5.9–12) for the slip associated with these two current components to produce a torque expression in terms of I_m and I_r. Although this approach is often used it has several disadvantages compared to using rotor flux; in particular, the two current components are not orthogonal components and the decoupling for transient behavior to be demonstrated in the next chapter is not complete. In effect, what happens is that a torque increase in this system requires a build up of the rotor leakage flux whereas the rotor flux orthogonal component system already has the proper flux level in the rotor circuits. The controlled air gap flux system can be considered as somewhat like a synchronous machine angle control scheme in which $\gamma \neq 0$; there will be a torque transient in response to a torque command instead of the instantaneous torque response associated with systems having $\gamma = 0$. It is, however, possible to design a compensated field oriented controller based on air gap flux as will be shown in Chapter 6.

5.9.4 Terminal Behavior in Terms of $I_{s\phi}$ and I_{sT}

While torque control focuses on the relation between the current components and the flux and torque, the terminal behavior of the machine also must be evaluated. The required terminal voltage and stator power factor are important quantities with respect to machine and converter ratings. Figure 5.23 is a pha-

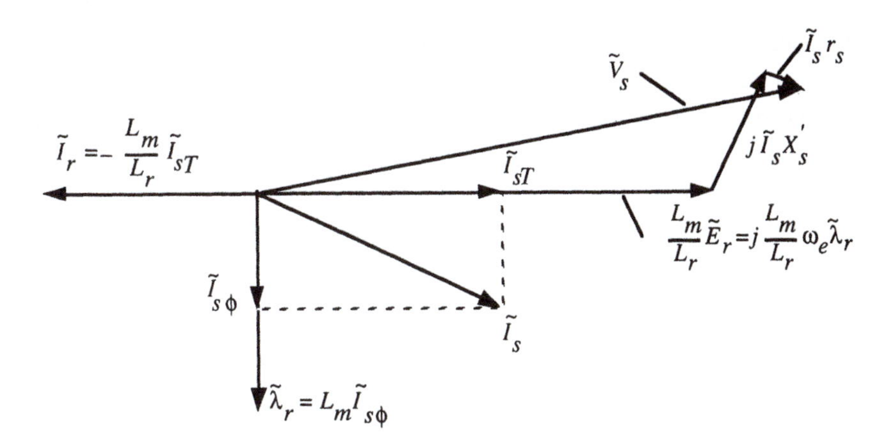

Figure 5.23 Phasor diagram showing rotor induced voltage and terminal voltage

sor diagram based on the equivalent circuit of Figure 5.18. The negative of the rotor induced voltage E_r multiplied by L_m/L_r is shown 90° ahead of the rotor flux vector. To obtain the terminal voltage, it is only necessary to add the stator $\tilde{I}_s r_s$ and $\tilde{I}_s X_s$ voltages exactly as is done in a synchronous machine. However, there is an important difference; in the induction motor the reactance is X'_s whereas in the synchronous machine the reactance is X_s. The transient reactance X'_s is approximately equal to the total leakage reactance

$$X'_s \cong X_{ls} + X_{lr} \tag{5.9–13}$$

and is usually of the order of about 0.2 per unit (pu). The synchronous reactance of a synchronous machine is a self–reactance and can be as large as 1–1.5 pu[†]. Thus, a typical induction machine will have an $I_s X'_s$ voltage of relatively small size even at base speed and torque while a synchronous machine can have a much larger reactance voltage. As can be seen from Figure 5.23, a larger reactance voltage leads to a larger terminal voltage and poorer power factor. Thus the inherently low value of X'_s is an advantage for the induction machine. In synchronous machines, a low value of X_{qs} is desirable for the same reason; better power factor and lower stator voltage.

Note that although the torque and flux relations involving $I_{s\phi}$ and I_{sT} are independent of rotor speed and frequency (a requirement of torque control), the terminal voltage and power factor relations are obviously speed dependent. At low speed the $\tilde{I}_s r_s$ voltage becomes dominant and the reactance voltage $\tilde{I}_s X'_s$ and induced voltage E_r become very small. Torque control via $I_{s\phi}$ and I_{sT} is independent of these variations so long as the current controller can continue to supply the commanded currents at the required terminal voltage and power factor.

5.10 Induction Machine Steady State *d,q* Model

The circuits and phasor diagrams of Section 5.9 have thus far been interpreted as conventional (stator referred) phasor models of the machine. It is much more common to use *d,q* modeling in treating vector control and it is therefore useful to introduce the steady state *d,q* model of the induction machine at this point.

---

†. In a salient pole synchronous machine the appropriate reactance is X_{qs} which is smaller than X_{ds} (typical range $0.2 < X_{qs} < 0.8$). The per unit system of units is discussed in Appendix A.

The most appropriate *d,q* model uses a synchronously rotating frame with the *d*–axis fixed to the rotor flux. This model was developed in Chapter 4 and is repeated in Figure 5.24 using slightly different notation to correspond more

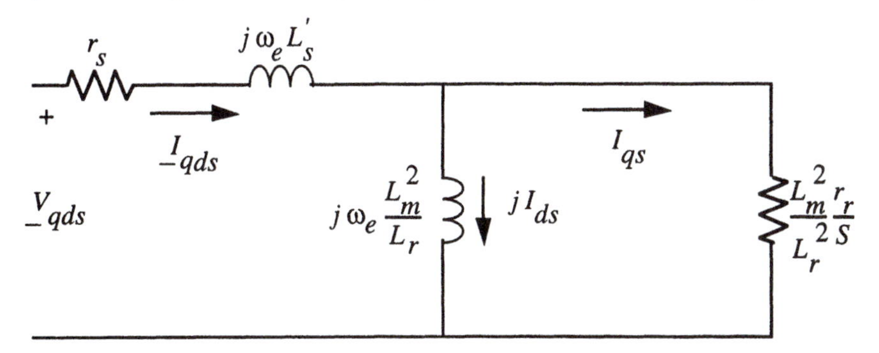

Figure 5.24 Equivalent circuit using complex steady state *d–q* currents – synchronously rotating reference centered on rotor flux vector

closely with the dynamic *d,q* analysis presented in later chapters. While this equivalent circuit has the same form as the stator referred circuit of Figure 5.21, the interpretation is quite different. In Figure 5.21, the circuit represents the relationships between conventional phasors; the phasors represent the magnitude and phase of stator frequency sinusoidally varying time functions. Figure 5.24 is a complex vector circuit diagram representing the relations between dc quantities in the *d*– and *q*– axes. This is, usually, a much more useful concept since the variables in this circuit can be directly associated with the dc control signals for I_{qs} and I_{ds} which are involved in the implementation of vector control. The vector diagram in Figure 5.25 shows these dc quantities and, using the obvious similarity between the stator referred and synchronous frame circuit models, illustrates the relationships between the *d,q* quantities I_{qs} and I_{ds} and the phasor magnitudes I_{sT} and $I_{s\phi}$*.

5.10.1 Example – Induction Machine Field Orientation

A 100 hp, 460 volt, induction machine is operated from an ideal controlled current electronic converter using field oriented control. The pu parameters of the machine are, in conventional per–phase symbols: (at rated frequency)

*. Phasors are usually expressed as rms quantities of sinusoids whereas *d,q* variables are usually peak values. Thus, the magnitudes of I_{sT} and $I_{s\phi}$ typically differ from I_{qs} and I_{ds} by $\sqrt{2}$.

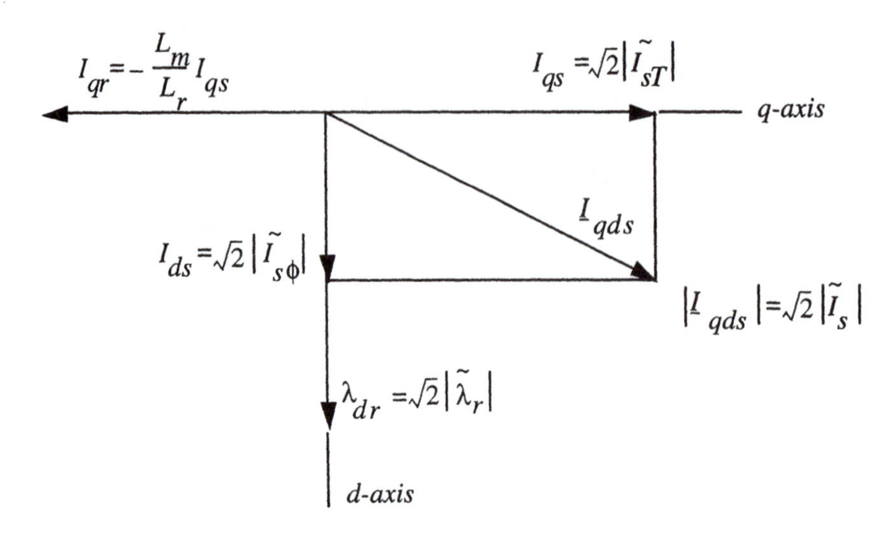

Figure 5.25 Steady state *d,q* currents – synchronously rotating reference centered on the rotor flux vector

$r_s = 0.015$ pu $\qquad x_{ls} = 0.10$ pu $\qquad x_m = 2.0$ pu

$r_r = 0.020$ pu $\qquad x_{lr} = 0.10$ pu

The rated slip is 0.0248. Express all answers in real units.

a) Find the required values of i_{qs} and i_{ds} and the slip relation relating $S\omega_e$, i_{qs} and λ_{dr} to produce operation at rated torque and rated speed if the terminal voltage and frequency are held at their rated values.

b) Find the final steady state torque and slip frequency if i_{ds} is reduced to one half its value in part a while the stator current amplitude is held constant at the value in part a. Find the terminal voltage and stator frequency if the new speed is twice that in part (a).

Solution – The solution assumes the following:

1) The machine has 4 poles

2) The per unit base is 100 hp, $460/\sqrt{3}$ V rms for phasors, $460\sqrt{2/3}$ V peak for *d-q* components.

The solution to part a also uses the standard steady state induction motor model (two leakages). It is definitely acceptable (and in many ways, easier) to use the motor model that separates torque and flux components. The most common error in so doing is failure to scale the components (r_r and L_m) appropriately.

a) Find the required values of i_{qs} and i_{ds} and the slip relation relating $S\omega_e$, i_{qs} and λ_{dr} to produce operation at rated torque and rated speed if the terminal voltage and frequency are held at their rated values.

The rotor flux linkage is: $\lambda_{dr} = L_r\, i_{dr} + L_m\, i_{ds}$ ($i_{dr} = 0$ in steady state). Hence, $\lambda_{dr} = L_m\, i_{ds}$. Therefore,

$$S\omega_e = \frac{r_r}{L_r}\frac{|I_{qs}|}{|I_{ds}|}$$

Rearranging,

$$|I_{qs}| = \frac{S\omega_e L_r |I_{ds}|}{r_r}$$

Also, by definition,

$$|I_s| = \sqrt{|I_{qs}|^2 + |I_{ds}|^2}$$

$$|I_{qs}|^2 = |I_s|^2 - |I_{ds}|^2$$

Combining expressions,

$$\left(\frac{S\omega_e L_r |I_{ds}|}{r_r}\right)^2 = |I_s|^2 - |I_{ds}|^2$$

To find I_s, use the steady state equivalent circuit

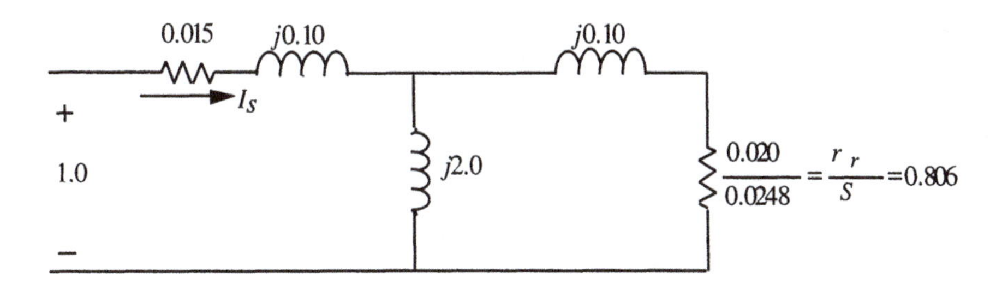

$$\tilde{I}_s = \frac{1 + j0}{(0.015 + j0.10) + [\,j2.0 \,||\, (0.806 + j0.10)\,]} = 1.053 - j0.7105$$

where '||' denotes the parallel combination of the two reactances on either side of the double vertical line

$$|I_s| = 1.27 \text{ pu}$$

Substituting,

$$\left(\frac{S\omega_e L_r |I_{ds}|}{r_r}\right)^2 = |I_s|^2 - |I_{ds}|^2$$

$$\left(\frac{(0.0248)\ (1.0)\ (2.0 + 0.1)\ |I_{ds}|}{0.020}\right)^2 = (1.27)^2 - |I_{ds}|^2$$

$$6.78|I_{ds}|^2 = 1.61 - |I_{ds}|^2$$

$$|I_{ds}| = 0.456\text{pu}$$

Base Current:

$$I_{base} = \frac{(100 \text{ hp})\ (746 \text{ W/hp})}{\sqrt{3}\ (460 \text{ V})} = 93.6 \text{ A rms } (132 \text{ A peak})$$

$$|I_{ds}| = (0.456 \text{ pu})\ (93.6 \text{ A}) = 42.7 \text{ A rms}$$

$$|I_{ds}| = 42.7\sqrt{2} = 60.3 \text{ A peak}$$

Then,

$$|I_{qs}| = \sqrt{(I_s)^2 - |I_{ds}|^2}$$
$$= \sqrt{1.27^2 - 0.456^2}$$

$$|I_{qs}| = 1.19 \text{ pu (rms)}$$

$$|I_{qs}| = (1.19)\ (93.6) = 111 \text{ A rms}$$

$$|I_{qs}| = 111\sqrt{2} = 157 \text{ A peak}$$

$$S\omega_e = (0.0248)\ (377\text{rad/s}) = 9.35 \text{ rad/s}$$

Checking,

$$S\omega_e = \frac{r_r}{L_r}\frac{|I_{qs}|}{|I_{ds}|}$$

$$= \left[\frac{(0.020)\,z_{base}}{(2.0+0.1)\,z_{base}/\omega_{base}}\frac{157A}{60.3A}\right] = 9.35 \text{ rad/s}$$

$$\omega_{base} = 377 \text{ rad/s}$$

b) Find the final steady state torque and slip frequency if i_{ds} is reduced to one half its value in part (a) while the stator current amplitude is held constant at the value in part (a). Find the terminal voltage and stator frequency if the new speed is twice that in part (a).

$$|I_{ds}| = \frac{1}{2}(0.456) = 0.228 \text{ pu} \quad (30.2 \text{ A peak})$$

$$|I_{qs}| = \sqrt{|I_s|^2 - |I_{ds}|^2} = \sqrt{1.27^2 - 0.228^2} = 1.25 \text{ pu} \quad (165 \text{ A peak}$$

Find z_{base}

$$z_B = \frac{V_B^2}{P_B} = \frac{(460)^2}{(100)\,(746)} = 2.84 \text{ ohms}$$

Find Torque

$$T_e = \frac{3}{2}\frac{P}{2}\frac{L_m^2}{L_r}I_{qs}I_{ds}$$

$$T_e = \left(\frac{3}{2}\right)\left(\frac{4}{2}\right)\frac{(2.0)^2\,(z_B/\omega_B)^2}{(2.1)\,(z_B/\omega_B)}(30.2)\,(165)$$

$$T_e = 214.8 \text{ N-m}$$

Find slip frequency

$$(S\omega_e) = \frac{r_r}{L_r}\frac{|I_{qs}|}{|I_{ds}|}$$

$$(S\omega_e) = \left[\frac{0.020}{2.1}\frac{(z_{base})}{z_{base}/\omega_{base}}\right]\frac{165 \text{ A}}{30.2 \text{ A}}$$

$$(S\omega_e) = 19.6 \text{ rad/s}$$

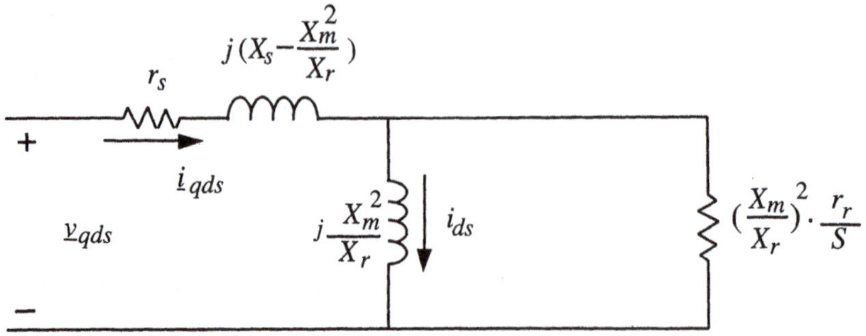

r_s $\quad j\left(X_s - \frac{X_m^2}{X_r}\right)$

$+$ $\quad \underline{i}_{qds}$

$\underline{v}_{qds}$ $\quad j\frac{X_m^2}{X_r}$ $\quad i_{ds}$ $\quad \left(\frac{X_m}{X_r}\right)^2 \cdot \frac{r_r}{S}$

$-$

Find terminal voltage and stator frequency when the speed is doubled.

$$\omega_r = 2\,(\omega_e - S\omega_e) = 2\,(377 - 9.35) = 735 \text{ rad/s}$$

The frequency is

$$\omega_e = \omega_r + (S\omega_e) = 735 + 19.0 = 754 \text{ rad/s}$$

$$\underline{V}_{qds} = \underline{I}_{qds}\left[r_s + j\left(X_s - \frac{X_m^2}{X_r}\right)\right] + \underline{I}_{ds}\left(j\frac{X_m^2}{X_r}\right)$$

Let $\underline{I}_{qds} = |I_{qs}| - j|I_{ds}|$ and $\underline{I}_{ds} = -j|I_{ds}|$. Remember the frequency is 754 rad/s (= 2.0 pu). This affects the reactance value proportionately. Substituting

$$\underline{V}_{qds} = (1.25 - j0.228)\left[0.015 + j\frac{754}{377}\left(2.1 - \frac{2^2}{2.1}\right)\right] + (-j0.228)\left(j\frac{2^2}{2.1}\right)\left(\frac{754}{377}\right)$$

$$\underline{V}_{qds} = 1.12 - j0.0256 \text{ pu}$$

$$|\underline{V}_{qds}| = |1.12 - j0.0256|\,(460)\,\sqrt{2/3} = 420.7\,\text{V}\ (514\ \text{V}\ l\text{–}l\ \text{rms})$$

5.11 Implementation of Field Orientation in Induction Machines

It should be apparent that the implementation of vector control and field orientation in induction machines can be carried out in the same manner as for synchronous machines if the position angle of the rotor flux is known. The basic implementation is illustrated in Figure 5.26. The only essential difference

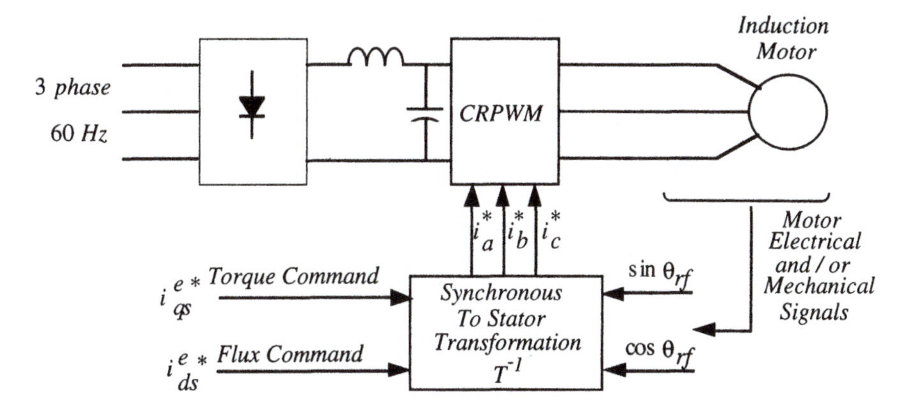

Figure 5.26 Basic induction motor field orientation system using a CPRWM

between the two types of machines is that the field winding (rotor flux) position angle in a synchronous machine is directly available as a measurable mechanical quantity while the position angle of the rotor flux in an induction machine is not mechanically measurable. Just as in the synchronous machine, two basic approaches to determining the rotor flux position angle have evolved; direct schemes which electrically determine the angle from flux measurements and indirect schemes which measure the rotor position and utilize the slip relation to compute the angle of the rotor flux relative to the rotor. The indirect method, which is analogous to rotor position measurement in a synchronous machine, is treated first.

5.11.1 Indirect Field Orientation

Indirect field orientation is based on the slip relation, eqn (5.9–13), repeated here for convenience in *d,q* notation

$$S\omega_e = \frac{r_r I_{qs}}{L_r I_{ds}} \tag{5.11–1}$$

This method makes use of the fact that satisfying the slip relation is a necessary and sufficient condition to produce field orientation, i.e. if the slip relation is satisfied, I_{ds} will be aligned with the rotor flux. At present this has only been shown for steady state operation but the dynamic analysis presented in the next chapter confirms that it is also true (in a slightly altered form involving the rotor flux linkage λ_{dr} instead of the current I_{ds}) under transient conditions.

Figure 5.27 illustrates the basic structure of an indirect field orientation

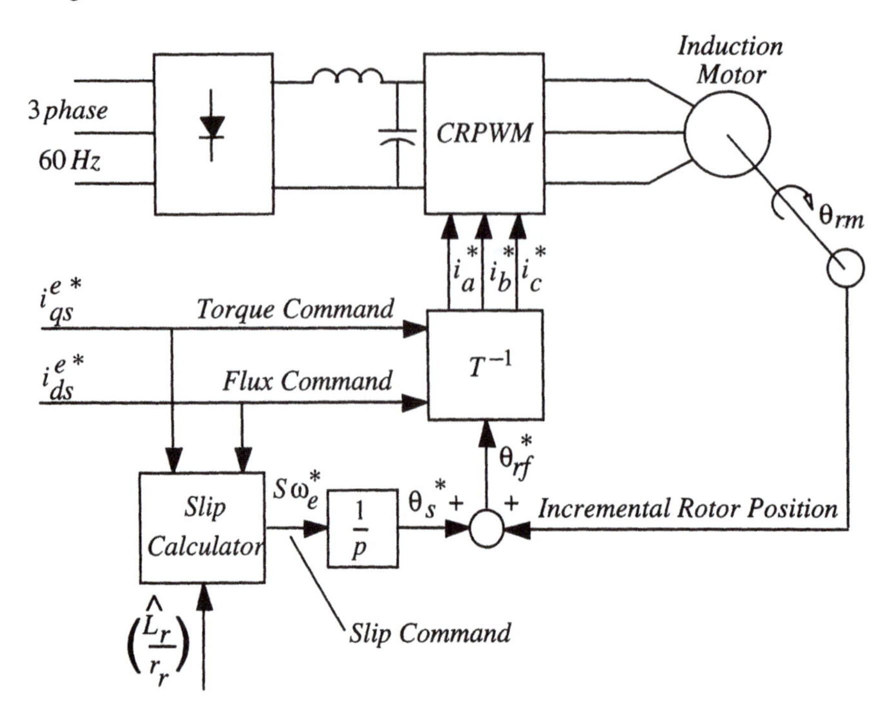

Figure 5.27 Indirect field orientation using a CRPWM

system using a current regulated PWM inverter. The commanded currents I_{qs}^* and I_{ds}^* are converted to stator referred reference currents by a synchronous to stator transformation just as in a synchronous machine. The field angle θ_{rf} is, however, generated by summing a rotor position signal and a slip position signal. Rotor position information is obtained directly from a shaft (incremental) encoder and the slip position is derived from a slip calculator implementing eqn (5.11–1). The conversion of the slip signal to an angle describing the position of the rotor field relative to the rotor (θ_s^*) and the summation of the two position angles θ_s^* and θ_r is normally implemented digitally to avoid drift and provide the needed accuracy. The transformation $\boldsymbol{T}^{-1}$ which appears in Figure 5.26 is defined in Figure 5.28 and this notation will be used through the remainder of the text.

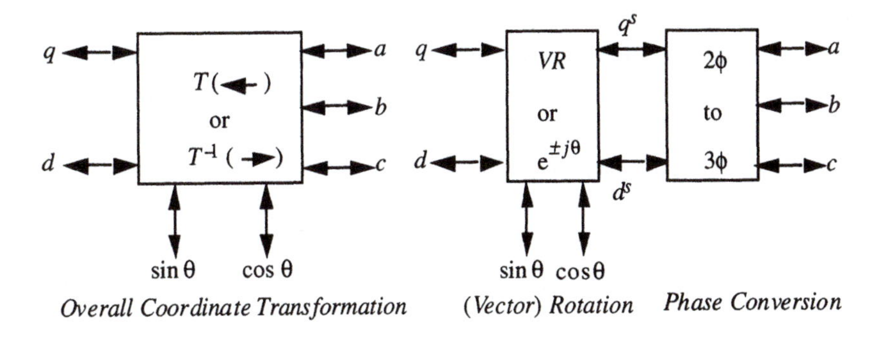

Overall Coordinate Transformation *(Vector) Rotation* *Phase Conversion*

$$f_q = f_a$$
$$f_d = -\frac{1}{\sqrt{3}}f_a - \frac{2}{\sqrt{3}}f_b$$

3ϕ *to* 2ϕ

$$f_a = f_q$$
$$f_b = -\frac{1}{2}f_q - \frac{\sqrt{3}}{2}f_d$$
$$f_c = -\frac{1}{2}f_q + \frac{\sqrt{3}}{2}f_d = -f_a - f_b$$

2ϕ *to* 3ϕ

PHASE TRANSFORMATION

$$\begin{bmatrix} f_q \\ f_d \end{bmatrix} = \begin{bmatrix} \cos\theta & -\sin\theta \\ \sin\theta & \cos\theta \end{bmatrix} \begin{bmatrix} f_q^s \\ f_d^s \end{bmatrix}$$

$$\boldsymbol{f}_{qd} = e^{-j\theta}\boldsymbol{f}_{qd}^{s}$$

Stationary to Rotating

$$\begin{bmatrix} f_q^s \\ f_d^s \end{bmatrix} = \begin{bmatrix} \cos\theta & \sin\theta \\ -\sin\theta & \cos\theta \end{bmatrix} \begin{bmatrix} f_q \\ f_d \end{bmatrix}$$

$$\boldsymbol{f}_{qd}^{s} = e^{j\theta}\boldsymbol{f}_{qd}$$

Rotating to Stationary

ROTATION TRANSFORMATION

Figure 5.28 Summary of transformations used in field orientation

The necessity to use angles in the summation in Figure 5.27 does not stem from the steady state theory presented in this chapter. In fact, the system will function to provide field orientation if the rotor position encoder is replaced by a tachometer and the calculated slip frequency is simply added to the tachometer signal to obtain a command for the stator frequency ω_e. The conversion to rotor flux angle would then occur just prior to the transformation to stator variables. This conclusion is also true in the transient state as will be demonstrated later. Such a system could be implemented using an analog tachometer and an analog slip calculating circuit and, except for offsets and drift in the analog circuits, would function as a field oriented system. The exception in the previous sentence is, however, an extremely important one. For example, an offset error in the tachometer (or in the analog circuitry) will immediately create an error in the resulting slip frequency in the motor. Recognizing the inherently small slip frequency associated with high efficiency induction motors, it is easy to see that the normal levels of offset and drift which occur in analog circuits (caused by temperature changes, for example) would create intolerable errors in the actual motor slip frequency. For these reasons, the addition of rotor speed and slip frequency is essentially always done with digital circuitry. This requires a digital signal describing the rotor speed. However, since it is ultimately an angle which is required in the transformation to stator variables, most systems use an encoder, convert the slip frequency signal to digital form and perform the addition of angles rather than of speeds.

Assuming the implementation is correctly (exactly) done, the system of Figure 5.27 is still subject to error because the rotor time constant τ_r directly affects the calculated slip. In the figure, the parameters L_r and r_r are entered with carets to indicate they are estimated values as opposed to actual values. To the extent that $\hat{L}_r/\hat{r}_r$ differs from L_r/r_r, the computed slip will be in error and correct field orientation will not be achieved. This can be a serious problem and will be examined in detail in Chapter 8.

The basic structure of an indirect field orientation system using a CSI is illustrated in Figure 5.29. As in Figure 5.27, the torque and flux command currents are used to generate a slip frequency command using a slip calculator. In addition, the current commands are converted to amplitude and phase commands using a resolver. The current amplitude command is directly employed as the reference for the CSI current regulator as shown in the figure. The phase command is passed through a lead circuit such that phase changes are added into the inverter frequency channel since these instantaneous phase changes

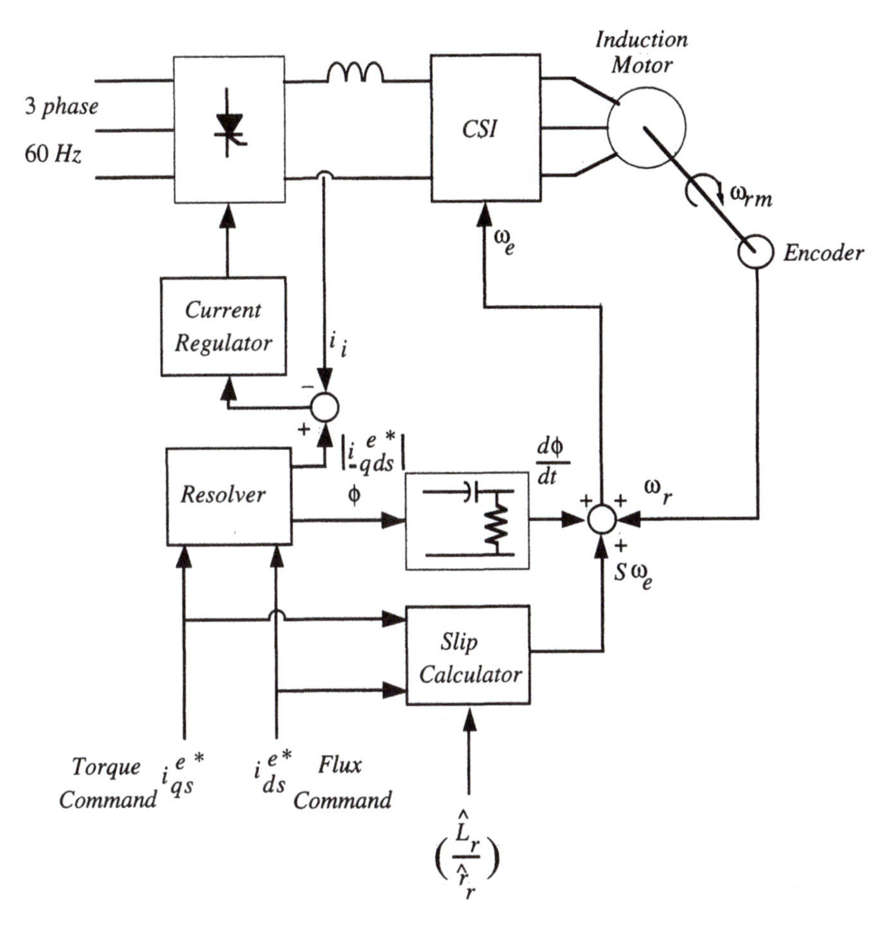

Figure 5.29 Indirect field orientation using a current regulated CSI

are not contained in the slip frequency command signal coming from the slip calculator.

Figure 5.30 illustrates the need for the separate addition of phase changes. For a change in the torque command from I^*_{qs1} to I^*_{qs2} (with a fixed I^*_{ds}) the current vector should instantaneously change from I^*_{qs1} to I^*_{qs2}. The slip frequency should also change instantaneously. The resolver does give the correct amplitude I^*_{qds} and the new slip frequency will be provided by the slip calculator. However, unless the phase change (frequency) $d\phi/dt$ is added as shown in Figure 5.29, the stator current command will correspond to the vector I^*_s in Figure 5.30 and there will be a phase error in the field oriented system. This

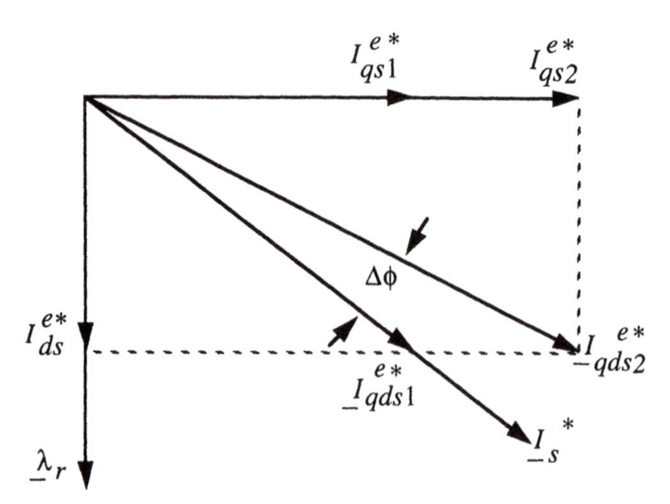

Figure 5.30 Phasor diagram showing required phase correction for system of Figure 5.29

would result in an instantaneous loss of field orientation which produces a very sluggish response (in both flux and torque). Such a system will ultimately attain the correct steady state torque and flux. The transient period is very long however, and is quite unacceptable for high performance applications. An analysis of the effects of incorrect orientation (detuning) is given in Chapter 7 and the instantaneous phase error described above can be considered as a form of detuned operation.

It is important to note that the necessity for separate consideration of phase changes does not exist in the PWM system of Figure 5.27. The synchronous to stator transformation used to generate the three phase current commands correctly includes the instantaneous phase changes associated with changes in torque command.

The indirect field orientation systems depicted in Figure 5.27 and Figure 5.29 are clearly very closely related to the slip command systems which have been employed for many years (especially in CSI drives). They differ primarily in that most other slip command systems are based on maintaining constant air gap flux while field orientation maintains constant rotor flux. They share with all slip command systems the basic problems of parameter sensitivity and the need for high accuracy in the tachometer loop. Digital techniques are adequate for the accuracy problem; the parameter sensitivity problem requires some form of adjustment of the control to account for motor parameter changes associated with flux level variations or temperature variations.

5.11.2 Direct Field Orientation

Direct field orientation in induction machines [3] is the exact counterpart of synchronous machine field orientation based on electrical measurement of the rotor field angle. As for synchronous machines, there are a variety of methods which can be employed to obtain the required rotor flux angle. Figure 5.31 illustrates the basic concept using a CRPWM. As in the case of the synchronous machine, if the measurement system also provides the rotor flux magnitude, additional (optional) loops can be added to regulate the torque and flux as shown in the figure. The consideration of the details of the measurement system is deferred to the next chapter on dynamics of induction machine field orientation.

Direct orientation can also be accomplished using a CSI. The basic system configuration is shown in Figure 5.32. Note that the phase angle of the stator current must be added to the rotor flux angle to properly orient the current (to avoid the problem illustrated in Fig. 5.27).

5.11.3 Torque Control Requirements

If we examine the induction machine field orientation systems of Figure 5.27 and Figure 5.29 or Figure 5.31 and Figure 5.32 to determine how the torque control requirements outlined in Section 5.3 are being met, the following observations can be made:

1) Independent current control of the stator current is employed to overcome the stator resistance, stator leakage inductance and induced voltage in the same manner as the dc machine.
2) Independent control of the flux is obtained by controlling $I_{ds} = \sqrt{2}I_{s\phi}$ independently of $I_{qs} = \sqrt{2}I_{sT}$. The stator current has two independently controlled components.
3) The spatial orientation of the rotor flux with respect to the current $I_{qs} = \sqrt{2}I_{sT}$ is maintained by using the rotor field angle θ_{rf} which is either obtained from flux measurement or computed from rotor position and slip frequency.

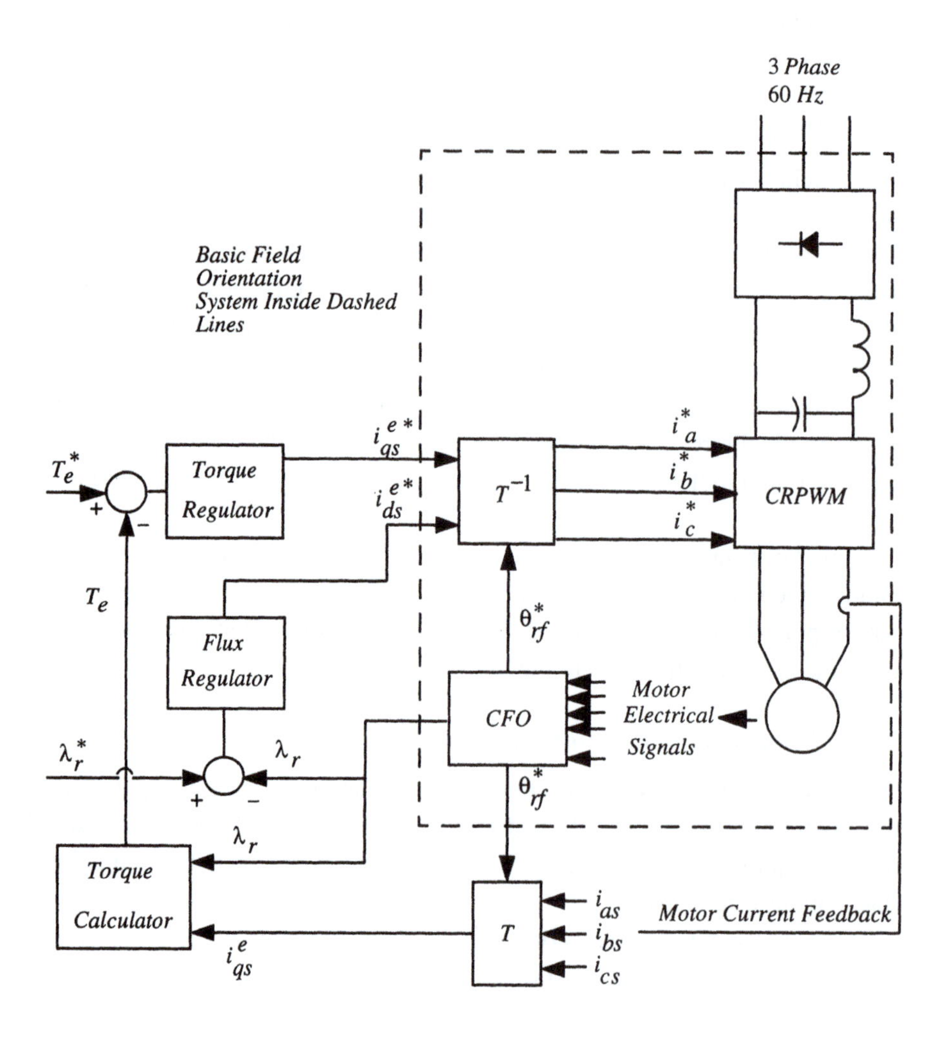

Figure 5.31 Direct implementation of induction machine field orientation using a CRPWM (torque and flux regulators optional)

Thus the electronic commutator concept introduced in connection with the synchronous machine drive carries over to the induction machine. The only major difference is that while the rotor flux position is directly measured in a synchronous machine, it is calculated in the induction machine drives of Figure 5.27 and Figure 5.29. Comparing these figures with Figure 5.11 and Figure 5.12 for the synchronous machine reveals the similarity.

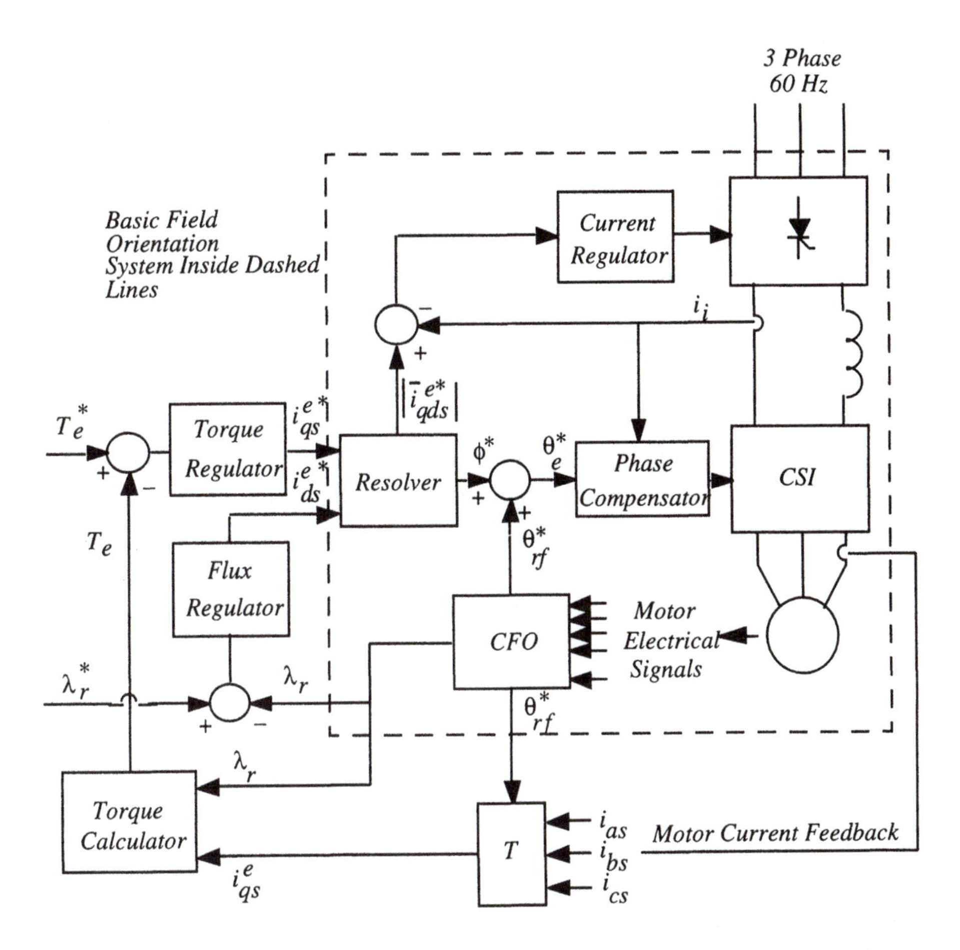

Figure 5.32 Direct implementation of induction machine field orientation using a CSI (torque and flux regulators optional)

5.12 A Basic Difference Between Synchronous and Induction Machine Field Orientation

While the similarity between synchronous and induction machine field orientation has been emphasized in this chapter, there is one important difference which should be noted. In a synchronous machine, the stator current creates a rotor magnetic field which is orthogonal to that produced by the field current (λ^r_{qr} in d,q notation). This effect is analogous to the cross magnetizing armature reaction in a dc machine and can cause a reduction in field flux as a result of saturation (demagnetizing effect of armature reaction as in a dc machine) [10].

In contrast, the induction machine under field oriented control always maintains zero flux in the axis orthogonal to the rotor flux (this is, in fact, the basic concept of induction machine field orientation). The field oriented induction machine is thus analogous to a dc motor with a compensating (pole face) winding. There is, therefore, no demagnetization caused by the torque component of stator current (*qs* component in *d,q* notation). Physically, the induced rotor current (I_r in Figure 5.23) functions as the compensating winding of a dc machine, preventing any net cross magnetization of the field (rotor) flux.

Thus, while the field oriented synchronous machine may exhibit a second order reduction of field flux at high torque, this effect is absent in field oriented induction machines. In surface magnet PM synchronous machines, the demagnetizing effect is likely to be small because of the long air gap associated with the magnets. The cross magnetizing ampere turns can, however, cause magnet demagnetization at high currents and must be carefully considered in motor design.

References

[1] K. Hasse, "Zur Dynamik Drehzahlgeregelter Antriebe Mit Stromrichter gespeisten Asynchron–Kuzschlublaufermaschinen" (On the Dynamics of Speed Control of a Static AC Drive with a Squirrel Cage Induction Machine), Ph.D. Dissertation, Techn. Hochschule Darmstadt, 1969.

[2] F. Blaschke, "The Principle of Field Orientation – the Basis for the Transvector Control of Three–Phase Machines", Siemens Zeitschrift, Vol. 45, No. 10, pp. 757–760, 1971.

[3] F. Blaschke, "Das Verfahren der Feldorientierung zur Regelung der Drehfeldmaschine" (The Method of Field Orientation for Control of Three Phase Machines), Ph.D Dissertation, University of Braunschweig, 1973.

[4] S. Miyairi and Y. Tsunehiro, "The Characteristics of Commutatorless Motor with SCR Inverter", Journal of IEE–Proceedings of Intermag Conference, April 1963.

[5] H. Sasajima, T. Tsuchiya and M. Naito, "Shunt Commutatorless Motor", Electrical Engineering in Japan, Vol. 88, No. 11, Nov. 1968, pp. 28–35.

[6] G. Kaufman, L. Garces and G. Gallagher, "High–Performance Servo Drives for Machine Tool Applications using AC Motors", IEEE–IAS Annual Meeting Conference Record, 1982, pp. 469–475.

[7] R. S. Colby, "Classification of Inverter Driven Permanent Magnet Synchronous Motors", IEEE–IAS Annual Meeting Conference Record, 1988, pp. 1–6.

[8] D. C. Hanselman, "Brushless Permanent Magnet Motor Design", (Book), McGraw–Hill,

Inc., 1994.

[9] B. V. Murthy, "Fast Response Reversible Brushless DC Drive with Regenerative Braking", IEEE–IAS Annual Meeting Conference Record, 1984, pp. 445–450.

[10] D.W. Novotny, and P.L. Jansen, "Field Oriented Control of Synchronous Machines," Journal of the Institution of Electronics and Telecommunication Engineers, Vol. 37, No. 1, 1991, pp. 46–56.

Problems

Problem 5–1 Constant hp Operation

For a 60 Hz induction motor with per unit parameters

$$r_s = 0.02 \qquad x_{ls} = 0.08 \qquad x_m = 2.2$$

$$r_r = 0.02 \qquad x_{lr} = 0.14$$

a) find the input current at base frequency, voltage and torque;

b) find the torque at $\omega_e = 1.5$ and $\omega_e = 2.0$ if V is held at 1.0 pu and I is held at the value found in (a);

c) find the approximate upper frequency limit of the constant power region of this machine.

Problem 5–2 Torque–Slip Frequency Characteristics

For the induction machine of problem 5–1:

a) find and plot the torque–slip frequency curve for operation at rated current for

$0 < S\omega_e < 10\pi$;

b) find and plot, on the same sheet, the torque–slip frequency curve for rated voltage, rated frequency operation;

c) repeat (b) at 5 Hz for rated volts/hz.

Problem 5–3 Synchronous Machine Field Orientation

A small permanet magnet synchronous machine has parameters,

$$E = 0.8pu \quad x_s = 0.4pu \quad r_s = 0.05pu \quad \text{at} \quad \omega_e = 1.0pu$$

The machine is operated from a controlled current electronic converter using rotor position feedback such that the internal power factor angle γ is always held at zero. Find:

a) the stator currents and terminal voltages to give output torques of 1.0, 0.5 and 0.0 when the stator frequency is 1.0;

b) repeat (a) at $\omega_e = 0.5$;

c) repeat "a" at $\omega_e = 0.0$.

Problem 5–4 Induction Machine Field Orientation

A 100 hp machine with the parameters below is operated from a controlled current electronic converter using field oriented control.

$r_s = 0.015$ pu $x_{ls} = 0.10$ pu $x_m = 2.0$ pu

$r_r = 0.020$ pu $x_{lr} = 0.10$ pu $r_m = 50$ pu at rated frequency

a) Find the values of I_{sT} and $I_{s\phi}$ and the slip relation relating $S\omega_e$, $I_{s\phi}$ and I_{sT} to produce operation at rated torque and rated speed if the terminal voltage and frequency are at rated values;

For each case below, find the current components, the slip frequency, the stator frequency and the terminal voltage:

b) 0.5 rated torque, rated speed, rated rotor flux (rated speed=0.9752 pu);

c) 0.5 rated torque, rated speed, 0.5 rated rotor flux;

d) zero torque, rated speed, rated rotor flux;

e) rated torque, 0.5 rated speed, rated rotor flux;

f) rated torque, zero speed, rated rotor flux;

g) zero torque, zero speed, rated rotor flux.

Problem 5–5 Field Oriented Operation of a PM Synchronous Motor with Voltage and Current Limits

For the PM synchronous motor of Problem 5–3 (parameters listed for convenience)

$$E = 0.8pu \quad x_s = 0.4pu \quad r_s = 0.05pu \quad \text{at } \omega_e = 1.0pu$$

operated above base speed subject to voltage and current limits of

$$V = 1.0pu \qquad I = 1.25pu$$

find:

a) the maximum speed attainable if field orientation is maintained;

a) the maximum speed attainable if field orientation is maintained;

b) the current and torque at $\omega_e = 1.20$ pu if field orientation is maintained;

c) the maximum speed attainable and the corresponding value of γ if phase advance of the current is utilized and γ is chosen to maximize the no load speed;

d) the maximum torque attainable and the corresponding value of γ if phase advance of the current is utilized to maximize the torque attainable.

Problem 5–6 Field Oriented Operation of an Induction Motor

Tests on a field oriented induction machine yield the following data:

	speed	frequency	stator current	stator voltage
	rad/s	Hz	amperes	volts
1) at rated flux–zero torque	188.5	60	21	200
2) at rated flux–rated torque	188.5	62.5	40	218

a) Use this data to find the rotor time constant, I_{sT} and $I_{s\phi}$ for rated torque and rated flux;

b) Find the stator current and frequency if the machine is adjusted to run at a speed of 377 r/s at 50 % of rated torque and 50 % of rated flux;

c) Estimate the stator voltage for part b. Explain your estimate.

Problem 5–7 Comparison of Field Orientation Drive Characteristics

Compare the steady state characteristics of field oriented control as applied to an induction machine and to a PM (permanent magnet) synchronous machine. Sketch vector diagrams in the synchronous frame comparing the two systems assuming each system has been adjusted to operate at rated conditions (rated rotor flux) at $\omega_r = \omega_{rB}$ and include diagrams at:

a) $\omega_r = \omega_{rB}$	b) $\omega_r = 0$	c) $\omega_r = 2\omega_{rB}$
$T_e = T_R$	$T_e = T_R$	$T_e = T_R$
and	and	and
$T_e = 0.0$	$T_e = 0.0$	$T_e = 0.0$

The diagrams should include $\underline{\lambda}_{qdr}$, $\underline{\lambda}_{qds}$, $\underline{I}_{qds}$ and $\underline{V}_{qds}$.

Use an induction machine with per unit parameters

$$r_s = 0.03 \qquad x_{ls} = 0.10 \qquad x_m = 2.0$$

$$r_r = 0.03 \qquad x_{lr} = 0.10$$

and let the PM machine have $x_s = 0.3$, $r_s = 0.03$, $x_{ls} = 0.10$ and an internal voltage such that it has the same power factor as the induction machine for rated conditions.

Discuss your results concerning the similarity (and differences) of the steady state operating points.

6 Dynamics of Vector Control and Field Orientation

6.1 Introduction

The steady state analysis in Chapter 5 has been used to demonstrate the basic concepts of vector control and field orientation in both synchronous and induction machines. The underlying motivation in Chapter 5 was to obtain torque (and flux) control independent of motor speed. Although the steady state arguments of Chapter 5 cannot provide information concerning the dynamic (transient) response, it was suggested that the same basic concepts as developed from steady state theory also produce dynamic decoupling and torque control in the dynamic state [1,2,3]. For high performance systems the dynamic response characteristics are as important as the steady state performance and this chapter is devoted to extending the concepts introduced in Chapter 5 to the dynamic state.

While it was convenient to treat the synchronous machine first in steady state analysis, the opposite is true for the general dynamic case. The symmetry of the induction machine results in a simpler dynamic model and will be treated first. It is also true that virtually all induction machine vector controllers employ a 90° spatial angle between the critical flux and MMF and are thus field oriented systems by our definition. There are, however, important practical applications of synchronous machines where other than a 90° spatial angle is employed in the vector controlled system. Such systems are not field oriented by our definition and do not result in transient state torque control. The analysis of such systems is more involved and is deferred until the end of the chapter.

6.2 Dynamics of Induction Machine Field Orientation

The vector control concepts leading to torque control and field orientation in induction machines as developed from steady state concepts in Chapter 5 will now be examined using a full d,q variable, transient state model. It will be shown that the basic concepts carry over to the transient state and that the only essential difference between steady state and transient behavior is the existence of a significant lag in the response of the flux to a flux command.

6.2.1 Induction Machine d,q Model with Axis Rotation at an Angular Velocity ω_e

The *d,q* axis model of an induction machine with the reference axes rotating at synchronous speed ω_e is

$$v_{qs}^e = r_s i_{qs}^e + p\lambda_{qs}^e + \omega_e \lambda_{ds}^e \tag{6.2–1}$$

$$v_{ds}^e = r_s i_{ds}^e + p\lambda_{ds}^e - \omega_e \lambda_{qs}^e \tag{6.2–2}$$

$$0 = r_r i_{qr}^e + p\lambda_{qr}^e + (\omega_e - \omega_r)\lambda_{dr}^e \tag{6.2–3}$$

$$0 = r_r i_{dr}^e + p\lambda_{dr}^e - (\omega_e - \omega_r)\lambda_{qr}^e \tag{6.2–4}$$

$$T_e = \frac{3}{2}\frac{P}{2}\frac{L_m}{L_r}(\lambda_{dr}^e i_{qs}^e - \lambda_{qr}^e i_{ds}^e) \tag{6.2–5}$$

where

$$\lambda_{ds}^e = L_{ls} i_{ds}^e + L_m (i_{ds}^e + i_{dr}^e) \tag{6.2–6}$$

$$\lambda_{qs}^e = L_{ls} i_{qs}^e + L_m (i_{qs}^e + i_{qr}^e) \tag{6.2–7}$$

$$\lambda_{dr}^e = L_{lr} i_{dr}^e + L_m (i_{ds}^e + i_{dr}^e) \tag{6.2–8}$$

$$\lambda_{qr}^e = L_{lr} i_{qr}^e + L_m (i_{qs}^e + i_{qr}^e) \tag{6.2–9}$$

The torque equation listed is one of many possible forms and has been selected for the specific purpose of this section.

6.2.2 Rotor Flux Referred Equations

The torque and flux control concepts developed in the previous chapter imply that the currents supplied to the machine should be oriented in phase and in quadrature to the rotor flux vector $\underline{\lambda}_{qdr}$. This can be accomplished by choosing

ω_e to be the instantaneous speed of $\underline{\lambda}_{qdr}$ and locking the phase of the reference system such that the rotor flux is entirely in the d–axis, resulting in

$$\lambda_{qr}^{e} = 0 \tag{6.2–10}$$

This expresses the field orientation concept in d,q variables. With this choice of reference, the d,q currents and rotor flux appear as shown in Figure 6.1. Note the obvious similarity of this d,q axis diagram to the phasor diagram

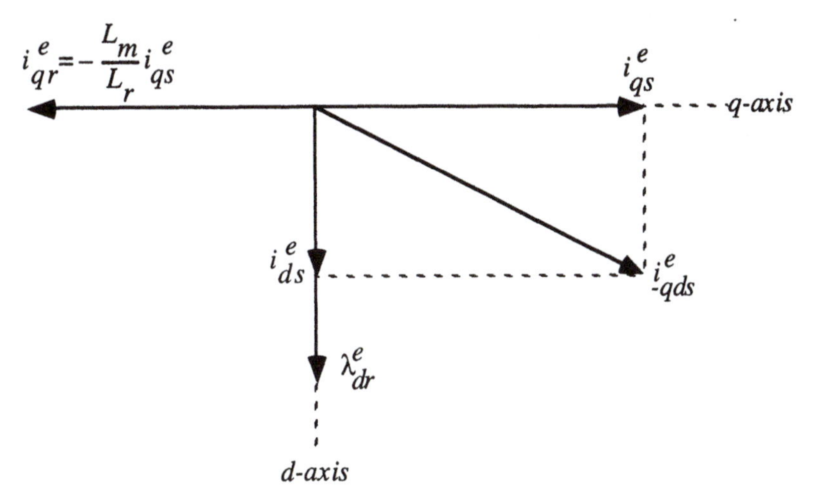

Figure 6.1 d,q currents with reference axes oriented to the rotor flux

showing the phasors $I_{s\varphi}$ and I_{sT} of steady state theory in Figure 5.22. Clearly the orientation of the d,q axes to the rotor flux produces a one to one correspondence between i_{qs}^{e} and I_{sT} and between i_{ds}^{e} and $I_{s\phi}$†. Assuming the machine is supplied from a current regulated source so the stator equations can be omitted, the d,q equations in a rotor flux oriented (field oriented) frame become:

$$0 = r_r i_{qr}^{e} + (\omega_e - \omega_r)\lambda_{dr}^{e} \tag{6.2–11}$$

$$0 = r_r i_{dr}^{e} + p\lambda_{dr}^{e} \tag{6.2–12}$$

†. Again it is common to use rms values for steady state phasors and peak values for d,q quantities.

$$\lambda_{qr}^{e} = L_m i_{qs}^{e} + L_r i_{qr}^{e} = 0 \tag{6.2–13}$$

$$T_e = \frac{3}{2}\frac{p}{2}\frac{L_m}{L_r}(\lambda_{dr}^{e} i_{qs}^{e}) \tag{6.2–14}$$

6.2.3 Dynamic Response of Field Oriented Induction Machines

Equations (6.2–11) to (6.2–14) describe the dynamic response of a field oriented induction machine. The torque equation, eqn (6.2–14), clearly shows the desired torque control property of providing a torque proportional to the torque command current i_{qs}^{e}. Note that when expressed in the form of eqn (6.2–14) (in terms of λ_{dr}^{e} and i_{qs}^{e}) the torque equation is the same for transient and steady state conditions.

A direct (ampere–turn) equilibrium relation between the torque command current i_{qs}^{e} and the rotor current i_{qr}^{e} follows immediately from eqn (6.2–13)

$$i_{qr}^{e} = -\frac{L_m}{L_r} i_{qs}^{e} \tag{6.2–15}$$

This is the same as the relation between I_{sT} and the steady state rotor current found in Chapter 5. The slip relation follows immediately from eqn (6.2–11)

$$\omega_e - \omega_r = S\omega_e = -\frac{r_r i_{qr}^{e}}{\lambda_{dr}^{e}} \tag{6.2–16}$$

or by using eqn (6.2–15),

$$S\omega_e = \frac{r_r}{L_r}\frac{L_m i_{qs}^{e}}{\lambda_{dr}^{e}} \tag{6.2–17}$$

which is the same as the steady state slip relation if the steady state result is expressed in terms of the flux λ_{dr}^{e}. The major departure from the steady state results of Chapter 5 occurs in the behavior of the rotor flux λ_{dr}^{e}. Equation (6.2–12) shows that in the steady state when $p\lambda_{dr}^{e}$ is zero, the rotor current component i_{dr}^{e} is zero. However, during flux changes, i_{dr}^{e} is not zero but is given by eqn (6.2–8)

$$i_{dr}^e = \frac{\lambda_{dr}^e - L_m i_{ds}^e}{L_r} \tag{6.2–18}$$

Combining (6.2–18) and (6.2–12) to eliminate i_{dr}^e yields the equation relating i_{ds}^e and λ_{dr}^e ($I_{s\varphi}$ and λ_r)

$$(r_r + L_r p)\,\lambda_{dr}^e = r_r L_m i_{ds}^e \tag{6.2–19}$$

or λ_{dr}^e can be eliminated to find the equation for i_{dr}^e

$$(r_r + L_r p)\,i_{dr}^e = -L_m p i_{ds}^e \tag{6.2–20}$$

This last relation clearly shows that i_{dr}^e exists only when i_{ds}^e is changing (changes in flux) and that in the steady state

$$\lambda_{dr}^e = L_m i_{ds}^e \qquad \text{(steady state)} \tag{6.2–21}$$

which demonstrates the correspondence between i_{ds}^e and $I_{s\varphi}$.

The close parallel to the steady state concepts of the preceding chapter is clear. Figure 6.2 emphasizes this correspondence in terms of torque production. In this figure the relation between the flux command current i_{ds}^e and the rotor flux λ_{dr}^e is shown as a first order transfer function with a time constant τ_r where,

$$\tau_r = \frac{L_r}{r_r} \tag{6.2–22}$$

In the steady state the relation between i_{ds}^e and λ_{dr}^e reduces to eqn (6.2–21) and the torque diagram of Figure 6.2 reduces to the steady state case of Chapter 5.

An alternative model for torque production is illustrated in Figure 6.3. Here the rotor flux is suppressed and the torque is represented entirely in terms of motor currents. Following transient changes in the flux command current i_{ds}^e, an induced *d*–axis rotor current is produced and has a transient influence on the torque. During steady state operation the *d*–axis rotor current is zero and the torque is proportional to the product of i_{ds}^e and i_{qs}^e.

The nature of the dynamic response for changes in torque command is illustrated at the left in Figure 6.4. With the flux command held constant, a change in i_{qs}^e is followed instantaneously by a corresponding change in i_{qr}^e. The result is an instantaneous change in the torque. The situation is more com-

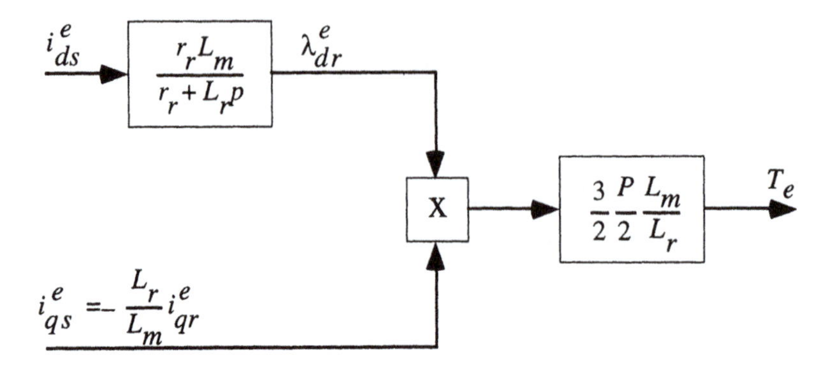

Figure 6.2 Torque production for field orientation in terms of rotor flux and q–axis stator current

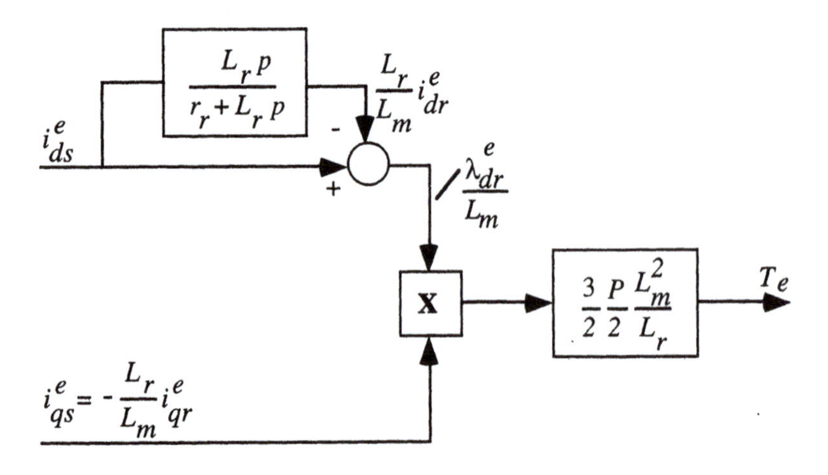

Figure 6.3 Torque production for field orientation in terms of currents

plex for a change in flux command. In this case, a change in i_{ds}^e induces a transient rotor current $i_{dr}^e(t)$ which subsequently decays with the rotor open circuit time constant L_r / r_r. The induced current prevents any sudden change in flux and the flux builds up to the new value exponentially with a time constant of L_r / r_r. This response is identical to that following a change in i_f in a dc machine; the induced eddy currents in the field poles being the equivalent of the induced rotor current i_{dr}^e.

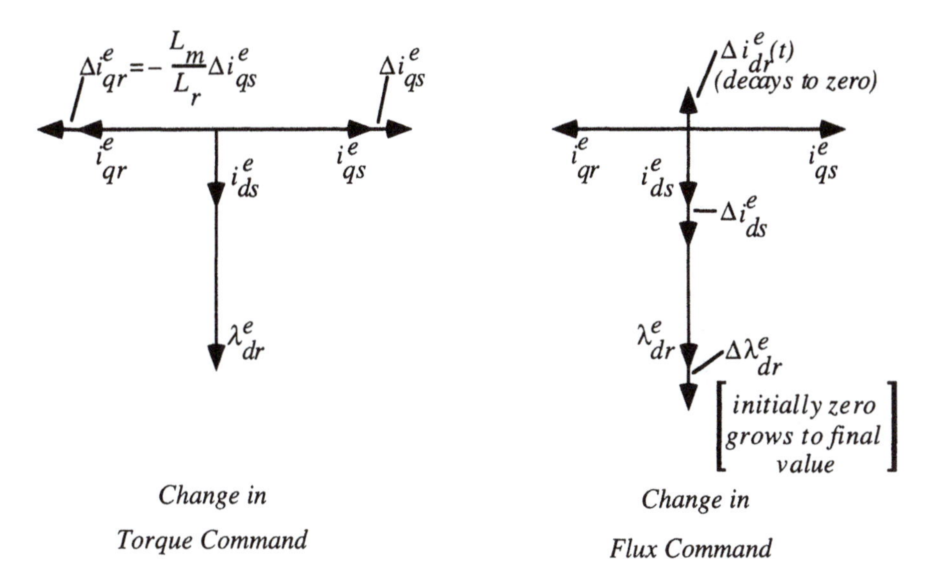

Figure 6.4 Illustration of response to step changes in torque command and flux command

In summary, the only difference between the steady state and transient behavior of a field oriented induction machine is the existence of a transient rotor *d*–axis current whenever flux changes are made. The flux accordingly follows an exponential change with time constant L_r / r_r as the transient *d*–axis current decays. Compensation for this effect can be included in the flux regulator but will require large pulses of *d*–axis stator current to attain adequate compensation. In effect, such compensation must overcome the tendency for the rotor flux to remain constant and this is a fundamental transient property of a closed circuit of low resistance.

6.2.4 Block Diagram of Field Oriented Induction Machine

Figure 6.5 shows a complete block diagram for a current fed induction machine in the rotor flux oriented reference frame defined by eqn (6.2–10). At the left side the input currents in a stator reference are indicated. A transformation to rotor flux oriented synchronous frame variables using the rotor flux angle θ_{rf} is shown. The rotor flux angle is obtained by adding the rotor speed ω_r and the slip speed $S\omega_e$ and integrating the result. The main portion of the diagram is based on eqns (6.2–15), (6.2–17), (6.2–19) and the torque equation, eqn (6.2–14). The difference between the electrical torque T_e and the load torque T_L is then shown as acting on the system inertia to produce the rotor

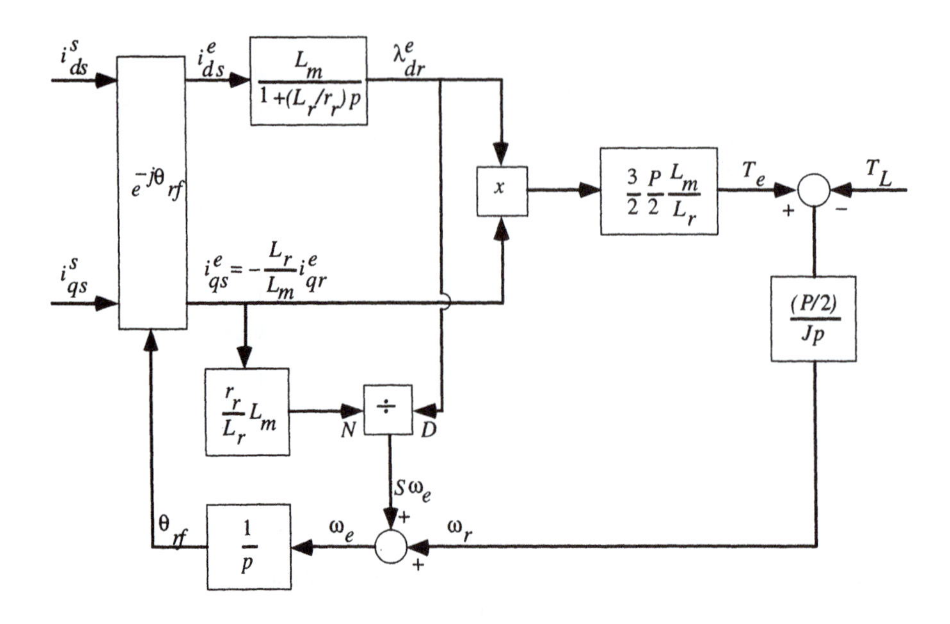

Figure 6.5 Complete block diagram of current fed induction machine in rotor flux oriented reference frame (field orientation)

speed ω_r. This signal is then used to complete the diagram. With the perspective used in constructing this machine model, the machine is viewed as internally consisting of a nonlinear slip frequency loop and a non linear torque loop. An indirect field oriented controller must perform the inverse calculations of the slip loop in order to obtain the field angle θ_{rf} from which the appropriate current commands for the machine (in a stator reference) can be obtained.

6.3 Indirect Controllers for Induction Machine Field Orientation

The indirect controller described in Chapter 5 is based on steady state considerations and will not properly maintain field orientation during transients which involve changes in the flux level. There are several versions of indirect controllers which correctly handle flux variations. These systems are described in this section [4].

6.3.1 *Indirect Controller with* i_{ds}^{e*} *and* i_{qs}^{e*} *as Inputs*

The simplest modification of the controller of Chapter 5 retains the currents i_{ds}^{e*} and i_{qs}^{e*} as inputs and only modifies the slip calculator to maintain orientation during flux changes. It employs the relation between the rotor flux λ_{dr}^{e} and the d–axis current given in eqn (6.2–19). In operational form, this relation is

$$\lambda_{dr}^{e} = \frac{L_m}{1 + p\tau_r} i_{ds}^{e} \tag{6.3–1}$$

Using this result in the slip relation of eqn (6.2–17) yields

$$S\omega_e^* = \frac{\left(\frac{1}{\hat{\tau}_r}\right) i_{qs}^{e*}}{\left(\frac{1}{1 + p\hat{\tau}_r}\right) i_{ds}^{e*}} \tag{6.3–2}$$

which shows that to maintain field orientation the lag in the flux response must be incorporated in a nonlinear slip calculator based entirely on current commands. The hat or caret in eqn (6.3–2) reminds us that the value used in the computation is not the actual value but only our best estimate of the actual rotor time constant. Figure 6.6 illustrates a controller based on this result.

Note that this controller is only a slightly modified form of the controller introduced in Chapter 5; the modification being the lag element in the slip calculator branch involving the flux command current i_{ds}^{e*}. For a fixed value of i_{ds}^{e*} (constant flux operation), this lag has no influence and the controller performs exactly as the simpler controller of Chapter 5. During the transient period following a change in i_{ds}^{e*}, the lag element delays the influence of the change in i_{ds}^{e*} to match the actual change in flux in the machine. The actual flux response follows eqn (6.2–19) and for this reason this controller is sometimes referred to as an *uncompensated flux response controller.*

6.3.2 *Indirect Controller with* λ_{dr}^{e*} *and* i_{qs}^{e*} *as Inputs*

A controller having λ_{dr}^{e*} and i_{qs}^{e*} as inputs has a fundamental difference from the preceding controller in that the controller must calculate the correct value of i_{ds}^{e*} as well as the slip. This is again accomplished using eqn (6.2–19); in this case the relation is used in the opposite sense as in eqn (6.3–1) by solving for i_{ds}^{e*} in terms of λ_{dr}^{e*}

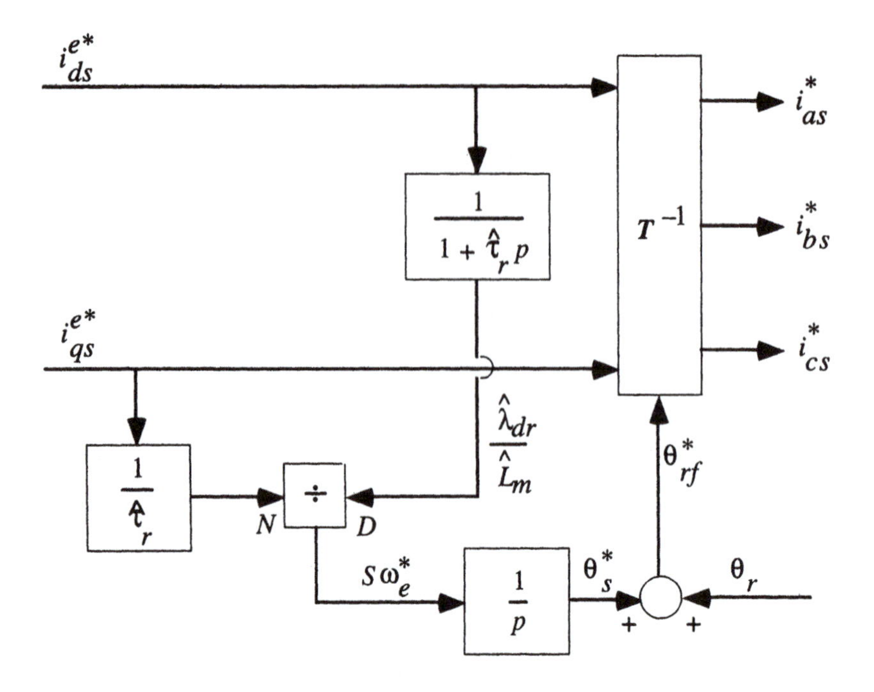

Figure 6.6 Indirect field orientation controller using input current commands (uncompensated flux response)

$$i_{ds}^{e*} = \frac{1}{\hat{L}_m}(1 + p\hat{\tau}_r)\,\lambda_{dr}^{e*} \tag{6.3–3}$$

The slip relation of eqn (6.2–17) is used directly and the resulting controller is illustrated in Figure 6.7.

This controller is substantially different from the controller of Figure 6.6 in that changes in flux command immediately alter the slip frequency calculation and also give rise to a *compensation component* of i_{ds}^{e*}. If the required i_{ds}^{e*} is within the capability of the stator voltage supply (and assuming ideal current control) the flux exactly follows the flux command. Clearly, the rate of change of the flux command must be limited in any practical system. Note that for a constant flux command this system is again identical to the simple system of Chapter 5.

The system of Figure 6.7 can be interpreted as an application of feedforward control theory to the block diagram of the induction motor given in Fig-

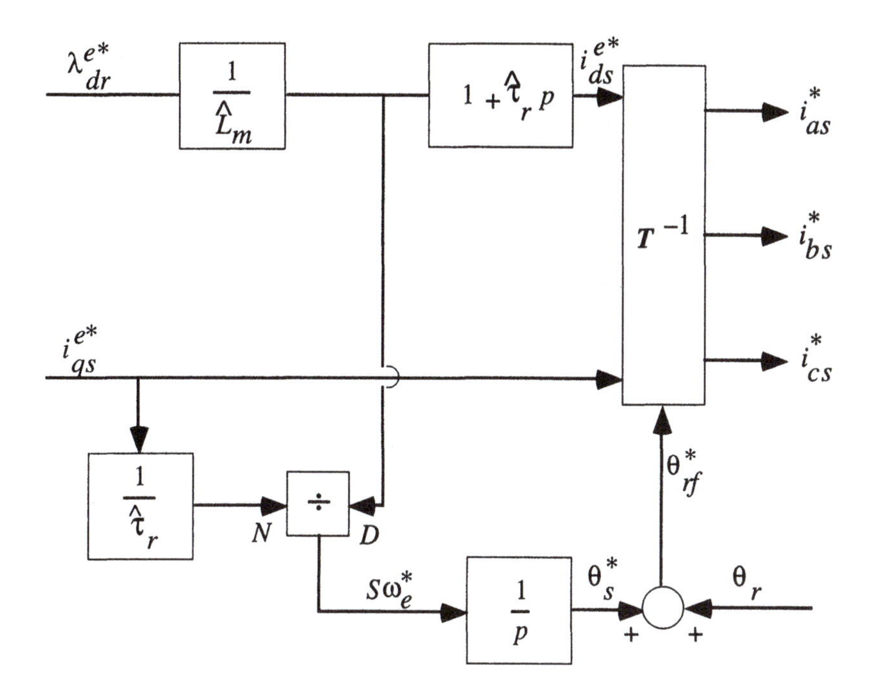

Figure 6.7 Indirect field orientation controller using flux and torque current commands (compensated flux response)

ure 6.5. With this interpretation the goal is to feedforward the proper θ_{rf}^{*} to control λ_{dr}^{e} and the torque (via i_{ds}^{e*}) in the presence of a system disturbance, in this case the rotor speed ω_r (or position θ_r). Since in this case the disturbance is measurable (via an encoder), feedforward theory would employ the inverse of the system transfer function and the measured disturbance to calculate the correct θ_{rf}^{*}. Thus, the inverse of the lag between i_{ds}^{e} and λ_{dr}^{e} is employed along with the slip frequency calculation to incorporate the rotor position (disturbance) in the feedforward controller. The result is the controller of Figure 6.7. Note that with this interpretation the rotor position signal is not part of a feedback controller but is instead a feedforward of information about a system disturbance.

6.3.3 Other Indirect Controllers

There are other forms which can be employed to implement indirect control. Most of these are variations of the systems in Figure 6.6 or Figure 6.7 altered to provide simpler implementations or modified to compensate for practical problems associated with real controllers. For example, some controllers patterned after the system of Figure 6.6 employ the measured currents in the slip calculator rather than the command currents. This allows extending the field oriented mode somewhat into the region where the current regulator is reaching its practical limits (either amplitude or frequency). A second example is the insertion of a limiter after the lead compensator for i^{e*}_{ds} in Figure 6.7 and shifting the input to the slip calculator to the output of this limiter. A lag element in the flux branch of the slip calculator is then required to maintain correct field orientation. This variation is shown in Figure 6.8.

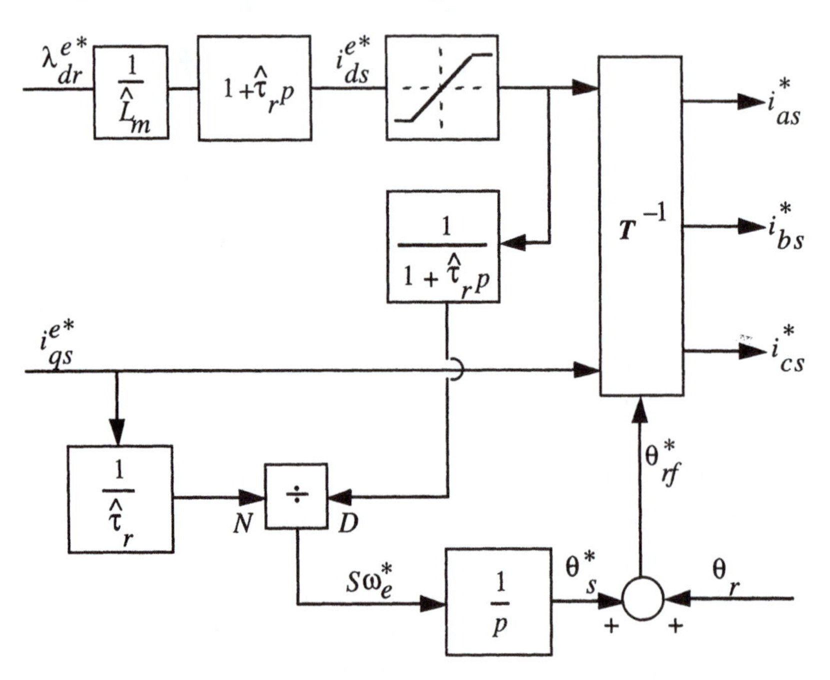

Figure 6.8 Indirect field orientation controller using flux and torque current commands with flux command current limiter

6.3.4 Indirect Control Using a CSI

Although the indirect controllers shown in Figure 6.6 to 6.8 are illustrated as developing command signals for a CRPWM inverter, they can also be employed using CSI systems. In this case the transformation block T^{-1} is replaced with a resolver and an angle summation as shown in Figure 6.9. The

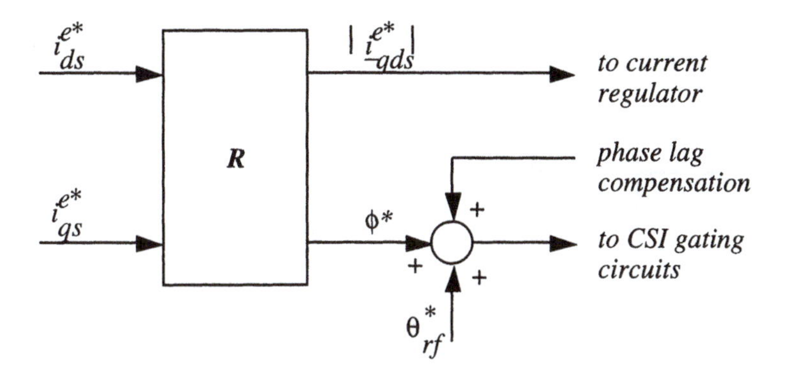

Figure 6.9 CSI implementation of field oriented control (replaces $\boldsymbol{T}^{-1}$ block of CRPWM systems)

input labelled *phase lag compensation* in the figure is included to emphasize that the commutation delay inherent in CSI operation must generally be compensated to achieve acceptable field oriented control. Note that the implementation shown in Figure 6.9 is equivalent to the scheme shown in Figure 5.26 where the addition of the angles is carried out in terms of frequencies. The arrangement shown in Figure 6.9 is well suited to digital implementation and provides more precise control of the angle of the stator current vector relative to the flux.

6.3.5 Indirect Field Orientation Start Up Transient

The issue of how the rotor flux builds up and becomes correctly oriented with the indirect field orientation controller from a start with no initial rotor flux and with only a relative sense of position is an interesting and informative subject for study [5]. A convenient means for carrying out the analysis is illustrated in Figure 6.10 in which the machine is modelled in the actual machine rotor flux frame and the controller is modelled in the controller's *estimated* rotor flux frame. This figure is essentially a composite constructed from the field oriented machine diagram of Figure 6.5 and the field oriented controller diagram of Figure 6.6. The error angle, $\Delta\theta_{rf}$, is the angle between the machine and controller axes and will become zero when the machine reaches the correct field oriented state. It is clear from Figure 6.10, and from the development in previous sections, that the transient flux build up process is independent of rotor speed and thus the analysis can be carried out at zero speed for simplicity.

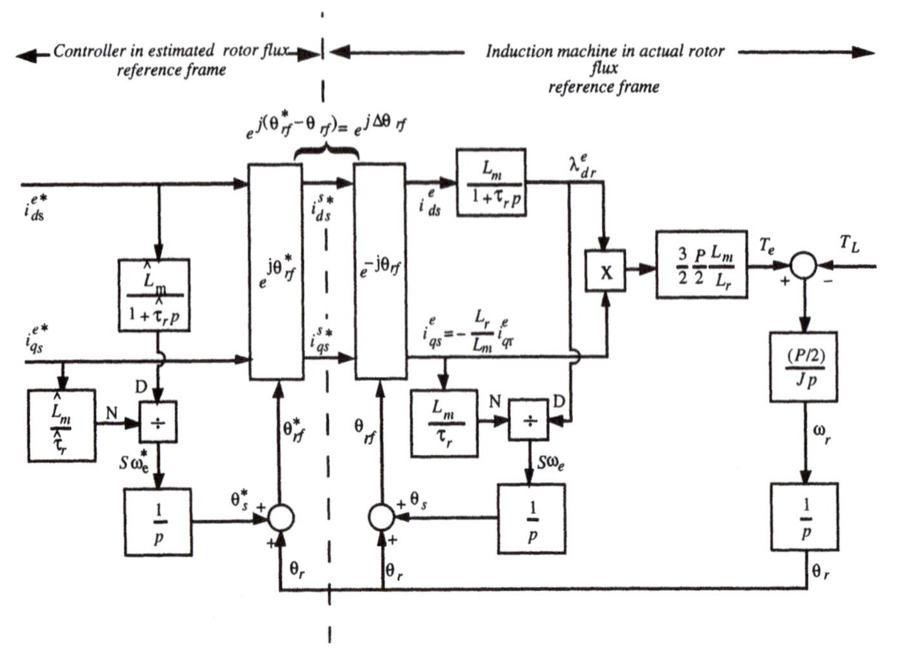

Figure 6.10 Induction machine indirect field orientation system for current source excitation showing error angle $\Delta\theta_{rf}$

In analyzing this transient process the ratio of the applied torque and flux command current components is an important parameter. It is convenient to define the commanded ratio of i_{qs}/i_{ds} as β, where

$$\beta = i_{qs}^{e*} / i_{ds}^{e*} \tag{6.3–4}$$

Case 1: Sudden application of flux command only, $\beta = 0$

Consider the start up condition where only a flux component ($i_{qs}^{e*} = 0$) of commanded current is suddenly applied to the machine. Initially both estimated and actual rotor flux angles are taken as zero and since $i_{qs}^{e*} = 0$ and speed = 0, dc is applied to the machine. The rotor flux is also initially zero and cannot change instantly. Mathematically at time zero there is a divide by zero and this results in an undefined '0/0' slip frequency. However, with the first increment of time a finite amount of rotor flux is built up and the slip frequency becomes zero, '0/finite = 0'. Physically, the rotor is at rest and the applied excitation is dc, therefore, the first application of current defines the machine *d*–

axis and results in a nearly equal and opposite induced rotor current. With zero slip frequency the applied dc current remains on the d–axis and the rotor flux builds up according to the rotor open circuit time constant and aligned with the flux command current. The build up of the flux is associated with the induced rotor current which continues to oppose the stator dc current and dies out as the flux grows. Ideally all of this occurs with no transient in torque, which says a free rotor would not have moved during the entire process.

Case 2: Sudden application of flux and torque command $\beta \neq 0$

Next consider the sudden application of stator current containing both torque and flux components. While this case is similar to that of only a flux command, there are distinct differences the most important of which is the existence of a finite commanded slip frequency. As in Case 1, at the first instant the imposed stator current excites a nearly equal and opposite rotor current vector as illustrated in Figure 6.11(a). The difference between the two collinear current vectors creates a small flux in the initial direction of the stator current vector which, therefore, becomes the initial motor d–axis. The error angle, $\Delta\theta_{rf}$, at the first instant is the angle between the current vector, which is also the machine d–axis, and the d–axis of the controller. This angle is illustrated in Figure 6.11(a).

As soon as this initial flux is established it interacts with the commanded dynamic slip frequency to create a rotor voltage and an induced current 90° away in space from this initial d–axis as a result of the cross coupling between axes in an induction machine, see eqn (6.2–11). As shown in Figure 6.11(b), this current interacts with the applied stator current and its induced, collinear rotor current and tends to rotate the resultant rotor flux and hence the machine reference frame in the direction to more nearly align it with the controller reference frame. At the end of the transient period the machine reference frame is aligned with the controller reference frame as shown in Figure 6.11(c) and only the q–axis component of rotor current remains. The actual transition can be overdamped or underdamped depending on the value of β and the transition time is governed by the rotor open circuit time constant as will be shown later.

Figure 6.12 and Figure 6.13 illustrate the results of simulations of the transient response following the sudden simultaneous application of q– and d–axis currents with varying ratios of i_{qs}/i_{ds} (β=0, 1/2, 1, 1.5 and 2) and with a constant 1.0 pu flux command. The rotor speed was held at zero throughout the simulations. Observe, in the expanded views of Figure 6.12, that from $t = 0$ to 20 μs

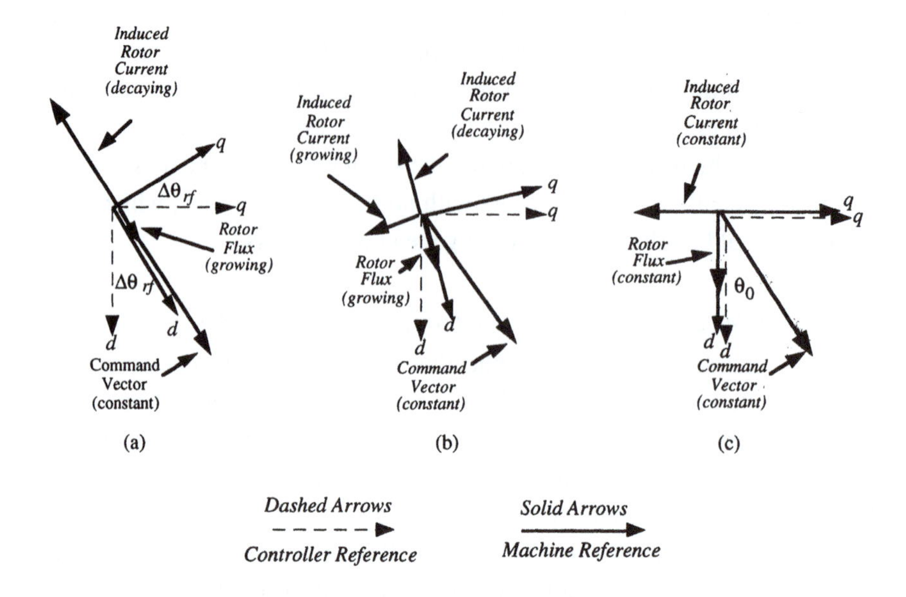

Figure 6.11 Motion of rotor flux and current vectors upon sudden application of flux and torque commands with the machine initially at rest

the rotor flux angle quickly builds up as a result of a very high dynamic slip frequency and then levels off with low slip frequency thereby first producing and then maintaining predominantly *d*–axis current. These curves (Figure 6.12) are the simulation equivalent of an impulse and a step function respectively and are a result of the specific initial conditions chosen for the simulation.

The mechanism creating these *switching transients* is as follows. Since both *q*– and *d*– components are commanded at the start, the commanded slip frequency is a non–zero value resulting in a ramp in commanded rotor flux angle. The initial conditions in the motor are all taken as zero and hence the rotor flux and rotor flux angle are initially zero. The actual slip frequency in the machine is initially an undefined '0/0'. This can be seen from the machine slip frequency calculation in the block diagram of Figure 6.10. Assuming that the controller initially has a zero commanded rotor flux angle, the applied current at the first instant is located exactly as commanded and the difference angle is initially zero. However the slip frequency in the machine immediately becomes very large (approximating an impulse) as finite *q*–axis current is applied with initially zero rotor flux. Given this large slip frequency, the actual rotor flux angle quickly integrates up (approximating a step increase) and

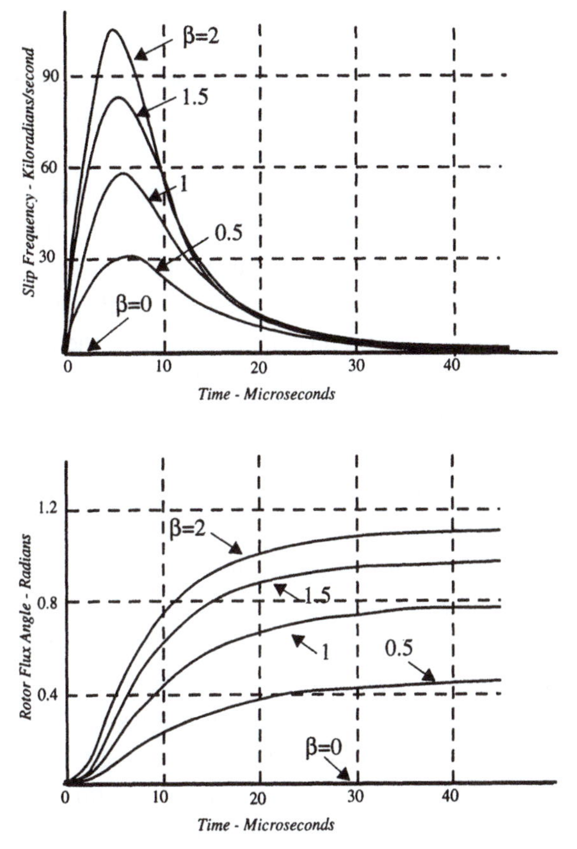

Figure 6.12 "Switching transient" associated with invalid initial conditions in simulation of field orientation start up transient, $\beta = i_{qs}^{e\,*}/i_{ds}^{e\,*}$

exceeds the commanded rotor flux angle. The difference angle then becomes negative and shifts the commanded stator current vector towards the machine's rotor flux d–axis. In doing so, the machine q–axis current decreases and the result is a reduced machine slip frequency. This shifting of the machine rotor flux angle is an inherent effect that causes the applied current to shift into the machine d–axis to create rotor flux.

As previously noted, this result occurs as a result of the choice of zero initial conditions for both the commanded and actual rotor flux angles. The dynamics shown in Figure 6.12 are therefore the dynamics of the simulation routine and not the true dynamics of the machine and controller. Note that with the more physical point of view of the previous paragraph, the initial position of the machine *d*–axis is immediately established as the direction of the initial current vector; no impulse of slip frequency or step in rotor flux angle is required. The difference in the two interpretations occurs as a result of the choice of the initial value of machine rotor flux angle; equal to zero in Figure 6.12 and equal to θ_0 in Figure 6.11(c).

Following this initial switching transient the actual dynamics of the initiation of field orientation occur. As illustrated in Figure 6.13, during this period the *q*–axis currents are held low, the rotor fluxes build up, and the commanded rotor flux angle catches up to the machine rotor flux angle after 1 to 2 seconds, all according to the rotor open circuit time constant. As the commanded rotor flux angle catches up, the applied currents gradually shift towards their commanded values, the commanded and the actual rotor flux angles equalize and steady state field orientation is ultimately achieved. It is important to note that although the commanded slip frequency (which is the actual applied frequency at zero speed) is constant, the actual dynamic slip frequency varies with time (Figure 6.13(d)). The difference between the two is a result of the transient swinging of the machine rotor flux axis. The correct definition of slip frequency is the difference between rotor flux speed and rotor speed, not between steady state synchronous speed and rotor speed. The distinction is crucial during transients while the two are identical in the steady state.

Note that the higher ratios of β exhibit some overshoot corresponding to the applied vector swinging past the commanded position and then ringing back in. The commanded slip frequency varies directly with the ratio β and it is thus reasonable to expect that the startup transient behavior would also depend strongly on this ratio. As shown previously, the case of $\beta = 0$ results in zero error and the flux builds up exactly where commanded. As the ratio is increased, the increased commanded slip frequency causes the rotor flux error angle to respond more quickly during the transient period. As a result the transient tends to become more oscillatory as the ratio β is increased, eventually showing substantial overshoot in flux and torque as illustrated in Figure 6.13. Except for this overshoot and ringing at high values of β the plots in Figure 6.13 are in basic agreement with the physical interpretation given previously and illustrated in Figure 6.11.

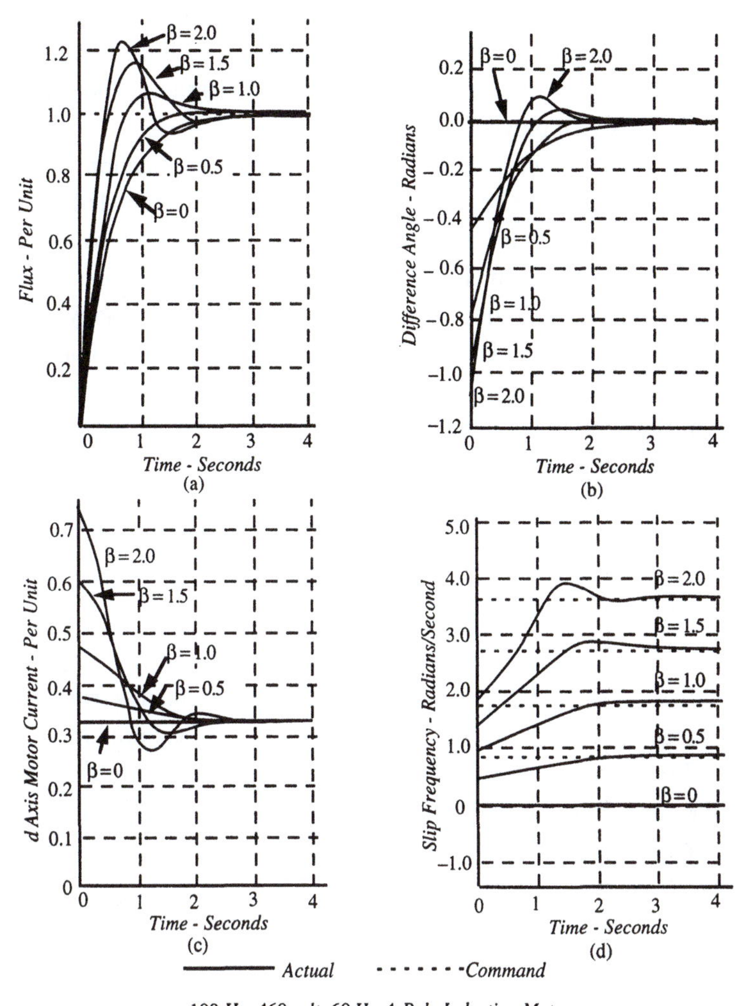

Per Unit Parameters

Stator Resistance = 0.10 *pu* *Stator Leakage Reactance* = 0.10
Rotor Resistance = 0.15 *pu* *Rotor Leakage Reactance* = 0.10
Magnetizing Reactance = 3.0 *Inertia Constant (H)* = 0.5 *Second*

Figure 6.13 Indirect field orientation start up transient – 100 hp machine

6.4 Direct Controllers for Induction Machine Field Orientation

As mentioned in Chapter 5, measurement (or calculation) of the rotor flux angle directly from machine electrical variables is also possible [2,3]. In these *direct controllers*, the measurement of rotor position is eliminated by making other measurements directly on machine electrical quantities. The slip relation is no longer directly employed.

The advantage of such systems is, of course, the elimination of the rotor position encoder. Unfortunately, this eliminates the direct knowledge of a significant disturbance to the system, i.e. the motor speed and rotor position. While there are schemes which can successfully overcome this loss of information, they themselves require new sensors which are in many ways less desirable than the encoder (Hall elements in the air gap, for example). While schemes which only require voltage and current measurements are also feasible, they are severely limited at low speed and introduce new and very troublesome parameter dependencies.

The following sections describe the basic principles of direct controllers and describe several important and useful variations of direct control.

6.4.1 Direct Determination of Rotor Flux Angle

Conceptually, the measurement of the flux angle requires two steps as illustrated in Figure 6.14; computation of the rotor flux components from measured quantities (rotor flux computer) and computation of the magnitude and angle of the rotor flux vector (field orienter). The field orienter is simply a resolver, exactly the same as the resolver introduced in Section 5.5 (Figure 5.13). The rotor flux computer takes on various forms depending on the motor quantities chosen as the inputs.

6.4.2 Measurement of Air Gap Flux

Figure 6.15 illustrates direct determination of λ_r and θ_{rf} by sensing air gap flux with flux sensing coils or Hall elements. The rotor flux equation in a stator reference is

$$\underline{\lambda}^s_{qdr} = L_m \underline{i}^s_{qds} + L_r \underline{i}^s_{qdr} \tag{6.4–1}$$

The measured air gap flux is given by

$$\underline{\lambda}_{qdm} = L_m (\underline{i}^s_{qds} + \underline{i}^s_{qdr}) \tag{6.4–2}$$

Eliminating the rotor current between these two equations yields

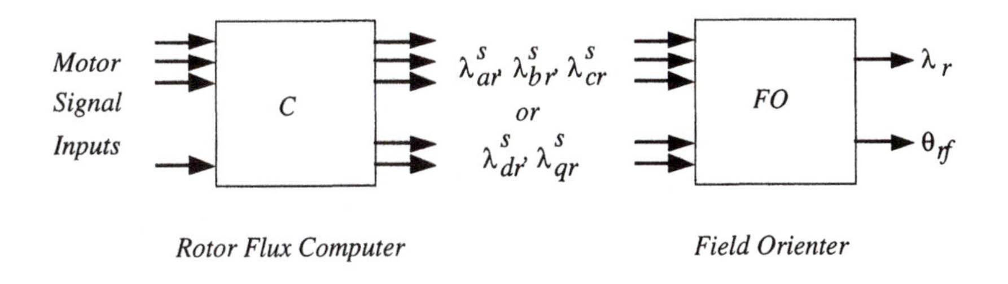

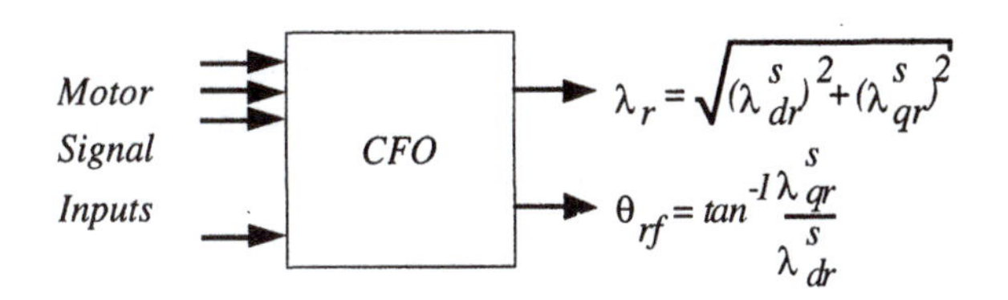

Figure 6.14 Rotor flux computer and field orientation

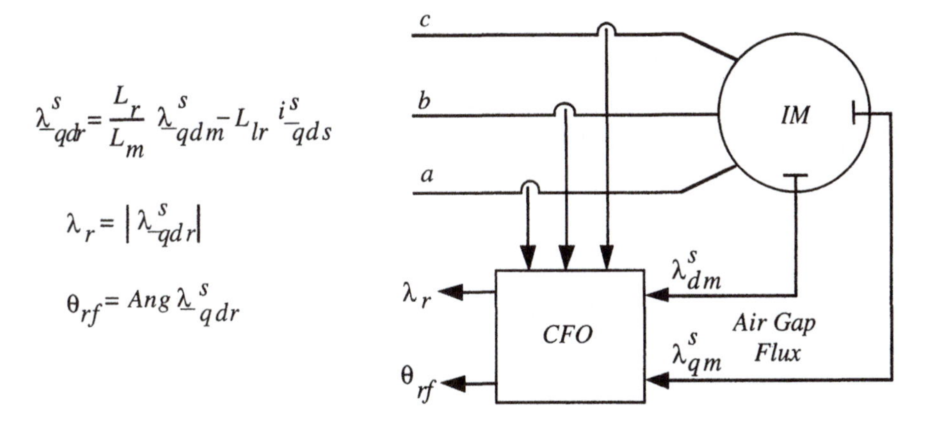

Figure 6.15 Field angle determination using flux sensors

$$\underline{\lambda}^s_{qdr} = \frac{L_r}{L_m}\underline{\lambda}^s_{qdm} - (L_r - L_m)\,\underline{i}^s_{qds} \tag{6.4–3}$$

$$\underline{\lambda}_{qdr}^{s} = \frac{L_r}{L_m}\underline{\lambda}_{qdm}^{s} - L_{lr}\underline{i}_{qds}^{s} \tag{6.4–4}$$

The only computation required is a correction for rotor leakage flux as indicated by the equation given in the figure (plus the resolution to magnitude and phase). This is the most direct approach to direct field orientation and has the considerable advantage of requiring only two motor parameters, the rotor leakage inductance L_{lr} which, (except for closed slot rotors), is substantially a constant value independent of temperature or flux level and L_r / L_m which is only moderately affected by saturation of the main flux paths in the machine. The disadvantage is, of course, the need for special sensing elements in the motor. If flux sensing coils are used there are also problems with integrating signals at low frequencies (near zero speed). A special problem occurs in machines with closed rotor slots (a typical construction in cage rotor machines) since the rotor leakage inductance then is strongly dependent on rotor current, especially at low values of rotor current. The resulting error in flux angle can be quite large if some form of adaptation is not employed. This parameter dependency is discussed in a later chapter.

6.4.3 Voltage and Current Sensing

Special sensors can be avoided by the scheme shown in Figure 6.16 [5]. Since the measurement now involves the stator voltage, the stator equation must be employed to solve for the stator flux linkage. In a stator reference, this equation is

$$\underline{v}_{qds}^{s} = r_s\underline{i}_{qds}^{s} + p\underline{\lambda}_{qds}^{s} \tag{6.4–5}$$

from which, in operational form

$$\underline{\lambda}_{qds}^{s} = \frac{1}{p}(\underline{v}_{qds}^{s} - r_s\underline{i}_{qds}^{s}) \tag{6.4–6}$$

The stator flux linkage expression is

$$\underline{\lambda}_{qds}^{s} = L_s\underline{i}_{qds}^{s} + L_m\underline{i}_{qdr}^{s} \tag{6.4–7}$$

which when solved for the rotor current yields

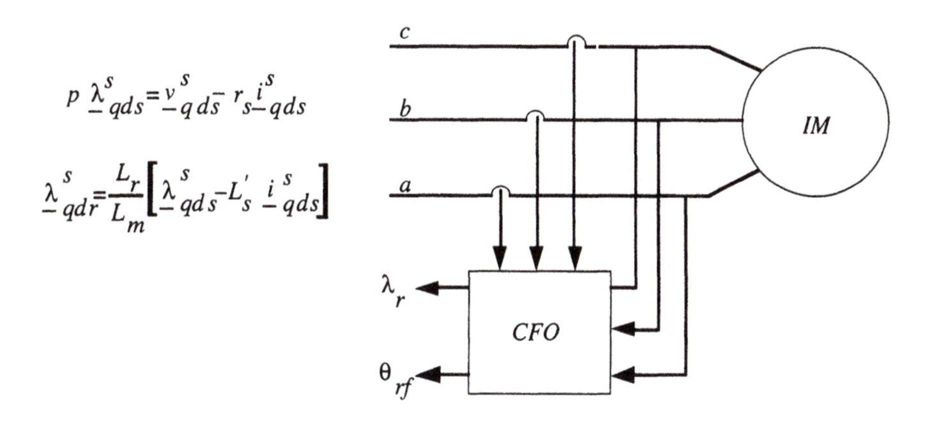

Figure 6.16 Field angle determination from terminal voltage and current

$$\underline{i}^s_{qdr} = \frac{\underline{\lambda}^s_{qds} - L_s \underline{i}^s_{qds}}{L_m} \tag{6.4–8}$$

The rotor flux linkage can then be written as

$$\begin{aligned}
\underline{\lambda}^s_{qdr} &= L_m \underline{i}^s_{qds} + \frac{L_r}{L_m}(\underline{\lambda}^s_{qds} - L_s \underline{i}^s_{qds}) \\
&= \frac{L_r}{L_m}\left[\underline{\lambda}^s_{qds} - \left(L_s - \frac{L_m^2}{L_r}\right)\underline{i}^s_{qds}\right] \\
&= \frac{L_r}{L_m}(\underline{\lambda}^s_{qds} - L_s' \underline{i}^s_{qds})
\end{aligned} \tag{6.4–9}$$

The major difficulty in this case is the need for three motor parameters, r_s, L_s' and L_r / L_m. The stator resistance is a significant problem because of temperature dependence (and skin effect in large machines), the two inductance parameters are only moderately affected by saturation. There are also problems with integrating low frequency signals and with the fact that the stator *IR* drop becomes dominant at low speed. These limitations preclude use of this scheme at low speed. It is, however, practical over a reasonable speed range and is used in many implementations. Again a special problem occurs in machines with closed slot rotors where L_s' becomes strongly dependent on rotor current because of the influence of saturation on L_{lr} as noted in the previous section.

A scheme for eliminating the dependence on the stator resistance has been proposed and implemented [6]. The method involves a winding tap at the mid–point of the phase belt as shown in Figure 6.17 [7]. Since the two winding sec-

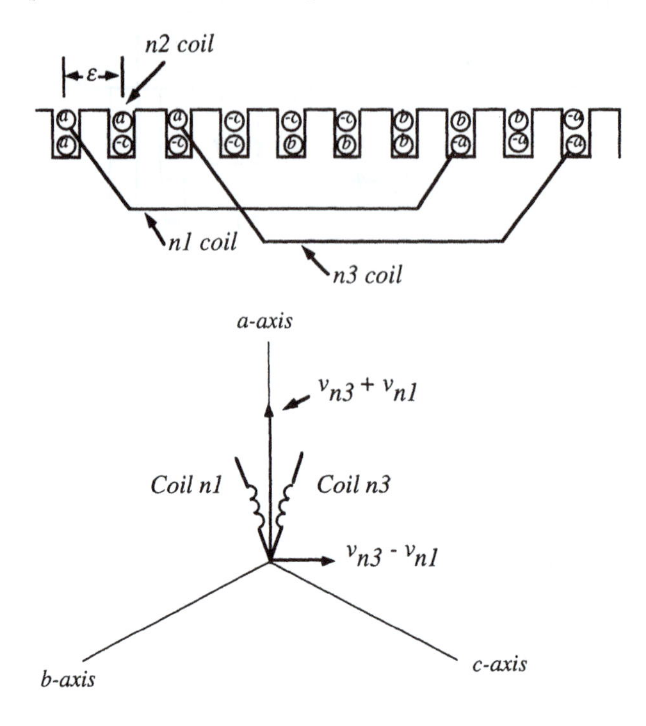

Figure 6.17 Field angle determination using tapped windings

tions are resistively identical, taking the difference eliminates the stator resistive drop. However, because of the spatial separation between the two winding sections, there is an induced voltage which can be related to a flux nearly equal to the stator flux. A relation similar to eqn (6.4–9) is then employed to evaluate the rotor flux linkage.

6.4.4 Implementation of Direct Field Orientation

The direct computation of the rotor flux angle as outlined in the previous sections eliminates the need for employing the slip relation and a measured rotor position signal. The complete controller thus only requires the flux compensation portion of the indirect controller plus the CFO to obtain the rotor flux angle as illustrated in Figure 6.18. For a CSI based system, the resolver and

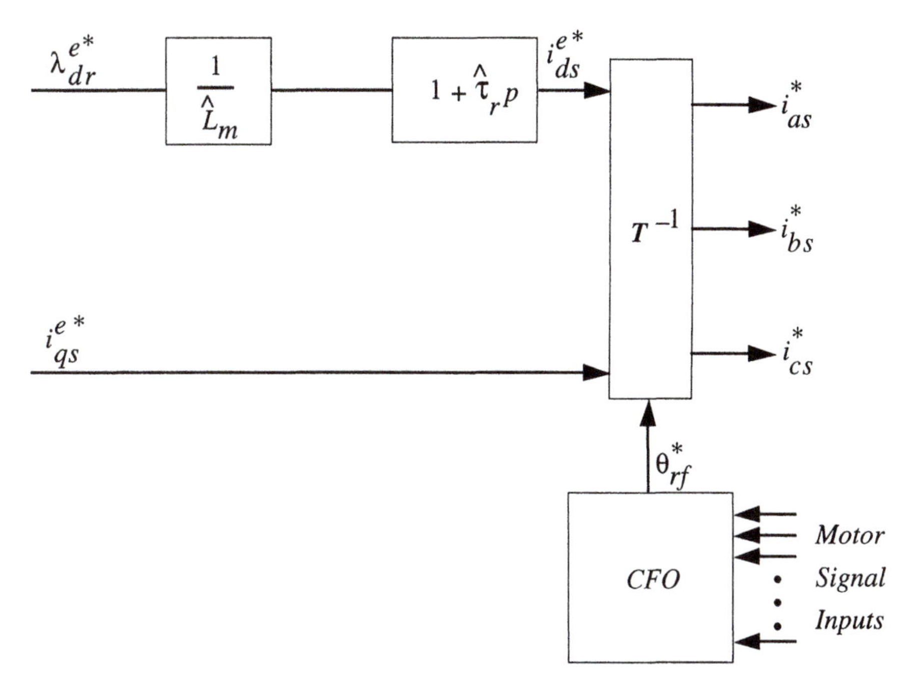

Figure 6.18 Direct field orientation scheme

angle adder of Figure 6.9 would replace the transformation block in Figure 6.18.

If the CFO employed provides rotor flux amplitude as well as the angle, the flux amplitude signal can be used as feedback for a conventional flux regulator. In this case the flux compensator lead element can be omitted as was illustrated in the direct field orientation block diagram in Chapter 5 (Figures 5.31 and 5.32). The flux amplitude and angle can also be combined with current feedback to provide a torque signal and a closed loop torque regulator can be implemented as shown in the Chapter 5 diagrams. Figure 6.19 illustrates the torque computation in block diagram form.

The direct field orientation controllers illustrated in Figure 6.18, 5.31 and 5.32 can utilize any of a wide variety of computational schemes in the CFO used to obtain the rotor flux angle. The three examples presented in this section are intended only as illustrations; new schemes will be developed as other sensors become available. Unfortunately, at present, only Hall sensors provide useful signals near zero speed; the others require integration and are unwork-

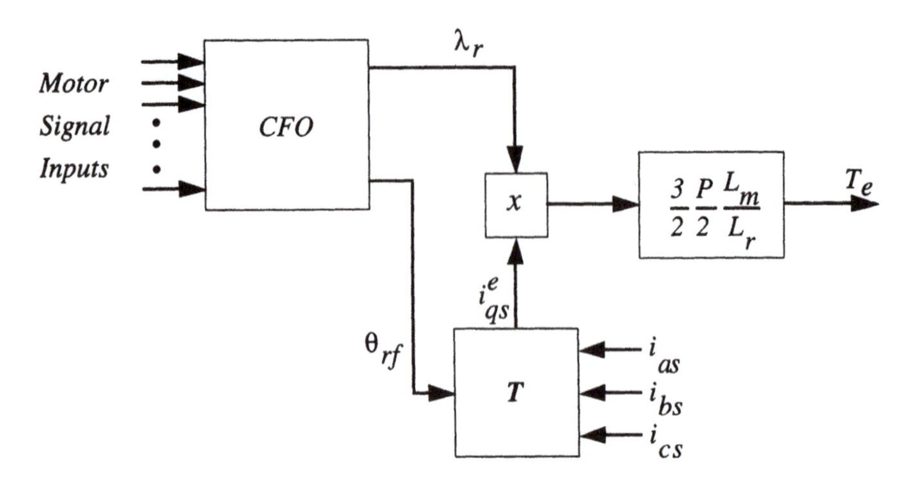

Figure 6.19 Torque computation using CFO output and current feedback

able at low speed. Since Hall sensors in the air gap are considered to be unreliable, direct field orientation has not been widely applied for servo drives where zero speed operation is necessary. However, where continuous operation at or near zero speed is not required, direct field orientation using voltage and current sensing is very attractive and will certainly see increasing application. Some of the problems associated with parameter variations when using this method are considered in Chapter 8.

6.5 Induction Machine Field Orientation Using Air Gap Flux

The field orientation controllers presented in Sections 6.3 and 6.4 obtain instantaneous torque control by controlling the spatial angle between the rotor flux and the torque command current. In the process, the amplitude of the rotor flux is also a controlled quantity. While this is the usual (and original) choice for field orientation, it is also possible to create systems which work with a flux other than the rotor flux [4].

Such systems are more complicated than the rotor flux based controller but have the advantage of working with flux quantities which are directly measurable, i.e. air gap flux or stator flux. It is also normal to relate saturation levels in the machine to air gap flux and hence a controller based on air gap flux is better suited to treating saturation effects. The study of a non rotor flux based controller is also useful in illustrating the existence of theoretical (and practical) limits to the torque capability and transient response of field oriented induction machines.

The following sections present the analysis and development of an air gap flux based field oriented controller. Stator flux based field orientation is treated in Section 6.8.

6.5.1 Air Gap Flux Referred d,q Equations

The basic synchronous frame *d,q* equations given in eqns (6.2–1) to (6.2–9) can be modified to incorporate the air gap flux by using the defining equations for the air gap flux

$$\lambda_{qm}^e = L_m(i_{qs}^e + i_{qr}^e) \tag{6.5–1}$$

$$\lambda_{dm}^e = L_m(i_{ds}^e + i_{dr}^e) \tag{6.5–2}$$

The stator and rotor flux linkages are then expressed as (in complex vector form)

$$\underline{\lambda}_{qds}^e = \underline{\lambda}_{qdm}^e + L_{ls}\underline{i}_{qds}^e \tag{6.5–3}$$

$$\underline{\lambda}_{qdr}^e = \underline{\lambda}_{qdm}^e + L_{lr}\underline{i}_{qdr}^e \tag{6.5–4}$$

where L_{ls} and L_{lr} are the stator and rotor leakage inductances

$$L_{ls} = L_s - L_m \tag{6.5–5}$$

$$L_{lr} = L_r - L_m \tag{6.5–6}$$

and the torque can be rewritten in terms of the air gap flux

$$T_e = \frac{3}{2}\frac{P}{2}(\lambda_{dm}^e i_{qs}^e - \lambda_{qm}^e i_{ds}^e) \tag{6.5–7}$$

If we now align the *d,q* axes such that the air gap flux is entirely in the *d*–axis, we can write

$$\lambda_{qm}^e = 0 \tag{6.5–8}$$

which is the counterpart of eqn (6.2–10) for a rotor flux oriented system. To distinguish the resulting equations from those of Section 6.2.2, we use a superscript *m* to indicate the reference is now the *main* or airgap flux instead of the rotor flux. Again assuming the machine is supplied from a current regulated source, the *d,q* equations in an air gap flux oriented frame are:

$$0 = r_r i_{qr}^m + pL_{lr} i_{qr}^m + (\omega_e - \omega_r)(\lambda_{dm}^m + L_{lr} i_{dr}^m) \tag{6.5–9}$$

$$0 = r_r i_{dr}^m + p(\lambda_{dm}^m + L_{lr} i_{dr}^m) - (\omega_e - \omega_r) L_{lr} i_{qr}^m \tag{6.5–10}$$

$$\lambda_{qm}^m = L_m (i_{qs}^m + i_{qr}^m) = 0 \tag{6.5–11}$$

$$\lambda_{dm}^m = L_m (i_{ds}^m + i_{dr}^m) \tag{6.5–12}$$

$$T_e = \frac{3}{2}\frac{P}{2}\lambda_{dm}^m i_{qs}^m \tag{6.5–13}$$

6.5.2 Dynamic Response of Air Gap Flux Controlled Induction Machine

Equations (6.5–9) to (6.5–13) describe the dynamic response of a machine field oriented to the air gap flux (compare with eqns (6.2–11) to (6.2–15) for rotor flux orientation). The torque equation clearly shows that instantaneous torque response is possible if the air gap flux is held constant.

A development very similar to that of Section 6.2.3 can be carried out to obtain the field orientation form of these equations with i_{qs}^m and i_{ds}^m considered as the input (controlled) quantities. From (6.5–11), we obtain

$$i_{qr}^m = -i_{qs}^m \tag{6.5–14}$$

and from (6.5–12)

$$i_{dr}^m = \frac{\lambda_{dm}^m}{L_m} - i_{ds}^m \tag{6.5–15}$$

Using these results, eqn (6.5–9) can be solved for $\omega_e - \omega_r$ (the slip relation)

$$\omega_e - \omega_r = S\omega_e = \frac{(r_r + L_{lr}p)\, i_{qs}^m}{\lambda_{dm}^m + L_{lr}\left(\dfrac{\lambda_{dm}^m}{L_m} - i_{ds}^m\right)} \tag{6.5–16}$$

$$= \frac{(r_r + L_{lr}p)\, i_{qs}^m}{\dfrac{L_r}{L_m}\lambda_{dm}^m - L_{lr} i_{ds}^m} \tag{6.5–17}$$

The flux relation is obtained from eqn (6.5–10), (6.5–14) and (6.5–15)

$$p\lambda_{dm}^m = -(r_r + L_{lr}p)\left(\frac{\lambda_{dm}^m}{L_m} - i_{ds}^m\right) - S\omega_e L_{lr} i_{qs}^m \tag{6.5–18}$$

or collecting terms and solving for $p\lambda_{dm}^m$

$$p\lambda_{dm}^m = -\frac{r_r}{L_r}\lambda_{dm}^m + \frac{L_m}{L_r}(r_r + L_{lr}p)\, i_{ds}^m - S\omega_e \frac{L_{lr}L_m}{L_r} i_{qs}^m \tag{6.5–19}$$

Note that unlike the rotor flux oriented system the slip relation (eqn (6.5–17)) and the flux relation (eqn (6.5–19)) are coupled equations. Note also that these equations reduce to the rotor flux oriented equations if the leakage inductance L_{lr} is taken as zero (but L_m and L_r must still be considered as unequal)!

Because of the coupling between the slip relation and the flux relation, a torque production block diagram analogous to Figure 6.2 must include both the slip and flux diagrams. Rewriting eqn (6.5–19) in the form

$$\lambda_{dm}^m = \frac{L_m}{1 + \dfrac{L_r}{r_r}p}\left[\left(1 + p\frac{L_{lr}}{r_r}\right) i_{ds}^m - S\omega_e \frac{L_{lr}}{r_r} i_{qs}^m\right] \tag{6.5–20}$$

in combination with eqn (6.5–17) leads to the block diagram in Figure 6.20. Clearly the coupling leads to a much more complicated conceptual picture of torque production based on air gap flux in comparison with the simple, decoupled conceptual diagrams associated with rotor flux control. However, instantaneous torque control can still be realized with a controller which maintains the air gap flux λ_{dm}^m constant as is easily seen from the diagram of Figure 6.20.

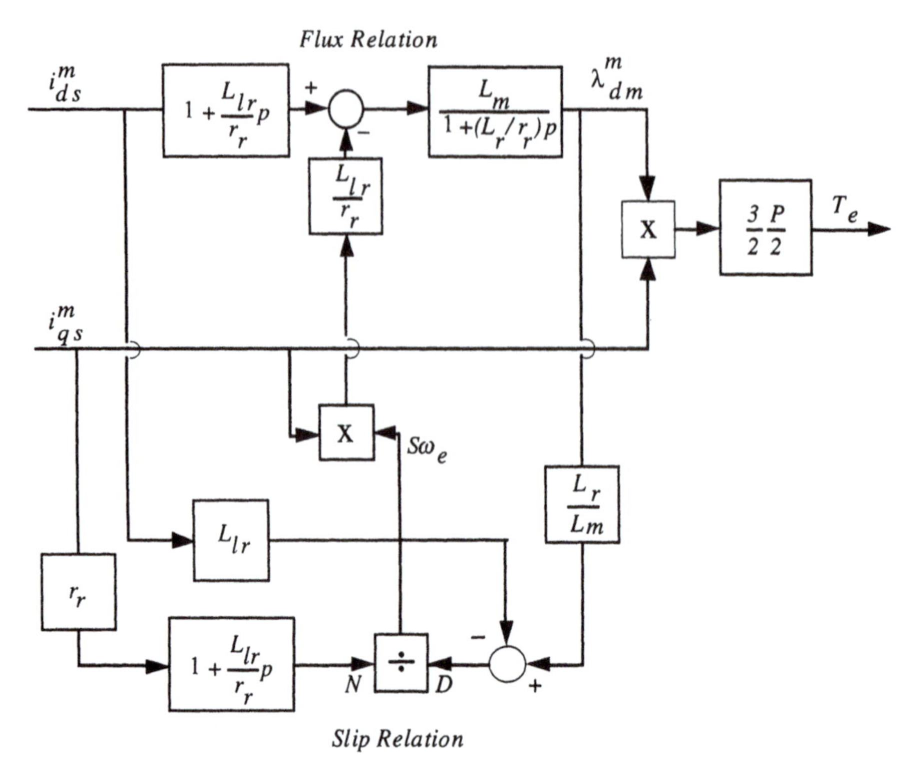

Figure 6.20 Torque production in air gap flux controlled induction machine

6.5.3 *Steady State Interpretation of Air Gap Flux Control*

Before outlining a controller for air gap flux control, it is useful to examine the steady state behavior of the system of Figure 6.20. The steady state forms of the slip and flux relations are

$$S\omega_e = \frac{r_r I_{qs}^m}{\frac{L_r}{L_m}\lambda_{dm}^m - L_{lr} I_{ds}^m} \tag{6.5–21}$$

$$\lambda_{dm}^m = L_m I_{ds}^m - S\omega_e \frac{L_{lr}}{L_r} L_m I_{qs}^m \tag{6.5–22}$$

Thus, even in the steady state, the coupling persists. The most significant fact is that the d–axis current I_{ds}^{m} does not independently control the flux as was the case for rotor flux control. This is easily seen to be the result of the fact that, in the steady state equivalent circuit of Figure 6.21, the rotor current has a *d*–axis component because of the leakage reactance of the rotor. The situation is illus-

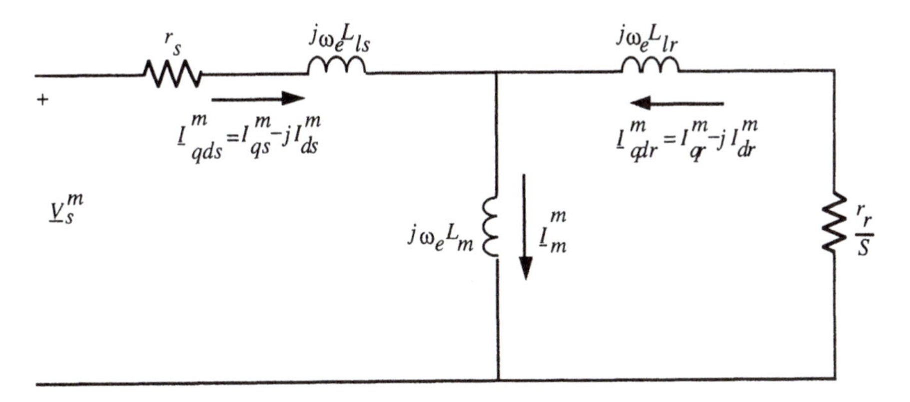

Figure 6.21 Steady state complex vector equivalent circuit for air gap flux control

trated in the vector diagram of Figure 6.22 where the magnetizing current I_{ds}^{m} clearly has two components, I_{ds}^{m} and jI_{dr}^{m}. Thus the stator *d*–axis current must supply the magnetizing current I_{m}^{m} plus a component to cancel the rotor *d*–axis component. This *d*–axis rotor current increases as the slip frequency increases and hence the stator *d*–axis current must also increase if the magnetizing current (and air gap flux) is to remain constant.

The steady state relations of eqns (6.5–21) and (6.5–22) can be easily derived from the steady state equivalent circuit of Figure 6.21. The exercise is left to the reader.

6.5.4 Indirect Field Orientation Controller for Air Gap Flux

An indirect (feedforward) controller for air gap flux control must clearly implement the slip relation of eqn (6.5–17). The second requirement is to implement a means of evaluating the *d*–axis control current i_{ds}^{m*} necessary to control the flux λ_{dm}^{m}. This can be accomplished by considering the flux λ_{dm}^{m*} as an input and solving the flux relation, eqn (6.5–20), for the required i_{ds}^{m*}. From eqn (6.5–20), the result is

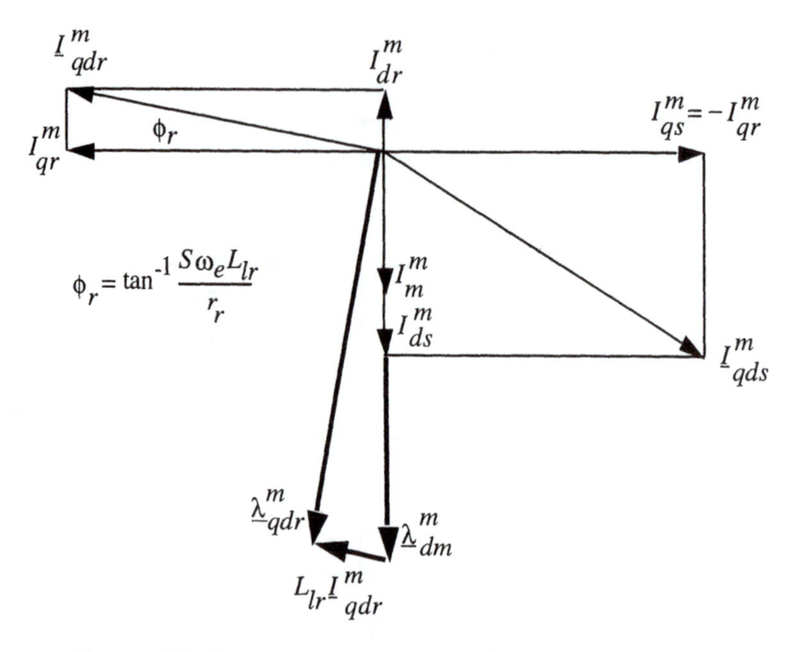

Figure 6.22 Complex vector diagram for air gap flux control

$$i^{m*}_{ds} = \frac{1}{1 + \frac{\hat{L}_{lr}}{\hat{r}_r}p}\left[\left(1 + \frac{\hat{L}_r}{\hat{r}_r}p\right)\frac{\lambda^{*}_{dm}}{\hat{L}_m} + (S\omega_e)^{*}\frac{\hat{L}_{lr}}{\hat{r}_r}i^{m*}_{qs}\right] \tag{6.5–23}$$

The resulting indirect controller is shown in Figure 6.23.

The air gap flux controller is clearly much more complicated than the rotor flux controller of Figure 6.7 and requires two additional parameters; the leakage time constant τ_{lr} and the ratio of the rotor leakage inductance to the rotor self inductance. Note that if these two parameters are set equal to zero, the air gap flux controller of Figure 6.23 reduces to the simpler rotor flux controller of Figure 6.7.

6.5.5 Direct Field Orientation Controller for Air Gap Flux

If direct control is implemented using air gap flux, the summation of slip angle θ^{*}_{s} with measured rotor angle θ_r is replaced by the output of the CFO producing the rotor angle θ^{*}_{rf}. In this case the slip relation is still required since it is

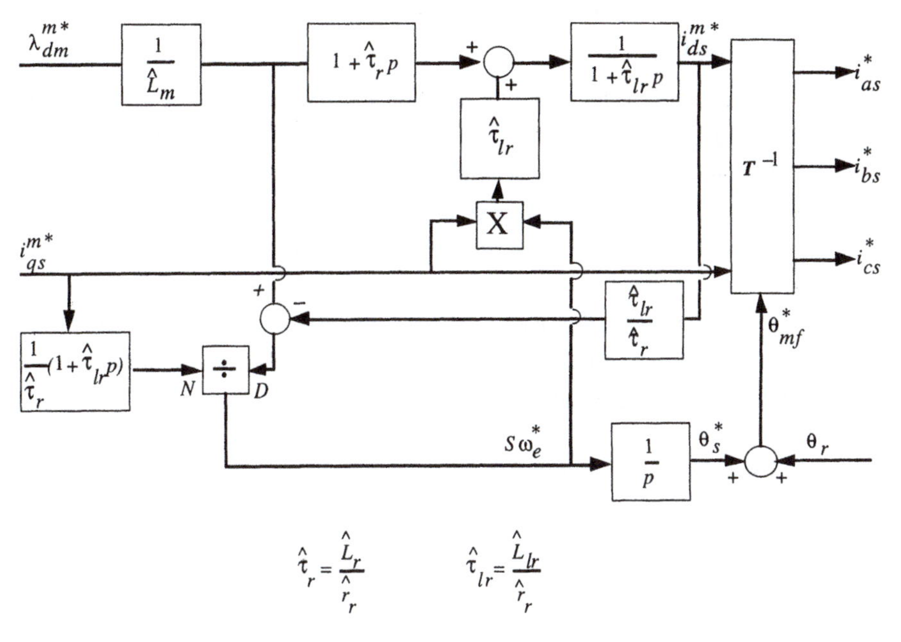

Figure 6.23 Indirect field orientation controller for air gap flux control

needed for the computation of the command value of d–axis current. The CFO computation, however, no longer requires the rotor leakage inductance; in the case of Hall sensors in the air gap, no calculation is required at all. For stator flux sensing, the air gap flux is obtained simply as

$$\underline{\lambda}_{qdm}^{s} = \underline{\lambda}_{qds}^{s} - L_{ls}\underline{i}_{qds}^{s} \tag{6.5–24}$$

Since the rotor leakage inductance is a special problem in machines with closed rotor slots, the air gap flux based controller has an advantage with regard to the CFO computation. The rotor leakage inductance does enter into the slip and i_{ds}^{m*} computation, but these computations are less important at light load where the rotor leakage inductance tends to be highly variable. A closed loop flux regulator utilizing the flux magnitude from the CFO as the feedback signal can also be employed to help reduce the sensitivity to changes in rotor leakage inductance.

6.6 Dynamics of Synchronous Machine Vector Control and Field Orientation

With the background of the previous sections and the basic vector control concepts presented in Chapter 5, the dynamic response of vector controlled synchronous machines will now be examined. Because there are practical systems which do not employ the 90° spatial relationship of flux and MMF used in field orientation, the influence of departures from true torque control will also be considered.

The dynamic model employed is the *d,q* machine model in a rotor reference frame (Park's equations). A full salient pole machine with dampers is used to allow the most general treatment.

6.6.1 d,q Model of Synchronous Machine

The *d,q* equivalent circuit model of a salient pole synchronous machine is illustrated in Figure 6.24. From Chapter 2 the circuit is defined by (the superscript *r* to denote the rotor reference frame and the prime to signify the turns ratio referral are omitted for simplicity)

$$\lambda_{qs} = L_{qs} i_{qs} + L_{mq} i_{qr} \tag{6.6–1}$$

$$\lambda_{ds} = L_{ds} i_{ds} + L_{md} i_{dr} + L_{md} i_{fr} \tag{6.6–2}$$

$$\lambda_{qr} = L_{qr} i_{qr} + L_{mq} i_{qs} \tag{6.6–3}$$

$$\lambda_{dr} = L_{dr} i_{dr} + L_{md} i_{ds} + L_{md} i_{fr} \tag{6.6–4}$$

$$\lambda_{fr} = L_{fr} i_{fr} + L_{md} i_{ds} + L_{md} i_{dr} \tag{6.6–5}$$

Note that each of the flux linkage equations as well as the voltage loop differential equations can be written by inspection of Figure 6.24.

The torque equation is

$$T_e = \frac{3}{2}\frac{P}{2}(\lambda_{ds} i_{qs} - \lambda_{qs} i_{ds}) \tag{6.6–6}$$

or by substituting for the flux linkages in terms of currents

$$T_e = \frac{3}{2}\frac{P}{2}[L_{md}(i_{fr} + i_{dr}) i_{qs} - L_{mq} i_{qr} i_{ds} + (L_{ds} - L_{qs}) i_{ds} i_{qs}] \tag{6.6–7}$$

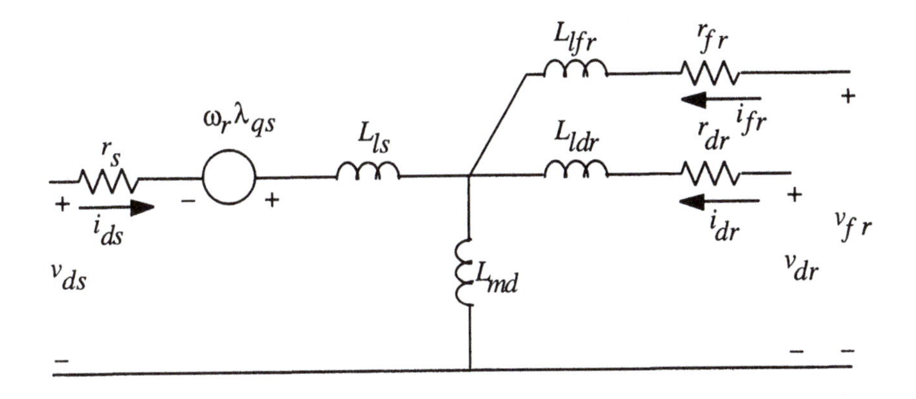

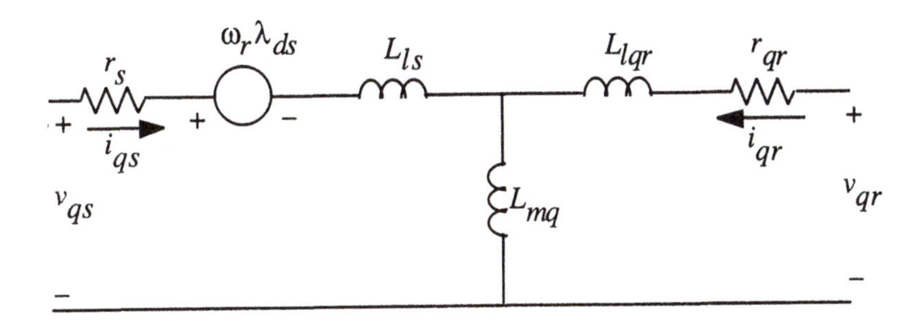

Figure 6.24 Equivalent circuits of three phase salient pole synchronous machine

which places in evidence the three torque components

1) $L_{md}\,(i_{fr} + i_{dr})\, i_{qs}$ – field plus d–axis damper winding torque

2) $L_{mq}\, i_{qr}\, i_{ds}$ – q–axis damper winding torque

3) $(L_{ds} - L_{qs})\, i_{ds}\, i_{qs}$ – reluctance torque (saliency torque)

6.6.2 Vector Control and Angle Control

The concept of rotor position feedback and vector control of the machine stator current to maintain the space angle between the field winding and stator MMF results in stator currents which translate to controlled values of i_{qs} and i_{ds} in the rotor reference frame. This is a result of the instantaneous control of the phase of the stator current to always maintain the same orientation of the stator MMF vector relative to the field winding in the d–axis of the d,q model. The resulting axis currents are illustrated in Figure 6.25. The concept of angle con-

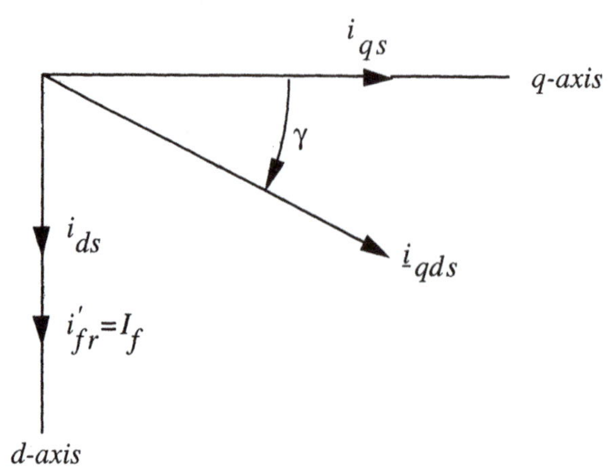

Figure 6.25 Synchronous machine currents in d,q axes

trol (and field orientation) is that the controlled current supply to the machine maintains this condition for transient changes in machine speed as well as under steady state conditions.

Compared to induction machine vector control, the currents i_{qs} and i_{ds} in the synchronous machine are analogous to i^e_{qs} and i^e_{ds} in the induction machine. The rotor flux axis in a synchronous machine is the field winding axis which is also the d–axis of the rotor referred Park equations. The current i_{ds} is, therefore, the stator current component in the rotor flux axis. The current i_{qs} is the *torque component* of stator current just as in the induction machine.

As shown in Chapter 5, under steady state conditions the currents i_{qs} and i_{ds} are constant quantities and the damper currents i_{qr} and i_{dr} are zero. During transients the currents may vary with time and, in general, the damper currents are non–zero. The field current can also vary with time.

In the remaining subsections of this section it is assumed in all cases that the stator current to the machine is directly controlled such that i_{qs} and i_{ds} are independent variables. The first case treated is field orientation; subsequent sections deal with the situation existing when the 90° spatial orientation between the d–axis and the stator MMF is purposely altered.

6.6.3 *Dynamics of Synchronous Machine Field Orientation*

The dynamics of the field oriented case ($\gamma = 0$) are extremely simple if we assume:

1) the stator current is the independently controlled input variable, and

2) the orientation of the stator *d,q* currents is maintained for all speeds including transient changes.

With γ=0, there is only a *q*–axis stator current component, i.e. $i_{ds} = 0$ for all transient conditions. Under these conditions the rotor voltage equations and flux linkage relations reduce to:

q–axis damper circuit

$$r_{qr} i_{qr} + p\lambda_{qr} = 0 \qquad \lambda_{qr} = L_{mq} i_{qs} + L_{qr} i_{qr} \tag{6.6–8}$$

d–axis damper circuit

$$r_{dr} i_{dr} + p\lambda_{dr} = 0 \qquad \lambda_{dr} = L_{md} i_{fr} + L_{dr} i_{dr} \tag{6.6–9}$$

field circuit

$$r_{fr} i_{fr} + p\lambda_{fr} = v_{fr} \qquad \lambda_{fr} = L_{md} i_{dr} + L_{fr} i_{fr} \tag{6.6–10}$$

and the torque expression reduces to

$$T_e = \frac{3}{2}\frac{P}{2} L_{md} (i_{fr} + i_{dr})\, i_{qs} \tag{6.6–11}$$

6.6.3.1 Constant Field Current Operation (Constant Torque Region)

With a fixed dc voltage applied to the field winding, the solution of Eq's. (6.6–9) and (6.6–10) is simply

$$i_{dr} = 0 \tag{6.6–12}$$

and

$$i_{fr} = \frac{v_{fr}}{r_{fr}} = I_f \tag{6.6–13}$$

This is the same as for steady state conditions and is a direct result of the absence of any *d*–axis stator current. The *d*–damper circuit and field circuit are simply a pair of coupled coils in the d–axis without any interaction with the stator as a result of the "field orientation". There is complete "decoupling" of the d–axis from the stator windings.

The q–axis equations can be combined and written in a form to emphasize i_{qs} as an input quantity to yield

$$r_{qr} i_{qr} + pL_{qr} i_{qr} = -pL_{mq} i_{qs} \tag{6.6–14}$$

or solving for i_{qr}

$$i_{qr} = \frac{-L_{mq} p}{r_r + L_{qr} p} i_{qs} \tag{6.6–15}$$

Thus, a change in i_{qs} will induce a transient q–axis damper current which will have an initial value of

$$\Delta i_{qr}(0) = -\frac{L_{mq}}{L_{qr}} \Delta i_{qs}(0) \tag{6.6–16}$$

where Δi_{qs} is the change in i_{qs}. This transient q–axis damper current will decay with the self–time constant (open circuit time constant) of the damper, L_{qr}/r_{qr}. There will, however, be no torque produced by this transient q–axis damper current because there is no d–axis stator current. The torque for the case of constant field excitation is simply

$$T_e = \frac{3}{2}\frac{P}{2} L_{md} I_f i_{qs} \tag{6.6–17}$$

a result identical to steady state (since $L_{md}\, I_f$ is equal to E / ω_r). Thus, the torque response for field orientation is instantaneous and follows the commanded value of i_{qs} exactly. This result is represented in block diagram form in Figure 6.26 [6]. Note that this is the same as for a dc machine — in fact it is a result of exactly the same physical phenomenon, the reaction between two fixed current distributions. In the dc machine both are stationary. In the field oriented synchronous machine both rotate in fixed relation to one another at rotor speed.

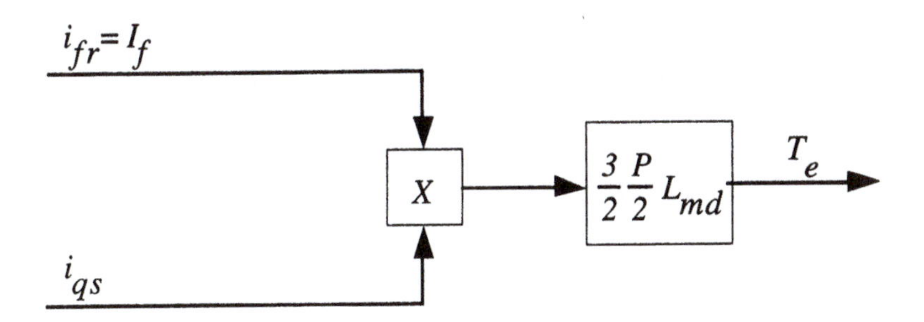

Figure 6.26 Torque production – field orientation ($\gamma = 0$) with constant field excitation

6.6.3.2 Variable Field Excitation (Field Weakening Region)

If the field current is changed, there will be an induced d–axis damper current which can be evaluated by solving eqn (6.6–10) for i_{dr} with i_{fr} treated as an input. The result is

$$i_{dr} = \frac{-L_{md}p}{r_{dr} + L_{dr}p} i_{fr} \tag{6.6–18}$$

This d–axis transient damper current will affect the torque since it will react with the stator q–axis current. A block diagram representation is given in Figure 6.27 along with a sketch of the response to a step increase in i_{fr}. Once again

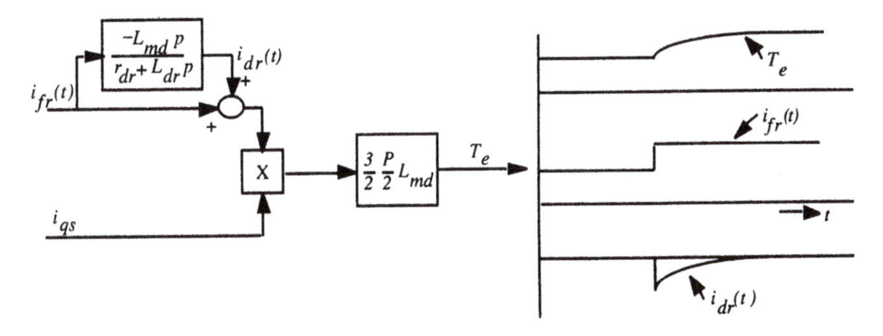

Figure 6.27 Torque production for a change in field current in a field oriented synchronous machine (γ=0,i_{ds}=0)

this is identical to the situation in a dc machine since a step increase in dc machine field current will induce currents in the field pole iron which must die away before the field flux can rise to its new value. This is precisely what happens in the field oriented synchronous machine except the induced current is treated as being in the damper circuit. Note that the lag in the torque response is in addition to any lag in buildup of the field current resulting from an imperfect field current regulator.

6.6.4 *Dynamic Response With $\gamma \neq 0$ (Angle Control)*

To illustrate the more complex situations that result if the 90° orientation of stator MMF and field MMF is not maintained, consider the case where γ is controlled to an angle other than zero [8]. In this case i_{ds} will not be zero. To keep things relatively simple, consider the case where i_{ds} is held at a constant value

$$i_{ds} = I_{ds} \tag{6.6–19}$$

and the torque is still to be controlled by controlling i_{qs}. Under these conditions, the expressions for i_{qr} and i_{dr} in eqns (6.6–15) and (6.6–18) are still valid. The torque expression, however, becomes

$$T_e = \frac{3}{2}\frac{P}{2}[L_{md}(i_{fr} + i_{dr})i_{qs} - L_{mq}i_{qr}I_{ds} + (L_{ds} - L_{qs})I_{ds}i_{qs}] \tag{6.6–1}$$

The q–axis transient current $i_{qr}(t)$ now does produce torque and the q–axis stator current produces reluctance torque $((L_{ds} - L_{qs})\, i_{qs}\, I_{ds})$. Figure 6.28 illustrates the torque production in block diagram form.

It is interesting to note that for the practical case of a negative I_{ds} (associated with operation at leading stator power factor for load commutation or field weakening for example) both the saliency torque and the q–axis damper torque would be negative causing a lag and a reduction in the torque response resulting from an increase in i_{qs}.

If the d–axis stator current is also considered as a variable, the torque production block diagram is even more interactive than shown in Figure 6.28. The additional term represents the transient d–axis damper current resulting from changes in i_{ds}. From eqn (6.6–9) with the d–axis stator current added in to the flux linkage expression for λ_{dr} (see eqn (6.6–4))

$$i_{dr} = \frac{-L_{md}p}{r_{dr} + L_{dr}p}(i_{fr} + i_{ds}) \tag{6.6–20}$$

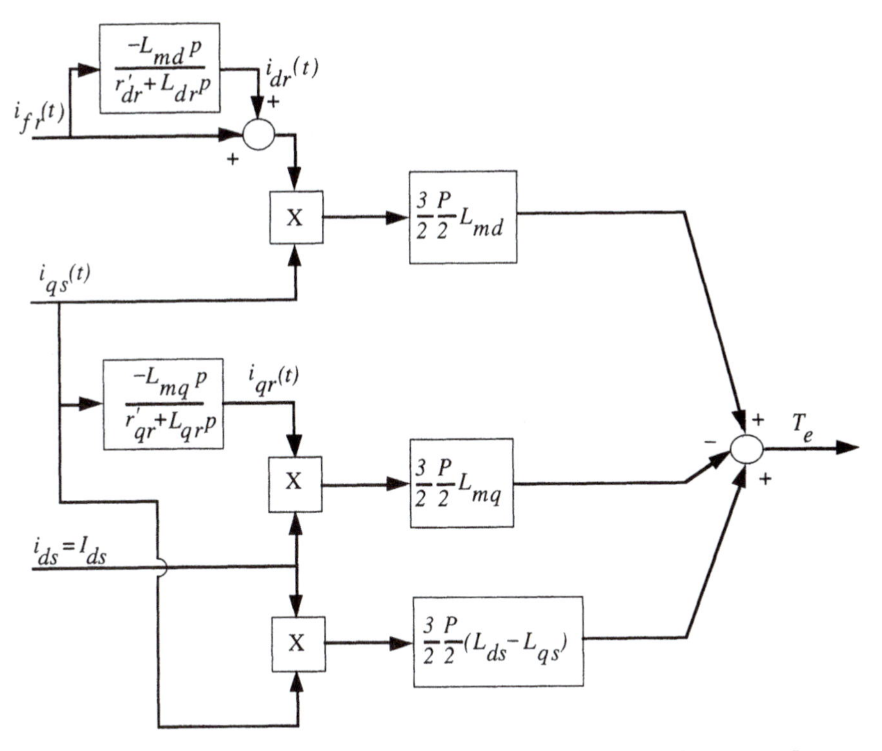

Figure 6.28 Torque production with $\gamma \neq 0$ for the case of constant I_{ds}

The resulting block diagram is shown in Figure 6.29. Note that this diagram can be considered as the general torque production model for a current fed synchronous machine; all of the other cases can be obtained by specializing this general model for the particular case of interest.

It is important to note that operation with $i_{ds} \neq 0$ results in a strongly non-instantaneous torque response only for changes in i_{ds} (or i_{fr}). When both i_{ds} and i_{fr} are held constant, the instantaneous torque becomes

$$T_e = \frac{3P}{22}\{[L_{md}I_{fr} + (L_{ds} - L_{qs})I_{ds}]\,i_{qs} - L_{mq}I_{ds}i_{qr}\} \qquad (6.6\text{–}21)$$

which is linear, except for the relatively small term involving i_{qr}, and yields nearly instantaneous response for changes in i_{qs}. Thus angle control ($\gamma \neq 0$) retains this aspect of field orientation when the d–axis currents are held constant, even for a salient pole machine.

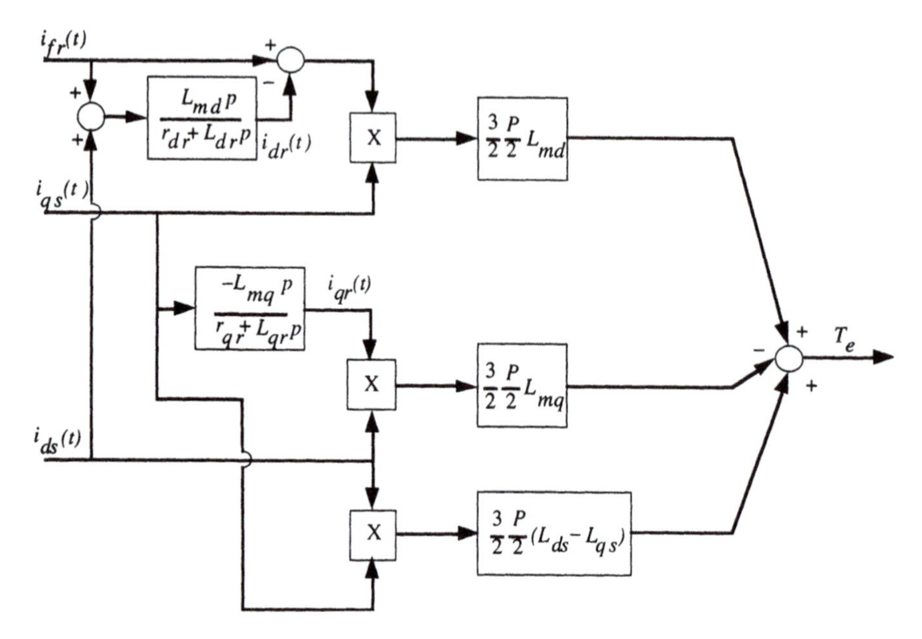

Figure 6.29 Torque production with $\gamma \neq 0$ for the general case

6.6.5 Example – Synchronous Machine Field Orientation

A 100 hp, 460 volt, 3 phase, 4 pole synchronous motor with the following per unit parameters

$X_{ls} = 0.1$	$X_{lf} = 0.2$	$r_s = 0.04$
$X_{md} = 1.1$	$X_{ldr} = 0.1$	$r_{dr} = 0.04$
$X_{mq} = 0.3$	$X_{lqr} = 0.15$	$r_{qr} = 0.08$
		$r_{fr} = 0.01$

is operated as a field orientation controlled machine. For this example assume an ideal current regulator.

a) For field orientation ($\gamma = 0$), find the rated current ($i^*_{qs} = I$), rated internal voltage E, and the terminal power factor for operation with rated terminal voltage, rated current, and rated frequency.

b) Assume the encoder is incorrectly aligned such that $\gamma_0 = 20°$. Find and plot T_e vs. I^*_{qs} from zero to rated value, also find the terminal voltage at rated $I = I^*_{qs}$.

c) Same as (b), but $\gamma_o = -20°$.

d) Find the torque $T_e(t)$ following a step input of i_{qs}^* (rated value) for cases (a), (b), and (c). Express time in seconds.

Solution:

a) At $\gamma = 0$, $i_{ds} = 0$

the steady state terminal relation is

$$E + jX_d I_{ds} + jX_{qs} I_{qs} + R_s I_{qs} + R_s I_{ds} = V$$

and the phasor diagram is

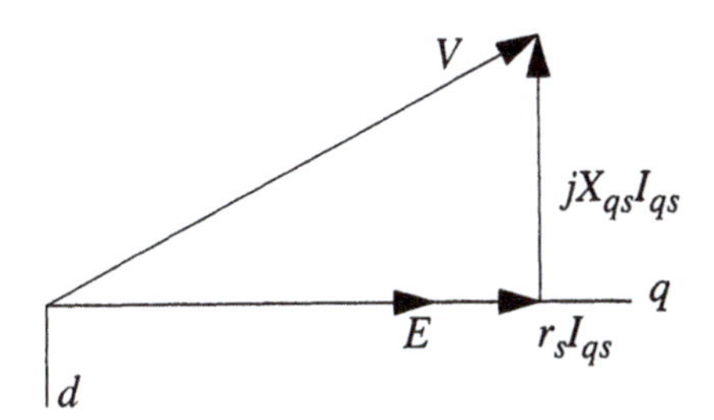

From the phasor diagram

$$|V|^2 = (X_{qs}|I_{qs}|)^2 + (|E| + r_s|I_{qs}|)^2$$

The general torque relation is, in per unit, $T_e = (\lambda_{ds}\, i_{qs} - \lambda_{qs}\, i_{ds})$
but, since $\lambda_{ds} = L_{ldr}\, i_{ds} + L_{md}\,(i_{ds} + i_{dr} + i_f)$
and $i_{dr} = 0$ in steady state $\Rightarrow \lambda_{ds} = L_{md}\, I_f$ (recall $I_{ds} = 0$)

$$\therefore T_e = L_{md} I_f I_{qs}$$

But $|E| = X_{md}\, I_f$ since $\omega = 1.0$ pu, $|E| = \omega\, L_{md}\, I_f \Rightarrow |E| = L_{md}\, I_f$

$$\therefore T_e = |E||I_{qs}|$$

Substituting,

$$|V|^2 = (X_{qs}|I_{qs}|)^2 + \left(\frac{T_e}{|I_{qs}|} + r_s|I_{qs}|\right)^2$$

Given: $V = 1.0$, $T_e = 1.0$, $X_q = \omega(L_{mq} + L_{ls}) = 1\ (0.3 + 0.1) = 0.4$

$$1 = (0.4|I_{qs}|)^2 + \frac{1}{|I_{qs}|^2} + 2(0.04) + (0.04)^2|I_{qs}|^2$$

$$0 = 0.1616|I_{qs}|^4 + (-0.92)|I_{qs}|^2 + 1$$

Solving,

$$|I_{qs}| = 1.22 \text{ pu}$$

Since, $\left(E = \frac{T_e}{|I_{qs}|}\right) \Rightarrow E = \frac{1.0}{1.22} = 0.820$ pu

The power factor is

$$pf = \frac{|E + r_s I_{qs}|}{|V|}$$

$$pf = \frac{0.820 + 0.04(1.22)}{1.0} \qquad pf = 0.869$$

[Note: Neglecting resistance will lead to a >10% error in E and in $|I_{qs}|$ in this example]

b) Assume the encoder is incorrectly aligned such that $\gamma_0 = 20°$. Find and plot T vs. I^*_{qs} from zero to rated value, also find the terminal voltage at rated $I = I^*_{qs}$.

I_f is the same (given) $\Rightarrow I_f = \left|\frac{E}{\omega L_{md}}\right| = \frac{0.820}{(1.0)(1.1)} = 0.751$ pu

$$I_{ds} = I^*_{qs} \sin\gamma_o \text{ and } I_{qs} = I^*_{qs} \cos\gamma_o$$

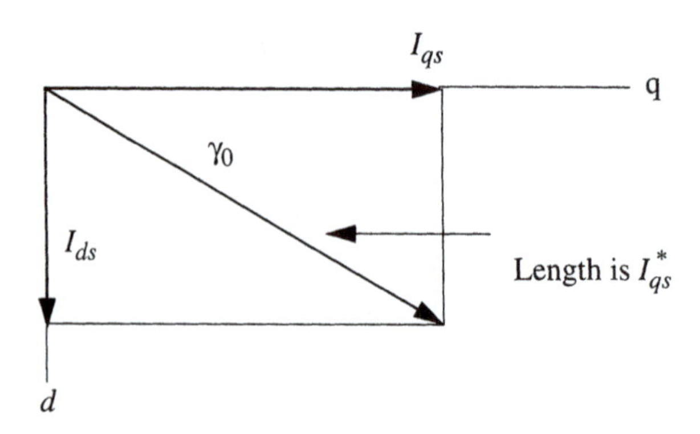

To find the torque expression,

$$T_e = L_{md} I_f I_{qs} + (L_{ds} - L_{qs}) I_{ds} I_{qs}$$

$$T_e = L_{md} I_f I^*_{qs} \cos\gamma_o + (L_{ds} - L_{qs}) (I^*_{qs})^2 \sin\gamma_o \cos\gamma_o$$

$$L_{md} = 1.1 \quad I_f = 0.751 \quad L_{ds} = 1.2 = L_{md} + L_{lds}$$

$$L_{qs} = 0.4 = L_{mq} + L_{lqs}$$

This can be plotted for $0 < I^*_{qs} < 1.22$ pu as required.

To find the voltage,

$$\underline{V} = \underline{E} + jX_{ds}\underline{I}_{ds} + jX_{qs}\underline{I}_{qs} + r_s\underline{I}_{ds} + r_s\underline{I}_{qs}$$

Substitute

$$\underline{V} = 0.82 + j(1.2)(-jI^*_{qs} \sin\gamma_o) + j(0.4)(I^*_{qs} \cos\gamma_o)$$
$$+ (0.04)(-jI^*_{qs} \sin\gamma_o) + (0.04)(I^*_{qs} \cos\gamma_o)$$

$$\underline{V} = 0.82 + (1.2)(I^*_{qs} \sin\gamma_o) + (0.04)(I^*_{qs} \cos\gamma_o)$$
$$+ j[0.4I^*_{qs} \cos\gamma_o - 0.04I^*_{qs} \sin\gamma_o]$$

For $\gamma_o = 20°$, $I^*_{qs} = 1.22$, $V = 1.44 \angle 17.9°$

(c) Same as (b), but $\gamma_0 = -20°$

$$T_e = L_{md} I_f + I^*_{qs} \cos\gamma_o + (L_{ds} - L_{qs}) (I^*_{qs})^2 \sin\gamma_o \cos\gamma_o$$

$$\underline{V} = 0.82 + 1.2I^*_{qs} \sin\gamma_o + 0.04I^*_{qs} \cos\gamma_o$$
$$+ j[0.4I^*_{qs} \cos\gamma_o - 0.04I^*_{qs} \sin\gamma_o]$$

$$\underline{V} = 0.599 \angle 52.5°$$

d) Find and sketch $T_e(t)$ following a step input of i^*_{qs} (rated value) for cases (a), (b), and (c). Express time in seconds.

The relevant equations are

$$v_{qs} = r_s i_{qs} + p\lambda_{qs} + \omega_r \lambda_{ds} \qquad \lambda_{qs} = L_{ls} i_{qs} + L_{mq}(i_{qs} + i_{qr})$$

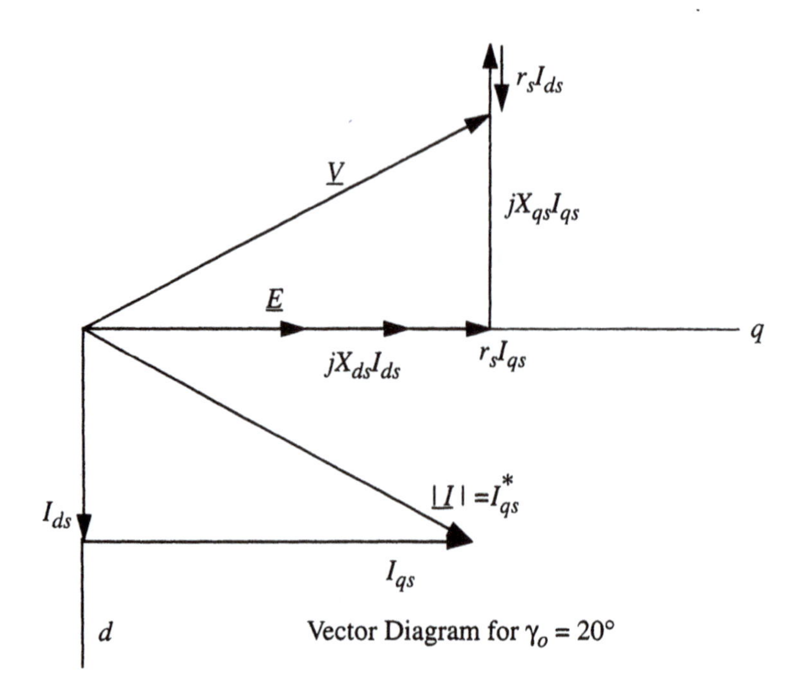

Vector Diagram for $\gamma_o = 20°$

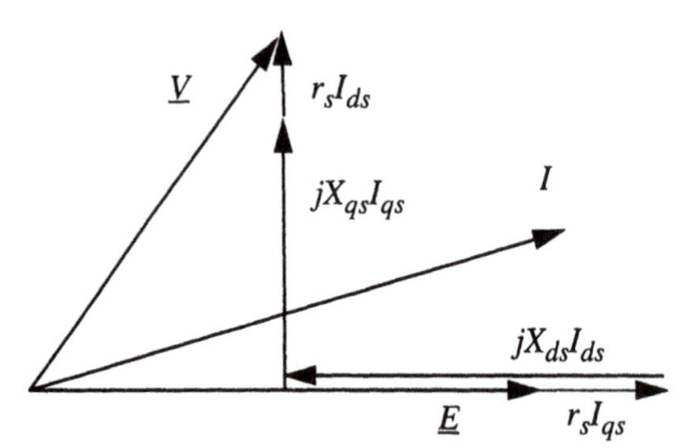

Vector Diagram for $\gamma_0 = -20°$

$$v_{ds} = r_s i_{ds} + p\lambda_{ds} - \omega_r \lambda_{ds} \qquad \lambda_{ds} = L_{ls} i_{ds} + L_{md}(i_{ds} + i_{dr} + i_f)$$

$$v_f = r_f i_f + p\lambda_f \qquad \lambda_f = L_{lf} i_f + L_{md}(i_f + i_{ds} + i_{dr})$$

$$v_{qr} = r_{qr} i_{qr} + p\lambda_{qr} = 0 \qquad \lambda_{qr} = L_{lqr} i_{qr} + L_{mq}(i_{qr} + i_{qs})$$

$$v_{dr} = r_{dr} i_{dr} + p\lambda_{dr} = 0 \qquad \lambda_{dr} = L_{ldr} i_{dr} + L_{md}(i_{dr} + i_{ds} + i_f)$$

Since

$$v_{qr} = 0 = r_{qr}i_{qr} + p\,(L_{lqr}i_{qr} + L_{mq}\,(i_{qr} + i_{qs}))$$

$$i_{qr} = \frac{-L_{mq}p}{r_{qr} + L_{qr}p}i_{qs} \qquad\qquad L_{qr} = L_{lqr} + L_{mq}$$

$$v_{dr} = 0 = r_{dr}i_{dr} + p\,(L_{ldr}i_{dr} + L_{md}\,(i_{dr} + i_{ds} + i_f))$$

$$i_{dr} = \frac{-L_{md}pi_{ds}}{r_{dr} + L_{dr}p} + \frac{-L_{md}pi_f}{r_{dr} + L_{dr}p}$$

In general, in per unit,

$$T_e = \lambda_{ds}i_{qs} - \lambda_{qs}i_{ds}$$

$$T_e = [L_{ls}i_{ds} + L_{md}\,(i_{ds} + i_{dr} + i_f)]\,i_{qs} - [L_{ls}i_{qs} + L_{mq}\,(i_{qs} + i_{qr})]\,i_{ds}$$

$$T_e = L_{md}\,(i_f + i_{dr})\,i_{qs} - L_{mq}\,(i_{qr}i_{ds}) + (L_{md} - L_{mg})\,i_{ds}i_{qs}$$

Let $i_{qs} = i^*_{qs}\ \cos\gamma_o u\,(t)$ and $i_{ds} = i^*_{qs}\ \sin\gamma_o u\,(t)$ and note that $pi_f = 0$. In Laplace Transform notation:

$$I_{qr}(s) = \frac{i^*_{qs}\ \cos\gamma_o}{s} \qquad\qquad I_{dr}(s) = \frac{i^*_{qs}\ \sin\gamma_o}{s}$$

Solve for $I_{qr}(s)$, $I_{dr}(s)$

$$I_{qr}(s) = \frac{-L_{mq}s}{r_{qr} + L_{qr}s}\ \frac{i^*_{qs}\ \cos\gamma_o}{s} \qquad I_{dr}(s) = \frac{-L_{md}s}{r_{dr} + L_{dr}s}\ \frac{i^*_{qs}\ \sin\gamma_o}{s}$$

$$I_{qr}(s) = \frac{-L_{mq}i^*_{qs}\ \cos\gamma_o}{r_{qr} + L_{qr}s} \qquad\qquad I_{dr}(s) = \frac{-L_{md}i^*_{qs}\ \sin\gamma_o}{r_{dr} + L_{dr}s}$$

Inverting these Laplace Transforms

$$i_{qr}(t) = -\frac{L_{mq}}{L_{qr}}i^*_{qs}\ \cos\gamma_o e^{-\frac{R_{qr}}{L_{qr}}t} \qquad i_{dr}(t) = -\frac{L_{md}}{L_{dr}}i^*_{qs}\ \sin\gamma_o e^{-\frac{R_{dr}}{L_{dr}}t}$$

This result fits the torque expression. Substituting,

$$T_e = L_{md} i_f i_{qs} + L_{md} i_{dr} i_{qs} - L_{mq} i_{qr} i_{ds} + (L_{md} - L_{mq})\, i_{ds} i_{qs}$$

$$T_e = L_{md} i_f i_{qs}^* \cos\gamma_o \qquad \text{Field Torque}$$

$$+ \left(\frac{-L_{md}^2}{L_{dr}} i_{qs}^{*2} \sin\gamma_o \cos\gamma_o \right) e^{-\frac{R_{dr}}{L_{dr}} t} \qquad \text{D–Damper torque}$$

$$+ \left(\frac{L_{mq}^2}{L_{qr}} i_{qs}^{*2} \sin\gamma_o \cos\gamma_o \right) e^{-\frac{R_{dr}}{L_{dr}} t} \qquad \text{Q–Damper Torque}$$

$$+ (L_{md} - L_{mq})\, i_{qs}^{*2} \sin\gamma_o \cos\gamma_o \qquad \text{Reluctance Torque}$$

The time constants are:

$$\frac{L_{dr}(\text{pu})}{r_{dr}(\text{pu})} \frac{z_{base}/\omega_{base}}{z_{base}} = \frac{L_{dr}(\text{pu})}{\omega_{base} r_{dr}(\text{pu})} = \frac{(1.1 + 0.1)}{(377 \text{ sec}^{-1})\,(0.04x)} = 79.6 \text{ ms}$$

$$\frac{L_{qr}(\text{pu})}{r_{qr}(\text{pu})} \frac{z_{base}/\omega_{base}}{z_{base}} = \frac{L_{qr}(\text{pu})}{\omega_{base} r_{qr}(\text{pu})} = \frac{(0.3 + 0.15)}{(0.08)\,(377 \text{ sec}^{-1})} = 14.9 \text{ ms}$$

6.7 Field Orientation Using Voltage as the Controlled Variable

Although the stator currents are the variables which must be controlled to implement field orientation, it is possible to employ the stator voltage equations and indirectly control the currents by controlling motor terminal voltage. To accomplish this task it is necessary to 'decouple' the voltage equations so as to be able to independently control the two components of stator current.

6.7.1 Stator Voltage Equations in Terms of Rotor Flux – Induction Machine

The first step in developing the voltage decoupling equation is to rewrite the stator equations given in eqn (6.2–1) and eqn (6.2–2) in terms of rotor flux. This is accomplished by solving the rotor flux linkage equations (eqns (6.2–8) and (6.2–9)) for the rotor currents and substituting the results into the stator flux linkage equations (eqns (6.2–6) and (6.2–7)) to yield

$$\lambda_{qs}^e = \left(L_s - \frac{L_m^2}{L_r}\right) i_{qs}^e + \frac{L_m}{L_r}\lambda_{qr}^e = L_s' i_{qs}^e + \frac{L_m}{L_r}\lambda_{qr}^e \qquad (6.7\text{–}1)$$

$$\lambda_{ds}^e = \left(L_s - \frac{L_m^2}{L_r}\right) i_{ds}^e + \frac{L_m}{L_r}\lambda_{dr}^e = L_s' i_{ds}^e + \frac{L_m}{L_r}\lambda_{dr}^e \qquad (6.7\text{–}2)$$

where the stator transient inductance

$$L_s' = L_s - \frac{L_m^2}{L_r} \qquad (6.7\text{–}3)$$

has been introduced to simplify the notation. Using these expressions in the stator voltage equations results in the desired voltage equations in terms of rotor flux linkages.

$$v_{qs}^e = (r_s + L_s' p)\, i_{qs}^e + \frac{L_m}{L_r} p\lambda_{qr}^e + \omega_e \left(L_s' i_{ds}^e + \frac{L_m}{L_r}\lambda_{dr}^e \right) \qquad (6.7\text{–}4)$$

$$v_{ds}^e = (r_s + L_s' p)\, i_{ds}^e + \frac{L_m}{L_r} p\lambda_{dr}^e - \omega_e \left(L_s' i_{qs}^e + \frac{L_m}{L_r}\lambda_{qr}^e \right) \qquad (6.7\text{–}5)$$

6.7.2 Decoupling Equations for Field Orientation

For a field oriented system these equations can be simplified since the field orientation forces

$$\lambda_{qr}^e = 0 \qquad (6.7\text{–}6)$$

With this constraint the stator voltage equations become

$$v_{qs}^e = (r_s + L_s' p)\, i_{qs}^e + \omega_e \left(L_s' i_{ds}^e + \frac{L_m}{L_r}\lambda_{dr}^e \right) \qquad (6.7\text{–}7)$$

$$v_{ds}^e = (r_s + L_s' p)\, i_{ds}^e - \omega_e L_s' i_{qs}^e + \frac{L_m}{L_r} p\lambda_{dr}^e \qquad (6.7\text{–}8)$$

Since the currents i_{qs}^e, i_{ds}^e and the flux λ_{dr}^e are command inputs or calculated from command inputs, these two equations can be used to calculate the required voltages. The computational scheme is illustrated in Figure 6.30.

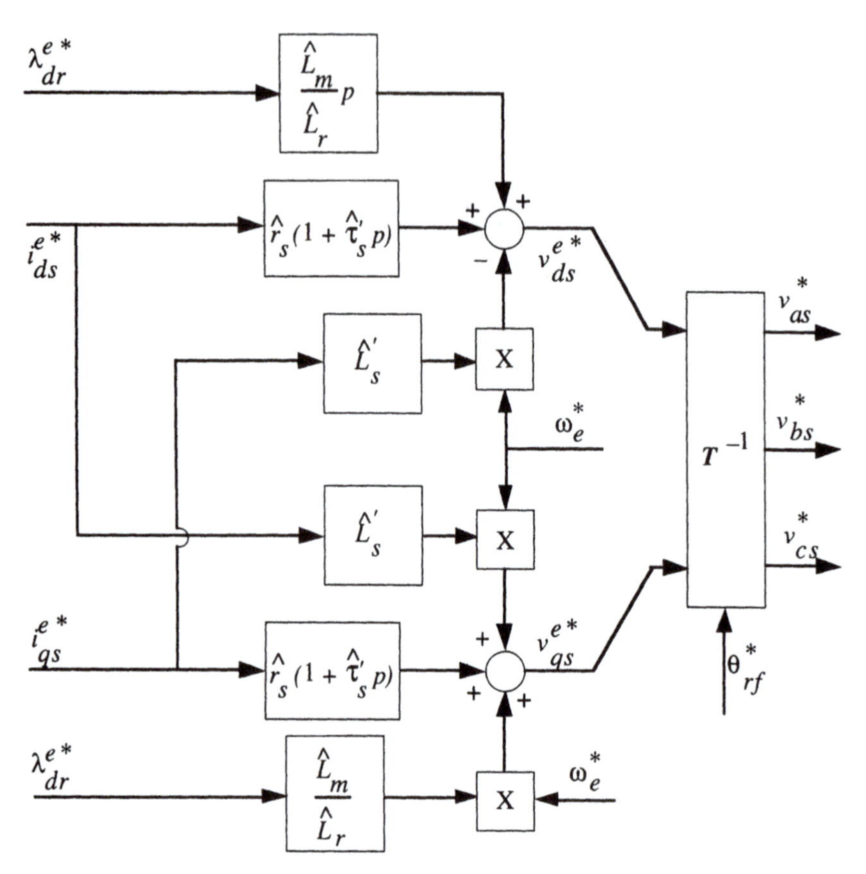

Figure 6.30 Computation of voltage commands for voltage controlled field orientation in induction machines (voltage decoupler)

The computation to decouple the voltages as shown in this figure is applicable to both direct and indirect field orientation. The only difference is in the method by which the various command quantities are calculated. In indirect control with λ_{dr}^{e*} and i_{qs}^{e*} as input commands, ω_e^* and θ_{rf}^* are obtained from the slip relation whereas in direct control they are obtained from the CFO. Thus, although hardware implementation will offer different practical problems, the general concept is independent of the method employed in the basic field orientation scheme.

For systems in which the flux is constant (no field weakening option), the decoupling equations are greatly simplified. With a constant flux, the steady state relation

$$\lambda_{dr}^{e} = L_m I_{ds}^{e} \tag{6.7–9}$$

can be used in eqn (6.7–7) to yield

$$v_{qs}^{e} = (r_s + L_s' p)\, i_{qs}^{e} + \omega_e \left(L_s' + \frac{L_m^2}{L_r} \right) I_{ds}^{e} \tag{6.7–10}$$

$$= (r_s + L_s' p)\, i_{qs}^{e} + \omega_e L_s I_{ds}^{e} \tag{6.7–11}$$

and eqn (6.7–8) reduces to

$$v_{ds}^{e} = r_s I_{ds}^{e} - \omega_e L_s' i_{qs}^{e} \tag{6.7–12}$$

The corresponding computational block diagram is shown in Figure 6.31. Note

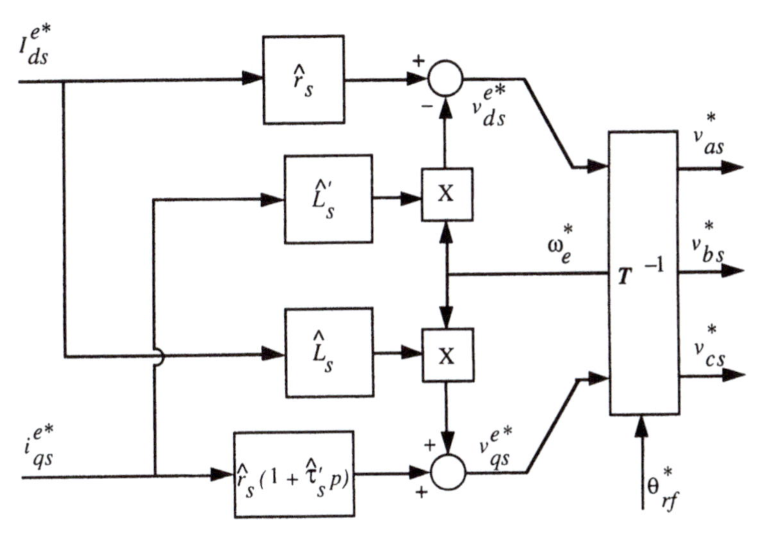

Figure 6.31 Computation of voltage commands for voltage controlled field orientation in induction machines with constant flux (constant flux voltage decoupler)

the close similarity of this model to a dc machine operated with constant field current; the only difference is the cross coupling term $\omega_e L_s' i_{qs}^{e}$ in the d–axis equation.

6.7.3 Examples of Field Orientation Using Voltage Controlled Inverters

To illustrate the overall system configuration of field orientation controllers using voltage controlled inverters, two examples are presented. Figure 6.32 is a block diagram of an induction machine indirect controller combining the basic indirect field orientation controller of Figure 6.7 and the voltage decoupler of Figure 6.30. The flux compensation block and the slip calculator are retained to

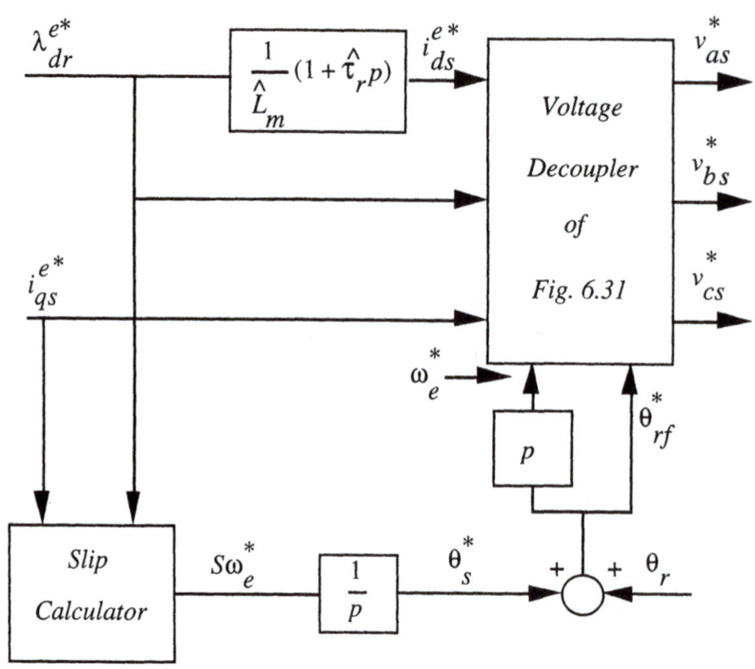

Figure 6.32 Indirect field orientation controller using voltage controlled inverter

provide the required input quantities for the voltage decoupler. If only constant flux operation is desired, the flux compensation block can be omitted and the simpler voltage decoupler of Figure 6.31 can be used. The system of Figure 6.32 could be used with a six step VSI inverter controlled to provide fundamental frequency output voltages corresponding to the voltage commands v_{as}^*, v_{bs}^* and v_{cs}^*.

The second example, shown in Figure 6.33, is an induction machine direct controller with external flux and torque regulators. Because of these external loops, the decoupler can be simplified by omitting the stator resistance–transient time constant blocks in the current feedforward paths thus eliminating the need for knowledge of the stator resistance.This system could again be implemented using a six step VSI inverter rather than a PWM inverter.

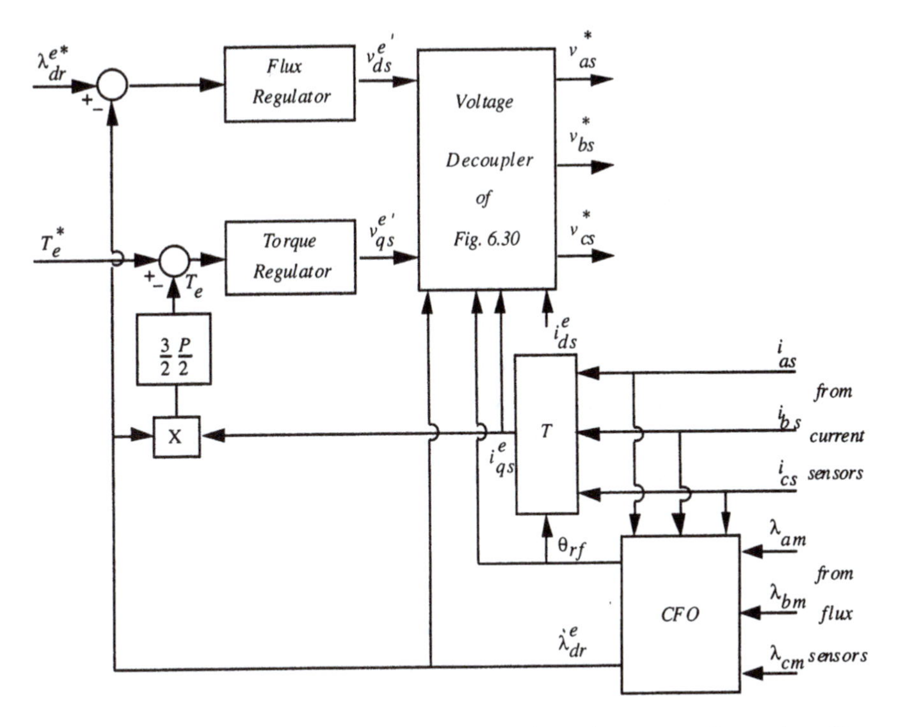

Figure 6.33 Induction machine direct orientation controller using flux sensors and a voltage controlled inverter

6.8 Stator Flux Based Field Orientation

Rotor flux field oriented induction machine drive systems have been shown to offer high performance as well as independent control on torque and flux. Efforts have been made to eliminate flux sensors in a direct field oriented (DFO) system or position sensors in an indirect field oriented (IFO) system without sacrificing performance. A commonly used method is to estimate rotor flux from terminal quantities. For example, the rotor flux can be estimated by the following equations:

$$\hat{\underline{\lambda}}_s = \int (\underline{v}_s - r_s \underline{i}_s)\, dt \tag{6.8–1}$$

$$\hat{\lambda}_r = \frac{L_r}{L_m}(\hat{\lambda}_s - \sigma L_s \underline{i}_s) \tag{6.8–2}$$

where $\hat{\lambda}_s$ is the estimated stator flux vector, $\hat{\lambda}_r$ is the estimated rotor flux vector and

$$\sigma = 1 - \frac{L_m^2}{L_s L_r} \tag{6.8–3}$$

is the leakage factor.

Equation (6.8–1) shows that the accuracy of the estimated stator flux depends on the accuracy of the estimated stator resistance. Usually, the stator resistance can be measured with reasonable accuracy and it is relatively easy to adapt to its slow variation with temperature. Therefore, it is reasonable to assume that the stator flux can be estimated accurately at least at relatively high speeds where the stator *IR* drop is relatively small compared to the amplitude of the applied voltage.

On the other hand, eqn (6.8–2) shows that estimation of rotor flux requires knowledge of the inductances of the machine, especially the leakage inductance. It has been reported that for a standard induction machine with closed rotor slots, the conventional no load and locked rotor tests do not give the correct values for machine inductances. It is also observed that due to the closed rotor slots, the inductances vary with operating conditions [9],[10]. Hence, it is difficult to tune the leakage parameters in the rotor flux estimator and adapt for their variations. Some error in the estimated rotor flux can always be expected. Structurally, the flux estimator is in the feedback path of the closed loop system. From the view point of control theory, the performance of a feedback control system relies on the accuracy of the feedback signal. To obtain accurate and robust control characteristics, the most accurate signal should be chosen as the feedback signal. Hence, it is natural to consider using the stator flux as the feedback signal, which leads to the implementation of a stator flux oriented system.

6.8.1 Mathematical Model of a Stator Flux Oriented Induction Machine

In the mathematical model of a rotor flux oriented system (eqns (6.2–11) through (6.2–14)), all variables are expressed in a reference frame oriented to the rotor flux. With the stator flux chosen as the measured quantity, it is more convenient to express the machine model in a reference frame oriented to the stator flux. Figure 6.34 illustrates the spatial relation between the two reference frames.

To obtain the machine equations in the stator flux reference frame, one can mathematically eliminate the rotor flux as a variable from eqns (6.2–1) through (6.2–9) and then force the q– component of the stator flux to be zero similar to the procedure used for air gap flux in Section 6.5. Assuming again that the machine is effectively supplied by a current source, the resultant equations can be written as follows:

$$(1+\sigma\tau_r p) L_s i_{qs}^{es} - S\omega_e \tau_r (\lambda_{ds}^{es} - \sigma L_s i_{ds}^{es}) = 0 \tag{6.8–4}$$

$$(1+\tau_r p)\lambda_{ds}^{es} = (1+\sigma\tau_r p) L_s i_{ds}^{es} - S\omega_e \tau_r \sigma L_s i_{qs}^{es} \tag{6.8–5}$$

$$T_e = \frac{3P}{4}\lambda_{ds}^{es} i_{qs}^{es} \tag{6.8–6}$$

Here, the superscript *es* is used to designate that the quantity is expressed in a synchronous reference frame fixed to the stator flux vector. The quantity σ is the *coupling factor* as defined by eqn (6.8–3).

6.8.2 Design of a Decoupler for Direct Stator Flux Oriented System

Comparing eqn (6.8–5) to (6.2–12) indicates that there now exists a coupling between the torque producing component of the stator current i_{qs}^{es} and the stator flux linkage. As a consequence, any change in i_{qs}^{es} without changing i_{ds}^{es} accordingly will cause a transient in the stator flux. In order to overcome this disadvantage, a *decoupler* can be designed to compensate the coupling effect. This approach is depicted in Figure 6.35.

The action of the decoupler can be expressed as follows,

$$i_{ds}^{es} = \left(K_p + \frac{K_i}{p}\right)(\lambda_{ds}{}^* - \hat{\lambda}_{ds}) + i_{dq}^{es} \tag{6.8–7}$$

where i_{dq}^{es} is the output of the decoupler and it is assumed that a proportional plus integral (P–I) regulator is used to control the stator flux linkage. Substituting eqn (6.8–7) into (6.8–5) yields:

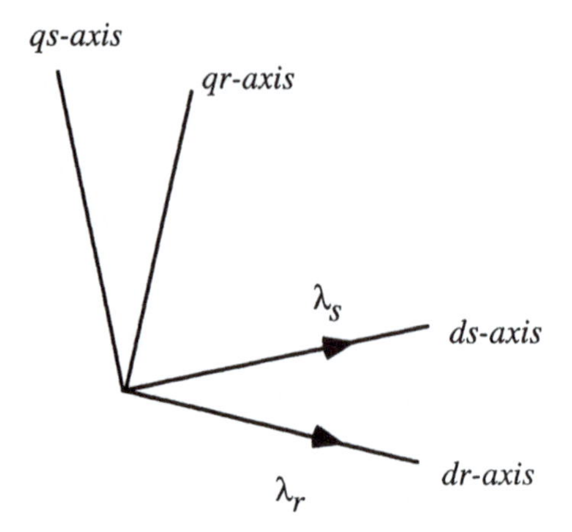

Figure 6.34 Rotor and stator flux linkage reference frames

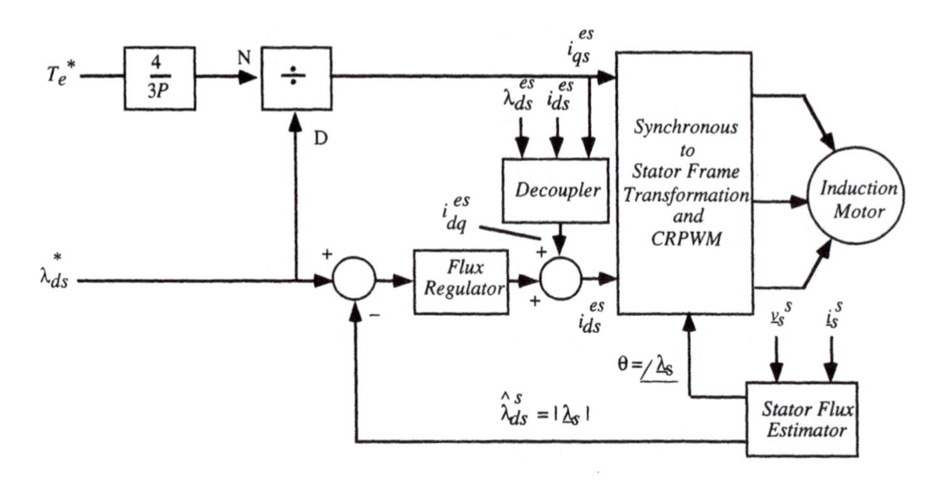

Figure 6.35 Block diagram of stator flux based field orientation system

$$(1+\tau_r p)\,\lambda_{ds}^{es} = (1+\sigma\tau_r p)\,L_s\left(K_p + \frac{K_i}{p}\right)(\lambda_{ds}{}^* - \hat{\lambda}_{ds}) + (1+\sigma\tau_r p)\,L_s i_{dq}^{es}$$
$$- S\omega_e \tau_r \sigma L_s i_{qs}^{es}$$

(6.8–8)

In order to decouple the d–axis stator flux from the q–axis stator current, it is required that

$$(1+\sigma\tau_r p)\,L_s i_{dq}^{es} - S\omega_e \tau_r \sigma L_s i_{qs}^{es} = 0 \qquad (6.8\text{–}9)$$

or

$$i_{dq}^{es} = \frac{S\omega_e \tau_r \sigma L_s i_{qs}^{es}}{(1+\sigma\tau_r p)\,L_s} \qquad (6.8\text{–}10)$$

where $S\omega_e$ is determined from eqn (6.8–4),

$$S\omega_e = \frac{(1+\sigma\tau_r p)\,L_s i_{qs}^{es}}{\tau_r\,(\lambda_{ds}^{es} - \sigma L_s i_{ds}^{es})} \qquad (6.8\text{–}11)$$

Equations (6.8–10) and (6.8–11) describe a decoupler which, when properly tuned, generates the correct value of i_{ds}^{es} so that for a specified value of i_{qs}^{es}, the flux linkage $\lambda_{ds}^{es} = \lambda_s$ is not altered when torque changes are demanded by the torque command $T_e{}^*$. The concept is similar to the decoupling relation in eqn (6.5–23) and Figure 6.23 for air gap flux field orientation. Note that since "p" occurs in the numerator of eqn (6.8–11), the decoupler requires the use of a differentiator which introduces noise into the current regulator. However, since most controllers are today implemented in digital form, modern numerical algorithms can be used to moderate this effect.

References

[1] K. Hasse, "Zur Dynamik Drehzahlgeregelter Antriebe Mit Stromrichter gespeisten Asynchron–Kuzschlublaufermaschinen" (On the Dynamics of Speed Control of a Static AC Drive with a Squirrel Cage Induction Machine), Ph.D. Dissertation, Techn. Hochschule Darmstadt, 1969.

[2] F. Blaschke, "The Principle of Field Orientation – the Basis for the Transvector Control of Three–Phase Machines", Siemens Zeitschrift, Vol. 45, No. 10, pp. 757–760, 1971.

[3] F. Blaschke, "Das Verfahren der Feldorientierung zur Regelung der Drehfeldmaschine"288288 (The Method of Field Orientation for Control of Three Phase Machines), Ph.D Dissertation, University of Braunschweig, 1973.

[4] R. DeDoncker and D.W. Novotny, "The Universal Field Oriented Controller," IEEE–IAS Trans., Vol. 30, No. 1, January/February 1994, pp. 92–100.

[5] N.R. Garrigan and D.W. Novotny, "Start–up and Sensor–Parameter Error Transients in

Field Oriented Induction Machines," Proc. IEEE–IAS Annual Meeting, October 1995.

[6] P.L. Jansen and R.D. Lorenz, "A Physically Insightful Approach to the Design and Accuracy Assessment of Flux Observers for Field Oriented Induction Machine Drives", Proc. IEEE–IAS Annual Meeting, October 1992, pp. 570–577.

[7] D.Zinger, F. Profumo, T.A. Lipo and D.W. Novotny, "A Direct Field Oriented Controller for Induction Motor Drives Using Tapped Stator Windings," IEEE Trans. on Power Electronics, Vol. 5, No. 4, October 1990, pp. 446–453.

[8] R. Lorenz and D.W.Novotny, "A Control Systems Perspective of Field Oriented Control for AC Servo Drives", Proceedings of Control Expo 88, June 1988, pp 18–1 to 18–12.

[9] T. Irisa, S. Takaka, R. Ueda and T. Sonoda, "On Reliability of Induction Machine for High Performance Based on Parameter Characteristics", IEEE–IAS Annual Meeting Conf. Record, 1983, pp. 547–554.

[10] R. Ueda, T. Sonoda, K. Fujitani, Y. Yoshida and T. Irisa, "Investigation of Induction Motor Characteristics by Means of Vector Control", IEEE–IAS Annual Meeting, Conf. Record, 1985, pp. 578–585.

Problems

Problem 6–1 Induction Machine Field Orientation

A 100 hp, 460 volt, induction machine is operated from an ideal controlled current electronic converter using field oriented control as illustrated in Fig. 6.6. The pu parameters of the machine are:

$r_s = 0.015$ pu	$x_{ls} = 0.10$ pu	$x_m = 2.0$ pu (at rated freq.)
$r_r = 0.020$ pu	$x_{lr} = 0.10$ pu	$r_m = 50$ pu (at rated freq.)

The rated slip frequency is 0.0248. Express all answers in real units.

a) Find the required values of i_{qs} and i_{ds} and the slip relation relating $S\omega_e$, i_{qs} and λ_{dr} to produce operation at rated torque and rated speed if the terminal voltage and frequency are held at their rated values.

b) Find the final steady state torque and slip frequency if i_{ds} is reduced to one half its value in part a) while i_{qs} is held constant. Find the terminal voltage and stator frequency if the new speed is twice that in part a).

c) Find the transient response of the torque for part b) assuming the current regulator can change i_{ds} instantaneously.

Problem 6–2 Synchronous Machine Field Orientation

A 200 hp, 460 volt, 3 phase, 4 pole, 60 Hz synchronous motor with the following per unit parameters

$X_{ls} = 0.1$	$X_{lf} = 0.2$	$r_s = 0.03$
$X_{md} = 1.3$	$X_{ldr} = 0.1$	$r_{dr} = 0.03$
$X_{mq} = 0.4$	$X_{lqr} = 0.15$	$r_{qr} = 0.06$
		$r_{fr} = 0.02$

is operated from a current regulated variable frequency supply. For this problem assume an ideal current regulator.

a) For field orientation $0 = (\gamma)$ with rated terminal voltage V=1.0 pu, current $I = i^{s*}_{qs} = 1.0$ pu, and rated frequency $f = 1.0$ pu, find the torque, internal voltage E, and the terminal power factor.

b) For unity power factor operation at 60 Hz, V=1.0 pu, with the same torque as in part a), find the required E, I, i_{qs}^{s*} , i_{ds}^{s*} and γ.

c) Find and sketch $T(t)$ following a step input of i_{qs}^{s*} (step from zero to that value which yields rated torque) for parts a) and c). Give the torque in pu but express time in real units. Assume i_{ds}^{s*} remains constant in both cases.

7 Current Regulation in Power Converters

7.1 Introduction

The power converter in a high performance ac drive essentially functions as a power amplifier, reproducing the low power level control signals generated in the field orientation controller at power levels appropriate for the driven machine. Typically the input power to the converter is three phase, fixed frequency, fixed voltage power from the utility power grid. The output must be variable frequency, variable voltage power with the frequency and voltage range appropriate to the intended operating range of the drive.

At present, except for very low speed drives where line commutated cycloconverters are employed, the power converters used in high performance drives are dc link inverters. Developmental work on direct converters using forced commutation and bilateral switches as well as resonant link converters utilizing a high frequency single phase link rather than the usual dc link are underway. However, application to commercially available drives is several years away. The material in this chapter will be primarily concerned with dc link inverters with only passing mention of other types of power converter arrangements.

Although field orientation controllers which provide voltage commands are feasible (see Section 6.8), most systems provide current commands and thus require power converters which function as controllable current sources. Such systems contain inner current loops very much like the current loop in a high performance dc drive. The primary difference is that the current control required in an ac drive must provide both amplitude and phase control. This is, partly, the origin of the term *vector control* as applied to high performance ac drives.

7.2 Current Regulated Inverters

As noted in the introduction, high performance drives utilize control strategies which develop command signals for the ac machine currents. The basic reason for the selection of current as the controlled variable is the same as for the dc machine; the stator dynamics (effects of resistance, inductance and induced emf) are eliminated. Thus, to the extent that the current regulator functions as an ideal current supply, the order of the system under control is reduced and the complexity of the controller is significantly simplified.

Current regulators for ac drives are more complex than for dc drives because an ac current regulator must control both the amplitude and phase of the stator current. In addition, the steady state currents are ac currents not dc currents so the straight forward application of conventional proportional–integral control (PI control) as applied in dc drives cannot be expected to provide performance comparable to that of a dc drive current regulator. As in the dc drive, the ac drive current regulator forms the inner loop of the overall controller. As such, it must have the widest bandwidth in the system and should have zero or nearly zero steady state error. Achieving these goals in ac current regulators has proven to be a challenging task and it is only recently that satisfactory, fundamentally sound techniques have been developed.

Both CSI and PWM inverters can be operated in controlled current modes; the VSI is not readily applicable and would offer no basic advantage over a PWM inverter. The CSI is a 'natural' current supply and can readily be adapted to controlled current operation. The PWM inverter requires more complexity in the current regulator but offers potentially much higher bandwidth and elimination of current harmonics as compared to the CSI. Since the concepts involved are much simpler in the CSI it is treated first with the current regulated PWM inverter treated subsequently as a solution to some of the inherent problems associated with the CSI.

7.2.1 The Current Regulated CSI Inverter

The fundamental implementation of current regulation in a CSI is very simple in concept requiring only the addition of a dc link current regulator as illustrated in Figure 7.1. The command input I_s^* is the reference input corresponding to the amplitude of the stator current. The phase of the output current is determined by the firing times of the CSI thyristors and thus the phase current command, $\angle\tilde{I}_s^*$, is shown as a direct input to the CSI inverter block.

Implementation of the dc link current regulator is essentially the same as for a dc drive current regulator since the current being regulated is a dc current and regulation is accomplished via the firing angle of the input phase controlled rectifier. Ordinary PI–control can be applied since the regulated quantity is dc in the steady state. Regulator performance and design is adequately

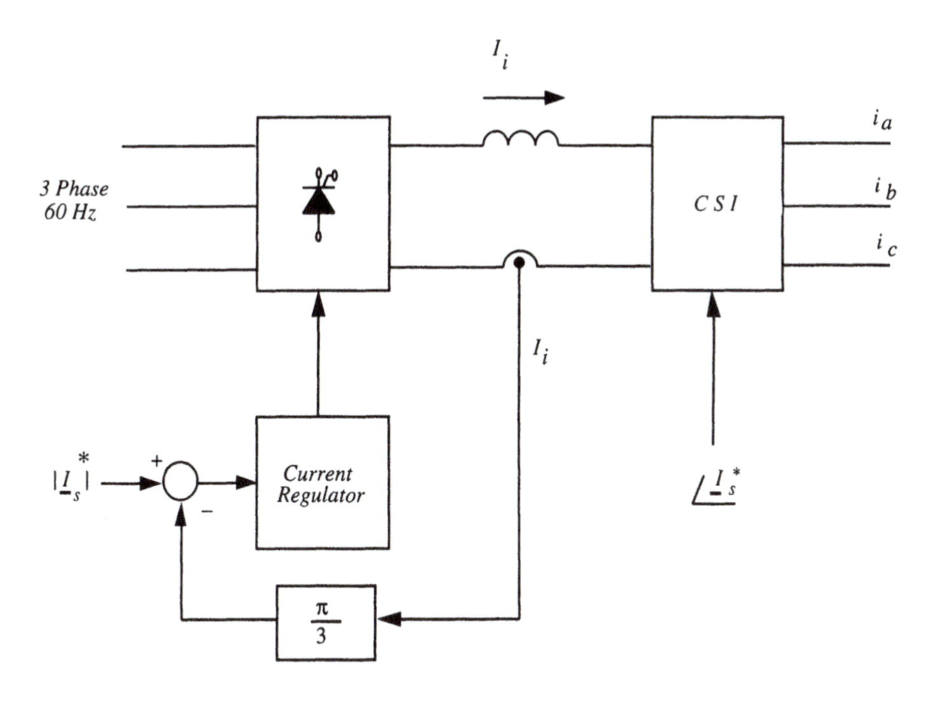

Figure 7.1 CSI inverter with dc link current regulator as a three phase regulated source

treated in many references and will not be considered in detail here. While a diode input rectifier and dc chopper could be used in place of the phase controlled rectifier, this would eliminate the inherent regeneration capability of the system. Although the chopper could potentially provide increased bandwidth, other inherent problems of the CSI would not be overcome and the loss of regeneration capability is a serious drawback. For these reasons, diode rectifier–chopper systems do not appear to be an important option.

While, in principle, phase angle control via thyristor firing time offers instantaneous phase control, there is a significant limitation imposed by the commutation delay inherent in the CSI. Unfortunately, this delay is both current amplitude and frequency dependent. Thus, implementation of compensation for this delay is a significant problem. The principle is illustrated in Figure 7.2 where compensation inputs for both amplitude and frequency are shown. Although reasonable compensation can be achieved, the non–linearity inherent in the commutation delay makes this task a difficult one.

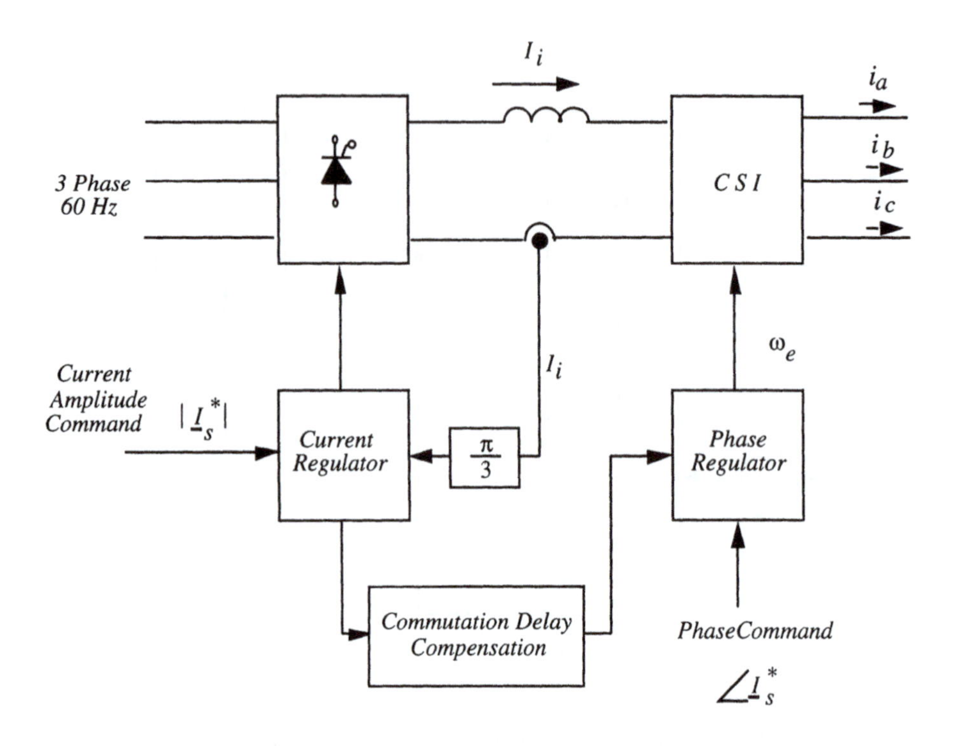

Figure 7.2 Current regulated CSI inverter with commutation delay compensation

The CSI system of Figure 7.2 also has the problem of producing significant harmonics in the output waveform since the CSI is basically a 6–step system. In effect, the essentially rectangular wave output current results in stepwise 60° phase advances in the stator mmf of the connected machine. While this may not have serious effects at higher frequencies, it will generally result in large torque pulsations and associated rotor speed oscillations or *cogging* at low frequencies. Although PWM techniques can be employed to reduce this cogging behavior at low speeds to acceptable levels, the fundamental problem of the 60° mmf steps and the associated torque pulsations remains at higher speeds. This is a result of the inherent CSI commutation delay and hence necessarily low PWM carrier frequency which can be employed. Above perhaps 10–20 Hz, the system must necessarily return to normal 6–step CSI operation with essentially rectangular (or trapezoidal) current waveforms. Twelve pulse, dual

CSI systems with six–phase (30° displacement) machines can be employed to produce 30° steps in the mmf wave.

Because of the commutation phase delay and the rectangular waveforms, the controlled current CSI has serious limitations as a high performance current supply. However, in high power applications where response time is less significant, it provides the possibility of achieving reasonable current control while retaining the practical advantages of CSI operation. This is especially significant in synchronous motor drives of high horsepower where load commutation is employed and the controlled current CSI is widely applied for such drives. Very high horsepower induction motor drives are also available utilizing the current controlled CSI.

7.2.2 The Current Regulated PWM Inverter

A current controlled PWM inverter operated with a switching frequency in the kHz range can function as a regulated current supply with the potential for good dynamic response and low harmonic content. It is the most commonly used power converter in the low to medium power range of high performance drives. Unlike the CSI, the PWM is a natural voltage source and conversion to current source operation requires closed loop control with feedback directly from the controlled ac currents. The feedback current sensors must therefore have a wide bandwidth, from the lowest fundamental frequency to be controlled (usually dc) to somewhat above the PWM carrier frequency. Figure 7.3 illustrates the basic system. Usually only two current sensors are employed since in the absence of a neutral connection the three currents must add to zero. It is, in fact, advantageous to use only two current sensors since this avoids error signals containing a zero sequence component resulting from sensor errors. The nature of the controlled output current is illustrated in Figure 7.4. As shown, the output consists of a reproduction of the reference current with high frequency PWM ripple superimposed. This system has several fundamental differences from the CSI system including:

1) The requirement of an actual time domain (waveform) reference instead of the amplitude and phase references required in the CSI,

2) The possibility of reducing harmonic content to arbitrarily small values as the PWM frequency increases,

3) The need for a current regulator operating with ac signals as opposed to the dc regulator of the CSI,

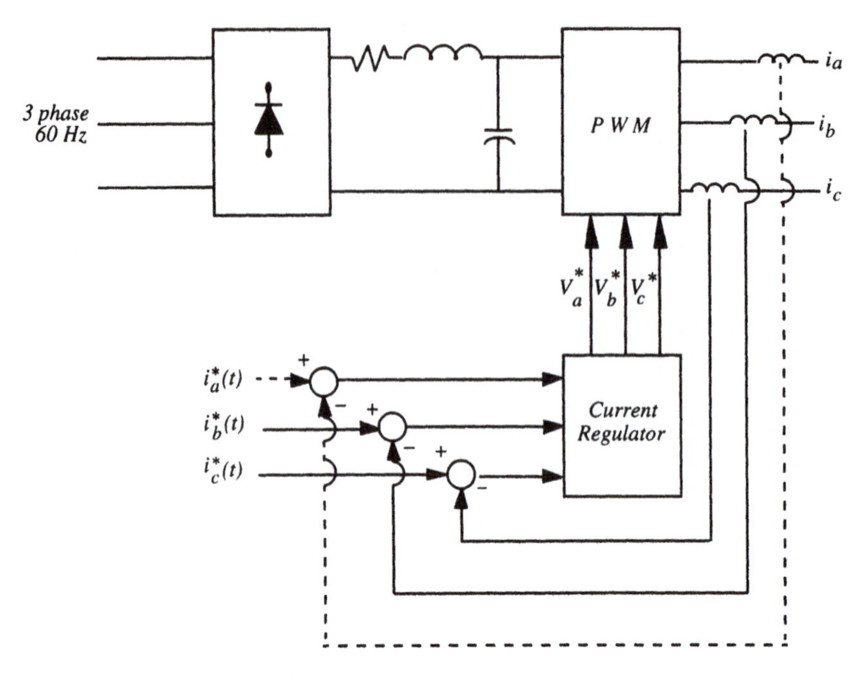

Figure 7.3 PWM system with current regulation to produce a controlled three phase current source

4) The need for wide bandwidth current sensors as described earlier.

While the sensor requirement and the development of good regulators has required considerable developmental effort, the excellent response and low harmonic content of the PWM system make it the best currently available regulated current supply. A number of different types of regulators have been proposed and used and will be discussed separately. Basically the regulator must function to convert the time domain current error signals to firing signals for the PWM inverter. The dual function of error processing and subsequent conversion to gate drive signals is required in all types of regulators.

The subject of current regulators has been the subject of intense activity. In general, current controllers can be classified into three groups: hysteresis regulation, ramp comparison and predictive controllers.

7.3 Hysteresis Regulators

The earliest and simplest type of PWM current regulator to implement is the hysteresis regulator [1],[2]. The basic concept of this type of controller is illus-

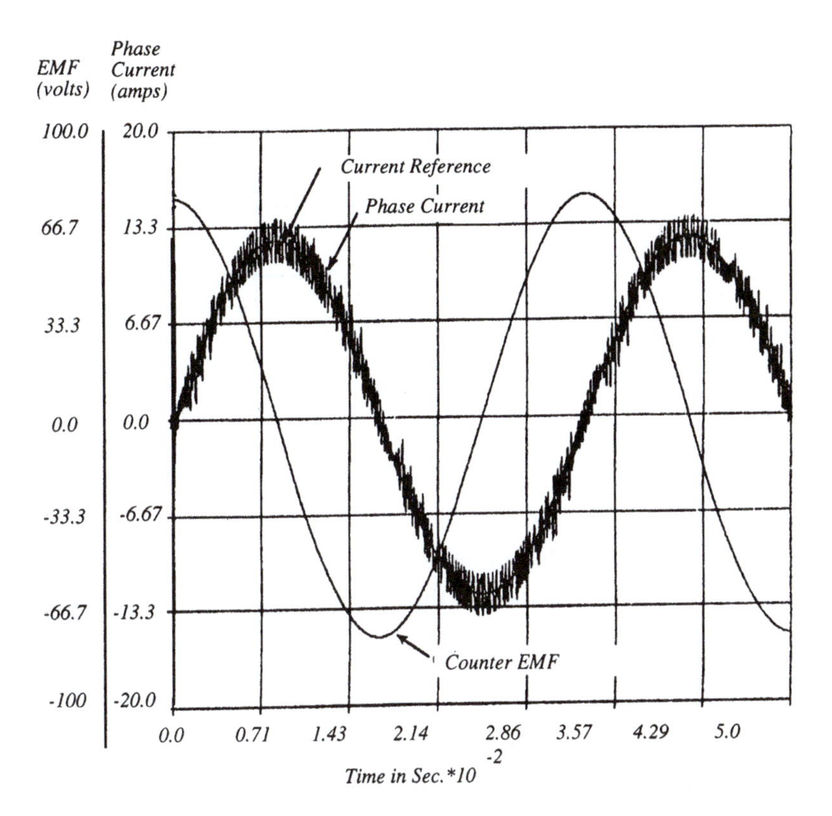

Figure 7.4 Simulated waveforms of a PWM inverter with current regulation using three independent hysteresis controllers

trated in Figure 7.5. The controller simply applies the current error signal to a hysteresis element, the output of which supplies the logic signal to gate the positive or negative inverter switching element. A lockout circuit is normally incorporated to allow for inverter switch recovery time and thus avoid short circuits across the dc link.

While this system is very simple and provides good current amplitude control, hysteresis regulation has the major disadvantage of producing a highly variable PWM switching rate. Low frequencies appear in the spectrum regardless of the switching frequency. Unfortunately, the variation of the switching rate is also opposite to the needs for good current control with the highest switching rates associated with the lowest reference frequencies. As will be discussed subsequently, the hysteresis controller also has the somewhat unex-

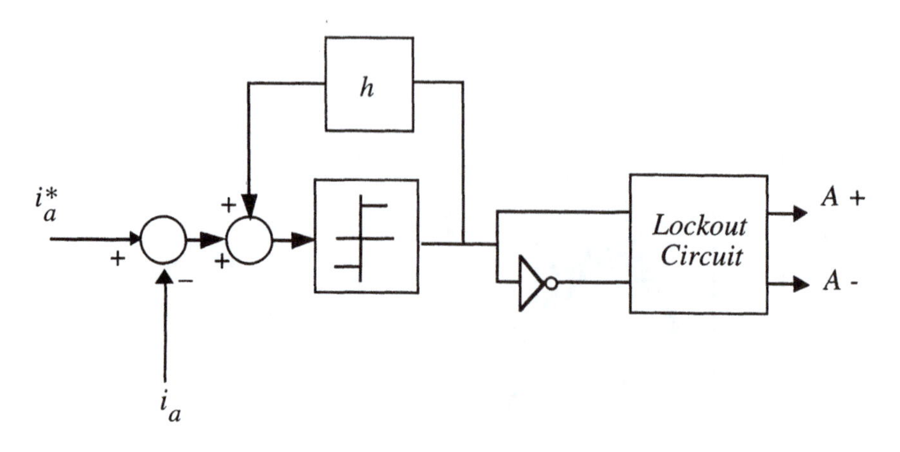

Figure 7.5 Hysteresis current controller for one phase

pected property of limiting the current error to twice the hysteresis band rather than to the band itself. Nonetheless, this controller is in active use in many applications.

The behavior of the hysteresis controller can be explained in terms of a complex plane switching diagram [2]. Figure 7.6(a) shows the reference current vector $\underline{i}_s^*$ the actual current vector $\underline{i}_s$ and the current error vector $\Delta\underline{i}_s$ in the complex plane along with the a, b and c axes of a three phase reference system. The line current errors Δi_{as}, Δi_{bs} and Δi_{cs} are the projections of $\Delta\underline{i}_s$ on these three axes. The hysteresis controller will switch the *a*–phase inverter leg when Δi_{as} exceeds the hysteresis band as represented in Figure 7.6(b) by the two switching lines drawn perpendicular to the *a*–axis. The switching lines are located a distance *h*, equal to the hysteresis band, from the tip of the current reference vector. Similar switching lines can be drawn for phases *b* and *c*; the resulting complete switching diagram is shown in Figure 7.6(c). The entire diagram moves with the current reference vector with its center remaining fixed at the tip of the vector.

The typical or expected behavior of the controller is to confine operation to the interior, hexagonal region of the switching diagram. Thus, whenever the current error touches one of the switching lines, that inverter leg is switched driving the current error in that leg in the opposite direction. Note, however, that the current error can be carried to one of the switching lines by motion of the current vector $\underline{i}_s$, or by motion of the current reference vector $\underline{i}_s^*$ on the switching diagram. Thus, the situation depicted in Figure 7.7 can occur, where

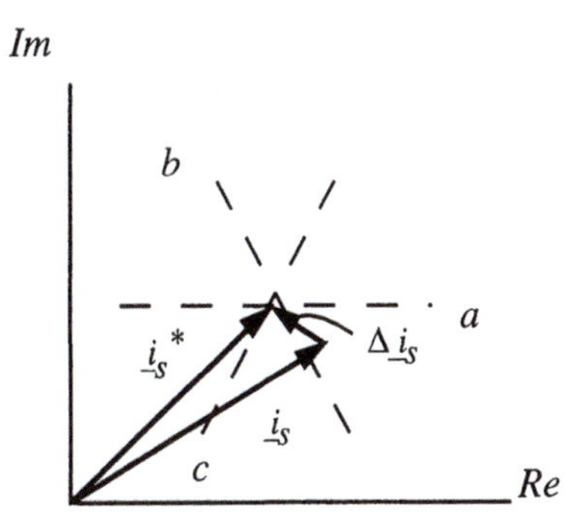

a) Current Vectors in the Complex Plane.

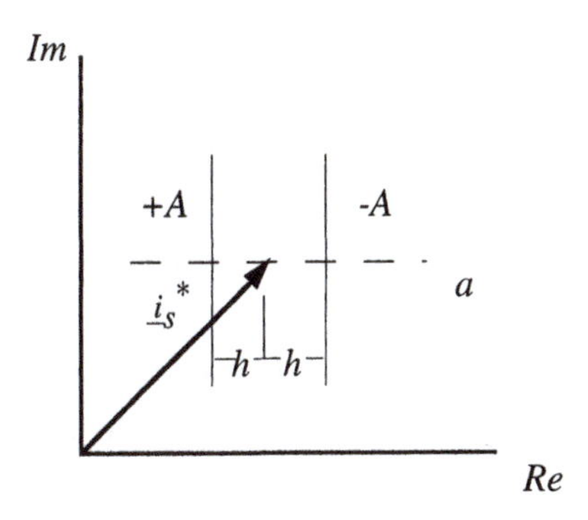

b) Switching Lines for Phase a.

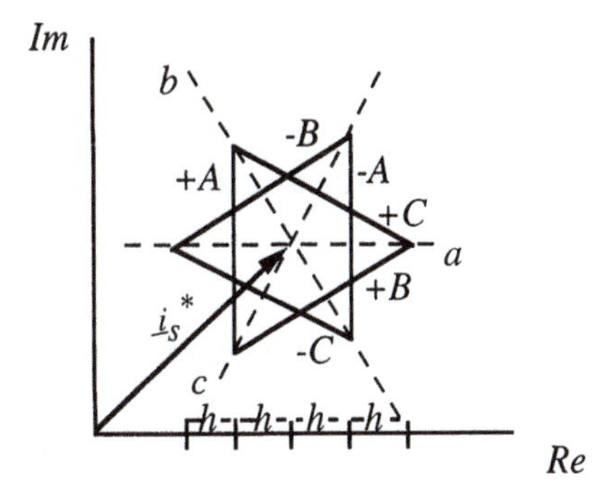

c) Complete Switching Diagram.

Figure 7.6 Hysteresis controller switching diagrams – three independent controllers

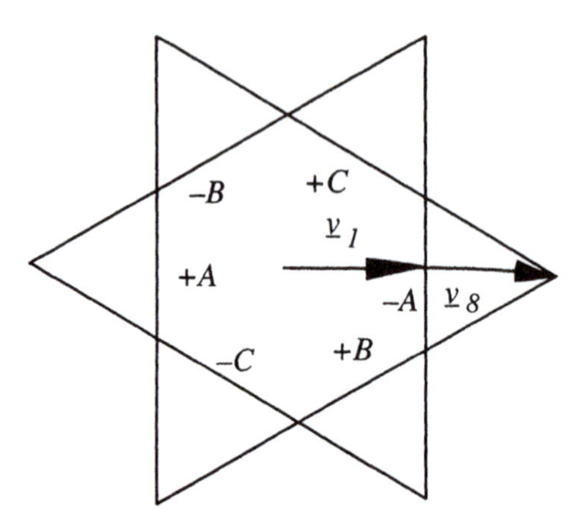

Figure 7.7 Hysteresis controller current trajectory resulting in error equal to twice the hysteresis band

the (−*A*) switching line is encountered with the inverter in the state (*a*+, *b*−,*c*−). The system then can 'coast' out to the tip of the switching diagram before a new switching of the inverter occurs. This demonstrates that an error of 2*h*, twice the expected value equal to the hysteresis band, can occur on an almost random basis.

The switching diagram can also be used to demonstrate the existence of high frequency limit cycle oscillations when the load counter emf is low (low frequency operation). In general, the vector voltage equation relating the inverter voltage $\underline{v}_s$, the load current $\underline{i}_s$ and the load counter emf $\underline{e}_s$ (voltage behind transient inductance of the induction machine) is

$$\underline{v}_s = r_s \underline{i}_s + L'_s p \underline{i}_s + \underline{e}_s \tag{7.3–1}$$

where r_s and L'_s are the load resistance and transient inductance per phase. If $\underline{e}_s$ is small and the resistance is neglected, this equation becomes

$$\underline{v}_s \approx L'_s p \underline{i}_s \tag{7.3–2}$$

which indicates that the change of the current vector is in the direction of the voltage vector.

Figure 7.8 shows a current trajectory, indicated by the solid line, which can occur and represents a high frequency limit cycle. The initial voltage vector $\underline{v}_1$ (*a*+, *b*−, *c*−) (mode 612 in Figure 1.5), forces the tip of the current vector to travel in the same direction as the voltage vector since the resistance and

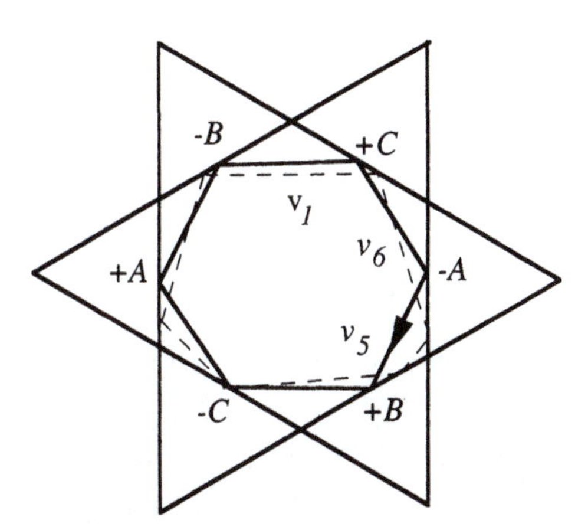

Figure 7.8 Hysteresis controller trajectory illustrating possible high frequency limit cycles

counter emf are assumed to be small. The current vector hits the $+c$ switching line causing inverter leg c to switch and produce the new voltage vector $\underline{v}_6$ (a+, b–, c+). Next the current vector will hit the $-a$ switching line producing the voltage vector $\underline{v}_5$ (a–, b–, c+). Continuing this reasoning, the six possible non zero voltage vectors are applied repeatedly and a high switching frequency results if the inductance is low and the hysteresis band small. Note that the magnitude of the current error vector is not reduced to zero during the limit cycle. The dashed line in Figure 7.8 represents a possible limit cycle when there is a non–zero counter emf.

The limit cycle will be occasionally interrupted by the intermittent occurrence of a zero voltage vector. A zero voltage vector will occur when one of the switching lines in the sequence is skipped as a result of the load counter emf or resistance, or the movement of the switching diagram caused by changes in the reference current. The application of a zero voltage vector will greatly reduce the inverter switching frequency when the counter emf is low since the velocity along a trajectory is proportional to the total applied voltage $\underline{v} - \underline{e}$ (see eqn (7.3–1)).

Figure 7.9 and Figure 7.10 illustrate the results of a simulation of a hysteresis controller which shows the behavior described above. Figure 7.9 is the current error signal for an unloaded induction motor operating at 0.5 pu speed. The occasional occurrence of a current error in excess of the 1.5 ampere hysteresis band is clearly shown; several of these are very close to twice the band. The controlled current in Figure 7.10, which is for a stalled motor, clearly

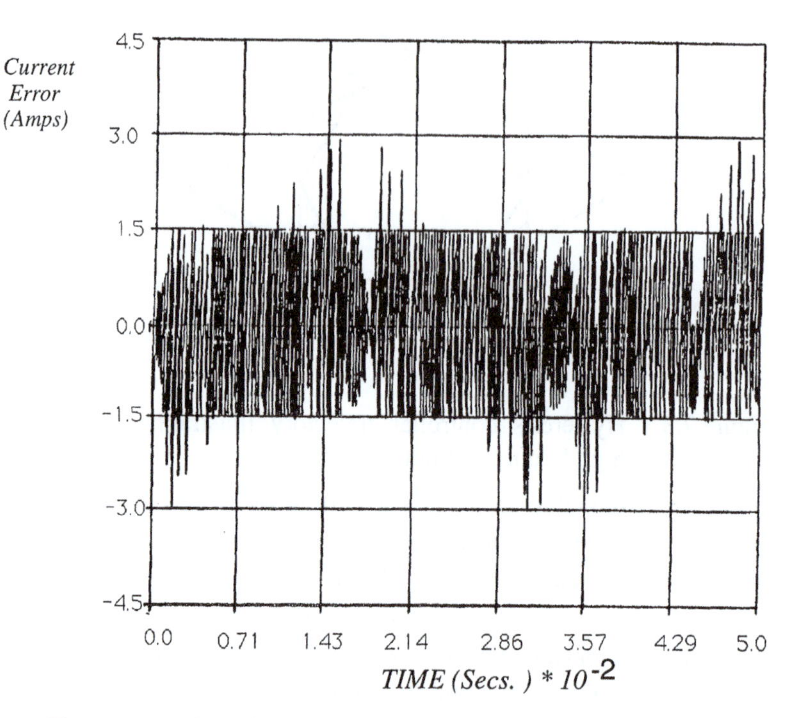

Figure 7.9 Simulation results for hysteresis controller driving a typical 10 HP induction motor at 30 HZ. and no load showing random large current errors

shows the sporadic limit cycles which can exist for low frequency operation. Note the nearly one half cycle of quite low switching frequency which also exists; the limit cycles are essentially random occurrences.

The oscillation frequency within the limit cycles shown in Figure 7.10 can be estimated by dividing the velocity of the current trajectory by the distance travelled in one complete period of the limit cycle. For zero counter emf, the velocity in amperes per second is approximately

$$\text{velocity} = \frac{di_s}{dt} \approx \frac{2}{3}\frac{V_{dc}}{L'_s} \tag{7.3–3}$$

and the distance (in amperes) travelled in a complete limit cycle is approximately

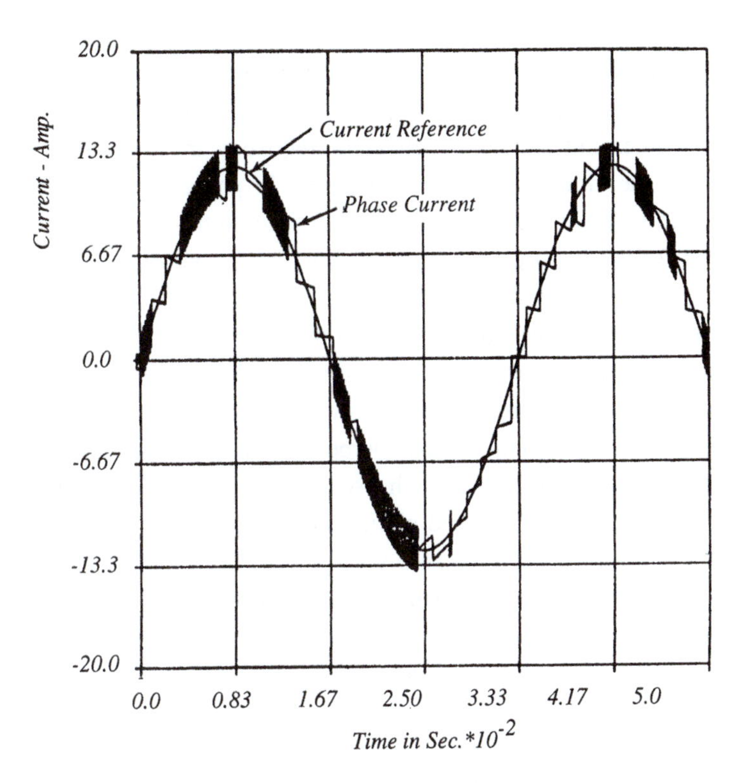

Figure 7.10 Simulation results for hysteresis controller driving a typical 10 hp induction motor at zero speed showing sporadic high frequency limit cycles in the motor phase current

$$\text{distance} = \Delta i_s \approx 6h \tag{7.3–4}$$

The inverter switching frequency is then

$$f_s = \frac{\text{velocity}}{\text{distance}} = \frac{V_{dc}}{9hL'_s} \tag{7.3–5}$$

For the system illustrated in Figure 7.10, the inverter bus voltage was 300 volts, the machine transient inductance was 3.36 mh and the hysteresis band was set at 1.5 amperes yielding a switching frequency of

$$f_s = \frac{300}{9\text{x}1.5\text{x}3.36\text{x}10^{-3}} = 6610\text{Hz} \tag{7.3–6}$$

which was found to be very close to the value obtained in the simulation. Clearly the highest switching frequency occurs during limit cycles with essentially zero counter emf and the inverter must be designed to operate successfully under this condition. Since high switching frequencies are not needed when the counter emf and hence reference frequency are low, this is a serious limitation of the simple hysteresis controller. In particular, the difficulty of working with variable switching rates with respect to the filtering of measured signals for feedback purposes argues against such controllers and has been addressed by several authors [3]–[5].

In Ref. [4] switchings still occur when the error vector crosses the boundary of the hexagon. However, the voltage vector that is selected differs from what a conventional hysteresis controller would select. When a segment of the hexagon is crossed the voltage vector perpendicular to that segment and pointing toward the origin is selected. Such a scheme would produce high frequency limit cycles of the type discussed earlier under all conditions if some means were not used to introduced the zero voltage state occasionally. In Ref. 3, the zero voltage vector is selected when the magnitude of $\Delta \underline{i}_s$ is less than a certain fixed amount and when the value of the back emf indicates that the $\Delta \underline{i}_s$ vector will be driven in a favorable direction. The back emf is calculated using measurements of rotor position and frequency. As this paper concerned a synchronous motor drive, frequency and rotor speed are essentially the same so that this problem is less difficult than for the induction motor. The use of the zero voltage vector is disabled if the $\Delta \underline{i}_s$ vector is outside the hexagonal region for a long period of time to allow faster transient response.

In Ref. [4] switchings are again initiated when the error vector crosses the boundary of the hexagon. In this case, however, the voltage that is selected when a segment is crossed is modified such that the association between hexagon segments and voltage vectors is decided based upon the instantaneous direction of the back emf. For any one value of back emf only three of seven voltage vectors are used. These are the three vectors which are closest to the back emf vector, and automatically includes the zero voltage vector. Only a simple estimate of the emf angle and not amplitude is required. An estimator for this quantity is described in the paper. The controller incorporates regulation of the average switching frequency by a feedback controller that adjusts the hysteresis band size.

Reference [5] uses a fundamentally different approach to solving the problems associated with the hysteresis controller. In this case the interaction between phases is eliminated by making the load appear to have a neutral connection back to a mid point in the dc link capacitor. This trick is accomplished by calculating the zero sequence component of the stator voltage and then simulating the behavior of the zero sequence circuit. From Chapter 2 the zero sequence voltage is given by

$$v_{0s} = \frac{1}{3}[v_{as} + v_{bs} + v_{cs}] \tag{7.3–7}$$

The zero sequence circuit equation is

$$v_{0s} = r_s i_{0s} + \frac{d(L_{0s} i_{0s})}{dt} \tag{7.3–8}$$

where L_{0s} represents the zero sequence inductance and is ideally equal to, but in actuality somewhat smaller than the per phase stator leakage inductance L_{ls}. Solution of this equation enables the calculation of the value of the zero sequence current that would flow if a neutral were connected. This computed current is added to the current feedback signals as shown in Figure 7.11, so that the individual hysteresis controllers appear to control phase currents that would flow if there were a neutral, thereby *decoupling* the three current controllers.

7.4 Ramp–Comparison Controllers

At present, most systems employ other forms of current regulators in which the switching frequency is either nearly constant or at least known and under good control. Constant switching frequency can be most easily obtained by a so called ramp comparison controller. The ramp comparison controller is a direct carry over from dc machine current regulators employing the four transistor bi directional chopper or H bridge. Figure 7.12 illustrates the basic concept which is essentially to employ a synchronous sine triangle PWM with the current error providing the sinusoidal input. Although not shown in Figure 7.12, it is common in dc systems to employ a PI controller ahead of the triangle comparison to enable high dc gain and a controlled roll off of the high frequency response.

The basic operation of the ramp comparison controller is very simple; the processed current error is compared to a triangular waveform and if the error amplifier output is greater (less) than the triangle waveform the inverter leg is

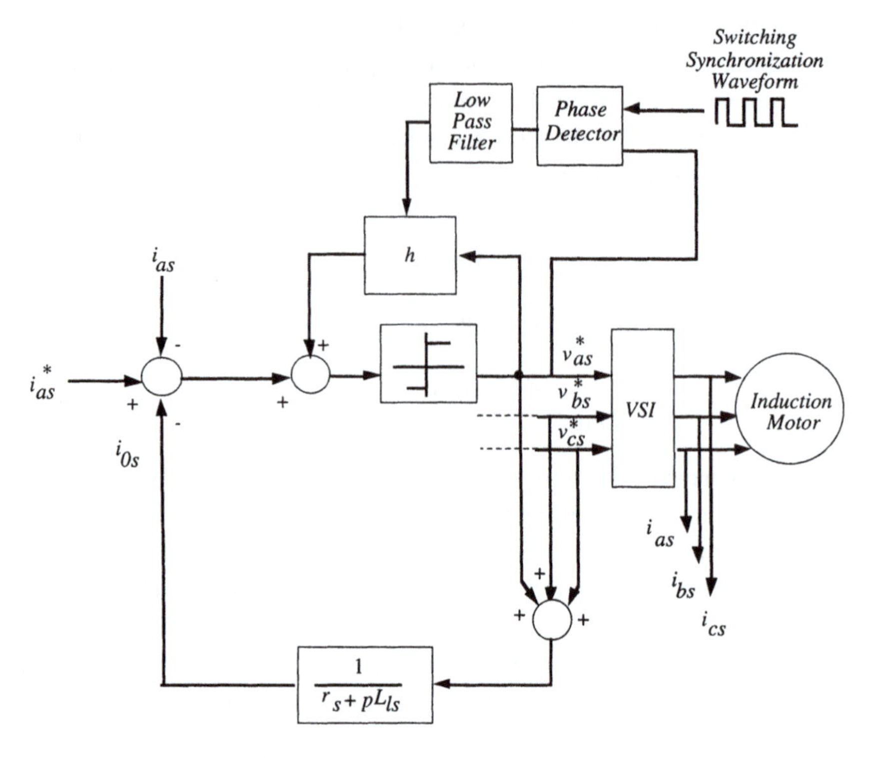

Figure 7.11 Modified hysteresis current controller of Malesani and Tenti [5]

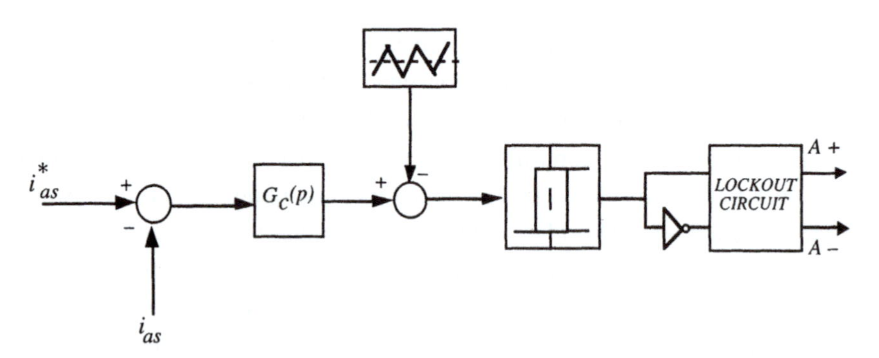

Figure 7.12 One phase of basic ramp comparison controller

switched to the positive (negative) polarity. The inverter switches at the frequency of the triangle wave and produces an output voltage which is proportional to the error amplifier output signal (assuming the output voltage is well below the inverter dc bus voltage) [6]. If the switching frequency is high compared to the frequency content of the input modulating signal (error amplifier output voltage), the proportional relationship between the input and the fundamental component of the output voltage across one phase of the motor can be expressed by

$$v_{out} = \frac{2}{\pi} V_i \frac{v_{in}}{V_\Delta} = K_\Delta v_{in} \tag{7.4–1}$$

where $K_\Delta = (2/\pi)(V_i/V_\Delta)$, V_i is the inverter dc input voltage and V_Δ is the peak value of the triangle signal. The input voltage v_{in} is, effectively, the voltage at the input of the summer leading to the hysteresis block. The ratio of the input voltage to the peak of the triangle voltage is typically termed the modulation index m_i. Hence, in terms of the modulation index eqn (7.4–1) can be written,

$$v_{out} = \frac{2V_i}{\pi} m_i \qquad m_i \le 1 \tag{7.4–2}$$

When the instantaneous amplitude of the input voltage exceeds the triangle wave, intersections of the input voltage and triangle wave are eliminated and consequently PWM pulses are dropped. In this case it can be shown that, the output voltage can be expressed by the describing function [6]

$$v_{out} = m_i \frac{V_{dc}}{\pi} [\sin^{-1}(1/m_i) + (1/m_i)\sqrt{1-(1/m_i)^2}] \qquad m_i > 1 \tag{7.4–3}$$

A sketch of the variation of the fundamental component of the output voltage as a function of m_i is given in Figure 7.13. Note that the amplitude of the input voltage v_{in} must exceed a factor of 5 times the amplitude of the triangle wave for full output voltage to be reached, i.e. square wave voltage operation. In Figure 7.14 the transfer function between the input voltage and the fundamental component of the inverter output phase voltage is expressed as a nonlinear gain. Note that the gain falls off rapidly as pulses are eliminated from the PWM pattern indicating a loss of control of the stator current as the PWM inverter approaches square wave operation.

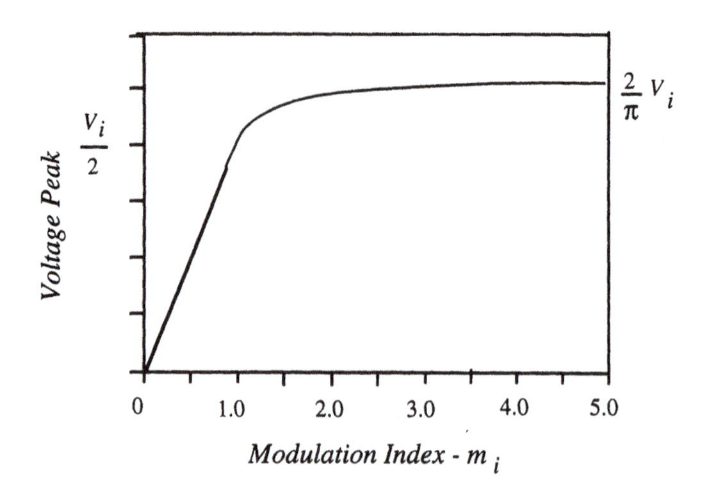

Figure 7.13 Fundamental component of phase voltage versus modulation index m_i

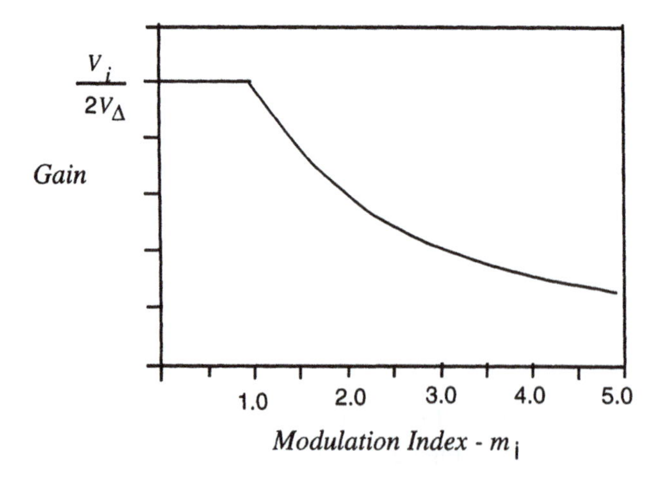

Figure 7.14 Fundamental component gain versus modulation index m_i

Although eqns (7.4–1) to (7.4–3) are derived for sinusoidal inputs, it is common to apply them to arbitrary variations of v_{in} provided the ratio of switching frequency to input frequency is sufficiently high. The hysteresis element shown in Figure 7.12 is included to prevent multiple switching of the inverter leg if the time rate of change of the input voltage exceeds that of the ramp. It is usually omitted when the input voltage rate is controlled.

In the typical dc current controller, a single ramp comparison controller is employed with the output used to control the four transistors of the H bridge in a complimentary fashion, i.e. either $a+$, $b-$ (positive output) or $a-$, $b+$ (negative output). The current error signal, as noted previously, is usually conditioned using a simple PI controller and very satisfactory static and dynamic performance can be attained. An attempt to extend this basic concept to three phase current controllers by using three ramp comparison controllers, one for each phase, immediately poses a problem since the three PI controllers result in an attempt to regulate three independent states when only two exist (because the three phase currents must add to zero). This immediate problem can be overcome by using only two PI controllers and slaving the third phase, an approach which is used in some cases. Alternatively, a zero sequence current can be synthesized in much the same manner as Figure 7.11 and fed back to the three regulators to decouple them. An intellectually pleasing as well as very effective approach is to view the current control problem in d,q coordinates, which immediately indicates the necessity of using only two regulators and algebraically establishing the three phase inverter gating signals.

7.5 Stationary Frame Regulators

Figure 7.15 illustrates the overall system diagram of a three phase ramp comparison controller based on stationary frame d,q variables. The regulator illustrated employs simple PI controllers and is a commonly used system as a result of dc machine experience. It is easily shown that this simple system has inherent regulation problems compared to its dc machine counterpart by considering the steady state response. In the dc machine case, the steady state response is characterized by zero current error because of the integration in the forward path. However, for the ac controller, the steady state condition requires sinusoidal output at the reference frequency and clearly the forward path integration in the PI controller does not produce zero current error because of the frequency dependent PI controllers.

The steady state error for the system of Figure 7.15 can be evaluated using the normal frequency domain steady state model derived in Chapter 4. From Figure 7.15, the frequency domain relation between the d,q axis output voltage and the current error can be written as

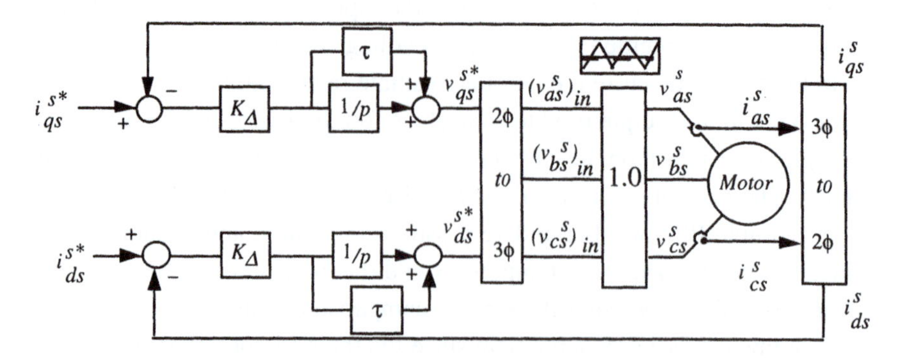

Figure 7.15 System diagram of stationary d,q frame ramp comparison current regulator using PI controllers

$$(\underline{i}_s^{s*} - \underline{i}_s^s)\, K_\Delta \underline{G}_c(j\omega_e) = \underline{v}_s^s \tag{7.5–1}$$

where, for generality, $\underline{G}_c(j\omega_e)$ is the frequency domain transfer function of the current error amplifier in Figure 7.12 rather than the PI regulator in Figure 7.15 and the complex quantities $\underline{i}_s^{s*}$, $\underline{i}_s^s$ and $\underline{v}_s^s$ are the steady state frequency domain values of the currents and voltages.

For steady state operation, the normal equivalent circuit of the induction machine (or synchronous machine) can be used to relate $\underline{i}_s^s$ and $\underline{v}_s^s$. It was shown in Chapter 4 that in the general rotating reference frame the equations for the stator and rotor circuits of a three phase, three wire induction machine can be written,

$$\underline{v}_s = r_s \underline{i}_s + (p + j\omega)\underline{\lambda}_s \tag{7.5–2}$$

$$0 = r_r \underline{i}_r + [p + j(\omega - \omega_r)]\underline{\lambda}_r \tag{7.5–3}$$

where the flux linkages are

$$\underline{\lambda}_s = L_s \underline{i}_s + L_m \underline{i}_r \tag{7.5–4}$$

$$\underline{\lambda}_r = L_m \underline{i}_s + L_r \underline{i}_r \tag{7.5–5}$$

Eliminating the rotor current in these equations yields

$$\underline{\lambda}_s = \left(L_s - \frac{L_m^2}{L_r}\right)\underline{i}_s + \frac{L_m}{L_r}\underline{\lambda}_r \tag{7.5–6}$$

where

$$L_s - \frac{L_m^2}{L_r} = L_s\left(1 - \frac{L_m^2}{L_s L_r}\right) = \sigma L_s = L'_s \tag{7.5–7}$$

Again, L'_s is the *stator transient inductance*. Substituting these results in the stator equation (eqn (7.5–2)) yields

$$\underline{v}_s = [r_s + L'_s(p + j\omega)]\,\underline{i}_s + (p + j\omega)\frac{L_m}{L_r}\underline{\lambda}_r \tag{7.5–8}$$

Eliminating the *ir* term in eqn (7.5–3) by use of eqn (7.5–5) results in, for the rotor voltage equation,

$$0 = \frac{r_r}{L_r}[\underline{\lambda}_r - L_m\underline{i}_s] + [p + j(\omega - \omega_r)]\,\underline{\lambda}_r \tag{7.5–9}$$

In the stationary reference frame eqn (7.5–8) together with eqn (7.5–9) can be written as

$$\underline{v}^s_s = [r_s + pL'_s]\,\underline{i}^s_s + \frac{L_m}{L_r}p\underline{\lambda}^s_r \tag{7.5–10}$$

$$0 = -\frac{L_m}{\tau_r}\underline{i}^s{}_s + \left[p - j\omega_r + \frac{1}{\tau_r}\right]\underline{\lambda}^s_r \tag{7.5–11}$$

where

$$\tau_r = \frac{L_r}{r_r}$$

Upon eliminating the rotor flux linkage from these two equations yields, finally

$$\underline{v}^s{}_s = \underline{i}^s_s\,\underline{Z}_{in}(j\omega_e, jS\omega_e) \tag{7.5–12}$$

where,

$$\underline{Z}_{in}(j\omega_e, jS\omega_e) = \frac{p^2L'_s + p(r_s + L_s/\tau_r - j\omega_e L'_s) + r_s[1/\tau_r - j\omega_e(1 - S)]}{p + 1/\tau_r - j\omega_e(1 - S)}$$

In eqn (7.5–13), the operator p should now be interpreted as the equivalent of the Laplace operator s. The use of another variable for the Laplace operator is clearly necessary to avoid confusion with the per unit slip S. In this context the operator p is frequently called the Heaviside operator after a famous British engineer who exploited many of the benefits of operator notation long before Laplace supplied the mathematical formalism.

The block diagram representing these equations is shown in Figure 7.16,

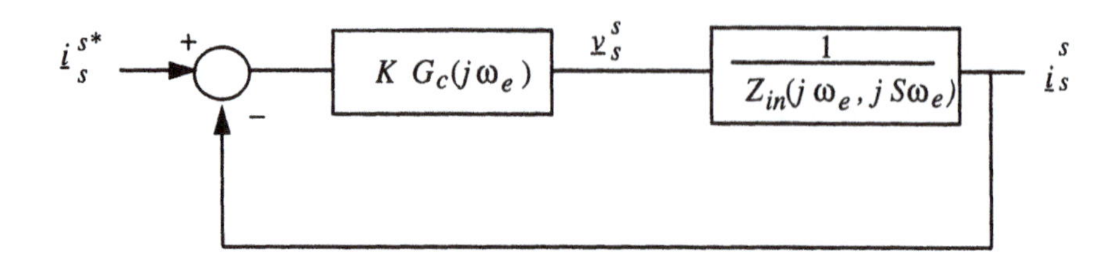

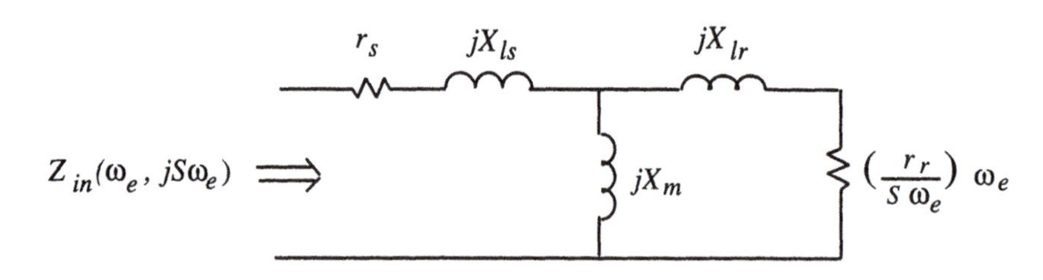

Figure 7.16 Steady state frequency domain block diagram for the current regulator of Figure 7.15 using the induction motor equivalent circuit model

from which the transfer function from current reference to output current can be written as

$$\frac{\underline{i}_s^s}{\underline{i}_s^{s*}} = \frac{1}{1 + \underline{Z}_{in}(j\omega_e, jS\omega_e)/K_\Delta \underline{G}_c(j\omega_e)} \tag{7.5–14}$$

Note that the circuit shown in Figure 7.16 represents the steady state version of the dynamic results expressed in eqn (7.5–14).

From this expression it is clear that the system of Figure 7.15 can result in significant phase and magnitude errors if the frequency curve of $G_c(j\omega_e)$ does not maintain sufficiently high gain at higher frequencies. It is also clear that this control problem quickly becomes very difficult since Z_{in} is essentially a

this control problem quickly becomes very difficult since Z_{in} is essentially a linearly increasing function of ω_e (for constant $S\omega_e$) and hence approximately a pure lead network.

An alternative representation of the machine in terms of the *transient model* (voltage behind transient reactance) can also be employed to illustrate the problem. This useful conceptual model can be obtained for this situation by again starting from the complex space vector form of the stator equations, in the stationary reference frame, that is

$$\underline{v}_s^s = [r_s + pL_s']\,\underline{i}_s^s + \frac{L_m}{L_r}p\underline{\lambda}_r^s \tag{7.5–15}$$

In the frequency domain we can write this equation as

$$\underline{v}_s^s = [r_s + j\omega_e L_s']\,\underline{i}_s^s + \underline{e}_s^s \tag{7.5–16}$$

This approach, which is also valid for the sub cycle behavior associated with the carrier frequency response, leads to the model shown in Figure 7.17. Here

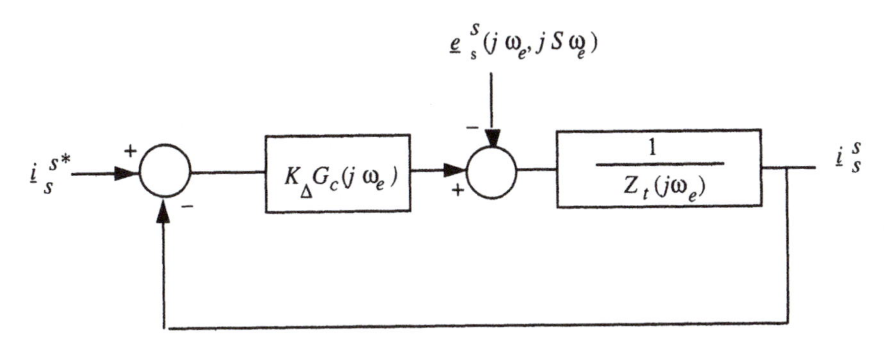

Figure 7.17 Steady state frequency domain block diagram for the current regulator of Figure 7.15 using the 'transient' model of the machine

the machine impedance is the stator transient impedance and the counter emf, $\underline{e}_s(j\omega_e, jS\omega_e)$, is the voltage induced in the stator by trapped rotor flux linkage. In Figure 7.17, the counter emf is shown as a steady state complex quantity representing the steady state voltage behind the equivalent transient impedance Z_t.

With the viewpoint of Figure 7.17, the system more closely resembles the dc machine case in which the machine counter emf enters as a disturbance to the controller. This suggests the possibility of employing feedforward compensation for the counter emf as is sometimes done in dc drives. In the dc case, the counter emf is calculated from knowledge of the field flux and rotor speed and added as a correction at the output of the error amplifier. Similar approaches are possible in ac drives; the details are left until the overall system has been described since implementation of the feedforward compensation is application specific.

7.6 Synchronous Frame Regulators

The inherent problems of regulating ac signals in the stationary frame were first recognized by Schauder and Caddy [7]. In their early work they showed that the stationary frame ramp comparison regulator is only one of an infinite collection of such regulators. In effect, regulators like that of Figure 7.15 but implemented in other reference frames exhibit quite different characteristics. That this is so is quite clear if one considers that the frequency of the current is different in different reference frames and hence the regulator performance (if it is frequency dependent) will also be different.

From this perspective, a synchronous frame seems especially appropriate since the steady state currents are dc currents and a simple PI controller will result in zero steady state error. A system diagram for a synchronous d,q frame regulator is shown in Figure 7.18. It is more complex and requires more hard-

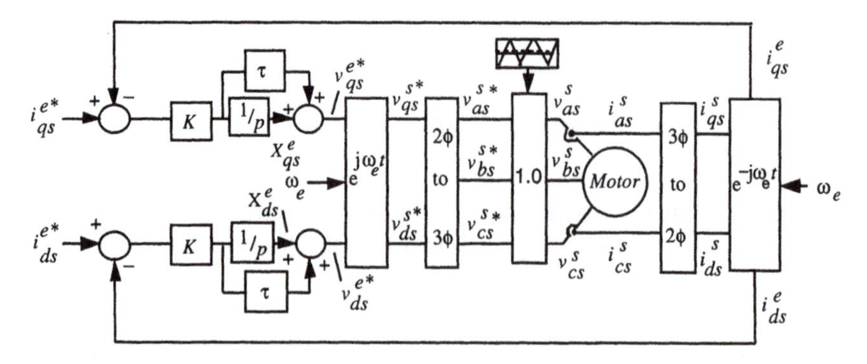

Figure 7.18 System diagram of synchronous frame ramp comparison current regulator using PI controllers

ware for implementation because of the requirement to transform the measured currents to a synchronous frame and subsequently to transform the error amplifier outputs back to a stationary frame to drive the ramp comparison controller. These transformations require explicit knowledge of the frequency ω_e.

To examine the differences between this synchronous frame regulator and the stationary frame regulator of Figure 7.15, it is convenient to transform the stationary frame regulator to the synchronous frame so a direct comparison can be made. The transformation is illustrated in Figure 7.19 which shows both the original regulator and its representation in the synchronous frame. The transformation is readily carried out using vector notation. From Figure 7.19(a), the equations of the stationary regulator in the stationary frame are

$$\underline{v}_s^{s*} = K\tau(\underline{i}_s^{s*} - \underline{i}_{qs}^s) + \underline{x}_s^s \tag{7.6–1}$$

$$p\underline{x}_s^s = K(\underline{i}_s^{s*} - \underline{i}_s^s) \tag{7.6–2}$$

Using the transformation equation from stationary to synchronous reference frame variables, (eqn 3.5–42), the stationary variables are replaced by

$$\underline{f}_s^s = e^{j\omega_e t}\underline{f}_s^e \tag{7.6–3}$$

to yield the synchronous frame variables. The result is

$$\underline{v}_s^{e*} = K\tau(\underline{i}_s^{e*} - \underline{i}_s^e) + \underline{x}_s^e \tag{7.6–4}$$

$$(p + j\omega_e)\underline{x}_s^e = K(\underline{i}_s^{e*} - \underline{i}_s^e) \tag{7.6–5}$$

Rewriting this last equation as

$$\underline{x}_s^e = \frac{1}{p}[K(\underline{i}_s^{e*} - \underline{i}_s^e) - j\omega_e\underline{x}_s^e] \tag{7.6–6}$$

suggests the synchronous frame equivalent of the stationary frame regulator as shown in Figure 7.19(b).

The most obvious aspect of the synchronous equivalent of the stationary frame regulator is the existence of frequency dependent cross coupling between the d– and q–axes. If the regulator is constructed in the synchronous frame as shown in Figure 7.18, no such cross coupling exists. There are three important effects of this cross coupling.

1) For steady state operation the cross coupling produces a response in the

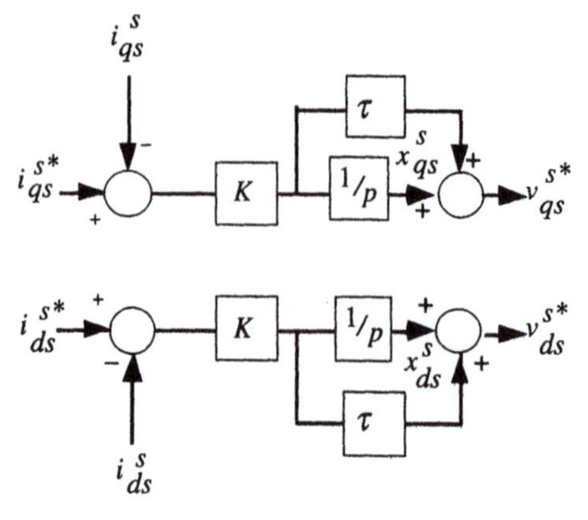

a) Stationary Frame

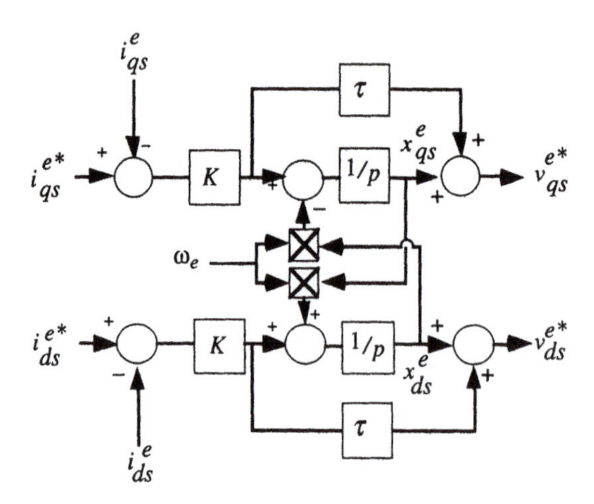

b) Synchronous Frame

Figure 7.19 Models of the stationary frame PI controller current regulator in a) the stationary frame and b) the synchronous frame

cross channel which increases in importance as the frequency increases, i.e., the q–axis current will not be zero even when the q–axis command is zero if there is a non zero command in the d–axis channel.

2) For steady state operation there is an error in the direct channel which increases with frequency since the cross coupling feeds back into the original channel as well, i.e., the q–axis current will not follow the q–axis command.

3) There is the possibility of strong resonance phenomena since the cross coupling terms themselves form a local oscillator at the frequency ω_ε.

The first two of these effects are the synchronous frame equivalents of the steady state error problems described in the previous section. In effect, they are previously anticipated results viewed from a different frame of reference. They are completely absent in the synchronous frame regulator of Figure 7.18 since no regulator cross coupling exists in the synchronous regulator.

The third effect is quite different in that it is not at all apparent from the stationary frame model. The existence of the possibility of local oscillation can be demonstrated by writing the equations around the cross coupling loop. These equations are, from Figure 7.19(b)

$$px^e_{qs} = -\omega_e x^e_{ds} \tag{7.6–7}$$

$$px^e_{ds} = \omega_e x^e_{qs} \tag{7.6–8}$$

from which, by eliminating x^e_{ds}, there results

$$p^2 x^e_{ds} = \omega_e x^e_{qs} \tag{7.6–9}$$

The solution to this linear oscillator equation is

$$x^e_{qs} = A\cos(\omega_e t + \varphi) \tag{7.6–10}$$

and

$$x^e_{ds} = A\sin(\omega_e t + \varphi) \tag{7.6–11}$$

If these synchronous frame sinusoidal oscillations are transformed back to the stationary frame, the result is

$$\underline{x}^s_s = e^{j\omega_e t}\underline{x}^e_s = e^{j\omega_e t}Ae^{-j(\omega_e t + \varphi)} = Ae^{-j\varphi} \tag{7.6–12}$$

which corresponds to a dc signal in the actual stationary regulator. Since a dc error will be regulated to zero in the actual stationary regulator, the oscillation does not exist in the steady state. However, during transients such oscillations can be initiated and they will die out according to the damping properties of the regulator. One would therefore expect the possibility of small signal resonances at low frequencies for the stationary frame regulator.

All of these effects are absent in the synchronous frame regulator since no cross coupling exists. The performance of the synchronous frame regulator is therefore greatly superior to the stationary frame regulator in all aspects. It is, however, as noted previously, considerably more hardware intensive, requiring two transformations between stationary and synchronous variables.

A much simpler hardware implementation has been suggested [6],[8] in which the synchronous regulator of Figure 7.18 is itself transformed to the stationary frame. This is, in effect, the opposite of the operation carried out to derive the circuit of Figure 7.19(b). It is, however, used to develop a simpler implementation of the synchronous frame regulator rather than as a means of analysis. Since the development is very much the same as before, only the result is given; the stationary frame version of the equations of the synchronous regulator in Figure 7.18 are

$$\underline{v}_s^{s*} = K\tau(\underline{i}_s^{s*} - \underline{i}_s^s) + \underline{x}_s^s \tag{7.6–13}$$

$$\underline{x}_s^s = \frac{1}{p}[K(\underline{i}_s^{s*} - \underline{i}_s^s) + j\omega_e\underline{x}_s^s] \tag{7.6–14}$$

The corresponding stationary frame equivalent of the synchronous regulator is shown in Figure 7.20. Once again, cross coupling terms are introduced. However, in this case the cross coupling is very useful and significant since the local oscillator formed in the cross coupling provides the required ac signal to allow the current error to go to zero in the steady state. The performance of the regulator is, of course, identical to the synchronous frame regulator of Figure 7.18. Note, however, that only two multipliers are required for implementation beyond that needed for the ordinary stationary regulator. Reference [8] provides an in depth study of the synchronous regulator and detailed comparison with the stationary regulator.

7.7 Feedforward Compensation

From the perspective of control theory, regulators can be grouped into three categories;

- Series compensation,
- Feedback compensation,
- Feedforward compensation.

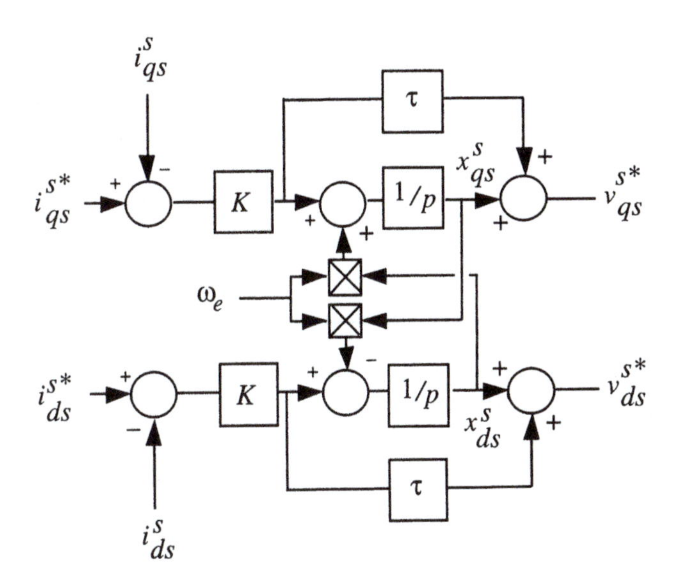

Figure 7.20 The stationary frame equivalent of a PI controller synchronous frame regulator

Thus far, all of the regulators that have been discussed fall into the category of series compensation. That is, whether implemented in the stationary or synchronous frames, both hysteresis and ramp comparison methods rely on a compensator placed between the error of the regulated quantity (motor current) and the output of the amplifier representing the PWM inverter. These regulators can be viewed generically as shown in the block diagram of Figure 7.21. Note that, neglecting the practical realities of scaling, the PWM inverter can be considered as an ideal unity gain amplifier. This assumption, of course, becomes invalid as pulses are dropped when the inverter ceiling voltage is approached and the inverter changes to six step square wave operation. Note again, that for purposes of analysis, the induction motor can be viewed as a counter emf in series with the stator resistance and stator transient time constant. The feedback block giving $e_s(p)$ in terms of $i_s(p)$ follows directly from eqns (7.5–10) and (7.5–11).

Other types of current regulation using feedback or feedforward compensation, either acting either alone or in conjunction with a series compensation method have been proposed in the literature. However, it is not necessary to concern ourselves with feedback compensation (which is not frequently used)

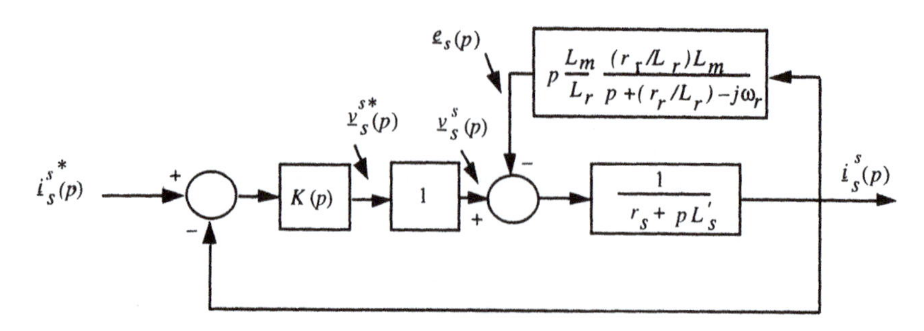

Figure 7.21 Block diagram of series compensated current regulated induction motor

since a nonunity feedback compensator $H(p)$ can be replaced by a series compensator $K(p) = H(p)$ and a feedforward compensator $1/H(p)$.

When all of the parameters of the machine are accurately known, it is in principle possible to regulate the stator currents without any feedback of the motor current by using a feedforward compensator that is the exact inverse of the motor transfer function. Such a scheme is shown in Figure 7.22. Unfortu-

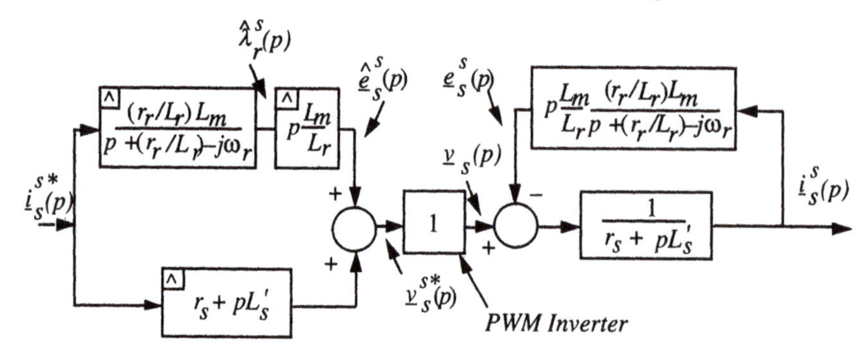

Figure 7.22 Block diagram of induction motor current controller with ideal feedforward compensation. The carot symbol '^' denotes an estimate of the quantity

nately, regulation of the currents will typically be unacceptably poor due to uncertainties in the motor parameter values. In Figure 7.22 the upper block following the current command synthesizes the back emf while the lower block accounts for the stator voltage drop. Note that the major difficulty with this scheme is the computation involved in the block to compute the rotor flux linkage from the stator current. This portion of the problem can be eliminated by calculating the rotor flux linkage from some other variable. For example, the time rate of change of air gap flux linkage can be measured from flux coils placed around the slots of the stator. When this quantity is integrated and processed by the complex vector equation,

$$\underline{\lambda}_r^s = \frac{L_r}{L_m}\underline{\lambda}_m^s - (L_r - L_m)\underline{i}_s^s \tag{7.7-1}$$

the rotor flux linkage can be obtained. Since this equation is algebraic, it is much simpler than the calculation shown in Figure 7.22.

7.8 Augmented Feedforward Compensation

The method of feedforward compensation is a type of open loop control since the quantity to be regulated is not measured or fed back to the input. While such an approach cannot, in itself, qualify as a high performance controller, an estimate of the current can be used to advantage in conjunction with one of the many series compensators as shown in Figure 7.23. In this case, the feedfor-

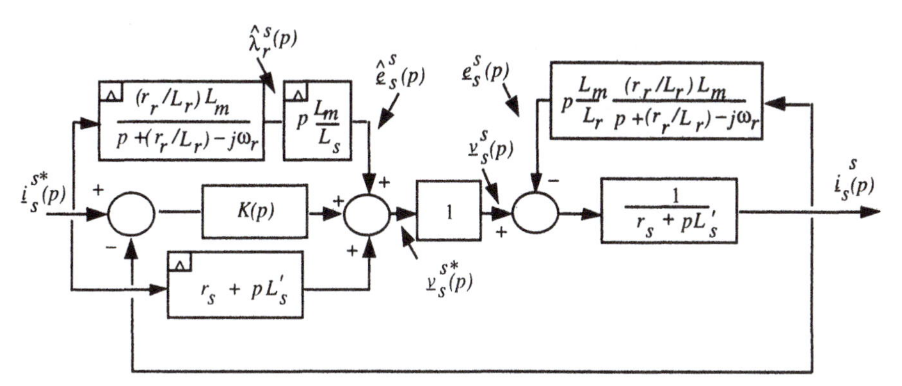

Figure 7.23 Block diagram of current regulator using series compensation augmented by feedforward compensation. The symbol '^' denotes an estimate of the quantity

ward compensator is used to provide only a rough estimate of the voltage command signal. The remainder of the required voltage is then contributed by the series compensator. Since the major portion of the voltage command is provided by the feedforward compensator, the series compensator is not called upon to slew as rapidly, thereby permitting a reduced gain in the series compensator. This reduced gain allows the circuit to become less sensitive to random errors and injected noise, thereby increasing the reliability, as well as the accuracy, of the overall control system. In practical applications, feedforward of only the counter emf is sufficient and the feedforward of the stator transient impedance drop is frequently omitted [9]. Alternatively, since the stator resistance is small and not reliably known due to heating, the feedforward of the transient impedance drop can be replaced by only an estimate of the drop contributed by the transient inductance, i.e. by replacing $(r_s + pL'_s)\underline{i}^s_s$ by $pL'_s\,\underline{i}^s_s$ in the feedforward path [10].

7.9 Augmented Feedforward Compensation with Decoupling

As has probably been surmised by the reader, it is possible to combine the benefits of the cross coupled regulator of Section 7.6 with the augmented feedforward approach of Section 7.8. The combined controller can be represented as shown in Figure 7.24. In this case the use of complex vectors in the block dia-

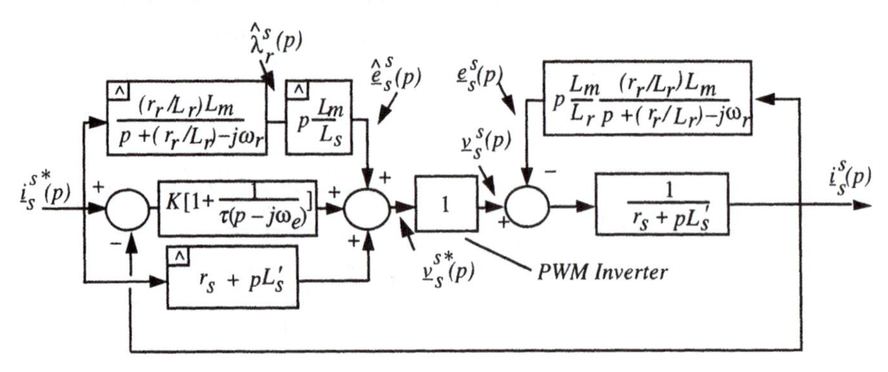

Figure 7.24 Block diagram of current regulator using cross coupled series compensation augmented by feedforward compensation. The symbol '^' denotes an estimate of the quantity

gram is retained so that the proportional plus integral block of the conventional series compensator, $K[1+1/(\tau\, p)]$, must be replaced with a *frequency shifted* compensator $K[1 + 1/(\tau(p - j\omega_e))]$. In effect, this series portion of the compensator represents the current regulator of Figure 7.20. It should be apparent that augmented feedforward can also be added to the current regulator in the stationary frame, Figure 7.18.

7.10 Predictive Control

Recently, use of powerful digital signal processors (DSPs) have become cost competitive with more conventional analog control circuits. While DSPs remain too slow to replace analog type hysteresis or ramp comparison methods for most applications at the present time, it will be only a matter of time before the full power of these computing chips can be utilized. In general, use of DSPs permits the on line computation of the condition of the circuit and enables the selection of more optimal strategies [11]. These methods are typically based on the complex space vector approach to the analysis of rotating machines presented in Chapter 2. If the supply voltages were sinusoidal, this vector would rotate with constant amplitude and fixed angular velocity. However, because of the discrete switching nature of the inverter, the vector can attain only six spatial positions plus the two zero vector positions (Figs. 3.3 and 3.4). Switching of the inverter results in discrete jumps of this voltage space vector rather than uniform rotation.

If the stator resistance is neglected, an approximation for the current trajectory is

$$\frac{d}{dt}\underline{i}_s^s(t) \cong \frac{1}{L'_s}[\underline{v}_s^s(t) - \underline{e}_s^s(t)] \tag{7.10–1}$$

If the switching interval T is sufficiently small, eqn (7.10–1) becomes

$$\underline{i}_s^s(\mathrm{T}) - \underline{i}_s^s(0) \cong \frac{\mathrm{T}}{L'_s}[v_s^s(0) - e_s^s(0)] \tag{7.10–2}$$

Consider now the case where the desired value of $\underline{i}_s^{s*}(\mathrm{T})$ is known at the beginning of the switching interval. Since it is desired to force the current error to zero over the interval [0,T], then clearly

$$\underline{i}_s^s(\mathrm{T}) = \underline{i}_s^{s*}(\mathrm{T})$$

so that, the required voltage over the interval is

$$\underline{v}_s^{s*}(0) = \frac{L'_s}{\mathrm{T}}[\underline{i}_s^{s*}(\mathrm{T}) - \underline{i}_s^s(0)] + \underline{e}_s^s(0) \tag{7.10–3}$$

where it is assumed that $\underline{i}_s^{s*}(\mathrm{T})$ is known at $t = 0$. The counter emf $\underline{e}_s^s$ can be calculated by a feedforward compensator in the same manner as in Figure 7.22.

Although eqn (7.10–3) defines the approximate vector value of inverter voltage needed to drive the current error to zero, the inverter can output only six discrete nonzero and two zero voltage vectors. Fortunately, minimum deviation of the current error is, in general, accomplished only if voltage states adjacent to the instantaneous position of the reference vector $\underline{v}_s^{s*}(0)$ are used to synthesize the required voltage vector $\underline{v}_s^s(0)$. Assume for example that the position of the command $\underline{v}_s^{s*}(0)$ is instantaneously located between vectors $\underline{v}_1$ and $\underline{v}_2$ as shown in Figure 7.25. Since all of the terms on the right hand side

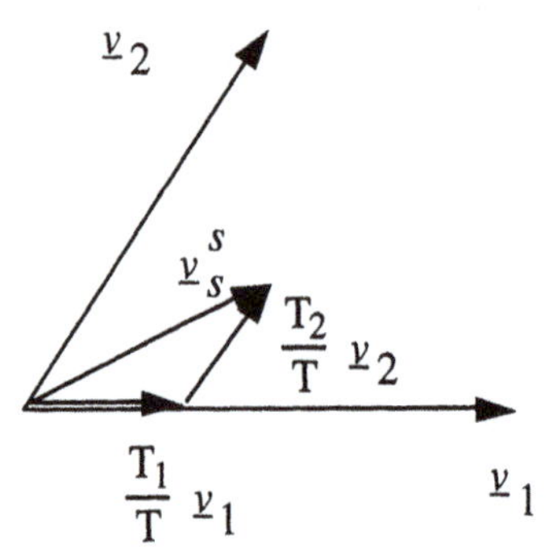

Figure 7.25 Space vector synthesis of a voltage $\underline{v}_s^s$.

of eqn (7.10–3) are constant it follows that over a switching interval T,

$$\int_0^T \underline{v}_s^{s*}(0)\,dt = \int_0^T \underline{v}_1 dt + \int_{T_1}^{T_1+T_2} \underline{v}_2 dt + \int_{T_1+T_2}^{T} \underline{v}_7 dt \tag{7.10–4}$$

where $\underline{v}_1$ and $\underline{v}_2$ are complex constants corresponding to the two adjacent switching states and $\underline{v}_7$ equals zero. Hence,

$$\underline{v}_s^{s*}(0) = \frac{T_1}{T}\underline{v}_1 + \frac{T_2}{T}\underline{v}_2 + \frac{T_3}{T}\underline{v}_7 \tag{7.10–5}$$

where $\underline{v}_7$ represents one of the two zero voltage switching states. The voltage necessary to drive the current to zero has thus been defined in terms of the inverter switching states over three subintervals. In effect, the two adjacent vectors are time weighted to produce the desired output voltage. In practice, the zero switching state selected is the one involving changing the state of the least number of semiconductor devices in order to minimize losses. A vector diagram representation of eqn (7.10–5) is shown in Figure 7.25.

7.11 Conclusion

Current regulation of inverters is clearly an interesting and important area of research in motor drives. This chapter has attempted to present only the highlights of a very dynamic and rapidly evolving technology. In addition to the concepts we have discussed, such advanced technologies such as expert systems and neural networks are beginning to be explored. We can look forward to many interesting new approaches to this problem in the years ahead.

References

[1] D.M. Brod and D.W. Novotny, "Current Control of VSI–PWM Inverters", IEEE Trans. Ind. Appl., Vol. IA–21, no. 4, pp. 562–570, May June 1985.

[2] D. M. Brod, "Current Control of VSI–PWM Inverters", M.S.E.E. Thesis, University of Wisconsin, 1984.

[3] G. Pfaff, A. Weschta and A. Wick, "Design and Experimental Results of a Brushless AC Servo–drive," IEEE Trans. Ind. Appl., Vol. IA–20, no. 4, pp. 814–821, July/August 1984.

[4] A. Nabae, S. Ogasawara and H. Akagi, "A novel Control Scheme for Current–controlled PWM Inverters," IEEE Trans. Ind. Appl. Vol IA–2, no. 4, pp. 697–701, July/August 1986.

[5] L. Malesani and P. Tenti, "A Novel Hysteresis Control Method for Current–controlled VSI PWM Inverters with Constant Modulation Frequency," IEEE Trans. Ind. Appl., Vol. IA–26, no. 1, pp. 88–92, Jan./Feb. 1990.

[6] T. Rowan, "Analysis of Naturally Sampled Current Regulated Pulse–width Modulated Inverters", Ph.D. Thesis, University of Wisconsin, 1985.

[7] C.D. Schauder and R. Caddy, "Current Control of Voltage–source Inverters for Fast Four–quadrant Drive Performance," IEEE Trans. Ind. Appl., Vol. IA–18, no. 2, pp 163–171, March/April 1982.

[8] T. Rowan and R. Kerkman, "A New Synchronous Current Regulator and an Analysis of Current–regulated PWM Inverters," IEEE Trans. Ind. Appl., Vol. IA–22, no. 4, pp. 678–690, July/August 1986.

[9] H. Nagase, Y. Matsusa, K. Ohnishi, H. Ninomiya and T. Koike, "High–performance Induction Motor Drive System Using a PWM Inverter," IEEE Trans. Ind. Appl., Vol IA–20, no. 6, pp. 1482–1489, Nov./Dec. 1984.

[10] R.D. Lorenz and D.B. Lawson, "Performance of Feedforward Current Regulators for Field–oriented Induction Machine Controllers," IEEE Trans. Ind. Appl., Vol. IA–23, no. 4, pp. 597–602, July/Aug. 1987.

[11] J. Holtz and S. Stadtfeld, "A Predictive Controller for the Stator Current Vector of AC Machines Fed from a Switched Voltage Source", Int. Power Electronics Conf. – Tokyo, March 27–31, 1983, pp. 1665–1675.

Problems

Problem 7–1 Stator Frame Ramp Comparison Current Controller

Consider a stator frame ramp comparison current controller as shown in Fig.7.15 used to supply a 100 hp machine with parameters

$r_s = 0.015, x_{ls} = 0.10, x_m = 2.0, r_r = 0.02, x_{lr} = 0.10$, all in pu.

Assume the machine is rated 460 volts (line to line) and that $K_\Delta = 10$.

a) Select the PI controller parameters K and τ such that the PI controller zero coincides with the motor stator transient time constant L'_s/r_s and that the steady state phase error of the current controller is 3 degrees at a stator frequency of 10 hz. and rated slip frequency.

b) For the PI controller parameters of part a, calculate the steady state amplitude and phase error of the current controller for frequencies of 0, 10, 20, 40 and 60 hz for rated slip frequency and for no load.

8 Parameter Sensitivity and Saturation Effects in Indirect Field Orientation

8.1 Introduction

In the indirect field oriented systems of Figures 6.6 or 6.7, the slip relation given in eqn 6.2–17 is employed to obtain the correct subdivision of the stator current into the torque and flux components i_{qs}^{e*} and i_{ds}^{e*}. If the rotor open circuit time constant L_r/r_r is not correctly known, or if it changes because of motor heating or other variations, this subdivision will not be correctly attained. The result will be detuning of the controller and a loss of correct field orientation [1],[2].

The overall effect of this detuning is to incorrectly calculate the rotor field angle. In general, this causes the actual stator current components to be incorrect with the result that:

(1) the flux level is not properly maintained,

(2) the resulting steady state torque is not the commanded value, and

(3) the torque response is not instantaneous.

These variations are explored in the following sections.

8.2 Phasor Diagram for Detuned Operation

The steady state effect of detuning is easily visualized by writing the rotor loop equation in the conventional equivalent circuit

$$j\omega_e L_m \tilde{I}_s + \left[\frac{r_r}{S} + j\omega_e (L_m + L_{lr})\right]\tilde{I}_r = 0 \tag{8.2–1}$$

Solving this equation for $\tilde{I}_r$ yields

$$\tilde{I}_r = \frac{-j\omega_s L_m \tilde{I}_s}{r_r + j\omega_s L_r} = \frac{-j\omega_s L_m \tilde{I}_s}{Z_r \angle \varphi_r} \tag{8.2--2}$$

where $\omega_s = S\omega_e$ and $L_m + L_{lr} = L_r$. Figure 8.1 [3] illustrates the situation exist-

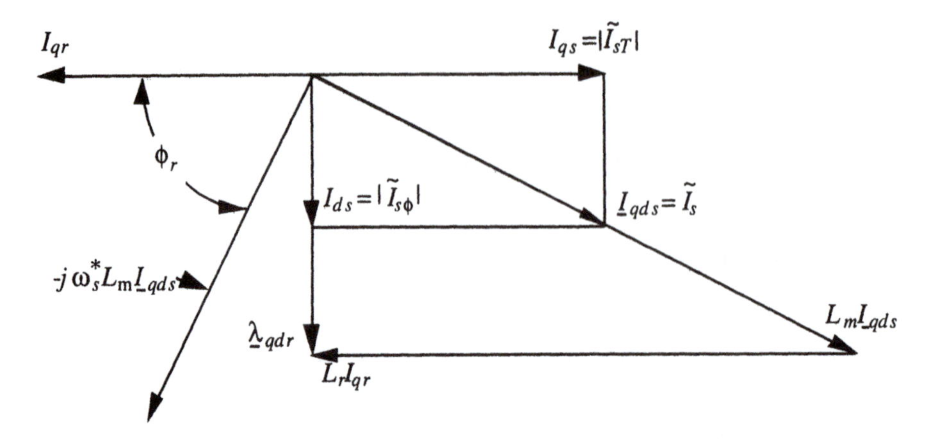

Figure 8.1 Indirect field orientation with correct value of slip frequency

ing when the slip calculator has the correct value of $\hat{L}_r/\hat{r}_r$. The specified values of $\tilde{I}_{sT}$ and $\tilde{I}_{s\varphi}$ locate $\tilde{I}_s$ which in turn locates the phasor $-j\omega_s{}^*L_m\tilde{I}_s$. If the slip is correct, the phase angle of rotor self impedance, $\varphi_r = \tan^{-1}\omega_s L_r/r_r$ will be exactly the value necessary to put $\tilde{I}_r = I_{qr}$ in phase opposition to $\tilde{I}_{sT}$ as shown in Figure 8.1 [3]. The rotor flux, which is orthogonal to the rotor current in the steady state, is therefore aligned with $\tilde{I}_{s\varphi}$. The figure is also labeled with the corresponding d,q axis currents (the $\sqrt{2}$ difference in scale between conventional phasors and conventional d,q axis currents is ignored) and the vector sum $L_m\underline{I}_{qds} + L_rI_{qr}$ is shown summing to $\underline{\lambda}_{qdr} = \lambda_{dr}$.

The situation existing when the slip is too small $(\hat{L}_r/\hat{r}_r > L_r/r_r)$ is illustrated in Figure 8.2. In this case, the phasor $-j\omega_s{}^*L_m\tilde{I}_s$ (or the vector $-j\omega_s{}^*L_m\underline{I}_{qds}$) has the same position as in Figure 8.1 but is shorter. The angle φ_r is, however, too small and the rotor current does not lie on the negative q–axis in opposition to I_{qs} (or $\tilde{I}_{sT}$). As a result the rotor flux $\underline{\lambda}_{qdr}$ is not entirely in the d–axis but is shifted by an angle δ as shown in Figure 8.2. Note that the magnitude of the rotor flux is also incorrect; in the example of Figure 8.2, $\underline{\lambda}_{qdr}$ is considerably larger than the correct value corresponding to the correct slip in Figure 8.1. Clearly the error in flux angle and amplitude will also cause an error in torque, i.e., the torque will not correspond properly to the torque command I_{qs} (or $\tilde{I}_{sT}$).

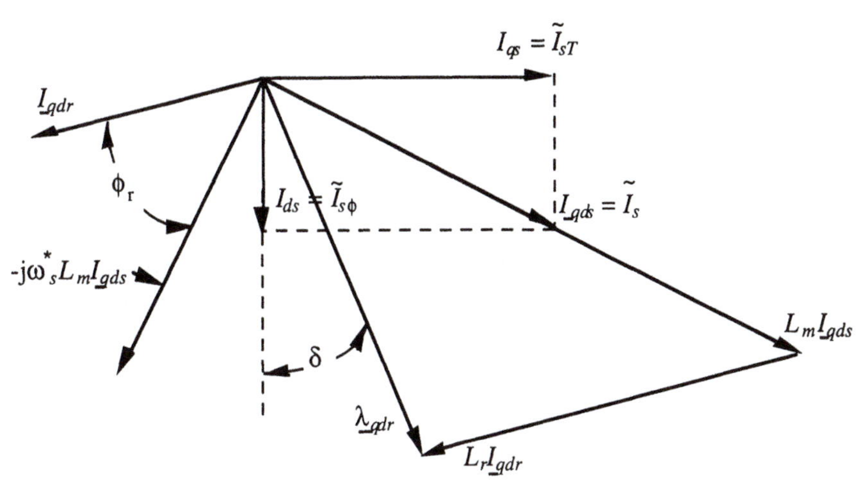

Figure 8.2 Indirect field orientation when calculated slip frequency is too small showing overexcitation and loss of orientation

8.2.1 Field Orientation Torque Characteristics

Determination of the field orientation torque characteristic for detuned operation is most conveniently made by returning to the conventional equivalent circuit of Figure 5.18. Since field orientation is a controlled current mode of operation, the important torque characteristic of the machine is the torque vs. slip frequency curve for a constant value of the stator current. From Figure 5.18, the torque can be expressed as [4]

$$T_e = \frac{3}{2}\frac{P}{2}\frac{I_r^2 r_r}{S\omega_e} \tag{8.2–3}$$

Using the rotor open circuit time constant

$$\tau_r = \frac{L_r}{r_r} \tag{8.2–4}$$

and the current division equation relating I_r to I_s the torque can be written as

$$T_e = \frac{3}{2}\frac{P}{2}I_s^2\frac{L_m^2}{L_r}\frac{S\omega_e\tau_r}{1+(S\omega_e\tau_r)^2} \tag{8.2–5}$$

Figure 8.3 illustrates this characteristic curve. It is important to note that this constant stator current torque curve:

1) is dependent only on slip frequency, it is independent of the actual motor speed,

2) has a peak value proportional to I_s^2 at the slip frequency where $S\omega_e\tau_r = 1$,

3) will exhibit strong saturation effects at low slip frequency where the flux becomes large as the constant stator current is forced into the magnetizing branch.

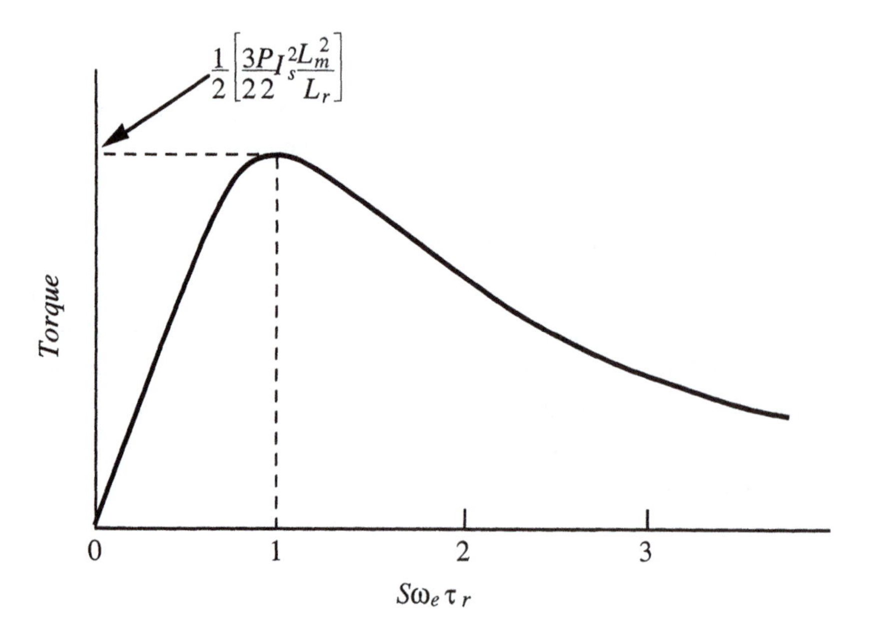

Figure 8.3 Torque vs. slip frequency characteristic for constant stator current

Field oriented operation may be examined by employing the steady state field orientation constraints which are the slip relation

$$S\omega_e\tau_r = \frac{I_{sT}}{I_{s\varphi}} \tag{8.2–6}$$

and the magnitude relation

$$I_s^2 = I_{sT}^2 + I_{s\varphi}^2 \tag{8.2–7}$$

Substituting the magnitude relation in the torque equation yields

$$T_e = \frac{3}{2}\frac{P}{2} I_{s\varphi}^2 \frac{L_m^2}{L_r}\left(1 + \left(\frac{I_{sT}}{I_{s\varphi}}\right)^2\right)\frac{S\omega_e\tau_r}{1 + (S\omega_e\tau_r)^2} \tag{8.2–8}$$

as a general torque expression describing field oriented operation. If the slip relation is used in eqn (8.2–8), the result is

$$T_e = \frac{3}{2}\frac{P}{2} I_{s\varphi}^2 \frac{L_m^2}{L_r}\frac{I_{sT}}{I_{s\varphi}} = \frac{3}{2}\frac{P}{2}\frac{L_m^2}{L_r} I_{s\varphi} I_{sT} \tag{8.2–9}$$

which demonstrates the correct linear relation between torque and torque command. This situation is illustrated in Figure 8.4 where the slip relation is used to draw *field orientation lines* which determine the field orientation operating points on the various constant stator current torque characteristics. Each specific choice of $I_{s\varphi}$ determines the slope of a unique field orientation line and also determines the unique value of I_{sT} which corresponds to each constant stator current torque characteristic where it intersects the field orientation line based on the magnitude relation. Two such lines are shown in the figure along with the appropriate scaling for I_{sT} on the torque curves.

If a normalized torque scale is used, the entire collection of field orientation lines can be reduced to a single line as illustrated in Figure 8.5. The normalization uses the coefficient of $I_{sT}/I_{s\varphi}$ in the first form of eqn (8.2–9) as the base for the torque. In this case changing $I_{s\varphi}$ does not alter the field orientation line but, instead, changes the value of the torque base, T_o, and hence the values of torque on the torque scale. The various constant stator current torque curves are now described by specific values of the ratio $I_{sT}/I_{s\varphi}$. Thus, a change in $I_{s\varphi}$ changes the torque scale and the numerical value of the ratio $I_{sT}/I_{s\varphi}$ for each torque curve, but there is only one field orientation line. It should be noted that these types of normalization are useful in the linear case (saturation neglected) but cannot be used when saturation is included.

Note that with the point of view of Figure 8.5, field orientation may be considered as a means of linearizing the highly nonlinear constant stator current, torque slip frequency curve. The cancellation of the $1 + (S\omega_e\tau_r)^2$ term in eqn (8.2–8) by the $1 + (I_{sT}/I_{s\varphi})^2$ term illustrates the means by which this linearization is achieved.

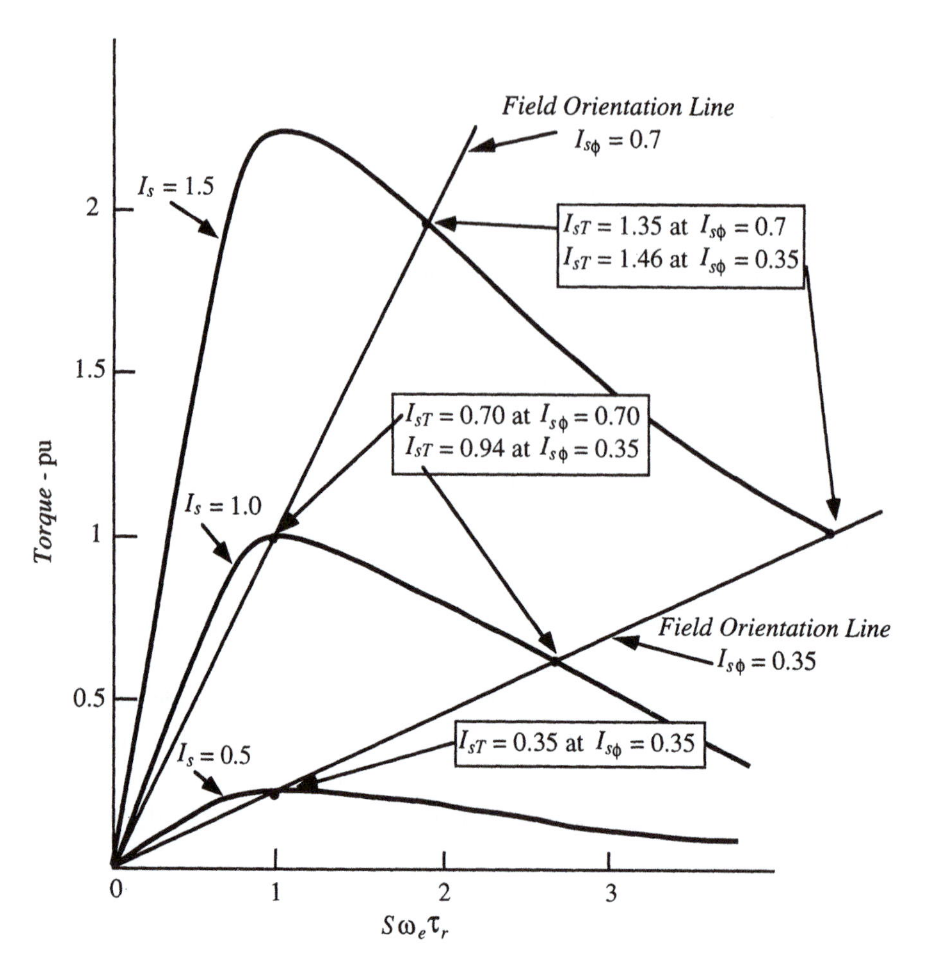

Figure 8.4 Field orientation lines superimposed on constant stator current torque characteristics

If the controller is detuned by having an incorrect value of $\hat{\tau}_r$ in the slip calculator, the field orientation lines on Figure 8.5 become curved lines. This can readily be observed by introducing an error into eqn (8.2–6) by rewriting it as

$$S\omega_e\tau_r = \alpha\frac{I_{sT}}{I_{s\varphi}} \tag{8.2–10}$$

where α is the error in slip gain and is equal to

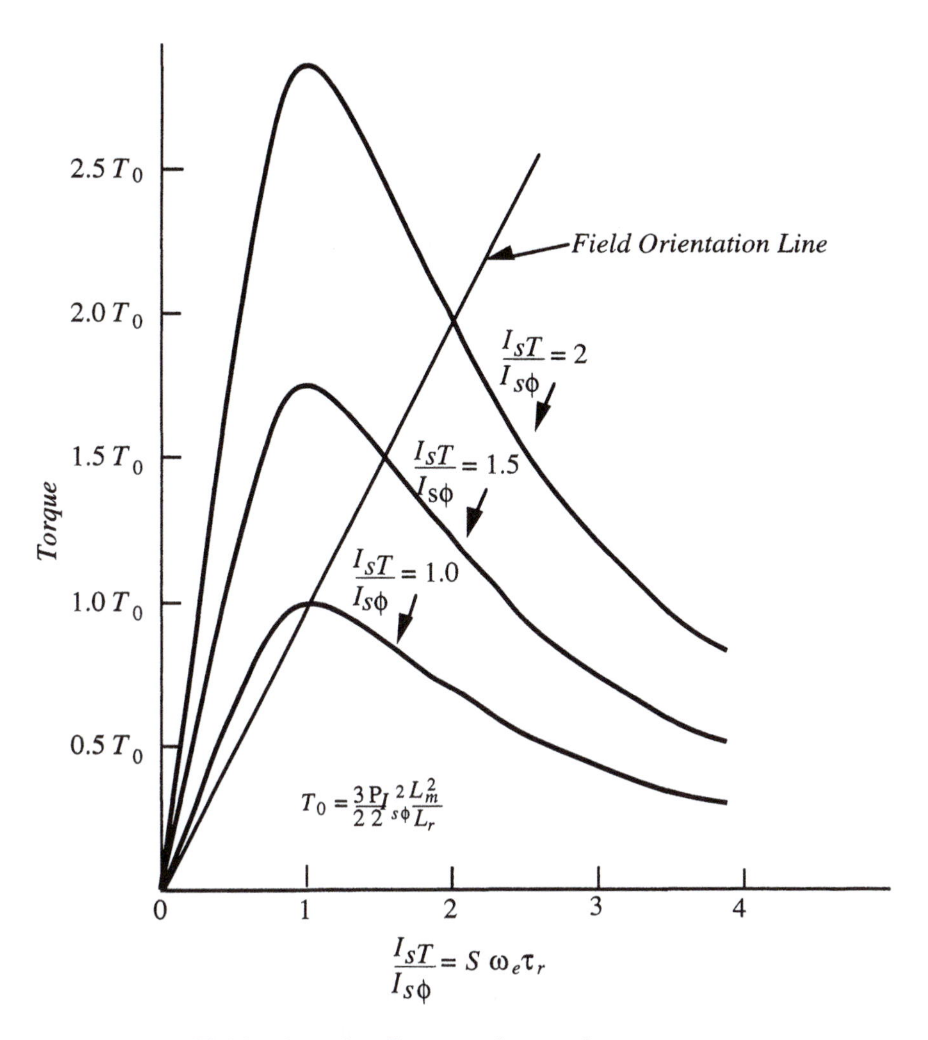

Figure 8.5 Field orientation line superimposed on constant stator current normalized torque characteristics

$$\alpha = \frac{\tau_r}{\hat{\tau}_r} \tag{8.2–11}$$

where τ_r is the correct time constant and $\hat{\tau}_r$ is the one used in the slip gain calculation. Using (8.2–10) in (8.2–8) yields

$$T_e = \frac{3}{2}\frac{P}{2} I_{s\varphi}^2 \frac{L_m^2}{L_r} \frac{1 + \left(\frac{I_{sT}}{I_{s\varphi}}\right)^2}{1 + \left(\alpha \frac{I_{s\varphi}}{I_{sT}}\right)^2} \alpha \frac{I_{sT}}{I_{s\varphi}} \tag{8.2–12}$$

which when plotted as a function of $I_{sT}/I_{s\varphi}$ is clearly not a straight line unless $\alpha = 1$. The nonlinearity is a result of the nonlinear relation between torque and slip frequency. The only condition which produces linearity is the original slip relation of (8.2–6) or (8.2–10) with $\alpha = 1$.

Since curved field orientation lines are both undesirable and inconvenient for analysis, it is useful to rearrange the method of viewing the results when the system is detuned. Equation (8.2–12) can be rewritten in a more compact form by defining

$$T_o = \frac{3}{2}\frac{P}{2} I_{s\varphi}^2 \frac{L_m^2}{L_r} = \text{Field oriented torque at } \frac{I_{sT}}{I_{s\varphi}} = 1 \tag{8.2–13}$$

$$\beta = \frac{I_{sT}}{I_{s\varphi}} \tag{8.2–14}$$

to yield

$$\frac{T_e}{T_o} = \frac{1 + \beta^2}{1 + (\alpha\beta)^2} \alpha\beta \tag{8.2–15}$$

If T_e/T_o is plotted as a function of α with β as a parameter, we have the very convenient result that the vertical line at $\alpha = 1$ is the field orientation line and all other vertical lines are detuned field orientation lines for various specific slip gain error multipliers α. Figure 8.6 [4] illustrates these results for slip gain errors between 0 and 3. The curves on this figure, which have the appearance of motor torque vs. slip frequency curves, are plots of eqn (8.2–15) for specific values of $\beta = I_{sT}/I_{s\varphi}$. The peaks of these curves occur at $\alpha = 1/\beta$ as can be seen in the figure. Note that along the vertical line $\alpha = 1$, the normalized torque is exactly equal to $\beta = I_{sT}/I_{s\varphi}$ as it should be for properly tuned field orientation. The variation of torque for any degree of detuning can be found by looking along the vertical line for the appropriate value of slip gain error α.

Each of the curves is also labeled with the value of relative stator loss which is simply the normalized value of the stator current squared given by $1 + \beta^2$.

Figure 8.6 clearly shows that the ratio of I_{sT} to $I_{s\varphi}$ (β, the normalized torque command) is the most significant parameter in determining the amount of impact a specific amount of detuning will have on torque production. Generally speaking, the smaller the value of β, the less effect detuning tends to have, although for errors in which the slip is too small there are some anomalies near the peaks of the curves. These curves can be viewed as representing a single machine at various levels of torque command or as a means of comparing different machines at, say, rated torque command. The following types of information can be obtained for any specific value of $I_{sT}/I_{s\varphi}$:

1) torque and slip frequency for correct field orientation,

2) relative stator I^2R loss,

3) torque and slip frequency for detuned operation for gain error ratios form zero to three,

4) required change in I_{sT} to compensate for a specified gain error.

8.2.2 Examples

To illustrate the application of the curves in Figure 8.6, consider the following examples.

1) A large machine with $I_{sT}/I_{s\varphi} = 4.0$ at rated torque. With correctly tuned field orientation, the normalized torque is 4.0, the slip frequency is $4.0/\tau_r$ and the relative stator loss is 17 (times the stator I^2R loss at zero torque). All of these values are read on the $\alpha = 1$ line using the $\beta = I_{sT}/I_{s\varphi} = 4.0$ curve

With a slip gain error of 2.0, the corresponding torque is 2.1 (53% of the commanded value), the slip frequency is $8/\tau_r$ and the relative stator loss is unchanged. These values are read at the $\alpha = 2$ line using the $\beta = I_{sT}/I_{s\varphi} = 4.0$ curve.

To bring the torque back up to 4.0 would require increasing the torque command β. Since no larger curve is available, eqn (8.2–15) can be used with $T_e = 4.0$, $\alpha = 2$

$$4 = \frac{1+\beta^2}{1+4\beta^2}2\beta$$

from which β must be increased to nearly 8.0 or about double the original value. Note that this would increase the relative stator loss to 65.

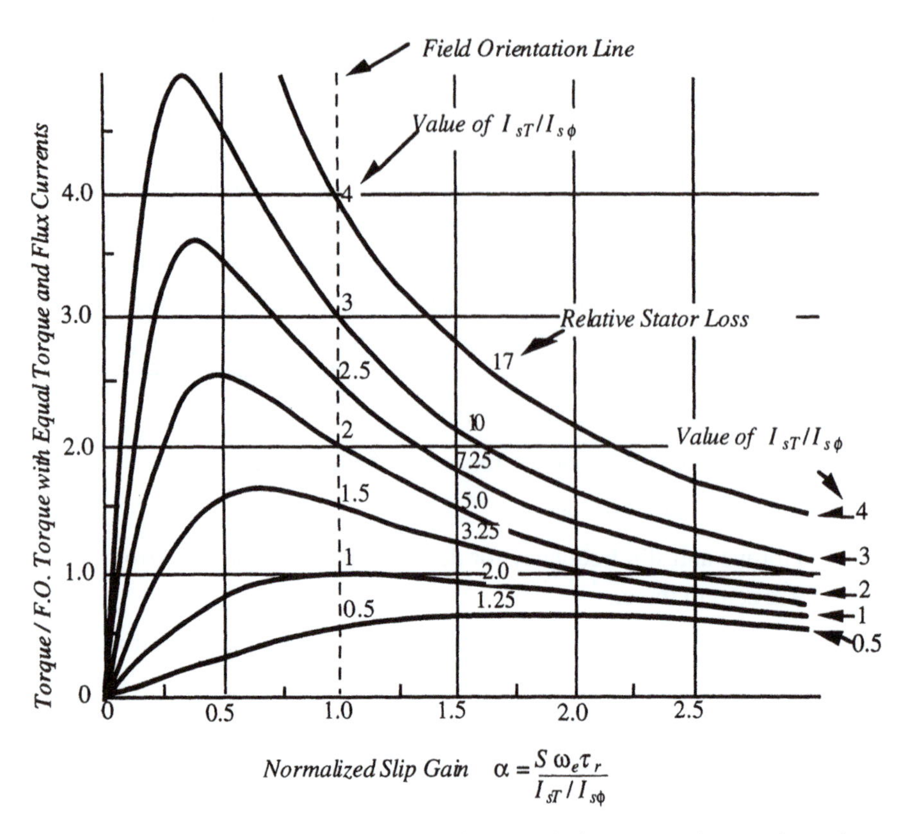

Figure 8.6 Field orientation torque characteristics (saturation neglected)

2) For comparison, consider a smaller machine having a larger magnetizing current and hence a value of $I_{s\varphi}/I_{sT} = 1.0$ at rated torque. With correct tuning the normalized torque is 1.0, the slip frequency is $1.0/\tau_r$ and the relative stator loss is 2.0.

With a slip error of 2.0, the corresponding normalized torque is 0.8 or 80% of the commanded value, the slip frequency is $2/\tau_r$ and the relative loss is unchanged.

To bring the torque back up to 1.0, the value β only needs to be increased to about 1.5 with the relative loss increased to 3.25.

3) To illustrate the complex nature of detuning, observe that for a slip error of 0.5 and a small $\beta = I_{sT}/I_{s\varphi} = 1.0$, the torque drops from 1.0 to 0.8 as a result of detuning. However, for the same detuning with $\beta = I_{sT}/I_{s\varphi} = 2.0$, the torque changes from 2.0 to 2.5 as a result of detuning. Thus, for small torque commands the detuning reduces torque but for large torque commands detuning

increases the torque whenever operation takes place in the vicinity of the peaks of the constant β curves.

It is easy to show from the curves of Figure 8.6 that large values of $\beta = I_{sT}/I_{s\varphi}$ (which are typical of larger, more efficient machines at rated torque) lead to:

(1) field oriented operation at slip frequencies larger than the slip frequency at peak torque,

(2) large torque loss (or gain) for detuned operation,

(3) large changes in stator I^2R loss for changes in I_{sT}.

Clearly, operation at lower values of $I_{sT}/I_{s\varphi}$ results in less sensitivity to detuning, less change in stator I^2R loss as I_{sT} changes and tends to increase the torque/ampere available from the machine for field oriented operation. This suggests that there are advantages to operating at higher values of $I_{s\varphi}$ (higher flux) but this conclusion is application dependent as will be demonstrated.

8.2.3 Field Orientation Flux Characteristics

As illustrated in Figure 8.2, detuning also results in a change in the flux amplitude. This change is, in fact, a primary cause of the torque changes described in the previous section. An expression for the rotor flux amplitude can again be readily derived from the equivalent circuit of Figure 5.18 following the same pattern as used to derive the torque relation of eqn (8.2–8). The result is

$$\lambda_r = L_m I_{s\varphi} \frac{\sqrt{1 + (I_{sT}/I_{s\varphi})^2}}{\sqrt{1 + (S\omega\tau_r)^2}} \qquad (8.2\text{–}16)$$

Once again, with proper field orientation $(S\omega\tau_r = I_{sT}/I_{s\varphi})$ the undesired variation with slip frequency cancels and the flux is constant at the value $L_m I_{s\varphi}$. Figure 8.7 illustrates the flux variation caused by detuning using the same slip frequency scaling as employed in Figure 8.6.

As can be seen from this figure, detuning such that the slip frequency is smaller than the correct field oriented value causes the flux to become larger than the correct value. This is a result of forcing more of the stator current to flow in the magnetizing branch as the effective rotor impedance increases due to the smaller slip. The curves are monotonic since an increase in slip results in more rotor current and less magnetizing current and flux. Note that, as for

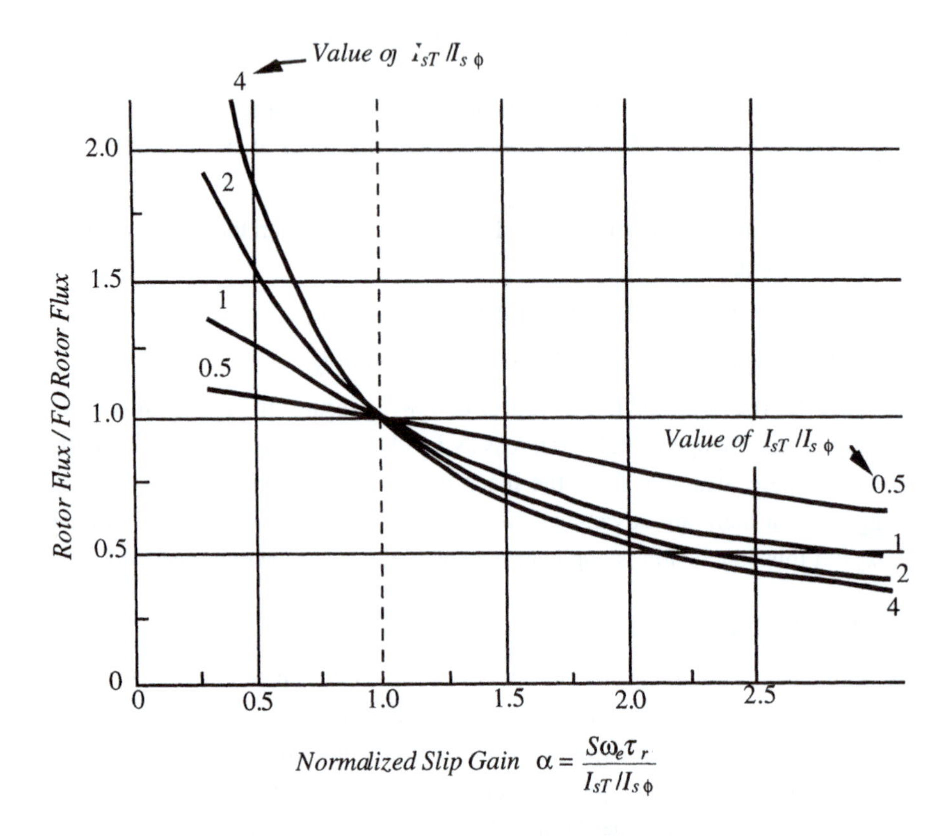

Figure 8.7 Field orientation flux characteristics (saturation neglected)

torque, larger values of $I_{sT}/I_{s\varphi}$ cause greater flux changes, i.e., greater sensitivity to detuning.

8.2.4 *Influence of Saturation*

The flux curves of Figure 8.7 show very large increases in flux for detuning to smaller than correct slip frequency values. Clearly these large flux changes will not occur in a real machine because the magnetic core will saturate and limit the flux change. Figure 8.8 illustrates the nature of the flux variation including saturation effects. The figure presents a set of calculated flux vs. slip frequency curves for a 7.5 hp, 230 V NEMA–A induction machine. The dashed curve for 100% current indicates the calculated curve with a constant value of X_m equal to the value at rated flux (rated slip equal to 0.022). The very pronounced influence of saturation is clearly evident [5],[6].

In addition to limiting the flux change, saturation also changes the value of L_m and hence of the rotor time constant τ_r. These changes in L_m and τ_r have important effects on the torque vs. slip frequency characteristics; reducing the peak torque and increasing the slip frequency at which the peak occurs. On the generalized curves of Figure 8.6, this results in lower peak torques for higher values of $I_{sT}/I_{s\varphi}$ and moves the peaks closer to the field orientation line at these higher values. The actual torque vs. slip frequency curves in Figure 8.6 would now only apply to a specific choice of $I_{s\varphi}$; a different set exists for every other value. In the saturated case it is therefore better to return to a characterization like that of Figure 8.4 and to use actual torque (not normalized to $I_{s\varphi}$). For any specific motor, one such set of curves exists representing the unique saturation characteristic of the machine.

Figure 8.9 represents the torque curves including saturation for 7.5 hp machine of Figure 8.8. Four field orientation lines are indicated on the figure representing four different choices of $I_{s\varphi}$. The field orientation lines are still very nearly straight lines even with saturation since field orientation produces constant rotor flux operation. Each line, however, has a different value of τ_r because of the change in L_m resulting from saturation. Note that, overall, the influence of saturation is to make the effect of detuning less pronounced; both the flux variation and torque variation are less severe. Note that saturation also tends to move the peak torque points more in line with a high flux field orientation line. This suggests it is possible to operate close to maximum torque per ampere, by operating at higher than normal flux, for nearly all values of torque command. In low speed applications (servo drives) this is generally an advantageous operating mode.

8.2.5 Power Loss Considerations

The changes in slip frequency and flux level associated with detuned operation will cause changes in the I^2R losses and core losses in the machine. In general, detuning to lower than correct slip frequency will reduce the I^2R loss and increase the core loss. The opposite effects occur if the slip frequency is too large. The quantitative changes depend upon the motor parameters and operating speed and can be quite different for different motors or different speeds of operation.

At low speed (servo applications), the core losses are relatively low and the major concern is the winding I^2R loss. The rotor I^2R loss clearly increases as the slip frequency increases; hence operation at greater than the correct field

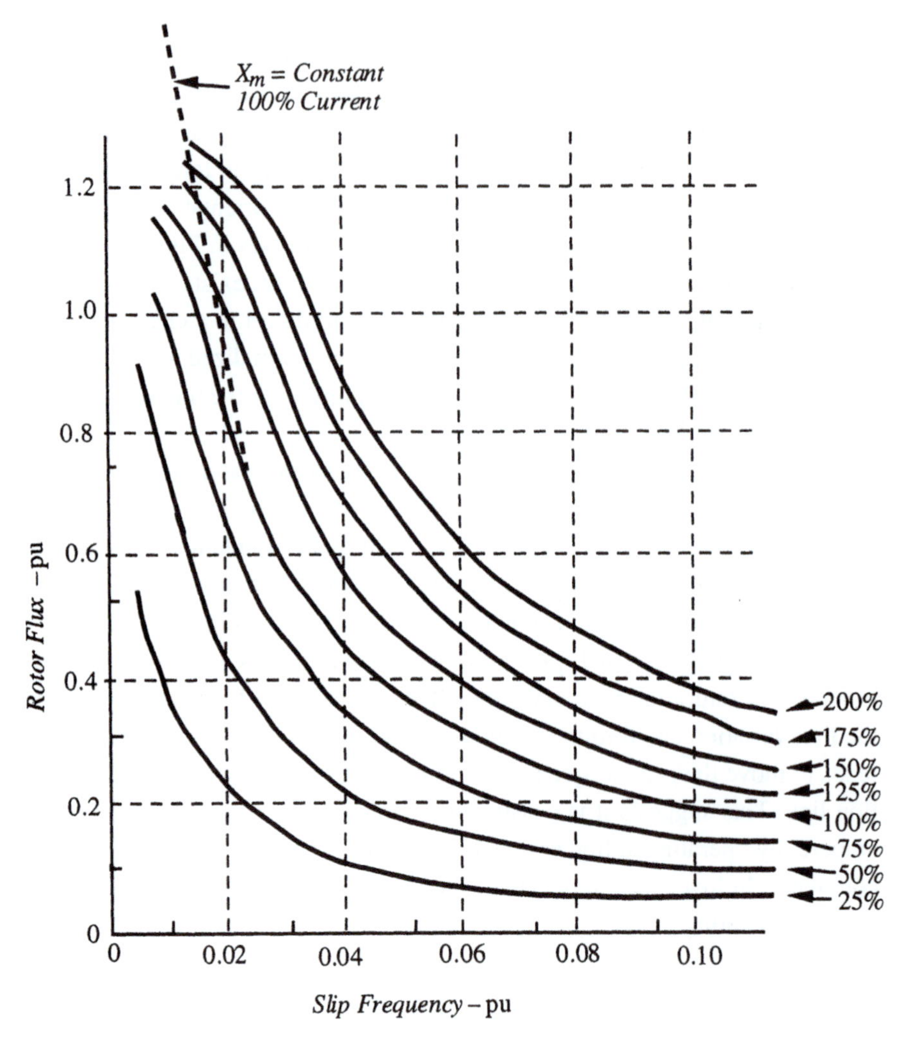

Figure 8.8 Flux vs. slip frequency for 7.5 hp induction machine — saturation included

oriented slip produces high rotor loss. The variation of rotor loss is shown in Figure 8.10 using the same normalization as in the general torque and flux characteristic curves of Figure 8.6 and Figure 8.7. Although the curves indicate relatively minor increases in rotor I^2R loss as the system is detuned to higher slip frequencies, the actual situation is considerably worse as a result of the torque loss for high slip frequency detuning. In most cases a low speed servo application must be examined at a specific torque output. Detuning to high slip fre-

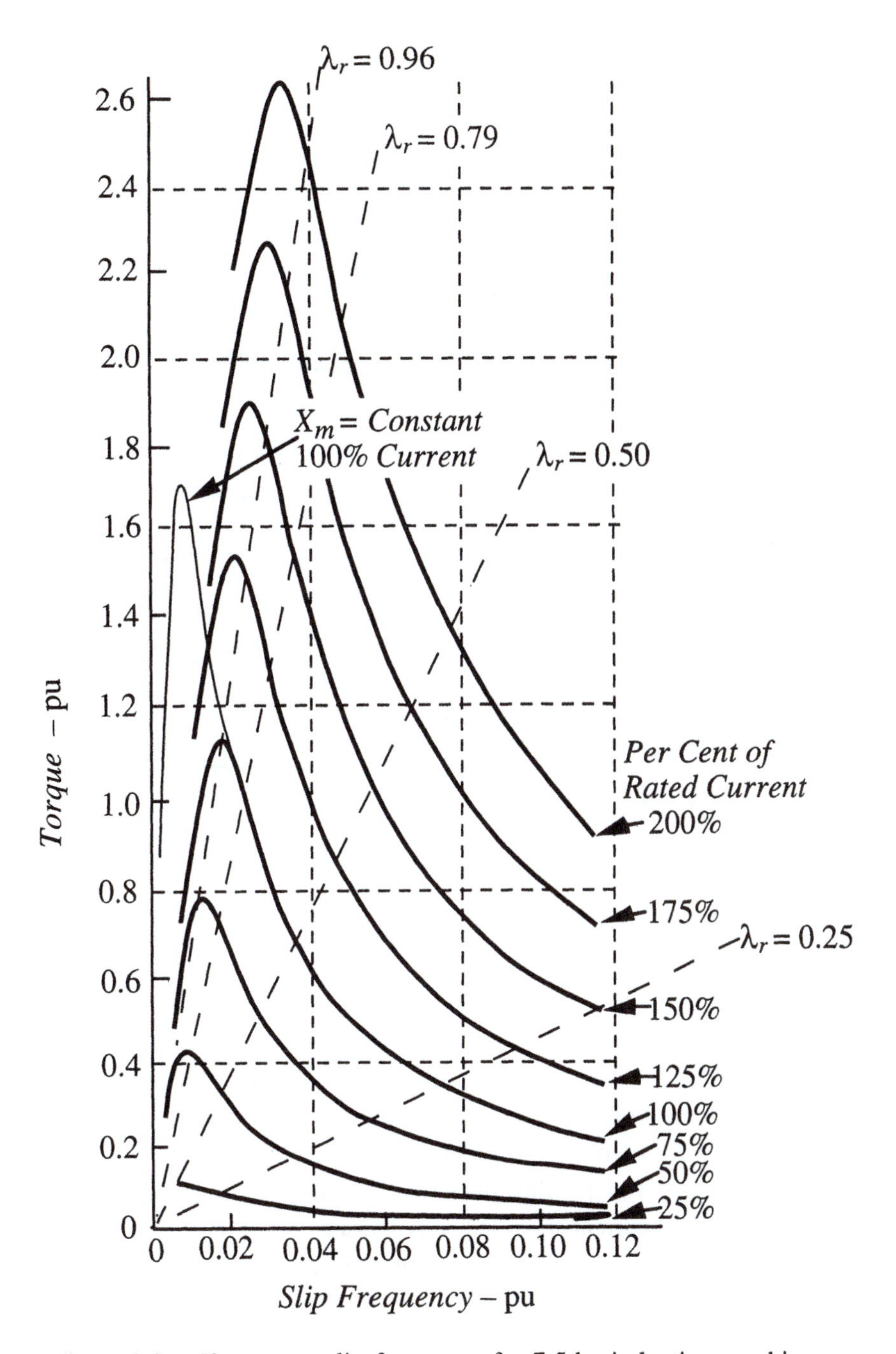

Figure 8.9 Torque vs. slip frequency for 7.5 hp induction machine — saturation included

quency will cause a torque/ampere loss and hence will require an increase in stator current. Thus, both stator and rotor losses will be increased and motor overheating is a serious problem. The converse is true for detuning to lower slip frequency until the detuning is sufficiently large to cause operation past the peak of the torque–slip frequency curve. Past the peak, the stator I^2R losses increase causing very large increases in total losses.

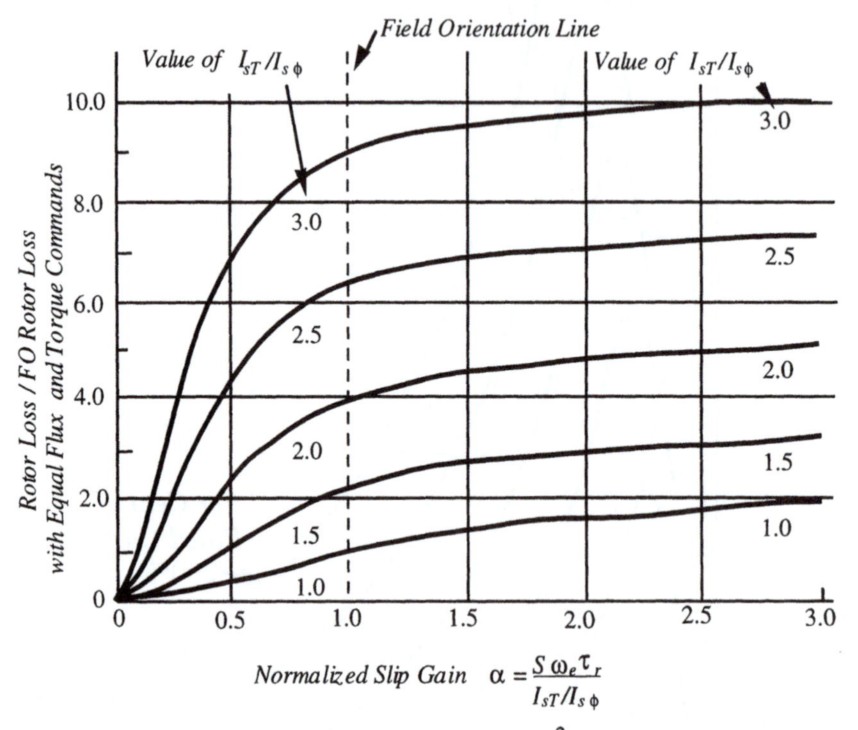

Figure 8.10 Field orientation rotor I^2R loss characteristics

For operation at substantial motor speed the core losses become important and the situation is more complex. In general, detuning to high slip frequency results in high I^2R loss but low core loss and the converse is true for low slip frequencies. Figure 8.11 [3] illustrates two typical cases for operation with a fixed flux command current and rated output torque. Since these curves include saturation effects, the rotor time constant is different for different amounts of detuning. The horizontal scale is, therefore, labelled in per unit values of the ratio of the rotor resistance used in the slip calculator $\hat{r}_r$ to the actual rotor resistance r_r. The curves show the variation of the rotor flux and the required torque command current as the system is detuned (low values of slip frequency which create high values of flux are to the left on these curves). The percent change in the motor input power is labeled along the axis. Since the curves are for rated torque and rated speed, the output power is constant and the changes in power input represent changes in motor losses. To the left of the tuned point the flux is high and the core losses become large. To the right, the flux is low and the I^2R losses become large. Note that in both cases there is a maximum efficiency point at a somewhat larger slip than the rated slip point where the systems of Figure 8.11 have been tuned to operate with field orientation.

The curves in Figure 8.11 illustrate several trends which are true in general for detuned operation. These include:

1) low slip detuning causes high core loss and low I^2R loss,

2) high slip detuning causes low core loss and high I^2R loss,

3) tuning to rated flux will not, in general, produce minimum losses since most machines have their maximum efficiency below full rated slip, and

4) small, low efficiency machines (high r/x ratios) are much less sensitive to detuning than large, high efficiency machines (low r/x ratios).

8.2.6 Example — Detuned Induction Machine Field Orientation

For the 100 hp, 460 volt, induction machine in the example in Chapter 5 with pu parameters:

$r_s = 0.015$ pu	$x_{ls} = 0.10$ pu	$x_m = 2.0$ pu at rated frequency
$r_r = 0.020$ pu	$x_{lr} = 0.10$ pu	$r_m = 50$ pu

The rated slip frequency is 0.0248. Express all answers in real units. Neglect r_m except when making efficiency calculations.

Assume the torques calculated in parts a and b of the example in Chapter 5 are the values required to supply the loads. Find the required value of i_{qs} to yield this torque if the slip calculator time constant $\hat{\tau}_r$ is equal to half of the correct value of τ_r in parts *a* and *b*. In each case calculate the copper and core losses and the efficiency when detuned and when correctly tuned.

Solution — The solution assumes the following:

1. The machine has 4 poles
2. Core losses are calculated by using the V_m (air gap voltage) found from a model that ignores core losses, $CL = V_m^2/R_m$.

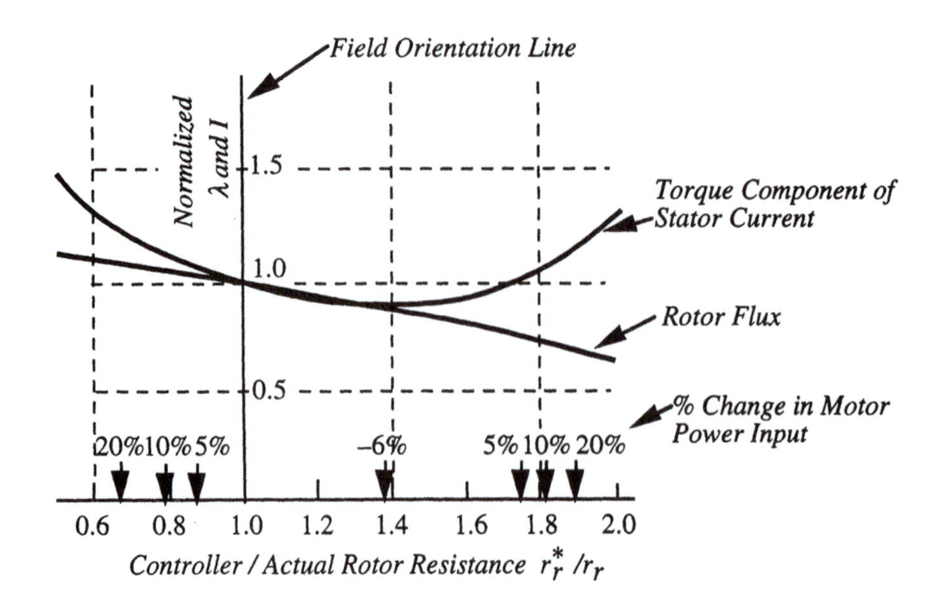

a) 1 hp Machine

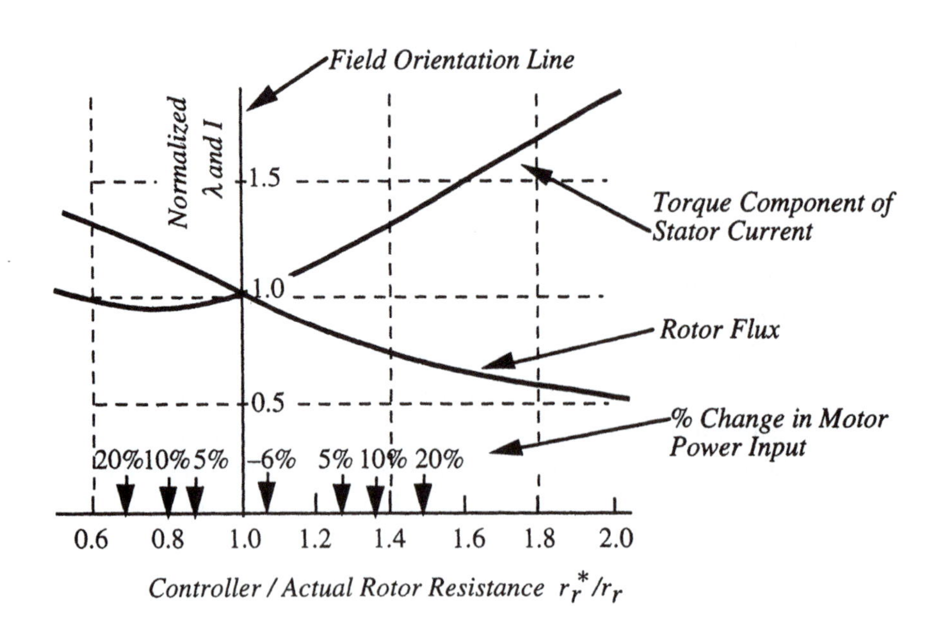

b) 100 hp Machine

Figure 8.11 Influence of detuning

3. Efficiency is dominated by core losses and I^2R losses.

Part a:

First compute values for the tuned case:

From the Chapter 5 example, $\tilde{I}_s = 1.053 - j0.7105 = 1.27$ angle $-34°$ pu

From current division in the rotor loop

$$\tilde{I}_r = -\tilde{I}_s\left(\frac{jX_m}{\frac{r_r}{S} + jX_r}\right) = (1.27 \text{ angle } -34)\frac{j2.0}{\frac{0.020}{0.0248} + j2.1}$$

$$\tilde{I}_r = 1.129 \text{ angle } -193.0°$$

Copper losses are

$$I^2R = I_s^2 r_s + I_r^2 r_r$$

$$= (1.27)^2(0.015) + (1.129)^2(0.020)$$

$$I^2R = 0.0497 \text{ pu}$$

Or, on a 100 hp base,

$$I^2R = (0.0497)(100 \text{ hp})(746 \text{ W/hp}) = 3.71 \text{ kW}$$

Core losses:

In per unit, core losses are assumed to be V_m^2/r_m, where V_m is the air gap voltage.

Find V_m:

$$\tilde{V}_m = \tilde{V}_1 - \tilde{I}_s(r_s + jX_{ls})$$

$$\tilde{V}_m = 1.0 \text{ angle } 0° - (1.27 \text{ angle } -34°)(0.015 + j0.10)$$

$$\tilde{V}_m = 0.918 \text{ angle } -5.9°$$

from which the core losses are,

$$P_m = CL = \frac{0.918^2}{50} = 0.0169 \text{ pu}$$

(The error caused by inserting r_m while keeping the same V_m is negligible. It affects the third significant figure).

or on a 100HP base,

$$CL = (0.0169)\ (100\ \text{hp})\left(746\ \frac{W}{\text{hp}}\right) = 1.26\ kW$$

If losses are assumed to be core loss and I^2R loss only,

$$\eta = \frac{Pout}{Pout + I_2R + CL} = \frac{74.6\ \text{k}W}{74.6\ \text{k}W + 3.71\ \text{k}W + 1.26\ \text{k}W}$$

$$\eta = 0.938 = 93.8\,\%$$

Part b:

The tuned case for part (b) is similarly computed. Recall that $\tilde{I}_s = \underline{I}_{qds} = (1.25 - j0.228)$ pu. With the change in frequency, the equivalent circuit becomes:

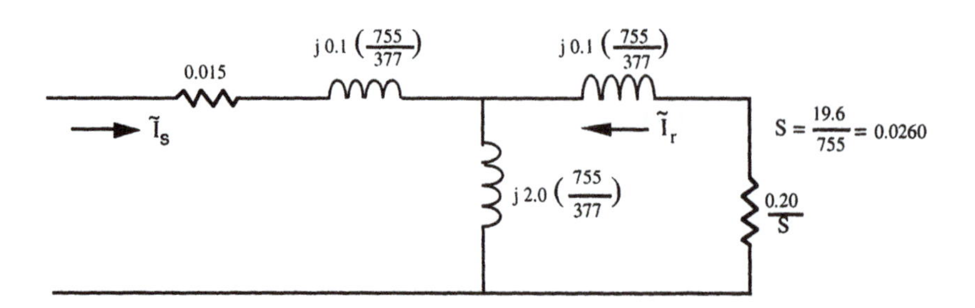

$$\tilde{I}_r = -\tilde{I}_s\left(\frac{j4.00}{j4.00 + j0.20 + \dfrac{0.020}{0.0260}}\right) = -1.19\ \text{angle}\ 0.040°\ \text{pu}$$

$$\tilde{V}_m = -\tilde{I}_r\left(\frac{0.020}{0.0260} + j0.1\frac{755}{377}\right) = 0.946\ \text{angle}\ 14.6°\ \text{pu}$$

I^2R losses: $I_s^2 r_s = I_r^2\ r_r = (1.27)^2\ (0.025)^2 + (1.19)^2(0.020) = 0.0525$ pu = 3.92 kW

Core losses:

$$CL = \frac{V_m^2}{r_m} = \frac{(0.944)^2}{50} = 0.0179 \text{ pu} = 1.34 \text{ k}W$$

Power output:

$$Pout = \frac{I_r^2\,(1-S)\,r_r}{S} = \frac{(1.19)^2\,(1-0.0260)\,(0.020)}{0.0260}$$

$$Pout = 1.061 \text{ pu} = 79.1 \text{ k}W$$

Efficiency:

$$\eta = \frac{Pout}{Pout + I^2R + CL} = \frac{79.1\text{k}W}{79.1\text{kw} + 3.92\text{k}W + 1.34\text{k}W}$$

$$\eta = 0.938 = 93.8\,\% \quad \text{(The same as part a)}$$

Detuned Operation: Part a

Torque is unchanged; speed is unchanged. The slip calculator value of τ_r is 1/2 of the actual value. Therefore,

$$\alpha = \frac{\tau_r}{\hat{\tau}_r} = \frac{\tau_r}{\frac{1}{2}\tau_r} = 2$$

Since,

$$\frac{T_e}{T_0} = \frac{1+\beta^2}{1+(\alpha\beta)^2}\alpha\beta \qquad \text{where } \beta = \frac{I_{sT}}{I_{s\phi}} = \frac{i_{qs}}{i_{ds}}$$

$$\text{and } T_0 = \frac{3}{2}\frac{P}{2}I_{s\phi}^2\frac{L_m^2}{L_r}$$

we need to find β to give the same value of T_e/T_0 as in the tuned case.

In the tuned case, $\alpha = 1$ and $\dfrac{T_e}{T_0} = \dfrac{1+\beta^2}{1+(1)^2\beta^2}(1)(\beta) = \beta$

Therefore,

$$\frac{T_e}{T_0} = \beta = \frac{I_{sT}}{I_{s\phi}} = \frac{i_{qs}}{i_{ds}} = \frac{157A}{60.3A} = 2.60$$

For the detuned case, we keep torque the same,

$$\frac{T_e}{T_0} = 2.60 = \frac{1+\beta^2}{1+(2\beta)^2}(2\beta)$$

Find new β

$$2.60\,(1+4\beta^2) = 2\beta + 2\beta^3$$

Therefore, $2\beta^3 + (-10.4\beta^2) + 2\beta + (-2.60) = 0$

Solving

$$\beta = 5.053,\ 0.147 \pm j0.502$$

Since β must be real, $\beta = 5.053$

Then, since $\beta = \frac{I_{sT}}{I_{s\phi}} = \frac{i_{qs}}{i_{ds}}$, $\quad i_{qs} = \beta\, i_{ds}$ with i_{ds} unchanged

$i_{qs} = (5.053)\,(60.3) = 305$ A peak

Since

$$S\omega_e\tau_r = \alpha\frac{I_{sT}}{I_{s\phi}} \qquad \tau_r = \frac{L_r\,\text{pu}}{r_r\,\text{pu}}\cdot\frac{Z_B/\omega_B}{Z_B}$$

then

$$S\omega_e = \frac{\alpha\beta}{T_r} \qquad \tau_r = \frac{2.1}{0.020\,(377)} = 0.279\ \text{s}$$

$$S\omega_e = \frac{(2)\,(5.053)}{0.279} = 36.3\text{rad/s}$$

$$\omega_r = \text{Same} = 377\text{–}9.35 = 367\ \text{rad/s}$$

$$\omega_e = 36.3 + 367.6 = 403.9\ \text{rad/s}$$

The slip is,

$$S = \frac{S\omega_e}{\omega_e} = \frac{36.9}{403.9} = 0.089$$

The equivalent circuit is:

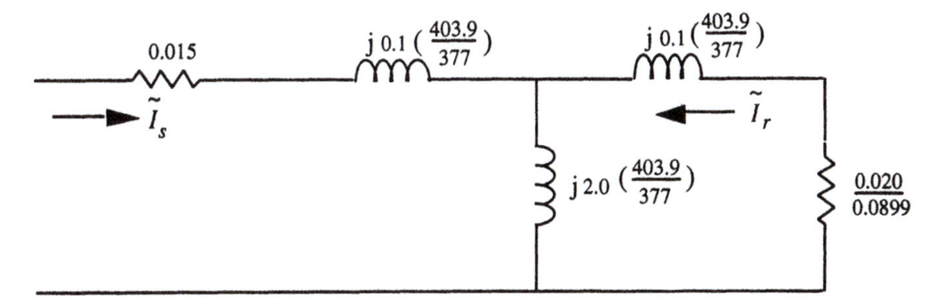

$$\text{Let } \tilde{I}_s = i_{qs} - j\,i_{ds} = \frac{(305 - j60.3)}{93.6\sqrt{2}} = 2.30 - j\,0.456 = 2.34\,\angle -11.2^\circ \text{ pu}$$

$$\tilde{I}_r = -\tilde{I}_s \frac{jX_m}{\frac{r_r}{S} + jX_r} = (0.948\,\angle 5.7^\circ)\,(I_s)$$

$$\tilde{I}_r = 2.22\,\angle -185.6^\circ \text{ pu}$$

$$\tilde{V}_m = \left[-\tilde{I}_r\left(\frac{r_r}{S} + jX_1\right)\right]$$

$$= (22.2\,\angle -5.6^\circ)\left(\frac{0.020}{0.0899} + j0.1\frac{403.9}{377}\right)$$

$$\tilde{V}_m = 0.549\,\angle\,20.1^\circ \text{ pu}$$

I^2R losses:

$$I_s^2 R_s + I_r^2 R_r = (2.34)^2\,(0.015) + (2.22)^2\,(0.020)$$

$$= 0.181 \text{ pu} = 13.5 \text{ k}W$$

Core losses:

$$CL = \frac{V_m^2}{R_m} = \frac{0.549^2}{50} = 0.00603 \text{ pu} = 450\ W$$

Efficiency:

$$\eta = \frac{Pout}{Pout + I^2R + CL} = \frac{74.6\text{k}W}{74.6\text{k}W + 13.5\text{k}W + 0.45\text{ k}W} = 0.842$$

Part b

The torque and speed are unchanged;

$\alpha = 2$ (Same as detuned Part a)

$$\beta \text{ tuned} = \frac{165}{30.2} = 5.46 = \frac{T}{T_0}$$

Therefore, $$5.46 = \frac{1+\beta^2}{1+(\alpha\beta)^2}(\alpha\beta))$$

Solve for β: $\beta = 10.83$

Since i_{ds} is constant = 30.2 A peak

$$i_{qs} = \beta i_{ds} = (10.83)(30.2) = 327 \text{ A peak}$$

Then, $$S\omega_e = \frac{2\beta}{\tau_r} = \frac{2(10.83)}{0.279\,\text{sec}} = 77.6 \text{ rad/s}$$

$$\omega_r = 735 \text{ rad/sec}$$

Therefore, $\omega_e = 735 + 77.6 = 813.6$ rad/s

$$S = \frac{77.6}{813.6} = 0.0955$$

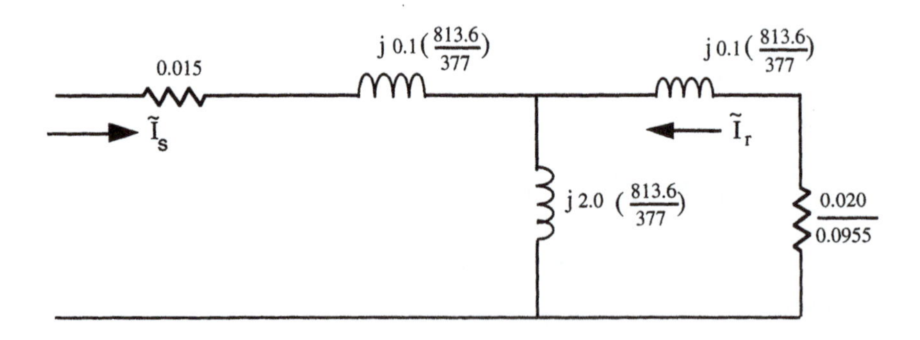

Let $\tilde{I}_s = I_{qs} - jI_{ds} = 327 - j30.2 = 328\angle{-5.3°}$ A peak $= 2.48\angle{-5.3°}$ pu

$$\tilde{I}_r = \tilde{I}_s\left(\frac{jX_m}{\frac{r_r}{S} + jX_r}\right) = 2.36\angle{-182.7°} \text{ pu}$$

$$\tilde{V}_m = (-\tilde{I}_r)\left(\frac{r_r}{S} + jX_1\right) = 0.731 \angle 41.5° \text{ pu}$$

I^2R losses:

$$I_s^2 r_s + I_r^2 r_r = (2.48)^2 (0.015) + (2.36)^2 (0.020)$$

$$I^2R \text{ losses} = 0.204 \text{ pu} = 15.2 \text{ kW}$$

Core loss

$$CL = \frac{V_m^2}{r_m} = \frac{(0.731)^2}{50} = 0.0107 \text{ pu} = 797 \ W$$

Efficiency

$$\eta = \frac{Pout}{Pout + I^2R + CL} = \frac{79.1\text{k}W}{79.1 \text{ k}W + 15.2 \text{ k}W + 0.797 \text{ k}W}$$

$$\eta = 0.832 = 83.2\%$$

8.3 Transient Response

In addition to the steady state errors described above, the loss of field orientation caused by detuning will result in a modification of the transient response. In particular, the instantaneous torque response to changes in torque command will no longer hold [1],[3],[7].

8.3.1 Torque Response with Slip Gain Error

To evaluate this situation, the transient equations describing the torque response for a change in torque command will be derived. The analysis assumes:

(1) the only parameter error is $\hat{L}_r/\hat{r}_r \neq L_r/r_r$

(2) the flux command $i_{ds}^{e*} = I_{ds}^{e*}$ is constant

and the following error variables are defined:

$\Delta\lambda_{qr}^e = \lambda_{qr}^e - \lambda_{qr}^{e*}$ = error in q–axis rotor flux linkage

$\Delta\lambda_{dr}^e = \lambda_{dr}^e - \lambda_{dr}^{e*}$ = error in d–axis rotor flux linkage

$\Delta\tau_r = \hat{\tau}_r - \tau_r = \dfrac{\hat{L}_r}{\hat{r}_r} - \dfrac{L_r}{r_r}$ = error in rotor time constant

ω_s^* = command slip frequency

Eliminating the rotor currents from the rotor voltage equations yields

$$p\lambda_{qr}^e = -\frac{1}{\tau_r}\lambda_{qr}^e - \omega_s\lambda_{dr}^e + L_m\frac{i_{qs}^e}{\tau_r} \tag{8.3–1}$$

$$p\lambda_{dr}^e = -\frac{1}{\tau_r}\lambda_{dr}^e + \omega_s\lambda_{qr}^e + L_m\frac{i_{ds}^e}{\tau_r} \tag{8.3–2}$$

Rewriting these equations, in a reference frame oriented to λ_{dr}^{e*}, in terms of the error variables results in (recognizing that ideally $\lambda_{qr}^{e*} = 0$)

$$p\Delta\lambda_{qr}^e = -\frac{1}{\tau_r}\Delta\lambda_{qr}^e - \omega_s^*\ (\lambda_{dr}^{e*} + \Delta\lambda_{dr}^e) + \frac{L_m i_{qs}^{e*}}{\tau_r} \tag{8.3–3}$$

$$p\Delta\lambda_{dr}^e = -\frac{1}{\tau_r}(\lambda_{dr}^{e*} + \Delta\lambda_{dre}) + \omega_s^*\ \Delta\lambda_{qr}^e + \frac{L_m i_{ds}^{e*}}{\tau_r} \tag{8.3–4}$$

The slip relation and the steady state flux equation provide the following control equations.

$$\omega_s^* = \frac{L_m i_{qs}^{e*}}{\lambda_{dr}^{e*}}\frac{1}{\hat{\tau}_r} \tag{8.3–5}$$

$$\lambda_{dr}^{e*} = L_m I_{ds}^{e*} \tag{8.3–6}$$

Using these two control equations to eliminate λ_{dr}^{e*} from eqns (8.3–3) and (8.3–4) yields

$$p\Delta\lambda_{qr}^e = -\frac{1}{\tau_r}\Delta\lambda_{qr}^e - \omega_s^*\ \Delta\lambda_{dr}^e + \frac{L_m i_{qs}^{e*}}{\tau_r}\frac{\Delta\tau_r}{\hat{\tau}_r} \tag{8.3–7}$$

$$p\Delta\lambda_{dr}^e = -\frac{1}{\tau_r}\Delta\lambda_{dr}^e + \omega_s^*\ \Delta\lambda_{qr}^e \tag{8.3–8}$$

as the final error equations for the error fluxes. The torque error is given by

$$\Delta T_e = T_e - T_e^* = -\frac{3}{2}\frac{P}{2}\frac{L_m}{L_r}(\Delta\lambda_{dr}^e\, i_{qs}^{e*} - \Delta\lambda_{qr}^e\, I_{ds}^{e*}) \quad (8.3\text{–}9)$$

The two equations for the flux errors are a pair of linear first order differential equations which exhibit oscillatory response. The eigenvalues which characterize the response are easily shown to be given by

$$p = -\frac{1}{\tau_r} \pm j\omega_s^* \quad (8.3\text{–}10)$$

and the natural frequency and damping ratio are therefore

$$\omega_n = \frac{1}{\tau_r}[1 + (\omega_s^* \tau_r)^2]^{1/2} \quad (8.3\text{–}11)$$

$$\zeta = \frac{1}{\sqrt{1 + (\omega_s^* \tau_r)^2}} \quad (8.3\text{–}12)$$

This last expression shows that the response becomes progressively more poorly damped as the torque command (and ω_s^*) increase. Note that the input to the error equations is the i_{qs}^{e*} term in eqn (8.3–7)

$$\frac{L_m \Delta\tau_r}{\tau_r \hat{\tau}_r} i_{qs}^{e*}$$

and that with $\Delta\tau_r = 0$ or $i_{qs}^{e*} = 0$ there is no error response. For any specific torque command the flux errors are proportional to $\Delta\tau_r$ although the presence of ω_s^* in the equations also influences the error amplitudes. Note also that the settling time of the torque response is proportional to τ_r only; it is independent of the error and the slip frequency ω_s^*.

A small signal transfer function relating the output torque to the torque command current can be derived using standard methods. The result is

$$\frac{\Delta T_e}{\Delta i_{qs}^*} = \frac{3}{2}\frac{P}{2}\frac{L_m}{L_r}\left[\frac{a_1 s^2 + a_2 s + (a_3 + a_4)}{s^2 + a_5 s + a_6}\right] \quad (8.3\text{–}13)$$

where

$$a_1 = \lambda_{dro}$$

$$a_2 = \frac{1}{\tau_r}(2\lambda_{dro} + \lambda_{qro}\omega_{so}\tau_r - L_m I_{dso} + \lambda_{dro} b)$$

$$a_3 = \frac{\omega_{so}}{\tau_r}(I_{qso} L_m + \lambda_{qro} + \lambda_{qro} b)$$

$$a_4 = \frac{1}{\tau_r^2}(\lambda_{dro} - L_m I_{dso} + \lambda_{dro} b)$$

$$a_5 = \frac{2}{\tau_r}$$

$$a_6 = \frac{1}{\tau_r^2} + \omega_{so}^2$$

$$b = \frac{\tau_r}{\hat{\tau}_r}$$

and the quantities with zero as the last subscript represent the operating point at which the linearization is carried out.

The poles of this transfer function are those given in eqn (8.3–10). The zeros are also complex and exactly cancel the poles when the system is correctly tuned ($\tau_r = \hat{\tau}_r$) yielding instantaneous torque response. Detuning causes the zeros to move away from the poles and results in non–instantaneous response. Figure 8.12 illustrates this pole zero pattern for the same two machines as in Figure 8.11. Note that the pole zero separation increases as the torque output increases and that the larger machine (100 hp) has greater separation than the small machine (1 hp). Figure 8.13 gives several torque step responses for these same machines to illustrate the nature of the small signal transient caused by detuning.

Although detuning eliminates the desirable instantaneous torque response of a field oriented system, the presence of the outer control loop (speed or position control) will often compensate for the less responsive detuned behavior. Unless the detuning becomes very severe, the transient effects of detuning are less serious than the steady state power loss increases described in the preceding sections.

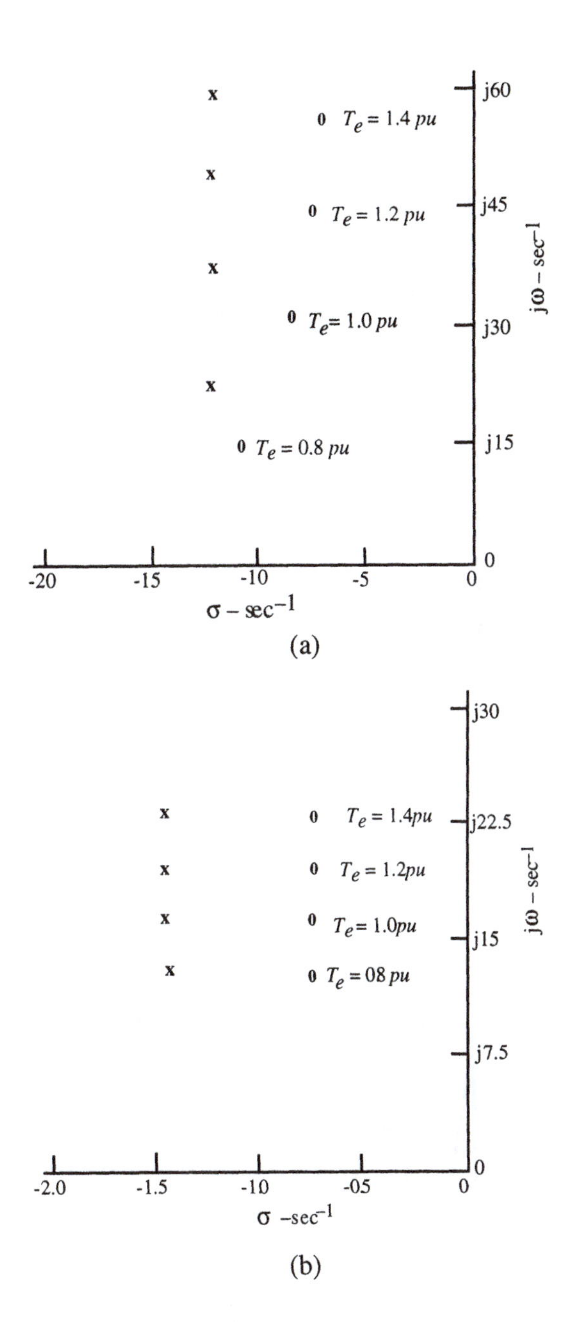

Figure 8.12 Pole – zero plot of the torque vs. torque command transfer function with a slip gain error of 0.50 a) for a 1 hp machine b) for a 100 hp machine

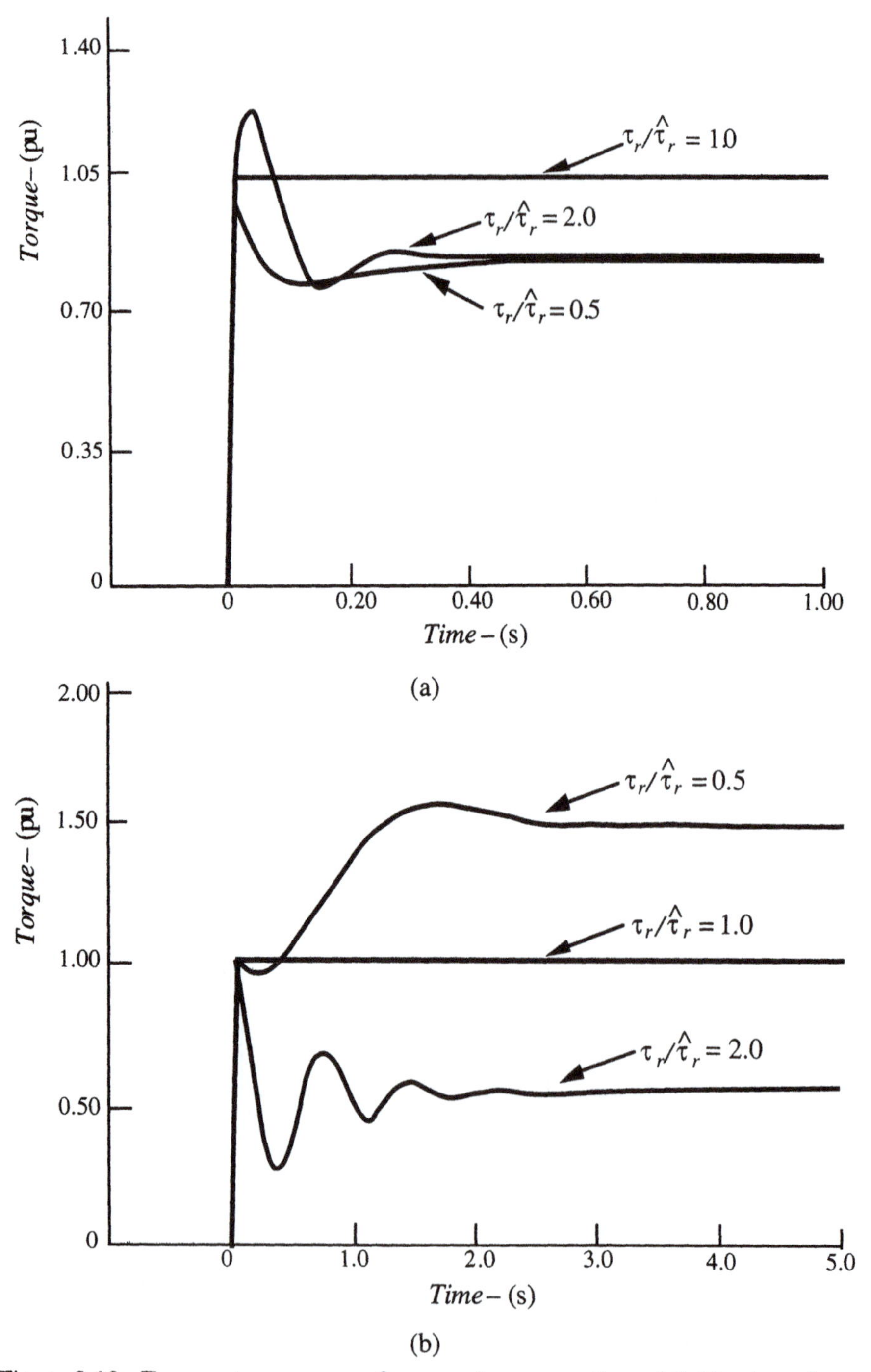

Figure 8.13 Torque step response for open loop operation of field oriented control a) with a 1 hp machine b) with a 100 hp machine

8.3.2 Block Diagram Representation of Slip Gain Error

An alternative formulation of the slip gain error problem in terms of a block diagram offers a different, more control system oriented means of visualizing the system behavior [8].

A block diagram with i^e_{qs} and i^e_{ds} as inputs and λ^e_{qr} and λ^e_{dr} as outputs can be constructed based on eqns (8.3–1) and (8.3–2). The result is shown in Figure 8.14 where the slip frequency ω_s is taken as a constant value about

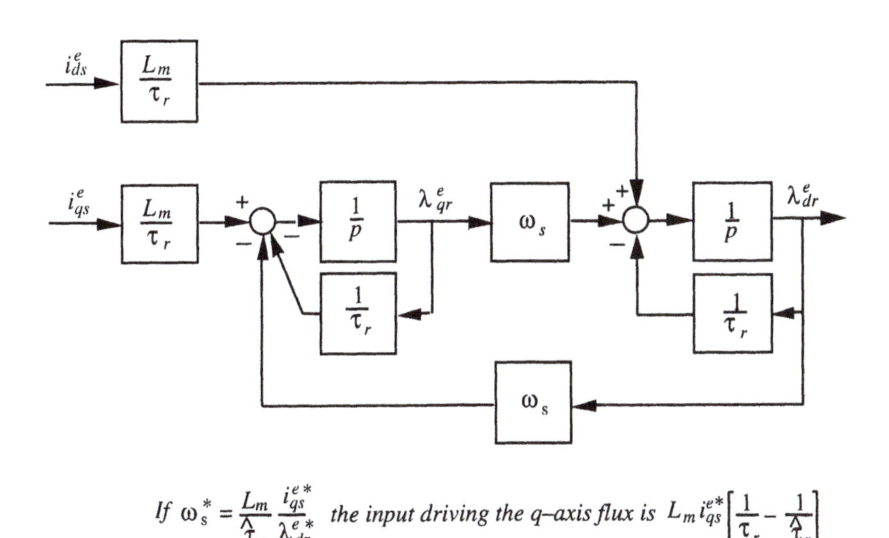

Figure 8.14 Block diagram representation of rotor voltage equations

which the system is linearized. As indicated in the figure, if the slip relation is utilized in an indirect field oriented controller, the net input into the q–axis rotor flux summing point is zero if the rotor time constant τ_r and the slip gain calculation time constant $\hat{\tau}_r$ are equal. In this correctly tuned case the outer loop containing ω_s is compensated and is effectively opened. The block diagram then reduces to the upper input path fed by i^{e*}_{ds} and the first order loop at the right side with the output λ^e_{dr}, as is expected based on earlier results in Chapter 6.

If, however, there is detuning in $\hat{\tau}_r$, the cancellation described above will not take place and the outer loop will not be rendered inoperative. With i^{e*}_{ds}, $\hat{\tau}_r$

and λ_{dr}^{e*} all constant, the slip frequency ω_s^* will also be constant. The characteristic equation associated with the outer loop in Figure 8.14 is

$$1 + \left(\frac{\tau_r}{1 + p\tau_r}\right)^2 (\omega_s^*)^2 = 0 \tag{8.3–14}$$

or solving for the roots

$$p = -\frac{1}{\tau_r} \pm j\omega_s^* \tag{8.3–15}$$

which agrees with the earlier result in eqn (8.3–10).

8.3.3 Simulation of Transient Response with Slip Gain Error

The transient response of a field oriented induction machine with an incorrectly tuned slip calculator can be conveniently simulated using the concept outlined in Figure 6.10. In this figure the machine is modelled in the actual rotor flux frame of the machine and the controller is modeled in the controller's *estimated* rotor flux frame. The error angle, $\Delta\theta_{rf}$, is the angle between the machine and controller axes and is zero when the machine is in a correct field oriented state. It is clear from Figure 8.10, and from the development in previous sections, that all transient or steady state behavior is independent of rotor speed and thus simulations can be carried out at zero speed for simplicity.

Figures 8.15 and 8.16 illustrate the behavior of a machine for four different values of the slip gain (time constant ratio) $\alpha = \tau_r/\hat{\tau}_r$, including the ratio 1.0 which represents correct tuning [9]. The machine is the same 100 hp, 460 volt machine used for the simulations in Figure 6.13 in Section 6.3.5. The simulations were made for a commanded torque current to flux current ratio β of 2.0 with a 1.0 pu flux command.

The torque step responses shown in Figure 8.15 are in agreement with the corresponding curves for the 100 hp machine in Figure 8.13. Note that higher values of the slip gain $\alpha = \tau_r/\hat{\tau}_r$ result in a more oscillatory response as was shown in the linearized analysis of Section 8.3.1. The flux responses in Figure 8.15 also show this behavior. The start up transients shown in the figure illustrate that the initial start up is also affected by the slip gain in the same way, with larger values yielding less well damped responses. Note also that unlike the step response behavior where field orientation yields a unique, transient free, response, there is nothing unique about field orientation during the start up transient. The field oriented case ($\alpha = \tau_r/\hat{\tau}_r = 1.0$) is simply one mem-

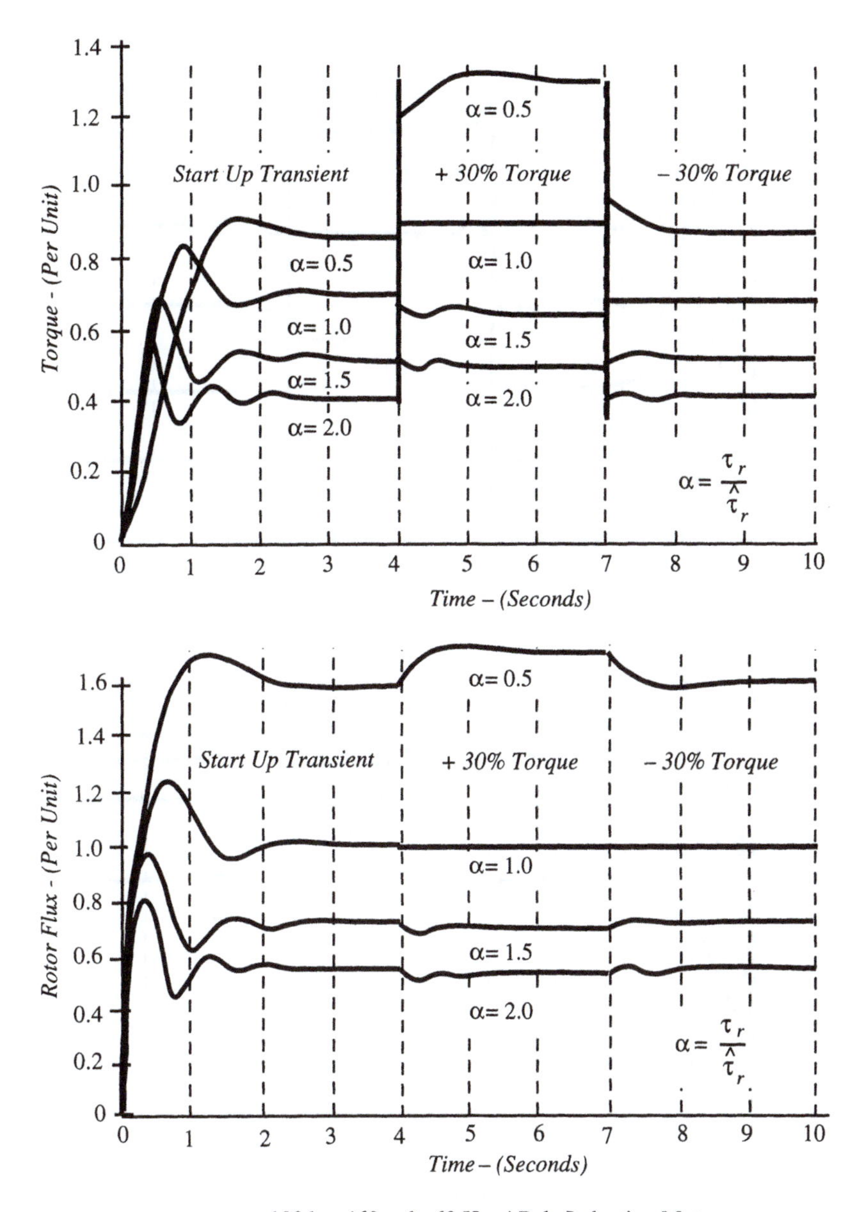

100 hp, 460 volt, 60 Hz, 4 Pole Induction Motor

Per Unit Parameters

Stator Resistance = 0.010 pu Stator Leakage Reactance = 0.10 pu
Rotor Resistance = 0.015 pu Rotor Leakage Reactance = 0.10 pu
Magnetizing Reactance = 3.0 pu Inertia Constant (H) = 0.5 Second

Figure 8.15 Influence of detuning on torque and flux – 100 hp machine

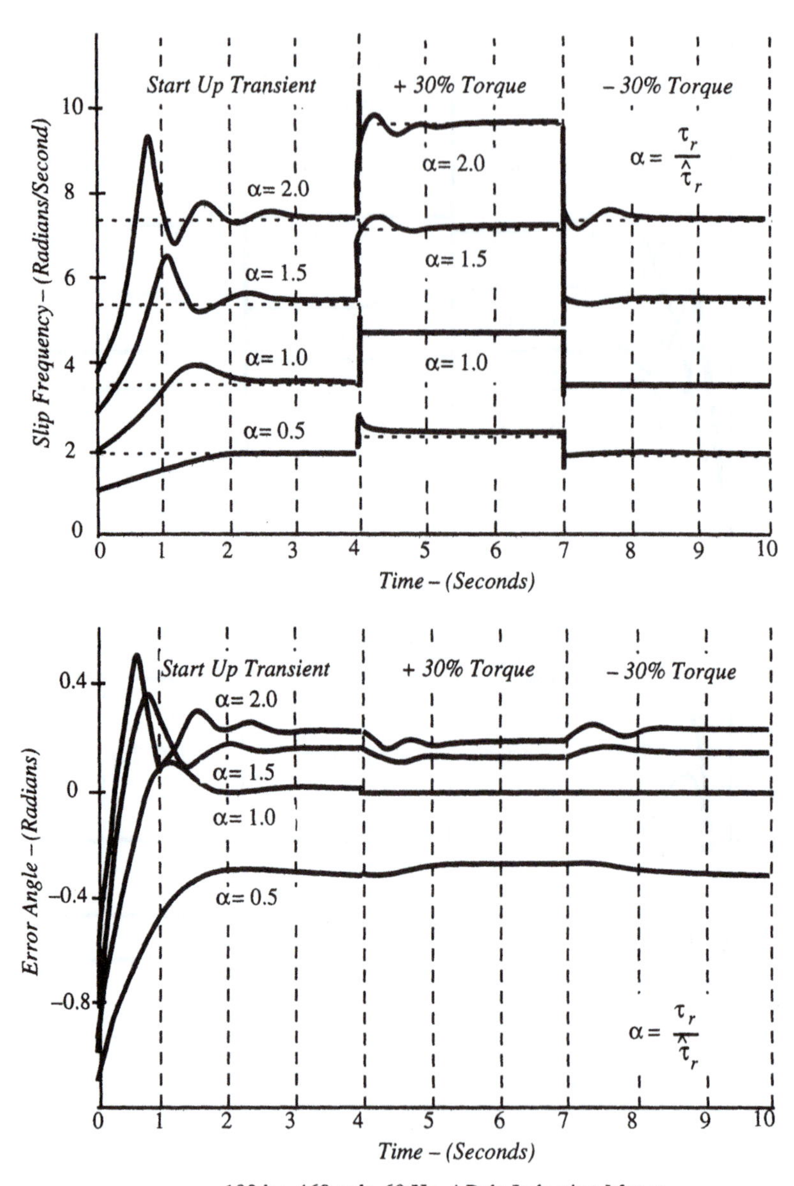

100 hp, 460 volt, 60 Hz, 4 Pole Induction Motor

Per Unit Parameters

Stator Resistance = 0.010 pu Stator Leakage Reactance = 0.10 pu
Rotor Resistance = 0.015 pu Rotor Leakage Reactance = 0.10 pu
Magnetizing Reactance = 3.0 pu Inertia Constant (H) = 0.5 Second

Figure 8.16 Influence of detuning on slip frequency and error angle – 100 hp machine

ber of a family of start up transients which fits into a continuum of responses with no special characteristic associated with field orientation. For the particular machine of Figure 8.15 the start up transient is underdamped for the choice of applied torque and flux command current components ($\beta = i_{qs}^{e*}/\, i_{ds}^{e*} = 2.0$) used in the simulation. For smaller values of β, the start up transient is better damped and becomes non–oscillatory when $\beta = 0.5$ as can be seen in Figure 6.13.

Figure 8.16 illustrates the slip frequency and error angle responses corresponding to the torque and flux responses of Figure 8.15. The slip frequency characteristics in Figure 8.16 clearly show the linearly proportional effect of the slip gain $\alpha = \tau_r/\hat{\tau}_r$ on the steady state slip frequency, which is, at least in part, the origin of the term *slip gain*. High values of the slip gain α clearly yield large values of overshoot in the slip frequency during the start up transient and this high gain can be viewed as the cause of the poor damping associated with the start up transient.

The error angle characteristics shown in Figure 8.16 illustrate the influence of slip gain errors on the alignment of the machine and controller reference axes. Again the oscillatory nature of the response increases with increasing values of the slip gain. Note that while there are zero steady state errors between the estimated and actual slip frequencies (as is forced by the controller), there are finite steady state errors in the rotor flux angles. This implies that the actual motor currents differ from their commanded values in such a way that the actual slip frequency, calculated with the actual value of τ_r, agrees with the commanded slip frequency calculated with the estimated value of $\hat{\tau}_r$. This requires that the actual ratio of i_{qs}^{e}/i_{ds}^{e} differ from the commanded value of $i_{qs}^{e*}/\, i_{ds}^{e*}$ by exactly the value of the slip gain error

$$\frac{i_{qs}^{e}}{i_{ds}^{e}} = \frac{i_{qs}^{e*}\tau_r}{i_{ds}^{e*}\hat{\tau}_r} = \beta\alpha \tag{8.3–16}$$

This result allows a solution for the steady state error angle in terms of the two parameters α and β

$$\tan\left(\Delta\theta_{rf}\right) = \frac{\beta(\alpha - 1)}{1 + \alpha\beta^2} \tag{8.3–17}$$

from which the steady state value of the actual motor currents can be found using

$$I_{qds}^{e} = I_{qds}^{e*} e^{j\Delta\theta_{rf}} \tag{8.3–18}$$

8.4 Optimal Selection of Flux Level

The preceding sections clearly show that the selection of flux level has important impacts on a variety of operating characteristics from efficiency to the effects of detuning. Optimal operation involves the proper selection of the flux level to satisfy the particular optimal requirements of specific applications. The influence of magnetic saturation in this selection of flux level is very significant and must be incorporated in the selection process to obtain valid and useful results [5].

While individual application requirements can create a large number of specialized optimization criteria, the following four optimal modes are of general interest and will be considered in some detail.

Peak Torque per Stator Ampere – In servo applications, when continuous operation at low speeds where core loss is not significant is required, peak torque per stator ampere operation is important. This allows obtaining minimum response time for the particular peak current available from the motor power supply.

Peak Drive Efficiency – Where long periods of continuous, relatively high speed operation are required, drive efficiency may become an important consideration.

Reduced Sensitivity to Controller Detuning – Temperature changes cause the motor resistance to vary and this can result in detuning of the slip calculator. Flux changes cause L_m and L_r to change which also affects the slip calculation. In some cases, reduced sensitivity to this detuning is an important issue.

Optimal Utilization of Converter – The power converter supplying the motor will always be limited in voltage and current capacity. Proper coordination of the motor and converter to meet the overall requirements of the drive is always a significant consideration. Each of these criteria will now be examined.

8.4.1 *Maximum Torque Per Stator Ampere*

Combining eqns (8.2–9), (8.2–6) and the flux equation $\lambda_{dr} = L_m i_{ds}$ allows rewriting the torque of a field orientation controlled motor as

$$T_e = \left[\frac{3}{2}\frac{P}{2}\frac{L_m^2\, i_{ds}^2}{r_r}\right] S\omega_e = \left[\frac{3}{2}\frac{P}{2}\frac{\lambda_{dr}^2}{r_r}\right] S\omega_e \tag{8.4–1}$$

Thus, for constant flux operation, a field oriented machine produces a torque which is linearly related to the slip frequency. The field orientation lines described by Eq. (8.4–1) can be superimposed on the motor torque–slip frequency characteristic to relate the field orientation and the motor characteristic as was done in Figure 8.4 and Figure 8.9. One such line can be drawn for each value of flux selected in the field orientation controller (via flux command). Figure 8.17 [5] illustrates this concept for three different values of flux.

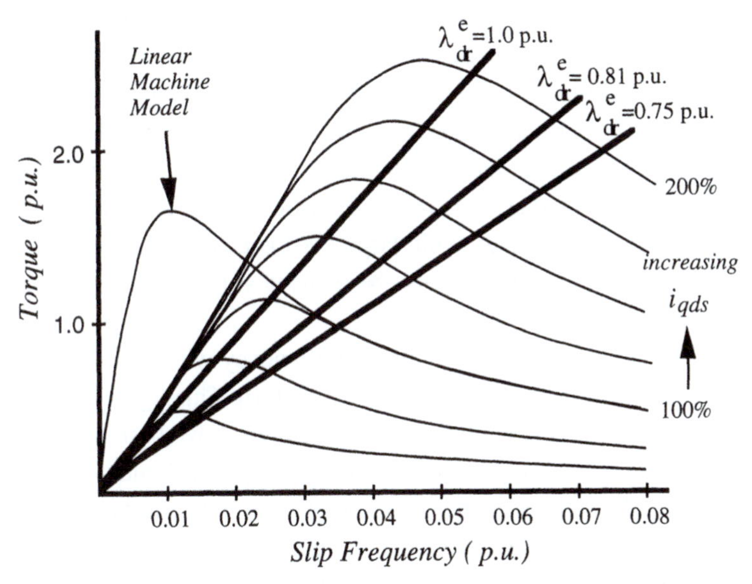

Figure 8.17 Field orientation lines of constant rotor flux plotted on constant stator current torque curves

Selection of the flux level for maximum torque per stator ampere is readily visualized from Figure 8.17. Maximum torque per stator ampere corresponds to the peak of the motor characteristic and hence the field orientation line should be selected to intersect the peak of the curve corresponding to the selected current level where peak torque is desired. Note that, because of saturation and the effect it has on the location of the peak of the motor characteristic, there is one field orientation line (and hence flux level) which is nearly optimal for all motor currents. This is in sharp contrast to the ideal, non–saturating motor where the peak torque per stator ampere always occurs at the same slip frequency. Saturation, therefore, tends to make one single flux level the optimal choice for all levels of stator current and in this sense is a 'good' effect. It also, of course, has the usual bad effect of reducing the peak torque for larger values of current.

8.4.2 Maximum Drive Efficiency

The efficiency of an induction motor is a relatively complex function of speed, torque, and flux. Optimal efficiency operation requires the proper balance between the motor I^2R losses and the core losses. This balance can be controlled by selecting the flux level in relation to the torque and speed of the machine. Thus, for any specific torque and speed, there is a specific flux level which minimizes the total losses and maximizes the efficiency. Unfortunately, this flux level is different for each combination of torque and speed.

In general the optimal flux level varies directly with torque and inversely with speed. At low torque the flux should be set low to minimize the core and magnetizing losses (and also the motor acoustic noise). At higher torque levels the flux must be increased to avoid large I^2R losses resulting from low torque per ampere. The flux level must also be reduced at high speed to avoid excessive core loss.

An illustration of the improvement in efficiency which can be attained by optimal flux selection is shown in Figure 8.18 [10]. This figure shows the efficiency of a high efficiency 5 hp machine operated above base speed with both conventional constant voltage control (flux varying as $1/f$) and with the optimal choice of flux. In both cases the power output was held constant at rated value (torque varying as 1/speed). Note that a significant efficiency improvement is possible, especially at high speed, and that there is a 'best optimal' speed for rated power at about 1.6 pu for this machine. Figure 8.19 [10] shows the inverter voltage and kVA required to achieve the optimal curve in Figure 8.18.

The results in Figure 8.18 and Figure 8.19 are typical of high efficiency machines. In general, the reduction of flux level with speed to attain optimal efficiency is less rapid than the $1/f$ variation normally suggested for constant hp operation. This is especially true for high efficiency machines. In fact, these machines typically have better efficiency at higher than rated flux for all

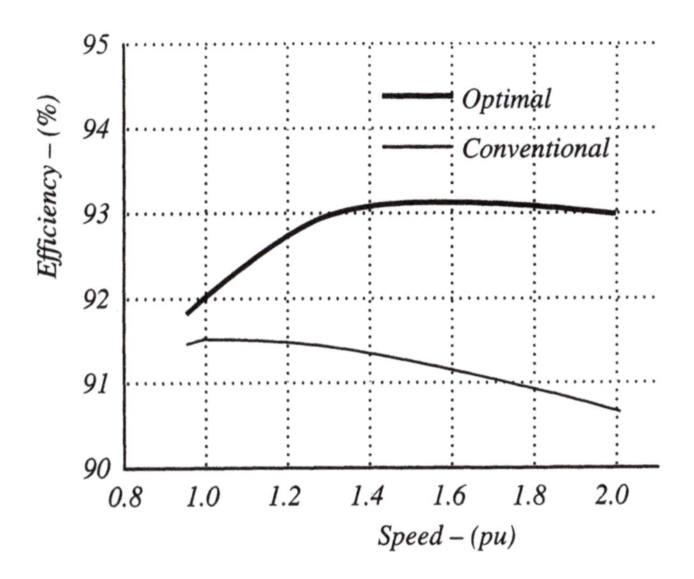

Figure 8.18 Efficiency comparison for conventional and optimal constant power operation — rated power, 5 hp high efficiency motor

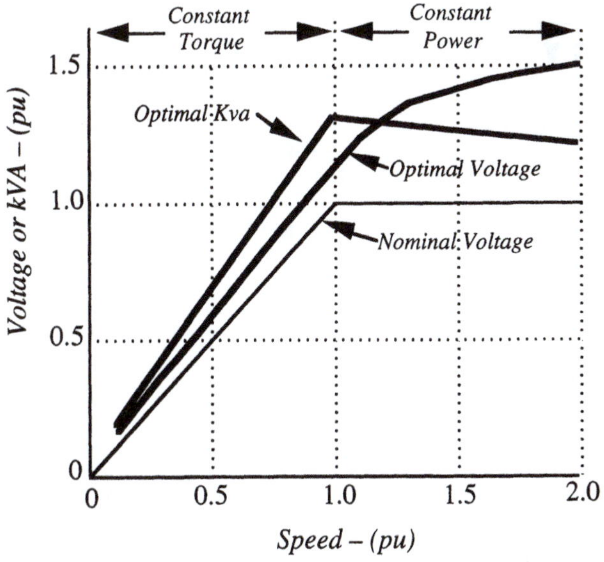

Figure 8.19 Voltage and kVA for optimal efficiency at rated load — 5 hp high efficiency motor

speeds. In contrast, standard design motors are often better at less than rated flux over a significant range of speeds. The existence of a 'best optimal' operating point is also of some interest since, given the choice, this point corresponds to the highest attainable efficiency for that particular output power. This speed always is well above rated speed for rated power output.

Unfortunately, there are no general analytical results which provide information on the best flux level for specific torque and speed conditions. The core and copper losses depend strongly on the saturation characteristic of the machine and hence the problem is strongly nonlinear. Calculations based on measured saturation and core loss are quite feasible, but direct measurement of motor performance is probably the best guide to proper selection of flux in each individual case.

A closed loop controller which automatically assures maximum efficiency operation has been demonstrated. Figure 8.20 is a block diagram of this system used in a speed regulated drive [11].

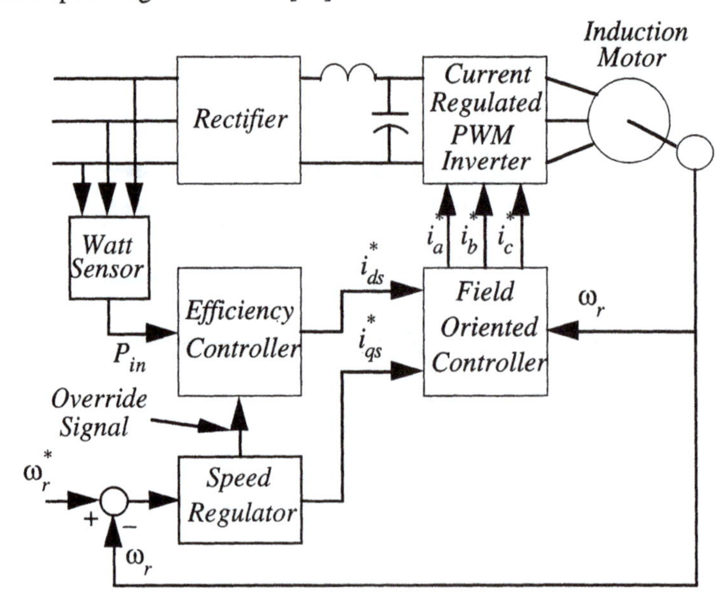

Figure 8.20 Block diagram of optimal efficiency controller

The system employs Hall effect power sensors to measure the drive input power. A microprocessor based controller makes small step changes in flux command and seeks the flux level which results in minimum power input. Since the power output is held constant by the speed regulator (for a fixed torque load), the system will always seek the maximum efficiency flux level. The flux level changes are made slowly relative to the response time of the drive except when a large speed error is detected. In the latter case, the flux is increased to a level which produces high torque per stator ampere to assist in responding to the speed error.

8.4.3 Reduced Sensitivity to Controller Detuning

The results presented in Figure 8.6 and Figure 8.7 clearly show that a small value of the ratio $I_{sT}/I_{s\phi}$ results in a low sensitivity to detuning. While this observation is generally useful in understanding how sensitive a specific machine is to detuning, the curves in Figure 8.6 and Figure 8.7 can also be used to help select a machine which exhibits less detuning sensitivity over a load range or how to select the flux level $I_{s\phi}$ or slip gain setting to reduce the detuning sensitivity. For example, with a machine having an $I_{sT}/I_{s\phi}$ ratio of 1.0 a detuned slip gain setting with $\alpha = 1.3$ on the cold motor will result in virtually no torque gain or loss as the motor heats and thus virtually no detuning effect. Similar gain settings can be found for machines with other values of the ratio $I_{sT}/I_{s\phi}$.

The single most significant factor which increases detuning sensitivity is to increase $I_{sT}/I_{s\phi}$. Thus, for example, detuning effects are much more significant in the field weakening region where $I_{s\phi}$ is much smaller and hence $I_{sT}/I_{s\phi}$ is much larger. Any changes which reduce $I_{sT}/I_{s\phi}$ will be helpful in reducing detuning effects.

Saturation of the air gap flux also helps to desensitize a field oriented controller to detuning. Generally, the lower slip gains which are needed to achieve high flux produce a reduced sensitivity to parameter errors. This is because although errors in the rotor time constant still cause incorrect positioning of the stator current vector, the amplitude of the rotor flux does not change appreciably. Thus, the output torque deviations are limited to the effects of vector mispositioning. Figure 8.21 [6] shows the effects on torque linearity for both the saturating machine and a constant parameter, non–saturating machine for a 3:1 detuning range.

While the 3:1 range of detuning is excessive, it demonstrates the beneficial effects of saturation (low slip gains). At very low flux levels (high slip gains), the differences between the saturating machine and the constant parameter machine are seen to be negligible.

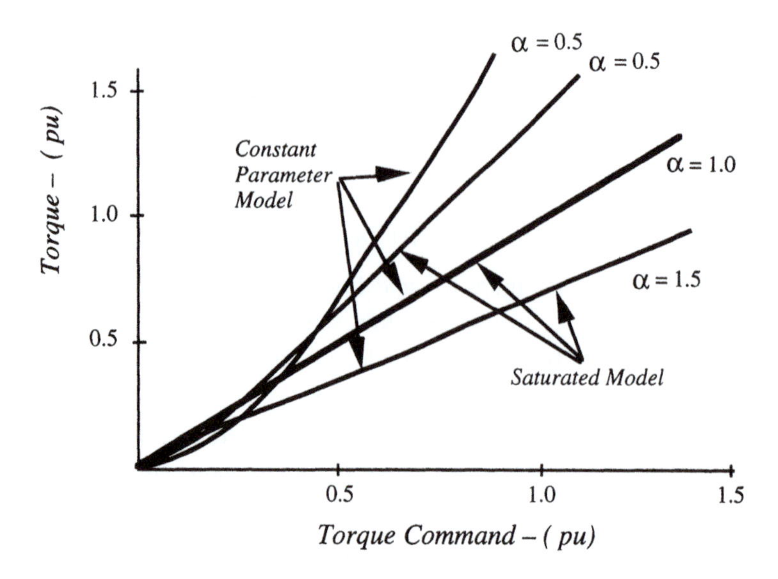

Figure 8.21 Detuning sensitivity characteristica showing that saturation model demonstrates less sensitivity to detuning than a constant parameter model, α = normalized gain, see eqn (8.2–11)

8.4.4 Optimal Utilization of the Converter

In general, flux selection may be determined by machine limits (primarily thermal) or by converter limits (primarily current and voltage). If the thermal limits of the machine dominate, then the operating flux may be chosen to optimize torque per ampere as presented in Section 8.4.1 above. However, the converter ratings for current and/or voltage are often below those required to operate at the machine limited values. Thus, it is appropriate to consider the selection of flux under converter limited conditions.

8.4.4.1 Converter Current Limits

Figure 8.22 shows the change in the maximum torque per ampere operating point corresponding to a converter current limit which is less than that the machine's thermal limit.

It is apparent that saturation reduces the effective change in torque/slip operating point as compared to the non–saturating machine model. Saturation thus makes the system less sensitive to this stator current limit.

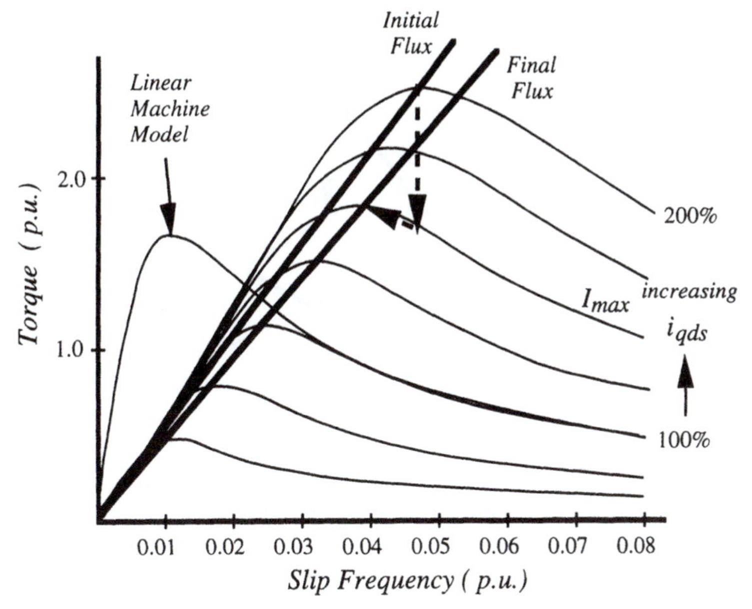

Figure 8.22 Operating point selection for an amplifier current limit which is lower than the machine's thermal limit

8.4.4.2 Converter Current and Voltage Limits

For high speed applications, the back emf voltage requirements of the machine may also limit the operating flux selection. Figure 8.23 [5] shows the change in maximum torque per ampere operating point as required by both a converter current and a voltage limit.

To evaluate this limit the stator voltage equations for field oriented induction machines must be invoked. In scalar, steady state, synchronous frame form they are:

$$V_{qs}^{e} = r_s I_{qs}^{e} + \omega_e L_s I_{ds}^{e} \tag{8.4–2}$$

$$V_{ds}^{e} = r_s I_{ds}^{e} - \omega_e L'_s I_{qs}^{e} \tag{8.4–3}$$

Given a maximum limit on the stator voltage and current amplitudes, and a value of slip frequency $S\omega_e$ as defined by eqn (8.2–6) it is possible to compute the magnetizing current and rotor flux iteratively. As a reasonable approxima-

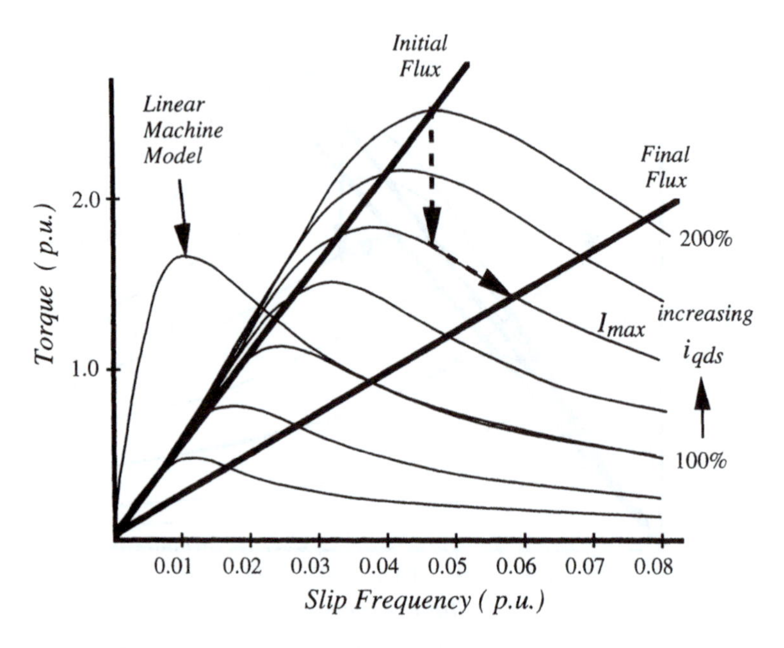

Figure 8.23 Operating point selection for an amplifier current limit which is lower that the machine thermal limit, and an amplifier voltage limit which is matched to the back emf voltage

tion, the magnetizing and maximum torque producing currents may be estimated respectively by the following equations:

$$I_{ds}^{e} \approx \frac{V_{max}}{\omega_e L_s} \tag{8.4–4}$$

$$I_{qs}^{e} = \sqrt{(I_{max})^2 - (I_{ds}^{e})^2} \tag{8.4–5}$$

where:

V_{max} = maximum amplifier voltage

I_{max} = maximum amplifier current

ω_r = rotor electrical speed

$\omega_e = \omega_r + S\omega_e$

This optimization may be viewed as using flux to impedance match the machine to the source (converter).

References

[1] L.J. Garces, "Parameter Adaptation for the Speed–Controlled Static AC Drive with a Squirrel Cage Induction Motor", IEEE–IAS Transactions, Vol. IA–16, No. 2, March/April 1980, pp. 173–178.

[2] M. Koyuma et al, "Effects of Parameter Changes on Coordinate Control System of Induction Motor", Conference Record of the International Conference on Power Electronics, March, 1983, pp. 684–695.

[3] K.B. Nordin, D.W. Novotny and D.S. Zinger, "The Influence of Motor Parameter Deviations in Feedforward Field Orientation Drive Systems," IEEE–IAS Transactions, Vol. IA–21, No. 4, July/Aug 1985, pp. 1009–1015.

[4] J. Loehrke, R.D. Lorenz and D.W. Novotny, "Torque Characteristics of Field Oriented Induction Machines," Conference Record, First Annual Conference on Applied Motion Control, Minneapolis, MN, June 1985.

[5] R.D. Lorenz and D.W. Novotny, "Optimal Utilization of Induction Machines in Field Oriented Drives," Journal of Electrical and Electronics Engineering, Australia, Vol. 10, No. 2, June 1990, pp. 95–100.

[6] F.M.H.Khater, R.D. Lorenz, D.W. Novotny and K. Tang, "Selection of Flux Level in Field Oriented Induction Machine Controllers with Consideration of Magnetic Saturation Effects," IEEE–IAS Transactions, Vol. IA–23, No. 2, March/April 1987, pp. 276–282.

[7] K.H. Bayer and F. Blaschke, "Stability Problems with the Control of Induction Motors Using the Method of Field Orientation", 2nd IFAC Symposium Conference Record, 1977, pp. 483–492.

[8] R. D. Lorenz and D.W. Novotny, "A Control System Perspective of Field Oriented Control for AC Servo Drives," Proceedings of Control Expo 88, pp. (XVIII–1)–(XVIII–12), June 1988.

[9] N.R. Garrigan and D.W. Novotny, "Start–up and Sensor–Parameter Error Transients in Field Oriented Induction Machines," Proc. IEEE–IAS Annual Meeting, October 1995.

[10] F.M.H. Khater and D.W. Novotny, "Efficiency Optimization for Constant Horsepower Operation of Induction Machines," Proceedings, International Conference on Evolution and Modern Aspects of Induction Machines, July 1986, Torino, Italy, pp. 9–16.

[11] D.S. Kirschen, D.W. Novotny and T.A. Lipo, "Optimal Efficiency Control of an Induction Motor Drive," IEEE Trans. on Energy Conversion, Vol. EC–2, No. 1, March 1987, pp. 70–76.

Problems

Problem 8-1 Detuned Field Orientation Operation in the Steady State

For an induction machine with per unit parameters $r_s = 0.03$, $x_{ls} = 0.10$, $x_m = 2.0$, $r_r = 0.03$, $x_{lr} = 0.10$ operated under indirect field oriented control:

a) evaluate the torque vs. torque command current (i_{qs}) characteristic if the machine is operated with a flux command current (i_{ds}) which yields rated flux. Plot the curve from zero to 150% of rated torque.

b) find the torque vs. torque command current (i_{qs}) characteristics which result under the conditions of part a) but for the cases where the slip calculator gain is 50% too large and 50% too small.

Problem 8-2 Volts/hz for Maximum Torque/ampere Operation

Use the results of problem 5-2 to find the required volts/hz to cause operation at maximum torque/ampere (peak of curve of part 5-2a) for the given machine at 60 Hz and at 5 Hz. Comment on how main flux saturation will affect these answers. Explain how these results illustrate the difficulty of obtaining good starting torque (maximum torque/ampere) in an open loop adjustable speed drive.

Problem 8-3 Detuned Induction Machine Field Orientation

Consider a 100 hp induction machine with parameters and performance data as follows:

460 volts (line to line)   100 hp   60 hz   4 pole
$Z_B = 2.84\ \Omega$ $V_B = 266$ volts $I_B = 93.6$ amp $T_B = 395.8$
$r_s = r_r = 0.015$ pu $= 0.0425\ \Omega$, $x_{ls} = x_{lr} = 0.100$ pu $= 0.284\ \Omega$
$x_m = 3.0$ pu $= 8.51\ \Omega$ $J = 2.0$ kg - m^2
Rated Operation $S_R = 0.0177$ pu $I_R = 1.17$ pu $\cos\theta_R = 0.880$

The machine is operated from a current regulated variable frequency supply. For this problem assume an ideal current regulator.

a) Find I_{ST} and $I_{S\phi}$ for operation with rated terminal voltage V_R = 1.0 pu, rated current I_R = 1.17 pu, rated frequency f_R = 1.0 pu, and rated torque T_R = 1.02 pu.

b) Find the terminal voltage V_s required to maintain the same I_{ST} , $I_{S\phi}$ and torque at 5 Hz. Compare this value with the voltage corresponding to constant volts/hz operation.

c) Find the value of the torque if I_{ST} is held constant at the value in part (a) and the new value of I_{ST} needed to bring the torque back to the value of part (a) if the slip gain is too small by a factor of 2. The value of $I_{S\phi}$ (command value) is held constant at the value in part (a). Find the rotor and stator fluxes for the original value of I_{ST} (express the fluxes as a multiple of the values for rated operation).

d) Find and plot (using simulation) the torque transient following a step change in I_{ST} from zero to the value in part (a) for the case of correct slip gain and for the slip gain error situation in part (c). Compare the predicted steady state torque with the result in part (c). The value of $I_{S\phi}$ (command value) is held constant at the value in part (a).

9 Field Weakening Operation

9.1 Introduction

So far we have examined carefully how torque can be accurately controlled in an induction motor drive. However, we have made little mention of the demands of the load. Clearly the electromagnetic torque and load torque must be equal in magnitude and opposing in direction during steady state operation. In general we can separate loads into four categories, 1) constant speed, variable or constant torque loads, 2) adjustable speed, constant torque loads, 3) adjustable speed, variable torque loads and 4) variable speed, variable torque loads.

In category #1 are included those typical applications which require only a fixed frequency supply such as constant speed pumps and fans, and compressors. This category is serviced by conventional induction motors operating from the utility 50 or 60 Hz supply in which the pole number is chosen to be compatible with the speed requirement of the application. (Note that constant speed is, in this case, slightly violated since the rotor speed changes by a few per cent with changes in load from zero to rated value.)

In category #2 applications such as elevators, fiber spinning, hoists, etc. are found. Note that here, *constant torque* is meant to imply that the torque changes only very slowly, such as when an elevator stops at each floor. By the term *adjustable* in contrast to the term *variable* is meant that the speed is adjusted externally by an operator, typically a human operator. In these applications, the management of acceleration or deceleration is also typically important.

In category #3 are found applications such as fans, pumps and compressors which are called upon to operate with changing speed. Because of the inherent flow versus pressure (or head) characteristics of such devices, the torque load follows square law characteristic as a function of speed. As a result the power requirement varies as the cube of speed. All three of the above applications are adequately served by the open loop speed controller of Figure 1.7 or Figure 1.13. With a correctly tuned slip frequency compensator (the gain box f_{slip}/I_i

in these figures), a speed regulation capability of 1% or less can be obtained at rated speed.

Finally, category #4 contains those applications in which the speed is variable rather than merely adjustable implying that dynamic requirements are attached to the speed and/or torque response. Category #4 contains those high performance applications addressed by this book, namely, traction motors, servos and the like.

9.2 Torque Demand and Capability of High Performance Drives

Prior to the modern ac inverter, variable speed, variable torque applications were the province solely of dc motor drives. The torque available from a dc motor was limited by three factors 1) the maximum armature and field currents which could be safely handled by the dc motor and dc power converter, 2) the maximum available voltage obtainable from the dc armature power supply and 3) the maximum current that could be handled by the brush/commutator mechanism at a particular speed [1].

If the field and armature currents are set to their maximum permissible values then, from eqn (5.2–4),

$$T_{e,\,max} = \frac{P}{2}\frac{L_{af}}{L_f}\lambda_{f,\,max}I_{a,\,max} \tag{9.2–1}$$

where $\lambda_{f,\,max} = L_f I_{f,\,max}$ and $I_{f,\,max}$ is the maximum field current. This constraint sets the limit curve #1 on Figure 9.1.

When the speed of the dc motor increases, the voltage applied to the motor must increase to match the back emf of the dc motor as defined by eqn 5.2–1. Eventually, at some speed, for any practical power supply, the voltage can not be further increased to maintain the current at the limit value $I_{a,\,max}$. If the field current is not changed at this point the current will rapidly drop to zero, i.e. at the speed where $V_{dc\,(max)} = E_a$, where $V_{dc\,(max)}$ is the dc output voltage of the power supply. Torque can continue to be maintained however, if the emf of the motor is held at a value lower than $V_{dc\,(max)}$. The minimum value of field current at any speed is set by the maximum permissible value of I_a. Hence, another boundary curve, curve #2 in Figure 9.1, can be identified which is set by the saturation of the dc power supply. This curve is usually labelled as the constant power region.

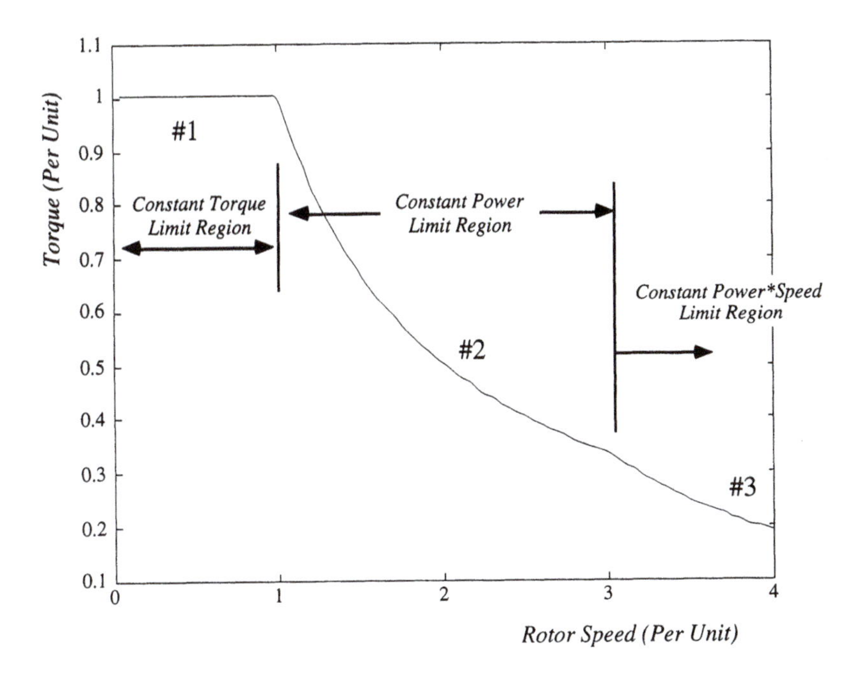

Figure 9.1 Torque vs. speed capability curve for dc motor drive

As speed continues to increase the minimum value of field current needed to supply rated armature current decreases in accordance with eqn 5.2–1 which specifies that as speed increases, the field flux linkage must decrease inversely with speed if E_a is to be maintained constant. However, as the field current is decreased, the distortion of the airgap flux caused by the armature current increases creating a very non-uniform distribution of bar to bar voltage on the commutator. This distortion and the high speed eventually cause bar to bar commutator voltages to reach unacceptable levels. Sparking begins to occur usually terminated by spectacular fireworks, a flashover of the brush and commutator rigging. Hence, above a certain speed the reduction of the field current to maintain constant armature current must be curtailed. The armature current must now begin to decrease. Since the need for commutating ability is proportional to the current to be commutated, the field current (as well as interpole current) now becomes proportional to the armature current resulting in a third mode of operation termed series field operation, curve #3. The corner point at which the machine reaches crossover between curve #2 and curve #3 is generally considered to be rated maximum speed. Operation along curve #3 is rarely used since commutating ability is marginal and the motor enters a speed region where physical integrity of the commutator bars becomes difficult to maintain. In addition the torque capability in this region is not much more than windage and friction losses making the additional complexity of operating in this region difficult to justify. The variation of the motor electrical variables needed to obtain Figure 9.1 is shown in Figure 9.2.

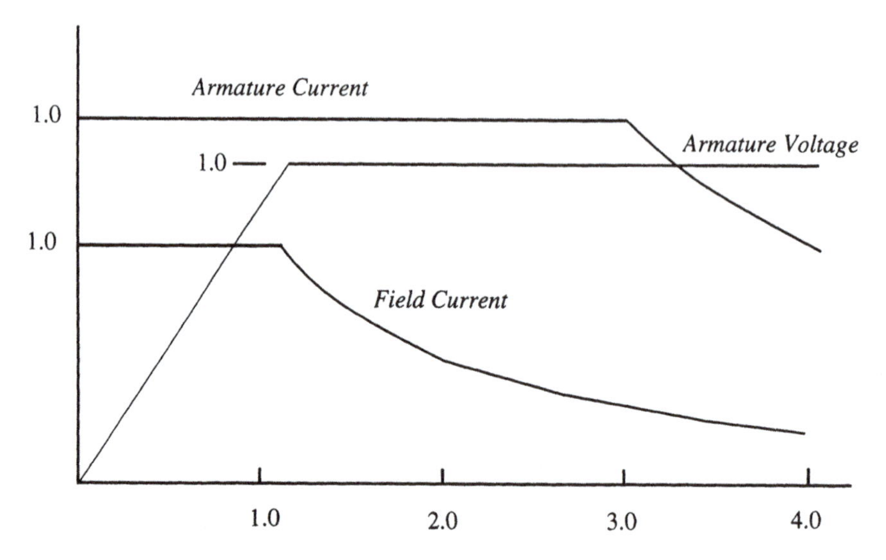

Figure 9.2 Electrical variables corresponding to Figure 9.1 as a function of speed

It should be noted that since the armature current and field current are constant the constraint set by curve #1 on Figure 9.1 inherently produces a constant torque limit on the speed–torque plane while constant armature current and constant back emf constraint of curve #2 imposes inherently a constant power limit. Neglecting armature resistance, Curve #3 imposes an inverse square law torque characteristic. Popular acceptance of these *capability* limitations of a dc machine has caused users of electrical equipment to define these same characteristics from the point of view of *torque requirements* not only to be imposed on a dc machine but any type of motor drive. It remains to be seen whether the maximum capability curves of an induction machine are consistent with torque requirements, as adopted from the maximum capability curves of a dc machine.

9.3 Induction Machine Control Strategy for Constant Power Operation

Referring back to Figure 5.21 it is quite easy to see that a maximum torque requirement imposed by curve #1 on Figure 9.1 is readily satisfied in the case of a vector controlled induction motor. In particular, when the slip frequency is held constant then both the reactance $j(L_m/L_r)X_m$ and the resistance $(L_m/L_r)^2[r_r/(S\omega_e)]\omega_e$ are proportional to the applied frequency. Hence if the stator current amplitude I_s is also held constant by means of current regulator, then the division of I_s between the magnetizing component $I_{s\phi}$ and the torque producing component I_{sT} remains unchanged and thus the torque is constant, as demonstrated by eqn 5.9–9. Except for the slightly different effect of the stator Ir drop the curves of the electrical variables are remarkably similar to the dc machine.

In the high speed range, the output torque capability of the induction machine depends on the field weakening strategy. The conventional method for the field weakening is to vary the rotor flux reference in proportion to the inverse of the rotor speed [2] as shown in Figure 9.3. The torque therefore

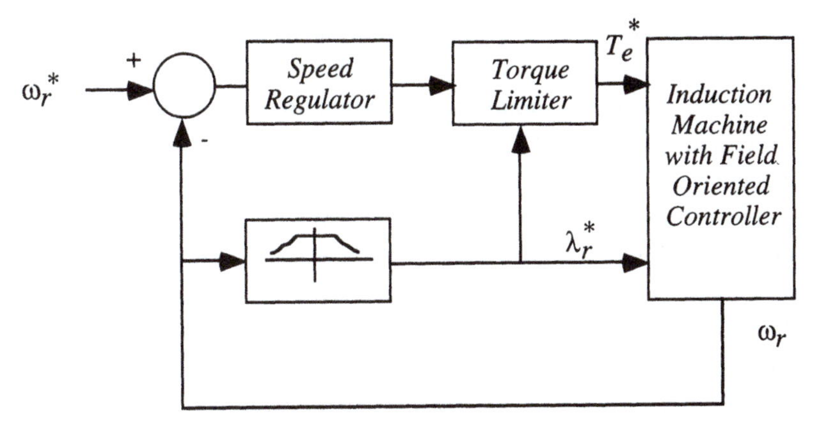

Figure 9.3 Block diagram of field weakening using rotor flux weakening

will decrease inversely in accordance with eqn (6.2–14). A plot of the corresponding electrical variables is shown in Figure 9.4. The parameters used to calculate these curves correspond to the parameters of the example in Section 5.10.1. However in this case, one per unit voltage is defined to be that voltage which produces a four to one field weakening range. The maximum torque in the torque limited region was selected to be 0.719 per unit.

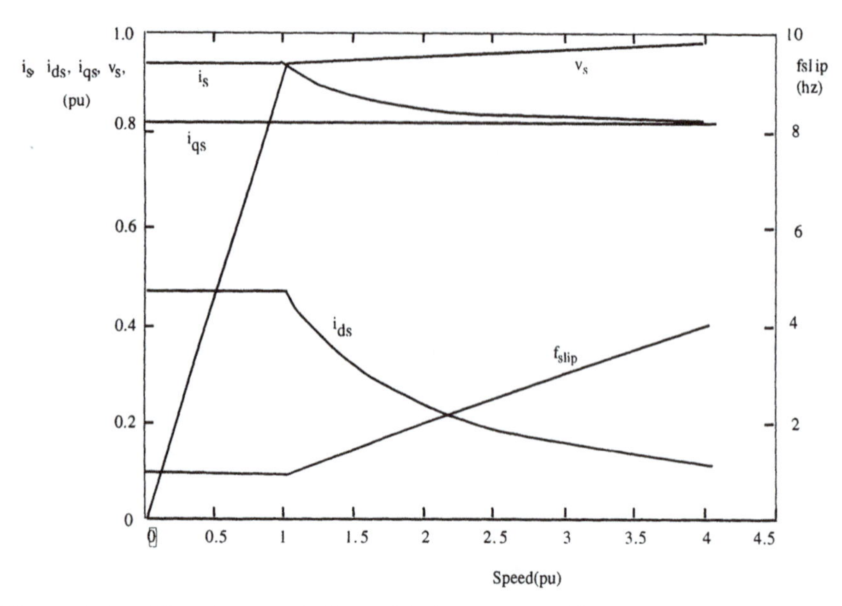

Figure 9.4 Electrical variables of induction machine satisfying the torque vs. speed requirements of Figure 9.1

Note that the constant power limit region can apparently be extended without limit. However, it is important to note that, in contrast to the dc machine, the voltage of the power supply must continue to increase during the power limited region to counteract the increased voltage drop across the stator leakage inductance. It should be noted that the price to be paid for an extended constant power region is a sacrifice in the torque producing capability. For example, from Figure 9.4 note that one per unit speed is reached at only 95% of the voltage capability of the inverter. Hence, 5% of the torque production capability is being sacrificed to meet the constant horsepower limit over the 4 to 1 speed range. In essence, the torque capability curve of an induction machine is, in fact, somewhat greater than the torque requirements imposed by the dc motor torque–speed characteristics. In the next section we will explore exactly how this additional torque capability of the induction machine can be extracted.

9.4 Inverter Imposed Voltage and Current Limits

The maximum output torque and power developed by the machine is ultimately dependent on the allowable inverter current rating and the maximum voltage which the inverter can apply to the machine. Therefore, considering the limited voltage and current capacities, it is desirable to consider a control scheme which yields the best possible torque per ampere over the entire speed range. In the high speed range, the output torque capability of the induction machine depends on the field weakening strategy. The conventional method for field weakening, discussed in the previous section, is to vary the rotor flux reference in proportion to the inverse of the rotor speed. In this method, if maximum inverter voltage is reached at one per unit speed, the voltage margin required to regulate current is not maintained. The problem can be overcome if we turn our attention to the stator flux linkage.

Assume that the voltage applied to the motor reaches its limit. The *d,q* axis model of the machine in the synchronous reference frame is, from eqns (6.2–1) and (6.2–2),

$$v_{qs}^{e} = r_s i_{qs}^{e} + p\lambda_{qs}^{e} + \omega_e \lambda_{ds}^{e} \quad (9.4\text{–}1)$$

$$v_{ds}^{e} = r_s i_{ds}^{e} + p\lambda_{ds}^{e} - \omega_e \lambda_{qs}^{e} \quad (9.4\text{–}2)$$

In the steady state the p becomes zero. In field orientation

$$\lambda_{ds}^{e} = L_s i_{ds}^{e} \quad (9.4\text{–}3)$$

and

$$\lambda_{qs}^{e} = L_s i_{qs}^{e} + L_m i_{qr}^{e} \quad (9.4\text{–}4)$$

where

$$i_{qr}^{e} = -\frac{L_m}{L_r} i_{qs}^{e} \quad (9.4\text{–}5)$$

Combining eqns (9.4–4) and (9.4–5),

$$\lambda_{qs}^{e} = \left(L_s - \frac{L_m^2}{L_r} \right) i_{qs}^{e} \quad (9.4\text{–}6)$$

or

$$\lambda_{qs} = L_s' i_{qs}^{e} \quad (9.4\text{–}7)$$

where L_s' is the stator transient inductance.

Equations (9.4–1) and (9.4–2) become:

$$V_{qs}^e = r_s I_{qs}^e + \omega_e L_s I_{ds}^e \tag{9.4–8}$$

$$V_{ds}^e = r_s I_{ds}^e - \omega_e L_s' I_{qs}^e \tag{9.4–9}$$

The maximum stator voltage, $V_{s,\,max}$ is determined by the available dc link voltage and the PWM strategy and must satisfy,

$$(V_{qs}^e)^2 + (V_{ds}^e)^2 \le V_{s,\,max}^2 \tag{9.4–10}$$

Hence, the voltage vector must remain on the locus of a circle of radius V_s.

Neglecting the stator *Ir* drop for high speed operation, the limit for the steady state currents is

$$(\omega_e L_s I_{ds}^e)^2 + (\omega_e L_s' I_{qs}^e)^2 \le V_{s,\,max}^2 \tag{9.4–11}$$

Equation (9.4–11) can be rearranged to form

$$\left(\frac{I_{ds}^e}{a}\right)^2 + \left(\frac{I_{qs}^e}{b}\right)^2 \le 1 \tag{9.4–12}$$

Hence, the locus of the current vector is constrained to the locus of a voltage limited ellipse with major axis $b = V_s / (\omega_e L_s')$ and minor axis $a = (V_s / (\omega_e L_s))$. The eccentricity of the ellipse is defined by,

$$e = \frac{\sqrt{b^2 - a^2}}{b} = \frac{\sqrt{L_s^2 - (L_s')^2}}{L_s} = \sqrt{1 - \left(\frac{L_s'}{L_s}\right)^2} \tag{9.4–13}$$

The shape of the voltage limited ellipse depends upon the leakage of the motor as a per unit of the stator self inductance. Figure 9.5 shows the voltage limited ellipse for several speeds in the field weakening region. Note that when the voltage is fixed the ellipse shrinks inversely with ω_e.

In general, the maximum stator current is also limited by the inverter current rating as well as the thermal rating of the induction motor. Hence, the *d*– and *q*–axis currents must also satisfy a limit in the stator current amplitude, i.e.

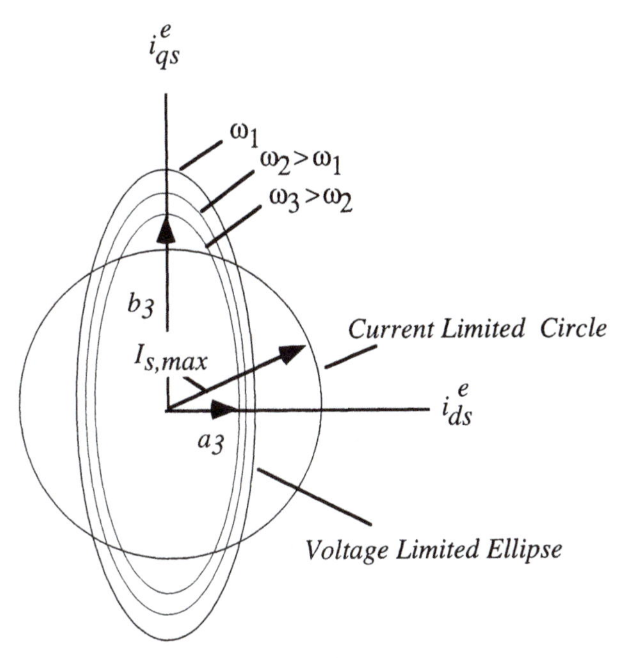

Figure 9.5 Voltage limited ellipse and current limited circle in the field weakening region

$$(I_{qs}^e)^2 + (I_{ds}^e)^2 \le I_{s,\,max}^2 \tag{9.4–14}$$

This expression represents a current limited circle the radius of which is the maximum allowable stator current as shown in Figure 9.5. Note that the current limited circle remains constant for any speed while the voltage limited ellipse begins to decrease as field weakening operation is entered.

9.5 Maximum Torque Capability Curve Using the Conventional Method

Consider now the commonly used means for field weakening operation in which the rotor flux reference is varied inversely proportional to ω_r. The d–axis reference current I_{ds}^{e*} is decreased so as to reduce the rotor flux linkage as in Figure 9.4. However, in this case, the q–axis reference current I_{qs}^{e*} can be increased in order to fully utilize the current rating of the power converter. The two current components are related to the maximum stator current $I_{s,\,max}$ by,

$$I_{ds}^{e*} = \frac{I_{ds,rated}}{\omega_{r(pu)}} \qquad (9.5\text{–}1)$$

$$I_{qs}^{e*} = \sqrt{(I_{s,max})^2 - (I_{ds}^{e*})^2} \qquad (9.5\text{–}2)$$

where $I_{ds,rated}$ is the rated d–axis current producing rated rotor flux and $\omega_{r(pu)}$ is the rotor speed in per unit and equals unity at the transition point. By combining (9.5–1) and (9.5–2), we have,

$$\left(\frac{I_{ds,\text{rated}}}{\omega_{r(pu)}}\right)^2 + (I_{qs}^{e*})^2 = I_{s,\,max}^2 \qquad (9.5\text{–}3)$$

Equation (9.5–3) indicates that when the rotor flux is reduced inversely with increasing speed, the trajectory of the reference current vector moves along a current limited circle in a counterclockwise direction. As long as the current vector for a given frequency is enclosed by the voltage limited ellipse for that frequency, field weakening operation can be maintained. However, as the voltage limited ellipse continues to shrink, a point is ultimately reached where the current vector lies on both the current limited and voltage limited curves shown as ω_3 in Figure 9.6. At this point the inverter is completely saturated and no voltage margin is available to regulate the current. Operation at any higher speed requires that the current vector now remain on the voltage limited ellipse.

9.6 Scheme for Achieving Maximum Torque Capability [4],[5]

When expressed in terms of stator currents, the torque of a field oriented induction machine can be expressed as,

$$T_e = \frac{3P}{4}\frac{L_m^2}{L_r} i_{ds}^e i_{qs}^e \qquad (9.6\text{–}1)$$

hence, the torque can be maximized by maximizing the $i_{ds} i_{qs}$ product. It should be mentioned here that this statement is true only if saturation is neglected. However, since this discussion pertains to field weakening the assumption is generally valid.

Figure 9.7 shows a sketch of this product superimposed on the voltage and current limited curves. The area within the intersection of these two curves cor-

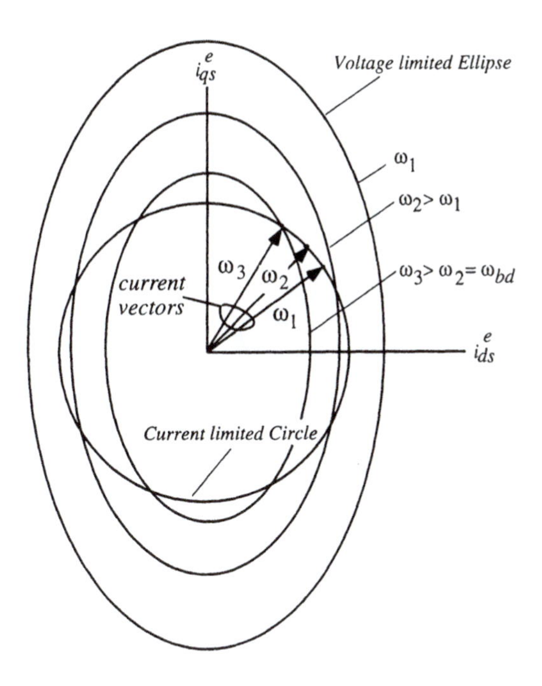

Figure 9.6 Locus of current vector for $1/\omega_r$ field weakening method showing operation reaching the voltage limit ellipse at $\omega_3 = \omega_{bd}$

responds to the allowable values of current at a certain speed. Clearly, the maximum torque is reached where the $i_{ds}i_{qs}$ hyperbola just touches the current limited curves at point A in Figure 9.7 which becomes the transition point from constant torque (region #1) to constant power operation (region #2).

9.6.1 Transition from Constant Torque to Constant Power Operation

When the maximum values of voltage and current are defined, the frequency at which the motor enters the field weakening region (region #2) is not an independent variable. If we combine eqns (9.5–1), (9.5–2) with (9.4–11), this results in an expression for the transition frequency, ω_{base}, is obtained

$$\omega_{base} = \frac{V_{s,\,max}}{\sqrt{L_s^2 I_{ds,rated}^{e\ 2} + L_s'^{\,2}(I_{s,\,max}^2 - I_{ds,rated}^{e\ 2})}} \tag{9.6–2}$$

Above this frequency the trajectory of the optimal current vector for maximum torque subsequently moves along the current limited boundary in the counter clockwise direction as the speed increases. As the speed continues to

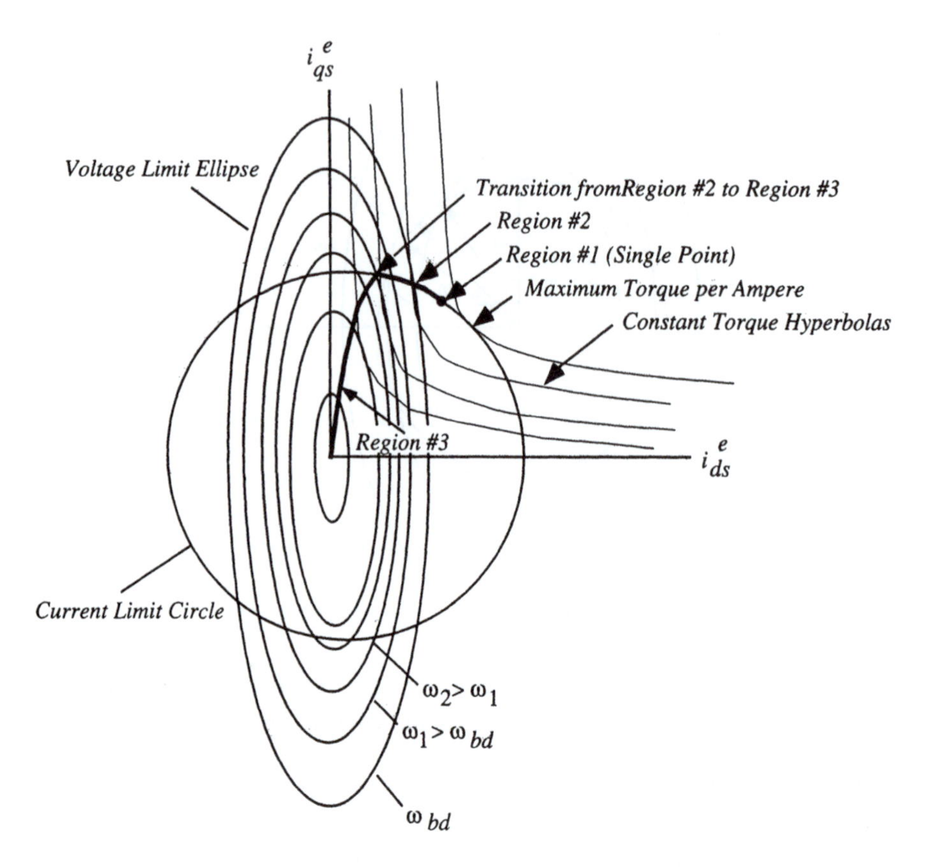

Figure 9.7 Illustrating intersection of the constant torque hyperbolas with the voltage limit curves and showing Region #3 operation

increase beyond ω_{base}, the slip frequency increases in order to keep the stator current amplitude at its prescribed maximum value.

9.6.2 *Region #2 – Field Weakening with Constant Output Power*

In region #2 the slip angular frequency $S\omega_e$ must be increased to maintain constant stator current. The slip frequency relationship, eqn (6.2–17) reduces to

$$S\omega_e = \frac{r_r i_{qs}^e}{L_r i_{ds}^e} \tag{9.6–3}$$

At the transition from region #2 to region #3 the maximum torque is limited by both the voltage limited ellipse and the current limited circle. Combining eqns (9.4–11) describing the voltage limit and (9.4–14) describing the current limit and assuming the equality sign holds, then solving for the d– and q–axis currents results in,

$$I_{ds}^{e} = \sqrt{\frac{(V_{s,max}/\omega_e)^2 - (L'_s I_{s,max})^2}{L_s^2 - L_s'^2}} \quad (9.6\text{–}4)$$

$$I_{qs}^{e} = \sqrt{\frac{(L_s I_{s,max})^2 - (V_{s,max}/\omega_e)^2}{L_s^2 - L_s'^2}} \quad (9.6\text{–}5)$$

The maximum value of slip is derived from eqn (9.6–4) and (9.6–5) subject to the current limit and voltage limit constraints, eqns (9.4–11) and (9.4–14), whereupon,

$$\omega_{sl,max} = \frac{L_s}{\tau_r L_s'} = \frac{1}{\sigma\tau_r} \quad (9.6\text{–}6)$$

This value of slip frequency corresponds to the slip for maximum torque for a voltage driven machine. As the machine moves through region #2, the slip frequency increases from the normal value in region #1 to the breakdown value of eqn (9.6–6). Increasing beyond this point produces less torque even though the current increases. This result indicates that a wide constant horsepower range is obtained by selecting a motor with as small a leakage inductance as possible. The frequency at the transition to region #3 is found to be

$$\omega_{e,1} = \sqrt{\frac{L_s^2 + L_s'^2}{2L_s^2 L_s'^2}}\left(\frac{V_{s,max}}{I_{s,max}}\right) \quad (9.6\text{–}7)$$

*9.6.3 Region #3 – Field Weakening with Constant Speed*Power*

As the frequency is increased beyond $\omega_{e,1}$, the ellipse continues to be reduced and eventually is included within the circle as shown in Figure 9.7. Thus the output torque is limited only by the voltage limited ellipse. The optimal currents for this case can be obtained from (9.6–1) and (9.4–11). The optimal d– and q– axis currents are, in this case,

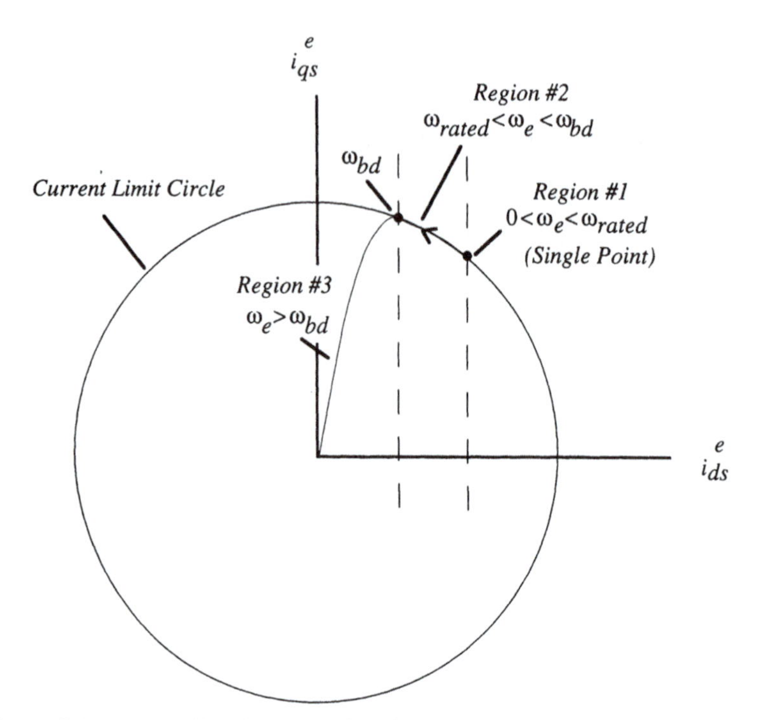

Figure 9.8 Overall trajectory of optimal stator current vector in the field weakening region

$$I_{ds}^e = \frac{V_{s,\,max}}{\sqrt{2}\omega_e L_s} \tag{9.6–8}$$

$$I_{qs}^e = \frac{V_{s,\,max}}{\sqrt{2}\omega_e L_s'} \tag{9.6–9}$$

The torque developed by the machine drops rapidly as roughly the square of the speed reproducing, to a remarkable extent the series field mode of operation of the dc machine. The trajectory moves along the constraint curve during regions#1, #2 and #3 as illustrated in Figure 9.8.

9.7 Control System Implementation

Figure 9.9 shows a block diagram of a possible implementation scheme [4]. The overall system consists of speed regulator, flux regulator, current regulator and indirect rotor flux oriented controller. A synchronous frame PI regulator

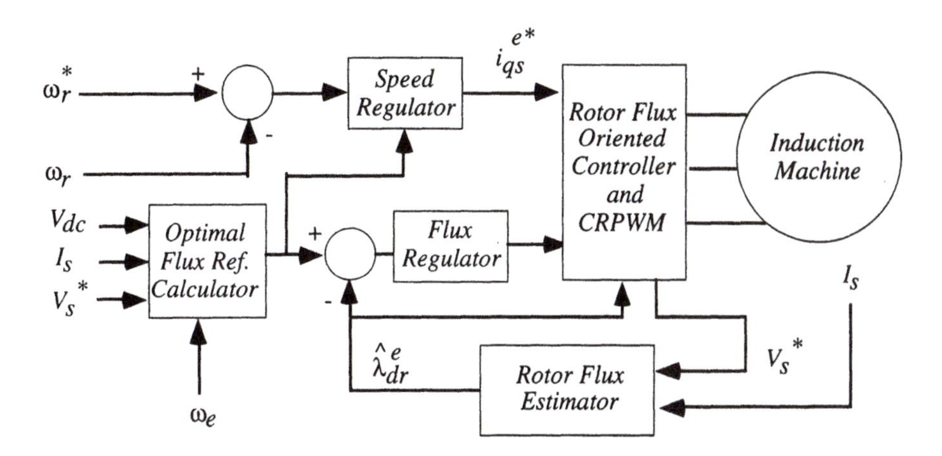

Figure 9.9 Block diagram of the overall control system for optimized operation in the field weakening range

with voltage space vector PWM is readily adopted as the current regulation scheme. The rotor flux is estimated from the terminal quantities as described by eqns (9.4–1) and (9.4–2) together with eqns (6.8–1) and (6.8–2). Feedforward compensation is added in order to decouple the *d,q* axes interactions as in Figure 7.23.

In practice eqn (9.4–1) is difficult to implement because it requires a pure integrator, which has initial value and drift problems. To avoid these problems, a pure integrator must be replaced by a low pass filter which results in good performance in the high speed region [3]. Figure 9.10 shows the flux reference and resulting torque vs. speed for the machine of Figure 9.4. From the comparison of the two methods it is apparent that the optimized method provides significantly improved torque production when compared with the simple algorithm which decreases rotor flux inversely with rotor speed.

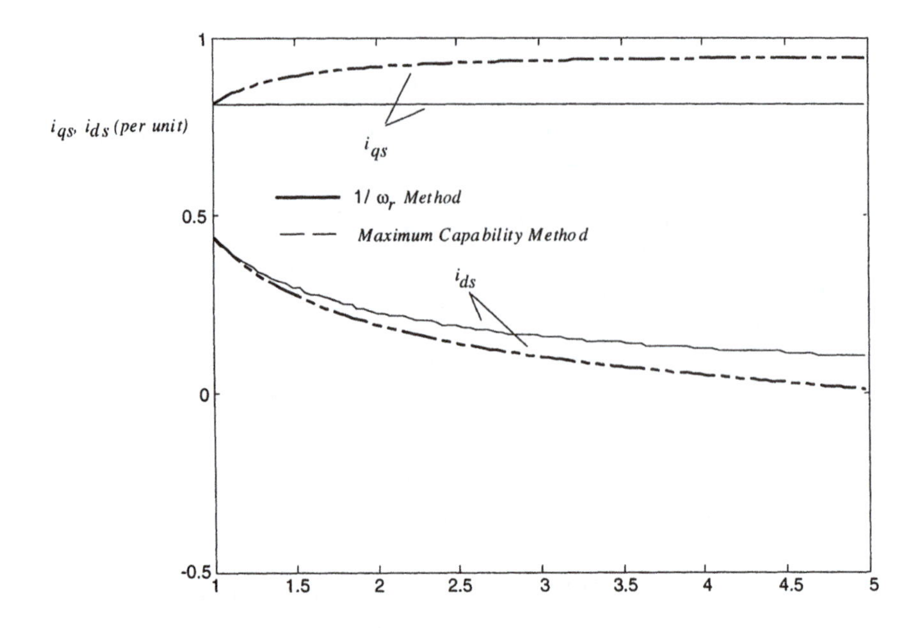

Figure 9.10 Flux reference and torque command currents vs. speed for $1/\omega_r$ method and optimized capability method

References

[1] D. W. Novotny and R D. Lorenz, "Introduction to Field Orientation and High Performance AC Drives", in Tutorial Course Rec., IEEE–IAS Annual Meeting Conf. Rec., 1985, Sections 1 and 6.

[2] R. Jotten and H. Schierling, "Control of the Induction Machine in the Field Weakening Range", in Proc. IFAC 1983, pp. 297–304.

[3] X. Xu, R. De Doncker and D. W. Novotny, "Stator Flux Orientation Control of Induction Machines in the Field Weakening Region", IEEE–IAS Annual Meeting Conf. Rec., 1988, pp. 437–443.

[4] S.–H. Kim, S.–K. Sul, and M.–H. Park, "Maximum Torque Control of an Induction Machine in the Field Weakening Region", IEEE IAS Annual Meeting Conf. Rec., 1993, pp. 401–407.

[5] S.–H. Kim and S.–K. Sul, "Voltage Control Strategy for Maximum Torque Operation of an Induction Machine in the Field Weakening Region", in IECON Conf. Rec., 1994, pp. 599–604.

Appendix
Per Unit Representation

Electric machine parameters and performance characteristics are often expressed in per unit terms rather than in actual units. This approach facilitates comparisons between machines and is helpful in clarifying the relative sizes of the machine parameters.

The most appropriate and commonly used per unit system is one based on the rated output power of the machine (shaft power if a motor, electrical terminal output power if a generator). The primary base quantities, normally taken as nameplate rated values, are:

$$V_B = V_R = \text{rated phase voltage (rms for phasors, peak for d-q)}$$

$$P_B = P_R = \text{rated output power}$$

$$\omega_B = \omega_R = \text{rated electrical radian frequency}$$

from which the following secondary base quantities are derived

$$I_B = \frac{P_B}{qV_B} = \frac{P_R}{qV_R}$$

$$Z_B = \frac{V_B}{I_B} = \frac{qV_R^2}{P_R}$$

$$\omega_{mB} = \frac{\omega_B}{P} = \frac{\omega_R}{P}$$

where q is the number of phases, P is the number of pole pairs and ω_{mB} represents the base mechanical angular velocity of the rotor shaft. The subscripts 'B' and 'R' denote 'base' and 'rated' respectively. All electrical and mechanical variables can now be expressed as a *per unit* quantity by taking the ratio of the physical quantity and dividing by the base. For example the per unit stator voltage, stator current and resistance are:

$$V_{s\,(pu)} = \frac{V_s}{V_B}$$

$$I_{s\,(pu)} = \frac{I_s}{I_B}$$

$$r_{s\,(pu)} = \frac{r_s}{Z_B}$$

In this per unit system, one per unit voltage denotes rated rms voltage while 1.41 per unit voltage would designate the corresponding peak voltage when the waveform is sinusoidal. It is interesting to note that the tendency to use rms voltage as the base loses its meaning when the supply voltage is non–sinusoidal, e.g. an inverter. In this case peak rated (sinusoidal) voltage is frequently used as the base. In this book rms bases are used for phasors and the corresponding peak sinusoid bases are used for *d-q* quantities.

Per unit quantities can also be employed in a system in which the frequency varies. The per unit reactance corresponding to an inductance L is then defined to be,

$$X_{pu} = \frac{\omega_e L}{Z_B}$$

where ω_e is the angular radian frequency of the supply. This equation can also be written as

$$X_{pu} = \frac{\omega_e}{\omega_B}\frac{(\omega_B L)}{Z_B} = \frac{\omega_e}{\omega_B}\frac{X}{Z_B} = \omega_{pu} X_{R,\,pu}$$

where X is the inductive reactance at the base frequency.

In *d-q* form the per unit power is

$$P_{pu} = \frac{\frac{3}{2}(V_{ds}I_{ds} + V_{qs}I_{qs})}{3V_B I_B} = \frac{\frac{3}{2}(V_{ds}I_{ds} + V_{qs}I_{qs})}{\frac{3}{2}(\sqrt{2}V_B)(\sqrt{2}I_B)} = V_{ds\,(pu)}I_{ds\,(pu)} + V_{qs\,(pu)}I_{qs\,(pu)}$$

Consequently, the per unit torque becomes,

$$T_{pu} = X_{m\,(pu)}\left[i_{qs\,(pu)}i_{dr\,(pu)} - i_{ds\,(pu)}i_{qr\,(pu)}\right]$$

It is important to note that in this per unit system:

1) All computations use the same mathematical expressions as used with actual units except that the number of phases and the number of poles no longer appear anywhere in the computations.

2) For a motor, the rated current is considerably greater than 1.0 because I_B does not incorporate the influence of the power factor or efficiency.

3) For an induction motor rated torque is slightly greater than 1.0 because T_B is based on synchronous speed and not on rated speed. Rated torque is equal to $\frac{1}{1 - S_R}$ in per unit.

Answers to Chapter 2 Problems

Problem 2–4

$$P_s = \frac{3}{2}Re\,[\underline{v}_{qds}\underline{i}^{\dagger}_{qds}] + 3v_{os}i_{os}$$

Problem 2–5

Stator Frame

$$v^S_{qs} = 375.6\cos(\omega_e t - 30°) \qquad v^s_{ds} = -375.6\sin(\omega_e t - 30°)$$

$$\underline{v}^s_{qds} = 375.6e^{j(\omega_e t - 30°)}$$

Rotor Frame

$$v^r_{qs} = 375.6\cos((\omega_e - \omega_r)t - 30°)$$

$$v^r_{ds} = -375.6\sin((\omega_e - \omega_r)t - 30°)$$

$$\underline{v}^r_{qds} = 375.6e^{j((\omega_e - \omega_r)t - 30°)}$$

Synchronous Frame

$$v^e_{qs} = 375.6 \qquad v^e_{ds} = 0 \qquad \underline{v}^e_{qds} = 375.6 + j0$$

Problem 2–6

Stator Frame

$$v^s_{qs} = 433.7\cos\omega_e t \qquad v^s_{ds} = 0 \qquad \underline{v}^s_{qds} = 216.8e^{j\omega_e t} + 216.8e^{-j\omega_e t}$$

Rotor Frame

$$v^r_{qs} = 433.7\cos\omega_e t\ \ \cos\omega_r t \qquad v^r_{ds} = 433.7\cos\omega_e t\ \ \sin\omega_r t$$

$$\underline{v}^r_{qds} = 216.8\,(e^{j(\omega_e - \omega_r)t} + e^{-j(\omega_e + \omega_r)t})$$

Synchronous Frame

$$v_{qs}^{e} = 216.8\,(1+\cos 2\omega_e t) \qquad v_{ds}^{e} = 216.8\sin 2\omega_e t$$

$$\underline{v}_{qds}^{e} = 216.8\,(1+e^{-j2\omega_e t})$$

Problem 2–7

a) $T_e = \frac{3}{2}\frac{P}{2} Im\,[\underline{\lambda}^{\dagger}{}_{qdm} i_{qdr}]$

b) $T_e = \frac{3}{2}\frac{P}{2} Im\,[\underline{\lambda}^{\dagger}{}_{qdm} \underline{i}_{qds}]$

c) $T_e = \frac{3}{2}\frac{P}{2}\frac{Lm}{Lr} Im\,[\underline{\lambda}_{qds} \underline{i}^{\dagger}{}_{qdr}]$

d) $T_e = \frac{3}{2}\frac{P}{2} Im\,[\underline{\lambda}^{\dagger}{}_{qds} \underline{i}_{qds}]$

e) $T_e = \frac{3}{2}\frac{P}{2} Im\,[\underline{\lambda}_{qdr} \underline{i}^{\dagger}{}_{qdr}]$

f) $T_e = \frac{3}{2}\frac{P}{2}\frac{L_m}{L_r} Im\,[\underline{\lambda}^{\dagger}{}_{qdr} \underline{i}_{qds}]$

g) $T_e = \frac{3}{2}\frac{P}{2}\frac{L_m}{(L_r L_s - L_m^2)} Im\,[\underline{\lambda}^{\dagger}{}_{qds} \underline{\lambda}_{qdr}]$

Problem 2–8

$$T_e = \frac{1}{\sqrt{3}}\left(\frac{P}{2}\right)[\lambda_a (i_b - i_c) + \lambda_b (i_c - i_a) + \lambda_c (i_a - i_b)]$$

Problem 2–9

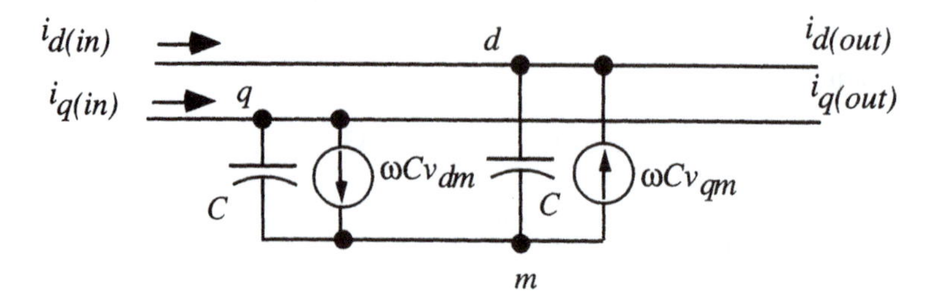

Problem 2–10

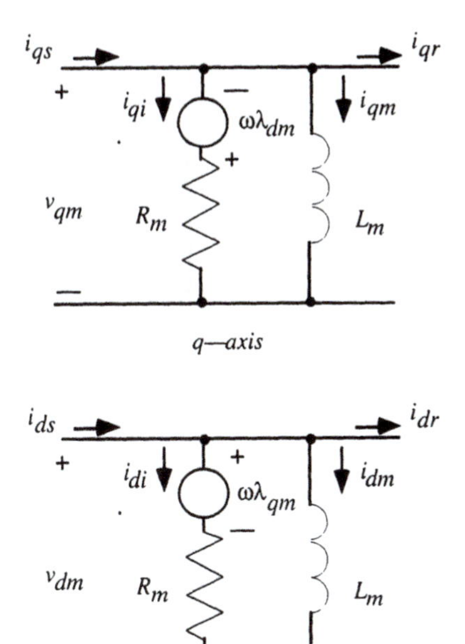

q—axis

d—axis

Answers to Chapter 3 Problems

Problem 3–1

$$\text{b)}\, v^s_{qs} = \frac{2}{\pi}\frac{3}{2}V_i\left[\cos\omega t + \frac{1}{5}\cos 5\omega t - \frac{1}{7}\cos 7\omega t - \ldots\ldots\right]$$

$$v^s_{ds} = \frac{2}{\pi}\frac{3}{2}V_i\left[-\sin\omega t + \frac{1}{5}\sin 5\omega t + \frac{1}{7}\sin 7\omega t - \ldots\ldots\right]$$

$$\text{c)}\, \underline{v}^s_{qds} = \frac{2}{\pi}\frac{3}{2}V_i\left[e^{j\omega t} + \frac{1}{5}e^{-j5\omega t} - \frac{1}{7}e^{j7\omega t} - \ldots\ldots\right]$$

$$\text{d)}\, \underline{v}^e_{qds} = \frac{2}{\pi}\frac{3}{2}V_i\left[1 + \frac{1}{5}e^{-j6\omega t} - \frac{1}{7}e^{j6\omega t} - \frac{1}{12}e^{-j11\omega t} + \ldots\ldots\right]$$

Problem 3–5

a)

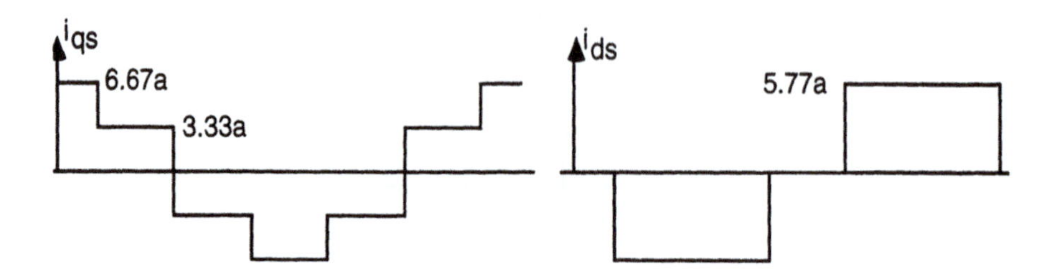

$i_i = 6.67$ A (constant)

b)

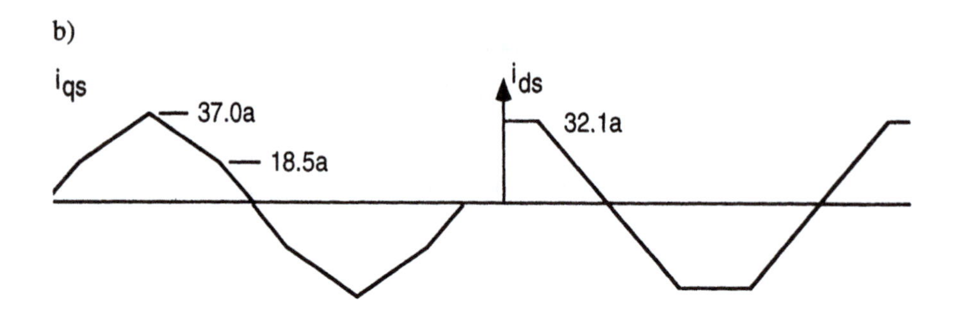

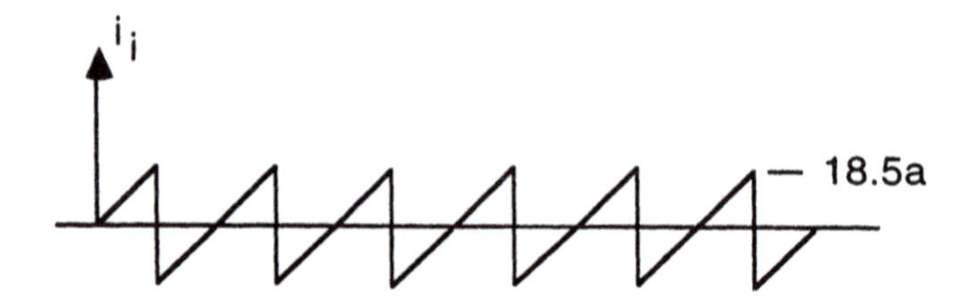

c)

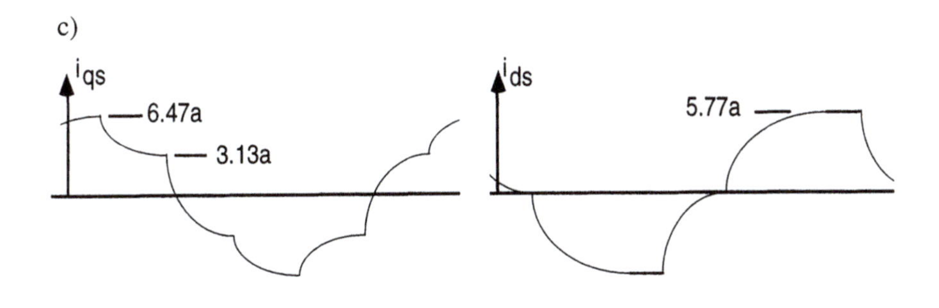

d)

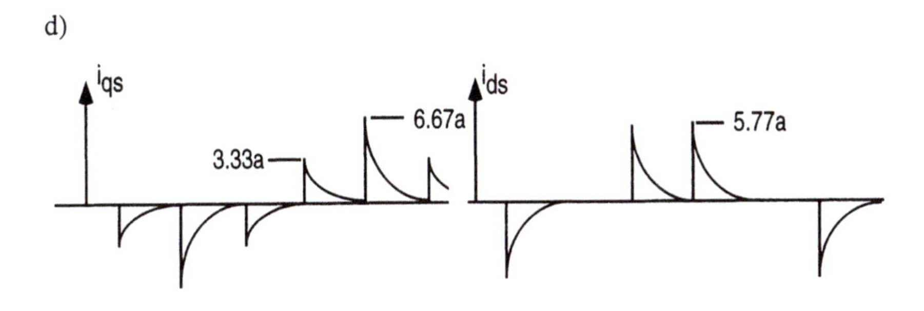

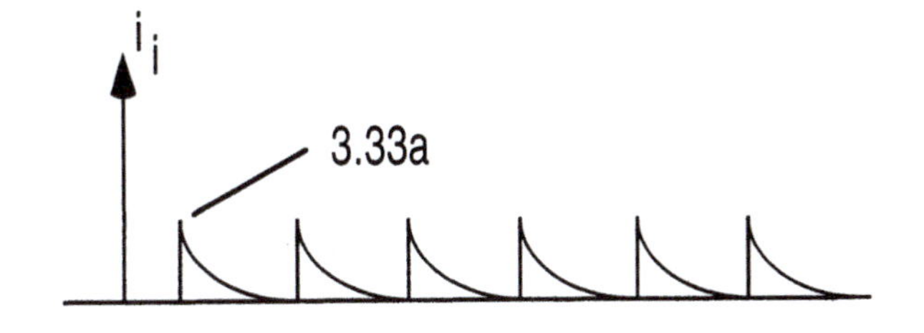

Problem 3–7

	(a)	(b)	(c)	(d)
i_{qs}^{s} {1} (A)=	6.37	33.8	6.26	1.18
i_{qs}^{s} {5} (A)=	1.27	1.35	0.927	0.873
i_{ds}^{s} {1} (A)=	6.37	33.8	6.26	1.18
i_{ds}^{s} {5} (A)=	1.27	1.35	0.927	0.873
i_{i} {dc} (A)=	6.67	0.0	6.06	0.597
i_{i} {6} (A)=	0.0	11.8	0.790	0.790

Problem 3–8

	(a)	(b)	(c)	(d)
v_{qs}^{s} {1} (V)=	110	577	112	595
v_{qs}^{s} {5} (V)=	22.1	115	30.3	32.2
v_{ds}^{s} {1} (V)=	110	577	112	595
v_{ds}^{s} {5} (V)=	22.1	115	30.3	32.2
v_{i} {dc} (V)=	200	0.0	200	200
v_{i} {6} (V)=	0.0	354	0.0	354

Problem 3–10

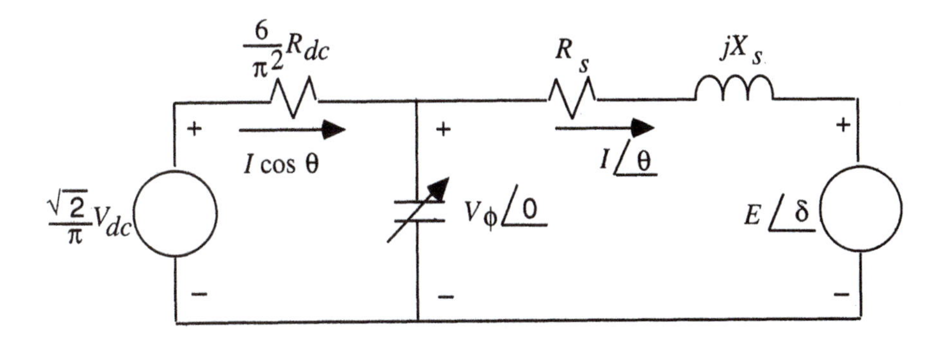

Torque characteristic similar to dc series machine.

Problem 3–13

Torque characteristic similar to dc shunt machine.

Problem 3–14

$$S(\omega_e L_m)^2 \approx -r_s r_r$$

Answers to Chapter 4 Problems

Problem 4–1

a) $i_{qs}^s = 0 \qquad i_{ds}^s = -\frac{2}{\sqrt{3}}I_{dc} \qquad i_{ns}^s = 0$

b) $$i_{qr}^s(ss) = -\frac{2}{\sqrt{3}}I_{dc}\frac{r_r\omega_{ro}L_m}{[r_r^2 + (\omega_{ro}L_r)^2]}$$

$$i_{dr}^s(ss) = \frac{2}{\sqrt{3}}I_{dc}\frac{\omega_{ro}^2 L_m L_r}{[r_r^2 + (\omega_{ro}L_r)^2]}$$

c) $v_{qs}^s(ss) = 0 \qquad v_{ds}^s(ss) = -\frac{2}{\sqrt{3}}I_{dc}r_s$

d) $T_e = -PI_{dc}^2 \dfrac{r_r \omega_{ro} L_m^2}{[r_r^2 + (\omega_{ro} L_r)^2]}$

Problem 4–2

a) $i_{qs}^r = \dfrac{2}{\sqrt{3}} I_{dc} \sin\omega_{ro} t$ $\qquad i_{ds}^r = -\dfrac{2}{\sqrt{3}} I_{dc} \cos\omega_{ro} t$

b) $i_{qr}^r(ss) = -\dfrac{2}{\sqrt{3}} I_{dc} \dfrac{\omega_{ro} L_m}{\sqrt{[r_r^2 + (\omega_{ro} L_r)^2]}} \sin(\omega_{ro} t + \phi)$

$i_{dr}^r(ss) = \dfrac{2}{\sqrt{3}} I_{dc} \dfrac{\omega_{ro} L_m}{\sqrt{[r_r^2 + (\omega_{ro} L_r)^2]}} \cos(\omega_{ro} t + \phi)$

where $\phi = \tan^{-1} \dfrac{r_r}{\omega_{ro} L_r}$

c) $v_{qs}^r(ss) = \dfrac{2}{\sqrt{3}} I_{dc} r_s \sin\omega_{ro} t$

$v_{ds}^r(ss) = -\dfrac{2}{\sqrt{3}} I_{dc} r_s \cos\omega_{ro} t$

d) Same as problem 4–1d

Problem 4–3

a) $i_{qs}^s = 0 \qquad i_{ds}^s = -\dfrac{2}{\sqrt{3}} I_m \cos\omega_e t \qquad i_{0s} = 0$

b) $\tilde{I}_{qr}^s = -\dfrac{2}{\sqrt{3}} \dfrac{r_r \omega_{ro} L_m I_m}{r_r^2 + (\omega_{ro}^2 - \omega_e^2) L_r^2 + j2 r_r \omega_e L_r}$ (phasor quantity)

d) $T_e = \dfrac{\sqrt{3}}{2} \dfrac{P}{2} L_m I_m I_{qr}^s [\cos\phi_q + \cos(2\omega_e t + \phi_q)]$

where $\phi_q = 180° - \tan^{-1} \dfrac{2 r_r \omega_e L_r}{[r_r^2 + (\omega_{ro}^2 - \omega_e^2)]}$

Problem 4–4

c) $\tilde{V}_x = j\frac{\tilde{V}_{bc}}{\sqrt{3}}$ $\tilde{I}_x = j\frac{\tilde{I}_b}{\sqrt{3}}$

Problem 4–5

a) $\tilde{I}_b = 1.77 \angle(-37.7)$

b) $T_p = 0.78$ $T_n = 0.0074$ $T_t = 0.77$

c) $T_{e2} = 0.802$

Problem 4–6

a) $r_s = 0.02836\Omega$ $L_s = 0.02332\text{H}$ $L_m = 0.02257\text{H}$

$r_r = 0.04254\Omega$ $L_r = 0.02332\text{H}$

b) $\lambda_1 = -19,300 \pm j0\,\text{sec}^{-1}$ $\lambda_2 = -1.82 \pm j370\,\text{sec}^{-1}$

c) $\underline{\lambda}^s_{qdr}(0) = -0.202 - j0.909\,\text{pu}$

$\underline{\lambda}^s_{qds}(0) = -j0.996\,\text{pu}$

d) $\underline{v}^s_{qds} = 4690e^{-19,300t} + 333e^{-j13.6°}e^{(-1.82 + j370)t}$

1.2 s to decay to 10%

Problem 4–7

a) $i_{qds}(t) = 5.08e^{-j(\omega_e t - 90°)} + 4.77e^{+j(\omega_e t + 90°)} + 10.2e^{-j90°}$

b) $T_e = 4.76\sin 2\omega_e t - 9.55\sin\omega_e t$

$T_{max} = 12.9\text{pu}$

Problem 4–8

$i^s_{qs}(1) = 0.865$ pu, $i^s_{qs}(5) = 0.0072$ pu, $i^s_{ds}(1) = 0.629$ pu,

$i_{ds}^{s}(5) = 0.205$, $i_{i}(dc) = 0.831$ pu, $i_{i} = 0.306$.

$i_{qds}^{s}(1) = 1.07 \angle -36°$, $i_{qds}^{s}(5) = 0.205 \angle -88°$

Answers to Chapter 5 Problems

Problem 5–1

a) $I_s = 1.26$ pu

b) $\omega_e = 1.5$ pu, $T_e = 0.671$ pu

$\omega_e = 2.0$ pu, $T_e = 0.469$ pu

c) $\omega_e = 1.92$ pu

Problem 5–3

a)		b)		c)	
$I_s = 1.25$	$V_s = 1.00$	$I_s = 1.25$	$V_s = 0.525$	$I_s = 1.25$	$V_s = 0.0625$
$= 0.625$	$= 0.867$	$= 0.625$	$= 0.449$	$= 0.625$	$= 0.0312$
$= 0$	$= 0.8$	$= 0$	$= 0.4$	$= 0$	$= 0$

Problem 5–4

a) $I_{sT} = 1.186$ $\quad I_{s\phi} = 0.455$ $\quad S\omega_e = 0.00952 I_{sT}/I_{s\phi}$

	I_{sT}	$I_{s\phi}$	$S\omega_e$	ω_e	V
b)	0.529	0.454	0.0124	0.988	0.958
c)	1.185	0.227	0.0497	1.025	0.558

d) 0	0.454	0	0.975	0.931
e) 1.185	0.454	0.0248	0.512	0.519
f) 1.185	0.454	0.0248	0.0248	0.042
g) 0	0.454	0	0	0.0068

Problem 5–5

a) $\omega=1.25$ pu

b) $I=0.410$ pu $\quad T=0.328$ pu

c) $\gamma = 90°$ $\quad \omega_e=3.33$ pu

d) $\gamma = 20.2°$ $\quad T_e=0.938$ pu

Problem 5–6

a) $\tau_r=0.103$ sec. $\quad I_{sT} = 34.0$ A $\quad I_{s\phi}=21.0$ A

b) $I_s = 35.6$ A $\quad f = 125$ Hz

c) $V_s = 265$ V

d)

	f_e	T_e	I_s	V_s
No Load	same	same	same	same
Loaded	same	increase	same	increase

Problem 5–7

All answers in pu with q–axis as reference for phase angles.

a) $\omega_r = \omega_{rB}$

	$\underline{V}_{qds}$	$\underline{I}_{qds}$	$\underline{\lambda}_{qds}$	$\underline{I}_{qdr}$	$\underline{\lambda}_{qdr}$
Machine/Cond.	Mag Ang	Mag Ang	Mag Ang	Mag Ang	Mag Ang
$IM/T = T_R$	1.00∠13.1°	1.31∠-20.0°	0.97∠-75.7°	1.17 ∠180°	0.89∠-90°
$SM/T = T_R$	1.00 ∠22.1°	1.30 ∠-11°	0.967∠ -66.6°	3.82 ∠-90°	0.814 ∠-90
$IM/T = 0$	0.9 ∠-0.8°	0.45 ∠-90°	0.94∠-90°	0.0 —	0.89 ∠-90°
$SM/T = 0$	0.89 ∠0.5°	0.25°∠ -90°	0.89 ∠-90°	3.82 ∠-90°	0.814 ∠-90°

b) $\omega_r = 0$

Machine/Cond.	$\underline{V}_{qds}$ Mag Ang	$\underline{I}_{qds}$ Mag Ang	$\underline{\lambda}_{qds}$ Mag Ang	$\underline{I}_{qdr}$ Mag Ang	$\underline{\lambda}_{qdr}$ Mag Ang
$IM/T = T_R$	0.074 ∠-3.1°	1.31 ∠-20°	0.97 ∠-75.7°	1.17 ∠180°	0.89 ∠-90°
$SM/T = T_R$	0.039 ∠11°	1.3 ∠-11°	0.97 -66.6°	3.82 ∠ -90°	0.81∠-90°
$IM/T = 0$	0.013 ∠-90°	0.45 ∠-90°	0.94 ∠-90°	0.0 –	0.89 ∠-90°
$SM/T = 0$	0.007∠-90	0.25∠-90	0.88 ∠-90°	3.82 ∠-90°	0.81 ∠-90°

c) $\omega_r = 2\omega_{rB}$

Machine/Cond.	$\underline{V}_{qds}$ Mag Ang	$\underline{I}_{qds}$ Mag Ang	$\underline{\lambda}_{qds}$ Mag Ang	$\underline{I}_{qdr}$ Mag Ang	$\underline{\lambda}_{qdr}$ Mag Ang
$IM/T = T_R$	1.08 ∠25.8	1.24 ∠-10.3	0.53 ∠-63.0°	1.17 ∠180°	0.45 ∠-90°
$SM/T = T_R$	not possible (flux constrained by magnet)				
$IM/T = 0$	0.9 ∠-0.4°	0.22 ∠-90°	0.47 ∠-90°	0.0 —	0.45 ∠-90°
$SM/T = 0$	not possible (flux constrained by magnet)				

Answers to Chapter 6 Problems

Problem 6–1

a) $I^e_{ds} = 60.6$ A (peak) = 0.457 pu
$I^e_{qs} = 157$ A (peak) = 1.186 pu
$S\omega_e = 9.35$ r/sec

b) $T_e = 204 N-m$
$S\omega_e = 18.7$ r/sec
$\omega_e = 754$ r/sec
$V_s = 496$ volts line to line (rms)

c) $T_e = 204\,(1 + e^{-3.6t})\,N-m$

Problem 6–2 Synchronous Machine Field Orientation

a) $T_e = 0.835$ pu, $E = 0{,}835$ pu, $\cos\theta = 0.865$

b) $E = 1.38$ pu, $I = 0.856$ pu, $i^*_{qs} = 0.785$ pu, $i^*_{ds} = 0.346$ pu, $\gamma = 23.8°$

c) Part a, $T_e(0) = 0$, T_e(final)=0.835 pu, Instantaneous response

Part b, $T_e(0)$=0.703, T_e(final)=0.835 pu, Time Constant =

$L_{qr}/r_{qr} = 0.0243$ sec.

Answers to Chapter 7 Problems

Problem 7–1

(a) K=16.1 τ=0.0345 sec

(b)

f(Hz)	|H|	∠ H (deg)
0.0	1.0	0.0°
10	0.963	–3.01°
20	0.910	–5.06°
40	0.821	–8.21°
60	0.747	–10.7°

Answers to Chapter 8 Problems

Problem 8–1

a) Tuned operation

Linear T_e vs. i_{qs} characteristic $T_e = 0.849 i_{qs}$

b) Detuned operation

Non–linear T_e vs. i_{qs} characteristic

Slip Gain Too Small		Slip Gain Too Large	
i_{qs}	T_e	i_{qs}	T_e
0.0	0.0	0.0	0.0
0.2	0.095	0.2	0.21
0.4	0.225	0.4	0.33
0.6	0.49	0.6	0.42
0.8	0.8	0.8	0.52

1.0	1.13	1.0	0.625
1.2	1.49	1.2	0.73
1.4	1.86	1.4	0.83

Problem 8–2

f=60 hz., V/hz=2.01pu

f=5 hz., V/hz=2.21 pu

Saturation will substantially reduce the V/hz values which yield maximum torque ampere.

Problem 8–3 Detuned Induction Machine Field Orientation

a) I_{sT} = 1.13 pu, $I_{s\phi}$ =0.309 pu

b) V_s = 0.0995 pu

c) T_e = 1.67 pu, I_{sT} = 0.790 pu, λ_r = 1.88 pu, λ_s = 1.78 pu.

d) T_e (steady state) = 1.67 pu

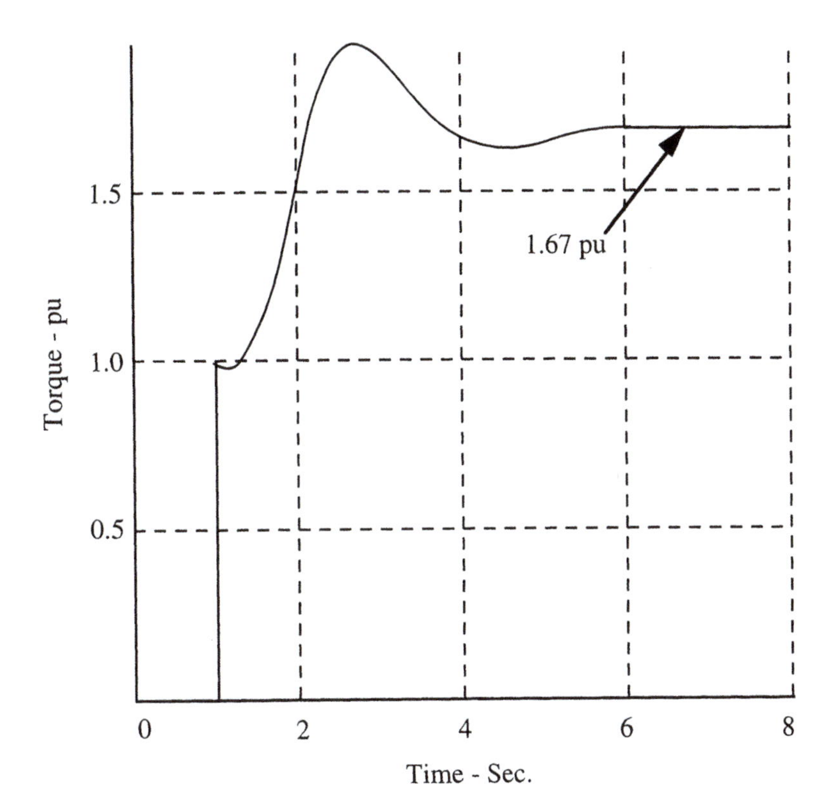

INDEX

A
air gap
  field intensity 39
  flux 39
  flux linkage 40, 75
  non-uniform 88
  packet of flux 40
  total air gap MMF 49

B
BJT (bipolar junction transistor) 11
brushless dc machine 224
  sine wave machines 223

C
complex vector
  definition 51
  equivalent circuit 287
  equivalent circuit in operational form
 153
  induction machine equivalent circuit
 152
  interpretation 61
  operational impedances 171
  voltage equations 52
complex vector diagram
  air gap flux reference 167
  for positive sequence 166
  rotor flux linkage reference 168
  round rotor synchronous machine ...215
  salient pole synchronous machine ...215
controlled rectifier 26
current limit circle 409
current regulation
  augmented feedforward compensation
 347
  augmented feedforward compensation
  with decoupling 348
  block diagram of series compensated
  346
  block diagram with feedforward compensation 346
  CSI 318
  CSI limitations 321
  dc link 318
  feedforward compensation 344
  hysteresis controller decoupling 331
  hysteresis controller switching diagram
 324
  hysteresis limit cycles 327
  hysteresis regulators 322
  predictive control 349
  PWM inverter 321
  ramp comparison 331
  stationary frame 335
  synchronous frame 340
current source inverter
  commutation delay 119
  current regulated 318
  d,q model 116
  dc link inductor 129
  duality with VSI 144
  switching functions 119
  synchronous frame switching functions
 123
  with synchronous machine 208

D
d,q model
  air gap flux referred form 283
  modified forms 155
  of synchronous machine 290
  ramp comparison controller 335
  rotor flux referred equations 258
  stator flux oriented induction machine
 310
  stator voltage equations in terms of rotor flux 304
  synchronous frame ramp comparison current regulator 340
  voltages for sinusoidal excitation 159
dc machine
  brushless dc machine 222
  capability curve 403
  commutating capability 6
  commutation 3, 4
  constant power limit 404
  constant torque limit 404
  dc drives 2
  efficiency 5
  field weakening operation 401
  requirements for torque control ... 206
  size and weight 7
  speed*power product 5
  torque 7
  torque control 203
detuning in indirect field orientation
  detuned field orientation lines 360
  eigenvalues 379

field orientation torque characteristics 362
flux characteristics 363
influence of saturation 364
loss trends 369
normalized torque command 361
phasor diagram 353
power loss 365
representation of slip gain error 383
rotor copper loss characteristics 370
slip gain error multipliers 360
small signal transfer function 379
torque response with slip gain error 377
torque step response 382
transient response 379, 384

direct field orientation
controller for air gap flux 289
decoupler for stator flux 311
implementation of 280
measurement of air gap flux 276
stator flux based 312
using flux sensors 277
voltage and current sensing 278

E

eigenvalues
constant speed 177
limiting cases 184
loci in the rotor reference frame 182
loci in the stationary reference frame 180
loci in the synchronously rotating reference frame 181
ranges of machine parameters for normal machines 186
rotor poles 179
stator poles 179
typical values 187

equivalent circuit
complex vector operational 153
constant rotor flux linkage model 190
for air gap flux control 287
fundamental component circuit of three phase six step CSI 143
fundamental component circuit of three phase six step VSI 142
fundamental component model of PWM-VSI 143
illustrating induction machine field orientation 230
induction machine complex vector form 152
invariance of the steady state 164
negative sequence 163
positive sequence 160
referred variables 154
rotor flux reference 169
salient pole synchronous machine 291
stator flux reference 170
stator referred 158
steady state 156
transient 191, 196
turns ratio 153

F

field 217, 410

field orientation
achieving maximum torque capability 410
block diagram for induction machine 263
constant flux voltage decoupler 307
direct 247
direct controllers 276
dynamic response 261
dynamic response of air gap flux 284
dynamics of 257
dynamics of synchronous machine 293
equivalent circuit for induction machine 230
field orientation lines 357, 389
field weakening operation 399
indirect 232, 241
optimized operation in field weakening range 415
phase correction for CSI system 246
rotor flux referred equations 258
slip expression 232
stator flux based 309
steady state d,q currents 236
steady state d,q model 234
synchronous and induction machine comparison 249
synchronous machine with constant field excitation 293
synchronous machine with variable field excitation 295
terminal behavior 233
torque characteristics 355
torque control requirements 247
torque production 261
using air gap flux 283
using voltage as the controlled variable 304
voltage decoupler 306

wide constant horsepower range 413
field oriented control 32
flux linkage
magnetizing component 40
self component of flux linkages 40
four quadrant drive 2
fundamental component model
CSI steady state current 137
CSI steady state voltage 140
d,q equivalent circuit of CSI drive in synchronous frame 136
d,q equivalent circuit of VSI drive in synchronous frame 134
equivalent circuit 141, 143
equivalent circuit of PWM-VSI inverter 143
PWM steady state voltage 139
three phase six step VSI 142
VSI steady state current 139
VSI steady state voltage 137

G
general purpose drives 31
GTO (gate turn off thyristor) 11

H
high performance drives 31

I
IGBT (insulated gate bipolar transistor) 11
indirect field orientation
compensated flux response 267
controller for air gap flux 288
detuning 353
field orientation lines 357
indirect control using a CSI 268
slip gain error 384
slip relation 260
start up transient 269
torque characteristics 355
using a CRPWM 242
using a CSI 244
using voltage controlled inverter ... 308
with d,q synchronous frame currents as inputs 265
with synchronous frame rotor flux and q-axis stator current as inputs 265

inductance
coupling factor 178
leakage 154
magnetizing 41, 45
mutual between phases 46
mutual from stator to rotor 47
rotor leakage 48
stator leakage 46
stator magnetizing 46
stator transient 191, 337
total self 46
induction machine
block diagram for current fed 263
cogging torques 25
complex vector equivalent circuit ... 152
complex vector model 151
constant power limit 406
direct field orientation 247
dynamics of field orientation 257
equivalent circuit 61
field weakening operation 401
idealized three phase 43
in rotor flux rotating frame 87
in rotor reference frame 83
in stationary axis d,q,o variables 81
in synchronously rotating reference frame 86
indirect field orientation 232, 241
indirect field orientation using a CSI 244
indirect field orientation using a CRPWM 242
inrush current 78
magnetizing inductance 55
negative sequence equivalent circuit 163
open loop speed controller 17
optimized operation in the field weakening range 415
positive sequence equivalent circuit 160
pulsation torque 25, 80, 162
rotor reference frame 84
short circuit of unloaded machine 187
simulated with phase variables 78
size and weight 7
squirrel cage winding 36
stationary frame d,q model with CSI 126
stationary frame d,q model with VSI 124
steady state complex vector equivalent circuit 287
steady state equivalent circuit derivation 156
steady state torque 161
synchronous frame d,q model with CSI 127
synchronous frame d,q model with VSI 125

torque capability curve 406
turns ratio 153
vector control in steady state 226
voltage decoupler 306
wide constant horsepower range 413
wound rotor 36, 61
inertia
ac machine 6
dc motor 6
inverter
commutation 10
CSI 25
commutation 27
induction motor 28
turn off devices 30
CSI (current stiff inverter) 9
current controlled PWM 321
current regulated 317
current stiff 9
d-q models in synchronous reference frame 120
feedback diodes 12
imposed voltage and current limits . 405
PWM
block modulation 19
changeover to six step mode 24
gear changing 22
natural modulation 22
phase locking 22
pulse dropping 24
sinusoidal modulation 20
PWM inverter 9, 18
resonant reversal 10
return current 12
ripple current 9
six step inverter 9
six step voltage inverter 12
snubbing 11
torque pulsations 25
voltage controlled oscillator (VCO) .. 18
voltage source 9
voltage stiff 9
Volts/Hertz control 15
VSI (voltage stiff inverter) 9
inverters
losses caused by 24

M

magnetic saturation.................. 393
MCT (MOS controlled thyristor) 11
MOSFET (metal oxide semiconductor field effect transistor) 11

N

negative sequence voltage 157

O

one quadrant converter 2

P

positive sequence voltage 157
power
electromechanical 74
in d,q circuits 69
three phase 68
power factor
input fundamental component 5
pulse width modulation
current regulated 321
hysteresis regulator 322
ramp comparison 331

R

ramp comparison PWM
augmented feedforward compensation 347
augmented feedforward compensation with decoupling 348
basic operation 331
block diagram 338
block diagram of series compensated 346
block diagram with feedforward compensation 346
dc current controller 335
describing function 333
feed forward compensation 344
modulation index 333
models of PI controller 342
predictive control 349
simpler synchronous regulator 344
stationary frame d,q variables 335
stationary frame regulator cross coupling 341
rectifier 9
reference frame
arbitrary 56
attached to vector 56
d,q axes 56
freely rotating 56
normal component 57
rotor 56
stationary 56, 63
stator 56

synchronously rotating 63
zero sequence component 56
rotor cogging 320

S

selection of flux level
best optimal 392
changes in magnetizing inductance and rotor time constant 365
effect of detuning 365
field orientation lines 389
maximum drive efficiency 390
maximum torque per stator ampere 389
optimal 388
optimal utilization of the converter 394
peak torque effects 365
reduced sensitivity to controller detuning 393
switching transients
ac component magnitude 189
complex eigenvalue 172
constant speed eigenvalues 177
coupling factor 178
dc component magnitude 189
general solution 177
ranges of machine parameters for normal machines 186
rotor poles 179
stator poles 179
stator transient inductance 191
sudden application of stator current source excitation 172
sudden short circuit 187
transient equivalent circuit 196
voltage behind transient reactance ..192
synchronous machine
angle control 292
as commutatorless dc machine 208
braking torque 103
choice of gamma 212
complex space vector 95
converter 8
CSI driven 208
CSI torque control 220
d,q variable vector diagram 214
damping torque 100
dynamic response with gamma not zero 295
dynamics of field orientation 293
excitation torque 100
field orientation 219
field orientation with a CSI 217
flux linkage equation 93
Görges Effect 100
induction motor torque 100
kVA requirement 8
line start of permanent magnet machine 104
mutual inductance between stator phases 90
Park's Equations 98
permanent magnet 102, 223
power input equation 99
reaction torque 100, 214, 216
reluctance torque 214, 216
rotor flux linkage equations 96
saliency torque 100
self inductance 89
steady state d,q model 213
three phase stator voltages 91
torque control 212
torque control requirements 221
torque control using a CRPWM 218
torque output equation 99
torque pulsation 103
vector control 208, 292
vector diagram for field orientation . 217
wound field 8

T

time constant
rotor open circuit 178
rotor short circuit 178
stator open circuit 178
stator short circuit 178
torque 103
adjustable 401
characteristics for field orientation . 355
complex vector expressions 77
constant 401
cross product form 76
electromagnetic 73, 74
response with slip gain error 377
variable 401
transformation
general complex form 65
general phase 65
general rotating complex vector 58
graphical interpretation of inverse ...66
inverse phase 66
inverse rotation 66
power invariant 56
turns ratio transformation 54

V

vector control 32
voltage behind transient reactance 192
voltage controlled oscillator (VCO) 17
voltage limit ellipse 408
voltage source inverter
  complex vector d,q form 114
  d,q model 115
  duality with CSI 144
  PWM switching functions 115
  switching functions 111
  synchronous frame switching functions 122
  zero voltage states 115

W

winding
  concentrated full pitch 39
  effective number of turns 38
  flux linkages 44, 46
  flux linkages of rotor 48
  leakage flux 45
  rotating 43
  rotor 43
  rotor flux linkages 47
  stationary 43
  stator 43
  voltage equations 43
  winding density distribution 37
  winding distribution 35

Z

zero sequence 66
  circuits 60
  equation 59